Situation Management for Process Control

Situation Management for Process Control

Situation Awareness and Decision-Making for Operators in Industrial Control Rooms and Operation Centers

Douglas H. Rothenberg

Notice

The information presented in this publication is for the general education of the reader. Because neither the author nor the publisher has any control over the use of the information by the reader, both the author and the publisher disclaim any and all liability of any kind arising out of such use. The reader is expected to exercise sound professional judgment in using any of the information presented in a particular application.

Additionally, neither the author nor the publisher has investigated or considered the effect of any patents on the ability of the reader to use any of the information in a particular application. The reader is responsible for reviewing any possible patents that may affect any particular use of the information presented.

Any references to commercial products in the work are cited as examples only. Neither the author nor the publisher endorses any referenced commercial product. Any trademarks or tradenames referenced belong to the respective owner of the mark or name. Neither the author nor the publisher makes any representation regarding the availability of any referenced commercial product at any time. The manufacturer's instructions on the use of any commercial product must be followed at all times, even if in conflict with the information in this publication.

Copyright © 2019 International Society of Automation (ISA)
All rights reserved.

Printed in the United States of America.
Version 1.0

ISBN-13: 978-1-945541-65-0 (Hardback)
ISBN-13: 978-1-945541-99-5 (EPUB)
ISBN-13: 978-1-945541-98-8 (MOBI)

No part of this work may be reproduced, stored in a retrieval system, or transmitted in any form or by any means, electronic, mechanical, photocopying, recording or otherwise, without the prior written permission of the publisher.

ISA
67 T. W. Alexander Drive
P.O. Box 12277
Research Triangle Park, NC 27709

Library of Congress Cataloging-in-Publication Data in process

Dedication

To my dearest wife, constant companion, champion, and best friend Kasia Gustaw Rothenberg who understood the potent draw of my work and encouraged my countless hours in the scriptorium writing, editing, and just thinking about everything that goes into such a project of unimaginable draw and dedication.

Pracę tą dedykuję mojemu najlepszemu przyjacielowi, mojej drogiej żonie Kasi Gustaw Rothenberg, która to wspierała moje zaangażowanie i niezliczone godziny spędzane w scryptorium nad formułowaniem i edytowaniem idei zawartych w tej książce.

I also dedicate this book to the men and woman of all walks of life who leave their homes, families, and friends and step into a control room or operation center for their assigned shifts. They experience periods of operational certainty and the unwelcome periods of upset and danger with responsible professionalism and keen sense of duty. I also want to recognize with respect and gratitude the surrounding staff that teaches and supports these dedicated people as they do the important work to make and distribute the many things of value for our grateful societies.

Contents

List of Figures xxv

List of Tables xliii

Foreword xlv

About the Author. xlvii

Acknowledgments. xlix

Part I   Operational Integrity 1

Chapter 1 Getting Started 3
    1.1   Key Concepts 4
    1.2   Introduction 5
    1.3   Cautions and Ground Rules. 8
            Design and Safety Notice 8
            Conflicts with Established Protocols or Statutory
               Requirements 9
            Exclusion of Special Systems and Responsibilities 9
            How to Read This Book 9
    1.4   Audience. 10
            Operators, Engineers, Technicians, and Support 10
    1.5   Contribution and Importance 11
    1.6   Situation Management 11

	Situation Management Is Straightforward................. 13
	Situation Management Fits Like a Glove 16
1.7	Structure of Situation Management 18
	Structure of Situation Management 19
	Essential Components of Control Room Management 20
	Essential Components of Situation Management 22
	Success Is Not a Trade-Off 23
1.8	Fundamental Tools for Situation Management 24
	Reminder about Management's Role 26
	Getting to Good Situation Management Can Be a Big Step ... 27
	TransAsia Crash....................................... 28
1.9	The Power of Time.. 30
	See-Understand-Decide-Act and Process Safety Time 30
1.10	Defensive Operating 32
	Five Principles 33
	Supplementary Guidance................................ 34
1.11	Situation Management Changes the Game 35
	Situation Management 37
1.12	The Overall Operational Setting 39
	Alarms (Something Is Wrong for Sure) 40
	Strong Signals (Something Is Wrong but Not Sure What).... 41
	Weak Signals (Clues Something Might Be Wrong, Find Out)... 41
1.13	Standards, Guidelines, and Practices 42
	Standard 43
	Recommended Practice................................. 44
	Best Practice 44
	Guideline... 44
	Recognized and Generally Accepted Good Engineering Practice 45
	OSHA 29 CFR 1910.119 45
	Best Available Technology Not Entailing Excessive Cost 46
	Duty of Care.. 47
	Discussion.. 47
1.14	Management's Role 47
1.15	The Human Operator 48
1.16	Frontline Supervisor's Role................................. 49
	Administrative Supervision............................. 50
	Development Supervision............................... 50
	Operational Supervision................................ 50
	Takeaway Message..................................... 50
1.17	How Situation Management Really Works 50
1.18	How Real Is All of This?................................... 53
	ABB 53
	Emerson.. 54
	Honeywell.. 54

		Rockwell Automation 54

           Rockwell Automation 54
           Schneider Electric (Foxboro, Invensys) 55
           Yokogawa 55
   1.19 Contribution and Importance 56
           Audience 56
           Suggestions for Reading 56
   1.20 Things to Keep in Mind 57
           To the Reader 58
           To Operators 58
           To Engineers and Technologists 59
           To Supervisors 59
           To Senior Management 59
           To Regulators and Inspectors 62
           Dual Responsibilities 62
   1.21 Review of Book 62
           Part I: Operational Integrity 63
           Part II: Situation Awareness and Assessment 64
           Part III: Situation Management 64
           Putting It All Together 65
   1.22 Suggestions for Using This Material 65
           Good Engineering Practice 65
           Evaluate and Improve Everything 69
   1.23 Actually Getting Started 69
           Suits and Coveralls 70
           Key Ingredients for Operational Success 70
           Some Causes of Poor Operation 71
           Knowing What Is Right 73
   1.24 Limitations of Situation Management 75
   1.25 Close 75
   1.26 Further Reading 76

Chapter 2 **The Enterprise 77**
   2.1 Key Concepts 78
   2.2 Introduction 78
           The Enterprise 79
           It is All About Culture 79
           Walk the Walk 80
           Understanding the Plant or Enterprise 81
           The Mindful Organization 81
   2.3 Silos 82
           Information Silos 84
           Functional Silos 84
           Communications between Silos 85
   2.4 Enterprise Capabilities 86
           International Association of Oil and Gas Producers on
               Process Safety 86

		Institutional Failure: PG&E San Bruno Pipeline 87

                Institutional Failure: PG&E San Bruno Pipeline 87
                Process Safety Begins in the Boardroom 87
                High-Reliability Organizations 88
2.5   Short-Term versus Long-Term 89
                Story of Two Enterprises................................... 89
                Valuable Message... 92
2.6   Delegation of Responsibility 92
2.7   Operational Integrity.. 93
                Operational Integrity Levels 95
                Plant Operability Components 95
                OiL as a Measurement Evaluation.......................... 97
2.8   Safety.. 99
                Characteristics of Safety 100
                Components of Safety.................................... 101
                Delivering Safety 103
2.9   Responsible Engineering Authority (REA) 103
2.10 The Magic of a Control Loop 105
                Temperature Control Example............................ 105
                Flow Control Example................................... 106
                Useful Message about Moving Disturbances 107
2.11 Selective Automation....................................... 107
                Advanced Basic Control 108
                Advanced Control 108
                Procedural Automation.................................. 110
2.12 The Automated Plant....................................... 112
                Selective Automation.................................... 113
                Operate Periodically with Automation Off 113
                Cease Operations during Significant Upsets............... 114
                Summarizing 114
2.13 Plant (or Enterprise) Area Model............................ 115
2.14 Decomposition 117
                Decomposition Basics 118
                Subsystem Boundary Attributes 118
                Subsystem Internal Attributes 119
2.15 Decomposition Underlying Situation Management 120
                Structure of Decomposition 120
                Looking for Abnormal Situations in Key Repeated
                    Elements 123
                Looking for Abnormal Situations in Key Repeated
                    Subsystems 125
                Summarizing 127
2.16 Transformational Analysis.................................. 127
                How It Works 127
                Using Transformational Analysis for Risk Assessment...... 129
                Using Transformational Analysis for Decomposition 130
                Use for Operational Area Division of Responsibility........ 131

		Identification of Structural Operational Issues 131
	2.17	Near Misses, Incidents, Accidents, and Disasters............ 132
		Pay Retail or Pay Wholesale............................ 133
		Hazard.. 134
		Abnormal Situation.................................... 134
		Near Miss 135
		Accident... 136
	2.18	Critical Failures 137
		Chains 138
		Disasters in Review.................................... 140
	2.19	Process Hazard Management 146
	2.20	Root and Other Causes................................... 148
		Definition of Root Cause............................... 148
		Explanatory Case for Root Cause 148
		Proximate Cause 149
	2.21	Layers of Protection...................................... 150
		Layer of Protection.................................... 150
		Independent Protection Layer 150
		Layers of Protection and Situation Management 150
	2.22	Close 151
	2.23	Further Reading.. 151

Chapter 3 | **Operators** **153**
	3.1	Key Concepts 154
	3.2	Operators' Creed 155
	3.3	Operators and Operations 155
		Definition of an Operator 155
		Peopleware 156
		Plants and Operations................................. 156
	3.4	Boundaries and Responsibilities 157
		Responsibility... 158
		Control.. 158
		Crossovers... 159
		Crossover Management 159
		Control Room Coordination with Its Operational Area Field.. 161
		Maintaining Responsibility 162
	3.5	Operator Readiness....................................... 164
		Understanding Fatigue................................. 164
		Managing Fatigue 168
		Understanding Impairment 168
		Managing Impairment 169
		Understanding Overload 170
		Managing Operator Load 171
		Operator Alertness 172
		Improving Operator Performance....................... 173

3.6	Operator Training	173
	Training and Skills	174
	Skills Training	176
	Competency Training	176
	"Personal" Tools	177
	Process Understanding	178
3.7	Qualified Operator	178
	Brief Glossary	179
	Message of Operator Qualification	181
3.8	Operator Tools	181
	Checklists	182
	Protocols	183
	Procedures	183
	Simulators	184
	Reports	185
3.9	Shift Handover	185
	Reasons for Shift Changes	186
	Beginning of the Operator's Shift Role	189
	Ending of the Operator's Shift Role	190
	Functional Components	191
	Noncontiguous Shifts	196
	Logs and Reports	196
	Handover	198
	Field Operators' Handover	202
	Shift Handover for Supervisors	204
	Special Case for Maintenance	206
3.10	Information Content of Shift Handover	208
3.11	The Mobile Operator	210
	Control Room Mobility	210
	Plant Area Mobility	211
	Large Geographical Mobility	211
	Requirements for Mobility Support	212
3.12	"Long Arm" of the Operator	213
3.13	Goals, Roles, and Culture	214
	Basics of Motivation	215
	Tenets of Operation	217
	Operator Objectives	218
	Message	222
3.14	Close	222
3.15	Further Reading	222

Chapter 4 High-Performance Control Rooms and Operation Centers ... 223

4.1	Key Concepts	224
4.2	Introduction	226
4.3	A Note about Scope	228

4.4	Control Room and Operation Center Requirements	229
	Physical Protection and Security	229
	Environmental Controls	229
	Information Support Tools and Technology	230
	Sufficient Process and Operational Controls	230
	Operational Support	231
	Control Room Access Management	231
	Special Operating Situations	232
	Permits, Personnel, and Visitors	232
	Scope Note	233
4.5	The Control Room	233
	A Control Room Is Remote (but Not Necessarily Distant)	234
	Design Evolution	234
	Architectural Aspects	235
4.6	Operation Centers	236
4.7	Collaboration Centers	237
	Requirements	237
	Configurations	238
	Importance	238
4.8	Advanced Technology Control Centers	239
4.9	Design of Effective Work Spaces	241
	The Concept of Space	241
	A Few Thoughts	244
4.10	User-Centered Design	244
	Human Factors Details	245
	Environment	245
	Scaling	246
	Compensation	247
	Understandability	247
	Implementability	248
	Unified Feel	248
	Example of Mixed Technology	248
4.11	Control Room Design	249
	Location	250
	Security	251
	Building Style	251
	Layout	251
	Design Considerations	252
	Principles and Ergonomics	252
	Console Design	252
	Life Cycle	253
4.12	The Mobile Control Room	253
4.13	The Role of the Control Room	254
4.14	Looking to the Future	255
4.15	An Architect Weighs In	256
4.16	Close	257

Chapter 5		**The Human-Machine Interface** **259**	
		Not a One-Stop Shop 260	
		Chapter Coverage 260	
	5.1	Key Concepts 261	
	5.2	Introduction 262	
		Physical Differences and Preferences 262	
	5.3	Nomenclature for Display Screens and Components. 263	
	5.4	Four Underlying Requirements for Operator Screens 265	
		Requirement 1: Purpose 266	
		Requirement 2: Understanding 266	
		Requirement 3: Use 266	
		Requirement 4: Complete 266	
	5.5	Principles of Display Screen Design 267	
		Wickens's 13 Principles 267	
		Engineering Equipment and Materials User Association's 10 Principles 269	
		Five Design Principles. 270	
		ISO 9241 Seven Design Principles and Five User Guidance Principles. 271	
	5.6	Seven Principles of Workspace Design 272	
	5.7	The Human-Machine Interface 274	
		A Wartime Story Sets a Stage. 275	
		Components of an HMI 277	
		HMI Design Philosophy 278	
		Style Guide 279	
		Graphics Library 280	
	5.8	Display Screen Design 281	
		Overview of Display Screen Design 281	
		Display Screen Structure 282	
		Flash. 291	
		Display Screen Design 292	
		Dynamic Page Assembly 295	
		Display Complexity and Minimum View Time 297	
		Color Blindness 297	
	5.9	Navigation. 298	
		Purpose of Navigating 298	
		The Navigating Cycle 299	
		Navigation Tools 302	
		The "Product" of Navigation 303	
	5.10	Glyphs, Icons, Dials, Gauges, and Dashboards 305	
		Relationship to Style Guides 305	
		Glyphs. 305	
		Icons. 308	
		Dials and Gauges. 311	

		Dashboards...	318
5.11		Design Fundamentals for Icons and Dashboards.............	325
		Foundations	326
		Design Types of Dashboards (and Dials and Gauges)	327
		Salience Requirements	328
5.12		Trend Plots	329
		Build-on-Demand Trends...............................	330
		Continuous Trends	330
		Pop-up Trends	330
		Complex Trends......................................	331
		Trend Components	332
		Special Types of Trend Charts	332
5.13		Example Display Screens	334
		Overview Page.......................................	335
		Secondary Page	339
		Tertiary Page...	343
		Pop-ups	345
5.14		Mass Data Displays	345
		Departure from Steady-State Value Mode	347
		Departure from Normal/Expected Value Mode............	347
		Things to Keep in Mind	347
		Extending Mass Data for an Overview....................	349
5.15		Multivariate Process Analysis	349
		What Is Multivariate Process Analysis?...................	349
		Bender Treater Example	351
		Important Note.......................................	353
5.16		Displays Large and Small...................................	353
		Workstation Displays	354
		Off-Workstation Large Displays	356
		Requirements for Large Off-Workstation (OWS) Displays....	358
		Illustrations of OWS Large Displays.....................	360
		Off-Workstation Small Displays.........................	361
		Head-Up Displays	363
5.17		Video Walls..	365
		Conventional Video Walls	365
		Video Walls for Control Rooms.........................	366
5.18		Paper versus Electronic Screens..............................	370
		Pros and Cons..	372
		Naturalness..	373
		Readability	373
		Following the Thread	374
		Personalization and Annotation	376
		Comparisons to Think About	376
5.19		Fire, Gas, Safety Instrumented Systems, and Security Systems....	377

	5.20	Sound, Audio, and Video 377
		Sound.. 377
		Cues 378
		Announcements... 379
		Video 380
	5.21	Evaluating Effectiveness............................... 380
	5.22	Loss of View and Key Variables..................... 383
	5.23	Building Effective Screens 385
	5.24	Further Reading... 386

Part II   Situation Awareness and Assessment 389

Chapter 6   Situation Awareness and Assessment 391

6.1	Key Concepts 392
6.2	Introductory Remarks.................................... 392
6.3	The Situation Management "Situation" 393
	The Operational Setting 394
6.4	Situations to Be Aware Of............................ 395
	Problematic Situations................................. 395
	Operational Situations................................. 396
6.5	Strong Signals... 396
	Indirect Strong Signals 397
	Direct Strong Signals................................... 398
6.6	Situation Management Roadmap 399
	The Process of Situation Awareness 400
	Active versus Passive Monitoring 401
6.7	Situation Awareness Tools 401
6.8	The Psychology of Situation Awareness................... 402
	Ownership 403
	"Relative" Prime Responsibility 403
	Accepting Reality... 404
	Leadership and Cooperation 404
	The Triple Package...................................... 405
	Intuition and "Raw" Information 406
6.9	Situation Assessment..................................... 406
	Situation Assessment Question 406
6.10	Surrogate Models... 407
	Sources for Surrogates................................. 408
6.11	The Four-Corners Tool 409
	Acting, Outcome Questions............................. 410
	Not Acting, Outcome Questions 410
	Discussion.. 410
6.12	Close 411

	6.13	Further Reading.. 411

Chapter 7 Awareness and Assessment Pitfalls........................ 413

	7.1	Key Concepts 415
	7.2	Introduction 416
	7.3	Readers' Advisory 416
	7.4	Why We Make Mistakes...................................... 416
		Looking without Seeing 418
	7.5	Dangers from Automation 418
		The Substitution Myth 419
		Automation Complacency 420
		Automation Bias... 421
		The Generation Effect 422
	7.6	Mental Models 422
		Expected Roles 423
		Failure Avoidance 425
		Logic-Tight Compartments.............................. 427
		Surrogate Models... 429
		The Surrogate Model Test............................... 430
		The Deception of Two Reasons 430
		Remembering... 431
		Good Is Not Really Good Enough..................... 431
	7.7	Doubt... 432
		Possible Doubt 433
		Probable Doubt... 433
		Reasonable Doubt 433
		Shadow of a Doubt 434
		Dealing with Uncertainty................................ 434
		Lingering Doubt.. 435
		Managing "Truths" 435
	7.8	How We Decide 436
		Short-Term versus Long-Term 436
		Loss Aversion 438
		Sixth Sense 440
	7.9	Biases... 441
		Confirmation Bias 442
		Continuation Bias.. 445
		Anchoring Bias.. 445
		Halo Effect 446
		Bandwagon Effect 446
		Diffusion of Responsibility.............................. 446
		Post Hoc Ergo Propter Hoc.............................. 447
		The "What Then" Question 448
	7.10	Inattention Blindness 448
	7.11	Partial Information 450

	7.12	Myth of Multitasking 452
		Setting the Stage 452
		Multitasking 452
	7.13	Personalities 454
		Accident-Prone Behavior 455
		The Quiet Ones 455
		Situation Management Points 456
	7.14	Geography of Thought 456
		Norms and Conventions 457
		Logic and Reason 460
		Individuality 463
		Handling and Reporting Problems 463
		Three Postal Codes 464
	7.15	Institutional Culture versus Individual Responsibility 464
		Polarity 464
		Alignment Failure 465
		Historical Incidents 465
	7.16	Close 469
	7.17	Further Reading 470

Chapter 8 Awareness and Assessment Tools 471

	8.1	Key Concepts 472
	8.2	Introduction 472
		Knowledge Fork 473
		Awareness and Assessment Situation 473
	8.3	Alarm System 477
		Alarm Fundamentals 477
		Anatomy of an Alarm 478
		Alarm Philosophy 479
		Alarms 480
		Alarm Management 480
		Alarm Rationalization 482
		Alarm Response Sheet 485
		Process Trouble Point 489
		Alarm Activation Point 491
		Alarm Priority 494
		Alarm Rationalization Step-by-Step 497
		Alarm Metrics 499
		Operator Alarm Loading 501
		Contribution of Alarms to Situation Management 503
	8.4	Operator Ownership Transfer at Shift Change 503
		Transferring Information 504
		Receiving Information 505
		Verifying Information and Taking Ownership 505
		In-Shift Handover Emulation 506
	8.5	Alerts, Messages, and Notifications 506

			Notifications as Weak Signals 506

      Notifications as Weak Signals 506
      Properties of Notifications 508
      General Design and Implementation Guidelines 509
      Notifications and Logs 511
  8.6 Putting It All Together 511
  8.7 Making Situation Awareness Happen 512
      Capable of Doing the Job 513
      Design and Implementation............................. 513
      Usability 513
      Selling Management 513
      Auditing 514
  8.8 Close 515
  8.9 Further Reading.. 515

Chapter 9 Weak Signals 517
  9.1 Key Concepts 521
  9.2 Introduction 522
      A Word to the Reader 522
  9.3 Weak Signals 524
      Weak Signals Announce................................. 526
      Finding Weak Signals 528
      Weak Signals for Situation Management 529
      Categories of Weak Signals 532
      Weak Signal Concepts................................... 533
      What Weak Signals Look Like 534
      Examples of Weak Signals 537
      Hunches and Intuition Might Be Weak Signals 540
      Expectations Will Interfere with Weak Signals............. 540
  9.4 Building and Displaying Weak Signals 541
      Characteristics of Weak Signals.......................... 541
      Weak Signals from Direct Measurements and
       Observations... 542
      Weak Signals from Indirect Measurements
       or Observations 545
      Weak Signals from Trend Plots 562
      The Role of Icons, Dials, Gauges, and Dashboards.......... 564
      Intuition and "Raw" Information 565
  9.5 Models for Weak Signal Analysis 566
  9.6 Weak Signal Management 566
      The Work Process....................................... 567
      Step 1: Identification 569
      Step 2: Forward-Extrapolation 570
      Step 3: Backward-Projection............................ 573
      Step 4: Evidence for Confirmation....................... 575
      Step 5: Resolution...................................... 581
      Never Assume the Problem 582

		Recapping the Steps.. 582
		Classifying Weak Signals—A Review 585
		Summary of Extrapolation and Projection................. 586
		Weak Signal Management: Before and After............... 586
	9.7	Digging into Weak Signals................................... 587
		Trouble Indicators Come in Sizes 588
		Actively Looking for Weak Signals...................... 591
		Special Case of a Weak Signal Mapped to a Specific Problem....................................... 593
		Prove True or Prove False............................... 594
		Weak Signals Observed by Experts...................... 595
		Collaboration and Consensus 595
		Weak Signals Do Not Escalate........................... 596
		Weak Signals as Flags................................... 598
		Two-Cycle Weak Signal Analysis 600
		Accentuate the Negative, Eliminate the Positive........... 602
		Weak Signals and Checklists............................ 602
		Persistent Weak Signals 603
		Weak Signals That (Seem to) Lead Nowhere.............. 603
		Weak Signals among Strong Signals..................... 605
		Actively Looking for Weak Signals...................... 606
	9.8	Weak Signals Might Not Persist Very Long.................. 607
		Weak Signal Life Cycle 608
	9.9	Other Weak-Signal-Type Extrapolations..................... 609
		Near Hits (Near Miss).................................. 609
		What-If and HAZOP................................... 610
		Root Cause Analysis 610
		Alarm Rationalization.................................. 611
	9.10	Weak Signal Templates?..................................... 611
	9.11	Retrospective Weak Signals Case Study...................... 612
		Texas City 612
	9.12	Relationship between Weak Signals and Alarms............. 615
		Situation Awareness Depends on Both Alarms and Weak Signals 617
		Few Weak Signals Lead to Alarms 618
		Precedence of Operator Activity 618
		Weak Signals from Alarm Activations................... 618
	9.13	Relationship between Weak Signals and Critical Variables 619
	9.14	Operator Weak Signals Are Tactical (Short-Term)............. 619
		Tactical versus Strategic 620
		Tactical Weak Signals 620
		Strategic Weak Signals 621
	9.15	The Dependence of Weak Signal Analysis on Model Quality... 622
		Model Fidelity 622
		Identifying Model Inadequacies 622
		Weak Signal Work Process.............................. 623

	9.16	Getting Weak Signals Working 623
		Proper Foundation... 624
		No Shortcuts for Weak Signal Management 625
		Weak Signal Overload... 626
		Selling Management 628
	9.17	Troubleshooting Guide... 628
		Finding Too Few Weak Signals 628
		Finding Too Many Weak Signals............................ 628
		Cannot Forward-Extrapolate a Weak Signal 629
		Finding Too Many Problems When Forward-Extrapolating ... 629
		Many Weak Signals Forward-Extrapolate to the Same Problem... 629
		Cannot Backward-Project a Potential Problem 629
		Many Potential Problems Backward-Project to the Same Evidence ... 629
		Confusing Evidence Obtained from Backward-Projections ... 629
		No Evidence Found from Backward-Projections 629
		The Weak Signals Procedure Does Not Seem to Work at All ... 630
		The Weak Signals Procedure Does Not Work All the Time ... 630
		The Weak Signals Procedure Is Too Hard to Use 630
		Finding Evidence of Potential Problems Actually Present ... 631
	9.18	Additional Thoughts about Weak Signals 631
		Artificial Intelligence for Weak Signal Analysis 631
		Identifying Gaps in Training and Procedures.............. 632
		Weak Signals and Incident Investigations 633
		Weak Signals Are Not the Only Way 634
		Skipping over Weak Signal Processing..................... 634
		Wrapping It Up.. 636
	9.19	Close 637
	9.20	Further Reading.. 637

Part III Situation Management........................ 639

Chapter 10 Situation Management 641
10.1 Key Concepts 642
10.2 Introduction 643
    The Situation Management Activity 644
    Step-by-Step Working Process 645
10.3 Lessons from Air France Flight 447 647
    Background... 647
    History of Failures... 648

	Lessons	648
10.4	Operations Safety Nets	649
10.5	Using Experts and Benefiting from Experience	651
	Experience	652
	Expertise	652
10.6	Safe Conversations	652
	First, a Word about Safety	653
	Safe Communication	653
	Mindful Conversations	655
	Mirroring, Acknowledging, and Tracking	656
10.7	Operational Drift	658
	Tenets of Operation	660
10.8	The Restricted Control Room	661
	Control Rooms in Normal Operations	661
	The Control Room in Abnormal Operating Conditions	662
	Restricted Control Room Conditions	663
	Initiators of a Restricted Control Room Condition	663
	Termination of a Restricted Control Room Condition	663
10.9	The Sterile Control Room	664
	Sterile Conditions	664
	Initiators of a Sterile Control Room Condition	665
	Termination of a Sterile Control Room Condition	665
10.10	Lessons of Defensive Operating	665
10.11	Managing Everyday Situations	666
	Fictional Illustration	666
	Operator Intervention Caution	667
	Operators Must Not Be Innovators	668
	Operator Duties	669
	Following the Rules	669
	Flow of Operator Activities	670
	General Flow of Operator Activities for Situation Management	670
10.12	Doubts and Concerns	673
10.13	Delegation	674
10.14	Collaboration	675
	Use the Knowledge Fork	675
	Caution	676
	Air Florida Flight 90 Crash	676
	Crew Resource Management	678
	Worst Case First	680
	10th Man Doctrine	680
	Triangulation	681
	Red Teams	684
10.15	Permission to Operate	684
	Management's Role	686

	Operating Situations 686
	Operational Modes 688
	How Permission to Operate Came to Be 689
	How Permission to Operate Works..................... 690
	Permission to Operate................................ 691
	Withdrawn Permission to Operate 692
	Alternate Methods for Having Permission 692
	Safe Operating Limits................................ 694
	Managing the Operator's Permission 694
10.16	Escalation 696
	Communication and Collaboration.................... 698
	Escalation 700
	Escalation Resources................................. 702
	Direction of Escalation 703
	First Duty of Escalation............................. 703
	The Process of Escalation 704
	When to Escalate 704
	Escalation Design................................... 706
	Escalation in Perspective............................. 706
10.17	Escalation Teams 706
	Escalation Team Composition 708
	The Escalation Activity.............................. 709
	Frontline Coaching and Mentoring.................... 709
	Readiness Evaluation Role 710
	Training.. 711
	Abnormal Situation Management Process Model........ 711
	Weak Signals for Abnormal Situation Management 712
10.18	Command and Control................................ 713
	Supervisor Not Fully Qualified to Be an Operator 713
	Supervisor Is Fully Qualified to Be an Operator......... 714
10.19	Alternate Safe Operating Modes 715
	Operator-Initiated Shutdown......................... 715
	Automated Shutdown 716
	Alternatives to Shutdowns 716
	Safe Park 718
	Keeping Perspective 719
10.20	Managing Major Situations 719
	Major Situations.................................... 720
	Operator Redeployment 720
10.21	Control Room Situation Codes.......................... 725
	Code Gray.. 727
	Code Orange.. 728
	Ending Codes....................................... 728
10.22	Operability Integrity Level for Online Risk Management...... 728
10.23	Managing Biases, Overcoming Pitfalls..................... 730

10.24 Safety and Protective Systems 731
10.25 Key Performance Indicators 731
10.26 Miracle on the Hudson 732
10.27 Achieving Situation Management 735
       Control Room Management Choices 737
       The Process Engineer 738
       The Controls Engineer 738
       The Operations Engineer 739
10.28 Training, Practicing, Evaluating, and Mastering 739
       Weak Signals 740
       Collaboration 740
       Escalation 740
       Permission to Operate 740
       Other Tools and Support 740
10.29 Closing Thoughts on Situation Management 741
10.30 Further Reading 741

Appendix: Definitions of Terms, Abbreviations, and Acronyms 743

Credits 761

Index 763

List of Figures

Figure 1-1	Our operator's first responsibility is to keep closely focused on the process to see everything unusual or abnormal. Everything observed will need to be followed far enough to either know that nothing important is going on or to correct things.	13
Figure 1-2	The operator has the important responsibility to find and remedy all abnormal situations that exist in his plant or operation. Unless he finds them before, the alarm system will announce those with alarms configured for them. He must find all the rest by whatever means that are available for him to use and that he knows how to use.	14
Figure 1-3	Alarms are the way process designers pass on to the operator for action all problems that should be identified by good engineering and a deep understanding of how the enterprise is supposed to operate. The rest must be found using something else.	14
Figure 1-4	Once alarms are taken care of, the operator's responsibility boils down to how to see all the rest of the abnormal conditions and situations he must understand and plan for. Anything missed or misunderstood usually causes trouble sooner or later.	15
Figure 1-5	Operators will need to find all the "likely real problems that could be happening" suggested by the "indications of abnormal operation" that they must see. This frames the requirements for operator interface design, as well as the situation awareness and assessment tools. All need to work.	15
Figure 1-6	Identified problems must be confirmed to be true; not assumed or just thought to may be true. This requires specific additional investigation by the operator. Only once they are confirmed, can they be worked on. And included in the "working" is the necessity to ensure that the steps taken actually did remedy the problem.	16
Figure 1-7	Framework for situation management requires effective control room management. Control room management is the competent foundation of management processes; everything in proper working order; proper equipment, processes, and training in place; and an operator interface that exposes the needed information and provides the necessary control handles. The situation	

	management structure on top requires the alarm system, the ability to detect early concerns (via weak signals) before they become problems, and the safeguard of moving the plant to a safe state (via permission to operate) when good operation cannot be assured.	19
Figure 1-8	Basic four-activity situation management process depicting the progression from "observe" to "confirm" to "remediate" with some possibility to move back a step when evidence is not there or not enough to accept. Escalation and collaboration may take place at any step.	25
Figure 1-9	TransAsia Flight 235 crashing into the Huandong Boulevard Bridge as a result of the pilot shutting off the only running engine, mistaking it for the other engine that had already shut down due to unknown causes (and therefore did not need attention during a takeoff). With no engines running, a crash was inevitable.	28
Figure 1-10	Process safety time and SUDA illustrating that everything takes time. (Illustration is not to scale.) It is only when the time required for proper operator management, "time to manage fault," is less than the time for the process to go into fault, "fault-tolerance time," that operator intervention is possible (and hopefully successful).	31
Figure 1-11	Normal, off-normal, and abnormal operating regions depicting the separation between normal (things are okay and need to be kept there), abnormal (something is wrong and should be visible), and off-normal (something might be headed toward a problem but is hard to see now).	36
Figure 1-12	Operation in the abnormal region is announced by appropriate alarms.	36
Figure 1-13	Operation in the off-normal operation region must be identified, confirmed, and remediated by operator activity not announced by alarms.	37
Figure 1-14	Operating regions identifying the "nested" nature of the differing regions as operation deteriorates from being right on target through upset and potentially damaging.	39
Figure 1-15	Operating region showing alarms announcing (protecting) likely transition out of the upset region toward worse operation. Operator intervention will be required in order to improve things.	40
Figure 1-16	Operating region showing strong signals announcing problems that were likely protected by alarms, but normal alarm response does not seem to be doing the job.	41
Figure 1-17	Operating region showing weak signals that must be found and processed by the operator, suggesting problems that were not considered serious enough to be protected by alarms but could eventually develop into problems if not recognized and handled.	42
Figure 1-18	Dilbert weighs in on best practices. They are the recognized way to do things by industry. In the United Kingdom they generally have the interpretation of due diligence.	44
Figure 2-1	Agricultural silos illustrating their individual significant size but showing no visible connection between them.	82
Figure 2-2	Pictorial depiction of a mechanical engineering silo.	83
Figure 2-3	Depiction of an enterprise with component organizational parts shown as silos. Three are illustrated. Of course, there are more. Within each is a complete set of organized components (let us also refer to them as subsilos, but each is a silo as well) that together make up a well-functioning component. Each silo and each subsilo has the essential properties within itself to properly function.	83
Figure 2-4	Two competing companies at year 1 showing a clear gain of pursuing short-term goals over long-term goals in the short term.	90

List of Figures

Figure 2-5	Two competing companies at year 2 still showing a gain of pursuing short-term goals over long-term goals but much less than that of the first year.	91
Figure 2-6	Two competing companies at year 5 showing a clear and dramatic loss by pursuing short-term goals over long-term goals in the long term.	92
Figure 2-7	Model reference design diagram illustrating how information from the real process is used to "build" a model that then permits a "look inside" to see what could not be seen in the real process. In this case, information about the alarm conditions inside.	109
Figure 2-8	Model reference adaptive control diagram where a model of the real process is used to constantly update a controller for the real process to ensure it works as needed despite significant changes in operation or situation.	110
Figure 2-9	Concept of operator as coordinator—no hands.	111
Figure 2-10	Conceptual plant area model showing the various levels of equipment and how it is "attached" in a diagram to the levels above and below. When such a structure is used for naming conventions, it is much easier to use automated or semiautomated procedures to modify operating conditions or alarm settings and the like.	116
Figure 2-11	Illustration of typical decomposition boundaries using natural (physical) relationships or functional relationships as decomposition boundaries. Shown are different colored parts of four separate functional components, such as a separator, a heater, a mixer, and an analytical processing component. Natural boundaries might be first stage, second stage, third stage, and finishing stage.	118
Figure 2-12	Input and output classification questions to accompany a decomposed element to choose the connectivity for each decomposed element with each other and the other components of the plant.	119
Figure 2-13	Generalization of how a decomposed element relates to other elements and the rest of the plant. This provides an effective but simplified model to use for alarming, human-machine interface (HMI) design, training, and more.	120
Figure 2-14	A few examples of key repeated elements. Each key repeated element represents a production module that is compact (clear boundaries, inputs, and outputs) and appears often in a plant. Managing them this way is efficient and promotes uniformity. Depicted are a vertical separation tower at the left, a significant flow loop in the center, and a stirred-tank reactor at the right.	121
Figure 2-15	Example of a key repeated subsystem that is constructed with key repeated elements defined earlier that fit the specific subsystem being described. Other repeated subsystems use other repeated elements as needed to adequately describe each. Managing them this way is efficient and promotes uniformity.	122
Figure 2-16	Entire plant or enterprise constructed of the building blocks defined earlier (key repeated elements and key repeated subsystems) and other pieces as needed to complete the description. Managing it this way is efficient and promotes uniformity.	123
Figure 2-17	Condenser and its operating regions. Each key repeated element is operated, alarmed, and understood in relationship to its own specific operating regions and associated risks.	124
Figure 2-18	Pump and its operating regions, where those operating regions will be specific to the pump.	124
Figure 2-19	Pump and tower connected by starting from a basic pump and vertical tower and customizing each to precisely fit the specific pump and tower here. Most of the basic element will remain unchanged to the degree that it had been carefully defined as a template.	125

Figure 2-20	The operating regions for a key repeated subsystem do somewhat reflect those of the individual elements from which they are built; however, by virtue of interconnections and operations, the regions are a new overall situation.	126
Figure 2-21	Transformational analysis diagram depicting the (left-to-right) flow of operations in a plant where each operation step is identified as to complexity (with simplest at the bottom and most complex at the top).	128
Figure 2-22	Transformational analysis diagram showing the degree of agreement between the number of important incidents and the degree of complexity with more incidents at the more complex step.	130
Figure 2-23	A chain is only as strong as its weakest link. When it stays intact, you can have whatever is attached to it.	138
Figure 2-24	Disaster chain represents an intended safeguard where each link that *does not break* represents a critical safety factor that *did not work*. If none of the links break, you pull the disaster right into your lap. Most serious industrial incidents have disaster chains more than three links long. This means that all three safeguards did not work as designed (and required) to ensure safe operation!	139
Figure 2-25	"Swiss cheese" model of how an incident (accident?) can occur despite several presumed safeguards in place. If each safeguard has significant weaknesses or faults, it is just possible for a single event to find a fault for each safeguard and therefore cause the sum total of all safeguards to fail. Therefore, it is the quality of each safeguard that is important, not just the number.	140
Figure 2-26	Fire and explosion at Milford Haven (1994) that was the "last straw" in serious industrial incidents in the United Kingdom. Following it, the HSE commissioned new regulations including the current best practices in alarm management.	141
Figure 2-27	Piper Alpha platform in the North Sea (1988) engulfed in a gas explosion and fire resulting in large loss of life caused by numerous operational and safety lapses.	142
Figure 2-28	Aftermath of the BP Texas City, Texas, refinery fire (2005) that resulted in large loss of life and was caused by numerous significant safety and operational lapses.	143
Figure 2-29	Deepwater Horizon drilling rig before and during the disaster (2010) caused by ignoring safety issues to keep drilling on schedule.	145
Figure 2-30	Olympic Pipeline disaster in Bellingham, WA, caused by a culture of neglect for pipeline monitoring and failed follow-up of problems and near misses.	146
Figure 3-1	Boundaries and responsibilities that identify who is responsible for trying to prevent operational excursions, from what region of operation, and who is in charge once an excursion has occurred.	157
Figure 3-2	Our circadian clock illustrating how our natural body cycle of alertness and functioning follows a daily rhythm. If our internal clock is much different from the actual time clock, we are not at our best.	166
Figure 3-3	Major incidents occur at our least-alert times. Notice that both Chernobyl and Three-Mile Island initiated near the "drowsy" level of the rhythm.	167
Figure 3-4	As operator loading changes from normal to upset situations, the amount of available time to attend to the problem not only disappears but without aid will overload.	172
Figure 3-5	Shift handover timing activities showing first communication to transfer knowledge followed by collaboration to transfer ownership.	186
Figure 3-6	Detailed board operator shift handover timeline identifying leaving operator preparing the handover report, sharing the report with the arriving operator,	

	List of Figures	xxix

	transferring the operational control, and sharing joint operation and collaboration leading to the arriving operator assuming full ownership of shift.	188
Figure 3-7	Example controller (operator) log entry tool that illustrates the capabilities and provides an example of a structure that is easy to use and promotes completeness and understanding.	197
Figure 3-8	Example of a shift handover tool designed to follow API Recommended Practice RP-1168, *Pipeline Control Room Management*, including a checklist of steps on the right and an area for a controller to make specific shift handover notes.	201
Figure 3-9	Detailed shift field operator handover timeline identifying the leaving field operator preparing the handover report, sharing the report with the arriving field operator, and transferring operational control resulting in the arriving field operator assuming full ownership of the shift.	202
Figure 3-10	Combined timelines for both board operator and field operator handovers. Green (barbell) line depicts how the alignment of the handovers relates to the handovers. This arrangement allows the control room to become fully settled before the field does its handover.	203
Figure 3-11	Detailed supervisor handover timeline noting that the leaving supervisor first monitors the operator shift handovers and then prepares the handover report. The arriving supervisor first monitors the fatigue levels of both arriving and leaving operators. Then both supervisors review the supervisor handover reports. Once that is done, the leaving supervisor departs, leaving the arriving supervisor in charge.	204
Figure 3-12	Combined timelines for both control room operator and supervisor handovers. The magenta (barbell) line depicts how the alignments of the individual handovers relate to each other. This arrangement allows the supervisors to monitor the joint operations phase of the control room handover, assess the fatigue levels of both operators, and assist in the handover in case of problems.	205
Figure 3-13	Combined timelines for control room operator, supervisor, and field operator handovers. The magenta (barbell) line depicts how the control room and supervisor handovers relate to each other. This arrangement allows the supervisors to monitor the joint operations phase of the control room handover, assess the fatigue levels of both operators, and assist in the handover in case of problems. The orange (barbell) line depicts how the supervisor and field operator handovers relate to each other. This permits the supervisor to monitor and assist with the field handover preparation and activity.	207
Figure 3-14	Personal needs hierarchy diagram that shows the importance of providing for the full self.	215
Figure 4-1	Early control room, circa 1950 design. Note the presence of important indicators yet few controller stations. The operator was expected to be able to gain a good overview of the entire process by "eyeing the board."	226
Figure 4-2	Contemporary control room circa 2014 illustrating the human-machine interface (HMI) configurations for each individual operator as well as the large shared displays for coordination and overview.	227
Figure 4-3	Meteorological control room for the European Organisation for the Exploitation of Meteorological Satellites with many displays arranged in a wide sweeping arc. Note that there are few operators.	228
Figure 4-4	Example simple control center design illustrating the presence of supporting documentation, individualized operator HMI displays, and overhead shared large displays for overview and coordination.	236

Figure 4-5	Example of a small bring-your-own workstation collaboration center showing the dedicated nature of it, electronic power and connectivity for individual work tools, and large shared display for collaboration.	239
Figure 4-6	Control room of the attack submarine USS *Seawolf* (1997). Note the proximity of coordinating operators and supervision located directly behind.	240
Figure 4-7	Human factors components of a successful control room design.	246
Figure 4-8	Control room of a Russian nuclear power plant illustrating the use of a wide variety of interface technologies deemed suitable to ensure situation awareness, assessment, and operator intervention for managing a critical plant.	249
Figure 5-1	ISO 11064-5 nomenclature diagram for graphical displays. This level of detail permits specific reference and discussion. Starting from the upper left, each successive component will be found as continuing a part of the component above.	264
Figure 5-2	Examples of the six basic instruments for all aircraft. From the top left (clockwise) are the basic "T," Schempp-Hirth Janus-C glider equipped for cloud flying, and Slingsby T-67 two-seater. By standardizing critical instruments this way, every pilot can fully orient to any aircraft in the world.	276
Figure 5-3	Pressure symbol icon depicting in one compact unit all important aspects for the variable being monitored. This ensures a proper frame for the information.	280
Figure 5-4	Pressure symbol (from Figure 5-3) showing an active medium-priority alarm (reserved color for medium priority alarms is yellow). Note the redundancy of the alarm rectangle around the icon in yellow and the solid yellow square with an "M" inside.	280
Figure 5-5	Solenoid valve icon. This valve is energized (note the filled "operator" portion) and closed (note the filled "butterfly" portion).	281
Figure 5-6	Three hierarchy levels for display screens. This structure permits all information to be located directly and then accessed without excessive "clicking." This structure also encourages uniformity and appropriate co-location of related information.	283
Figure 5-7	Responsibility-based display screen organization. This provides ready access to all information and controls within the given operator's specific area of responsibility. The arrangement of content of each level is governed by the relationship of the parts to the overall responsibility of the operator without undue regard to exactly where the equipment is physically located.	286
Figure 5-8	Risk-based display screen organization. This provides ready access to all information and controls within the given operator's specific area of responsibility focused only on operational risk without undue regard to exactly where the equipment is physically located. The arrangement of content for each level starts with the highest-level risks. Each subsequent level supports access to understanding and working needed problems or issues. This organization would be important during significant operational problems.	287
Figure 5-9	Task-based display screen organization showing two parallel paths (vertical portions) and how access to the next level down (horizontal links) provides detailed instructions and/or information to support understanding and management of the situation at hand. This arrangement allows direct focus on the task at hand and avoids undue navigation to locate needed supporting information.	288
Figure 5-10	Similarity-based display screen organization used to highlight operational differences, including subtle ones, between very similar process and operational units. This type of display makes it easy to understand where	

	problems may be developing by employing the same "mental model" of the process for all of them.	289
Figure 5-11	Geographically based display screen organization making it easy for the operator to see all the "moving parts" that should be working together harmoniously because they are colocated. It is presumed that any operational problems would quickly cascade to other closely located parts on the display screen.	290
Figure 5-12	Illustration of the flash cycle for a text field. The thin-lined red box identifies the flash-off cycle. The broad-lined red box identifies the flash-on cycle. A red line is used to indicate an alarm condition that uses the color red. During the entire period where the message is flashed, a lined box is always visible. If the message is not needed to be visible when it is not being flashed, then the line identified by "Message NO FLASH" will not be visible.	292
Figure 5-13	A flashing red light on a rural highway. The illuminated white outside ring is visible at all times. The inner red light cycles between being on (red) and being off (unlit), depicted as black. With this signal depiction, it would be very hard to miss the signal being present and working properly (flashing or not).	292
Figure 5-14	Example of the inappropriate use of color in an outdated HMI design. While color clearly separates the various parts of the screen and is interesting to view, it actually fails to highlight any but the most primitive abnormalities. Notice how the bright yellow of the heat exchangers appears to jump out of the view, but their actual operation is of very little interest. Even when the eye is looking for things, the colors get in the way by their distractions.	294
Figure 5-15	Example of the overuse of color, 3-D shading, and physicality. Notice how distracting all of this is. The attempts to add physical reality, in addition to color, do even more damage to the viewer's ability to see and understand the information. You see the picture. You do not see the information.	295
Figure 5-16	Illustration of unintended consequences of the specific way graphical artifacts are drawn. The two views represent the *same* process units. Notice how the connection from the pump to the vessel at the left has a drop in the leg. The view at the right does not. The very presence of the drop distracts by suggesting that it represents something important. We know it does not.	296
Figure 5-17	Glyph that means "Do not dispose of in ordinary trash." It is clear and easy to understand exactly what is intended.	306
Figure 5-18	Glyph traffic sign indicating to drivers to slow down vehicle speed on the road ahead.	306
Figure 5-19	Example tool tip showing "Hypertext Markup Language" displayed when the cursor is hovered over the HTML link. Where abbreviations or other short codes are used in notations or messages, tool tips can be used to provide more information for understanding should it be needed.	307
Figure 5-20	Example tool tips combined with glyphs. Only the appropriate one is displayed as they are used one at a time.	307
Figure 5-21	Glyphs as navigation buttons illustrating 12 different buttons. Only the appropriate one is displayed as they are used one at a time.	307
Figure 5-22	Social media glyphs illustrating 16 different types or links. Only the appropriate one is displayed as they are used one at a time.	308
Figure 5-23	Example icon illustrating a normal temperature. In addition to the label *Normal*, this icon depicts a normal temperature reading because it does not look abnormal. That is, nothing about the icon is visible except the blue bar	

	showing the relative value of the current reading. Contrast this with the icon in Figure 5-24.	309
Figure 5-24	Example abnormal temperature icon, often referred to as a *contextualized icon*. Note the change from the previous icon. Now showing is the relative value (Exception) regions in the temperature range showing the "sweet spot" for normal (the white region) and the increasingly abnormal regions (the gray areas of increasing darkness). The current value (wide blue bar) is in alarm (red region) with added alarm indication (red boxed value). The damaging region value (thin blue bar) is slightly above.	310
Figure 5-25	Sample progress bar used to depict both the current progress of a task (blue squares), the total progress the task will take (black rectangle), and a rotating arrow to show that the task is progressing despite any slow accumulation of blue squares.	310
Figure 5-26	Sample progress bar with more progress shown (more blue squares than the progress bar in Figure 5-25).	310
Figure 5-27	Example gauge to track the current state of all nuisance alarms (for a wastewater plant). At a glance you can see that the (hourly average) number of nuisance alarms is 22.5 (hourly average) and increasing (within the center circle); which are the oldest and newest ones; the relative number of times in alarm (the length of the arrow) with the longest at the top (decreasing in numerical number clockwise around the wheel); and whether the individual ones are increasing or decreasing in number (arrow out or arrow in).	311
Figure 5-28	Nominations gauge explanation of symbols. Though they might appear to be complex, they are not complicated. Once mastered, they will permit the viewer to see the overall progress and completion potential.	314
Figure 5-29	Design structure for nominations gauge identifying all component parts. Although none of these labels will be shown on the gauge they might be coded as tool tips if desired.	315
Figure 5-30	Nominations gauge illustrating that without mishap, the current nominations (345 MMB) will be met at the required time (15 hours and 43 minutes from now).	315
Figure 5-31	Nominations illustration requiring a flow decrease in order to satisfy the requirements. Unless a decrease is made, the actual delivery will be over by 2.7 MMB.	316
Figure 5-32	Nominations require increase of flow. Unless an increase is made, the actual delivery will be under by 3.5 MMB.	317
Figure 5-33	Nominations require an increase in flow, but no increase is possible. Without any resources or room to make the necessary increase, the actual delivery will be under by 2.9 MMB.	317
Figure 5-34	Nominations requiring a decrease in flow, but a sufficient decrease cannot be made. While there is room to make a flow reduction (red dotted box), no reduction level is available that can eliminate the overage. Without any change, the overage will be 4.3 MMB.	318
Figure 5-35	Social media respondents showing their physical location on a street map backdrop. The respondents are using various media engines (Linked In, Facebook, Google Plus, and Twitter) to support a local issue (unknown). The deeper gray region shows the center of mass.	319
Figure 5-36	Entire pipeline system nomination performance dashboard. The system contains six segments. One segment (Treseer Valley) is low and cannot make up the	

	needed difference. Another segment (Moreland Gap) is a bit over and has plenty of room to reduce flow to meet the requirements. All others are on target.	320
Figure 5-37	Deviation diagram showing an overview of an entire operator area. Shown are 36 critical operating values and their departure from the expected value. The arrangement is the order of the significant movement of product from area entry (at the left) to area exit (at the right).	321
Figure 5-38	Overloaded dashboard display screen. Clearly, there is a lot of information and it might all be related. There are many reference marks for normal and acceptable limits. However, there is so much that it is of too little use unless closely studied. Dashboards are not intended to be studied. All important inferences should be visible at a glance and readily understandable.	323
Figure 5-39	Example of excessively complicated dashboard screen. It is composed of three nuisance alarm gauges, but it is just too dense to be conveniently useful.	324
Figure 5-40	Clear and effective dashboard. Everything is related. Each dial, graph, glyph, and gauge is simple and understandable. Quickly, a viewer will gain all intended information.	325
Figure 5-41	Single trend. This is useful when the time-changing values of the variable (the history) are important to see in addition to the current value.	329
Figure 5-42	Overly complex trend combination. Not only is each one dense, the combination appears overwhelming. Any screen element that requires study and close attention is problematic: doing so takes a lot of time, and something important will almost always be missed or misinterpreted.	331
Figure 5-43	Trend plot components identified. All are important to ensure the viewer correctly understands the intended information.	332
Figure 5-44	Example of superimposed time-related trends. This type of graph is used to depict how several variables relate to each other in real time.	333
Figure 5-45	Illustrative fan chart (based on the Bank of England "Inflation Report") extremely useful to compactly depict where future values are likely to lie. The "fan" part begins at the beginning of the uncertainty period (which may actually be the present, before, or a bit later).	334
Figure 5-46	Sparkline chart showing key. This is an effective way to show a trend in a way that mimics an icon or gauge.	334
Figure 5-47	Overview display screen showing all the important aspects of an operator's entire area of responsibility. Key variables are shown as icons. Alarms are summarized so their locations and criticality are apparent. Trends are present for key inflowing variables coming from other plants. Similar trends are for the important product variables of this plant that flow to other plants or for shipment.	336
Figure 5-48	Identifying the components of the overview display page (shown before this as Figure 5-47).	337
Figure 5-49	Overview display screen now showing alarms that are summarized (see four-box stacks depicting one or more of the highest priority alarms) so their area locations and criticality are apparent. Note the pop-up dialog box with suggested actions in the lower center, and the "tool tips" values for certain variables.	338
Figure 5-50	Secondary page for Riser/Regenerator. This is one of the five major components of the entire plant. This level provides all details including alarms and access to controls. The icons and other components use the same convention as the overview.	340

Figure 5-51	Identifying the components of the secondary display page (shown earlier as Figure 5-50).	341
Figure 5-52	Riser/Regenerator subordinate secondary display at the riser level. Here all useful process variables and alarm conditions are shown.	341
Figure 5-53	A situationally based secondary page that is pre-engineered to provide information to help the operator understand what the situation is and how to manage it. This display combines the window (shown earlier as Figure 5-52) together with three additional windows to form a coherent view. It is constructed by the process of focus and yoking.	342
Figure 5-54	This figure explains the (on-the-fly) construction details that use focus and yoking to build a situationally based secondary page. This specific view was set up because of the (in red) low-temperature alarm in the regenerator. It adds reader-information labels to the information shown in an earlier screen (Figure 5-53).	343
Figure 5-55	Procedure tracking display screen. Each window depicts a different level of activity for the work being done. Figure 5-56 identifies each.	344
Figure 5-56	Identifying windows and glyphs for tertiary page.	345
Figure 5-57	Heat exchanger example of a mass data display. Depicted is a single measurement class of "heat efficiency" that is computed at every location shown by the arrows. The inclination of the arrow indicates whether the efficiency is increasing (the more up, the more increasing), steady (horizontal), or decreasing (the more down, the more decreasing).	346
Figure 5-58	Mass data display *format* included within an advanced control monitoring display. Certain locations in the furnace are control points, so the added elements provide deeper detail into the understanding of the particular performance of the controls.	348
Figure 5-59	Illustration of multivariate process monitoring construction. The center scatter diagram-like element shows the locations for the key information variables over time. Analysis permits the construction of a normal region of them. Movement of the "dots" in the diagram out of the normal region signifies that the process itself is moving away from normal.	351
Figure 5-60	Overview diagram of components for a Bender Treater process unit.	352
Figure 5-61	Bender Treater showing clusters of operation. Movements between clusters are important to observe and understand. The diagram makes it easy to see this movement.	352
Figure 5-62	Workstation configuration options depicting a mix of sizes of displays and their physical arrangements. The main operator working displays are at the lower center. The upper ones and the side ones are usually preassigned so the information is always where expected. Because all are on-workstation, the relative sizes are for convenience and preference.	355
Figure 5-63	Closely spaced main workstation displays combined with poor use of large off-workstation display. Note the excessive use of color, content density, and small visible sizes that make them difficult to read and identify what is important.	356
Figure 5-64	Workstation (WS) displays with a large off-workstation (OWS) display above.	357
Figure 5-65	Entire plant floor status mimic display arrangement.	358
Figure 5.66	Physical analog mimic panel. Key measurement points are there within a superimposed line drawing of the entire power plant. The dedicated alarm panel "windows" are arranged at the top. While appearing sparse, this design afforded an excellent overview of the entire plant.	359

List of Figures   xxxv

Figure 5-67	Two sets of flat panels arranged as OWS large displays with a third shared one in the middle. (The information content could be better selected.)	360
Figure 5-68	Operator using flat panels arranged as OWS large displays. This use is not considered a video wall because the displays do not depict a continuous view; rather, each display is a "tile" in a multi-tile mosaic view display. This has an enormous potential for significant distraction away from the operator tasks required in the control room.	361
Figure 5-69	Typical smartphone screens. The very small display area requires special attention to content and links in order to be useful. Special content and formatting is required.	362
Figure 5-70	Smartphone and tablet navigation icons. These are familiar to most of us, of course; however, this methodology is suitable only for small-format devices. Without careful attention, it is easy for an entire screen to be filled without proper grouping and arrangement.	362
Figure 5-71	A head-up display example for an automobile. Only the white content is the display portion. It appears directly in the driver's line of sight, making it less necessary for the driver to take his eyes off the road. This one includes the vehicle speed (90 kmh) and notification of an approaching turn (300 m) ahead.	364
Figure 5-72	Eyeglasses version of a head-up display. This has a very small display overlaid within each lens.	365
Figure 5-73	Video wall as background.	366
Figure 5-74	A view shows the operation room at the Meteo-France Toulouse site, named as Meteopole, outside the city of Toulouse, France, November 3, 2015. With two of the most powerful supercomputers in the world, French national meteorological service Meteo France participates in the international scientific work on climate change, on which is based the negotiations of the Climate Conference (COP21). Paris [hosted] the World Climate Change Conference 2015 (COP21) from November 30 to December 11. Picture taken November 3, 2015.	367
Figure 5-75	Video wall as the significant view into the process.	367
Figure 5-76	Futuristic video wall display as a main display.	368
Figure 5-77	Not a particularly good use of a video wall–type display. Depicted here is a personal workstation arrangement; however, it really is used as an extension to augment a very limited tablet display. The overview function is entirely missing. Note the sit-stand capability of height adjustment. Collaboration could be facilitated, but its provision must be supported with appropriate screen content in this configuration.	369
Figure 5-78	MBTA Boston Transit System main control room. There are several individual control pods as well as a separated interior area with a viewer's balcony above.	370
Figure 5-79	Another example of the use of a video wall. Here the shared element is the entire power distribution grid. Each operating pod manages a portion of this grid. A raised view and command center overlook the entire operation.	371
Figure 5-80	Video wall in an uncomplicated control room design. Shown here are main operators as well as 80 or support personnel at the rear. Notice this use of the video wall has content segmented. Doing it this way requires the content to be located in the area of local need and therefore not in view of all operators.	371
Figure 5-81	Original unimproved screen. There is excessive use of color, yet it fails to identify only abnormal aspects. There is the distracting use of 3-D. The mix of identifying labels, process values, and process icons is confusing and lacks focus.	382

Figure 5-82	Improved screen. Color is consistently used to represent abnormal (red and yellow for alarms, blue for equipment. There is a consistent use of icons (although the duplication of actual readings is unnecessarily distracting). The equipment depiction is clean and simple, thus providing just the proper backdrop for the other content.	383
Figure 6-1	Operator roles and responsibilities and how they differ depending on where he is during the shift.	393
Figure 6-2	This illustration identifies the different operating regions and their "nested" nature as operation deteriorates from being on target through upset and potentially damage. Also shown at the right side are the parts of the infrastructure that are intended to manage upsets or, if they happen, to protect against the worst.	394
Figure 6-3	The interactive flow from situation awareness to situation management showing how the operator's knowledge must build by first observing, then understanding, and finally managing. The bidirectional arrows depict the fact that each step is a collaboration between the one preceding and the one next; any concerns in one must be examined in light of both.	400
Figure 6-4	Four-corners decision framework that forces a proper balance between what might be good or bad for both the act of making a choice and the alternative of making no choice. With this information in hand, an informed decision is likely.	410
Figure 7-1	Asiana Flight 214 July 2013 crash caused by the flight crew failing to collaborate regarding approach problems due to cultural customs overriding safe flight requirements.	424
Figure 7-2	Aerial view of the April 2005 deadly Amagasaki, Japan, rail crash site caused by excessive speed due to the train operator trying to avoid excessively harsh punishment expected to come from minor operating delays on the trip.	426
Figure 7-3	The 80:20 Rule (effort versus benefit view) illustrating the seductive nature of early benefits that seem beneficial yet are not enough for responsible work. While the first 20% of the effort yields about 80% of the benefits, often the real and lasting value is derived primarily from the last 20%. Gaining that requires an extra 80% of effort.	432
Figure 7-4	Illustration of loss of scale showing a proper category of response (have a fire, then try to put it out) but one that is completely inadequate and dangerous for the true magnitude of the problem.	444
Figure 7-5	Passing the basketball (still frame) setting up the situation of a foreground task requiring an outside viewer's full focus of attention.	449
Figure 7-6	The gorilla, clearly in view, was missed by about half of the outside viewers of this basketball-passing exercise.	450
Figure 7-7	Partially obscured phrase that appears to be quite readable due to our ability to project meaning from incomplete causes. Often we are wrong, and without finding the truth, we act inappropriately.	451
Figure 7-8	World map delineating East, West, and Blended cultural frameworks. These regions (marked by the magenta and green dotted lines) exhibit different norms with regard to their approaches to life and thought.	457
Figure 7-9	Illustrative page from an HMI screen showing the usual component layout structures of windows, formats, and elements.	458
Figure 7-10	Eye flow on an HMI page (depicted in three windows, from left to right) illustrating different starting points and flow of view based on cultural convention. This suggests that layouts will need to guide the viewer's view. In	

	the West it is left to right, top to bottom. In the Middle East it is right to left, top to bottom. In the East it is top to bottom, right to left.	458
Figure 7-11	Nonverbal marketing message whose entire meaning is dramatically different when the individual frames are viewed in order from left to right (the intended flow by a Western designer) or right to left (typical for the culture where it was targeted).	459
Figure 7-12	Aquarium showing fish and the environment of the tank consisting of the pebbled and planted bottom, added natural features of the bottom, and a rear tank picture to complete the view.	459
Figure 8-1	Knowledge fork with the three "tines" describing how to qualify every bit of information available and intended to be used for problem identification and remediation (if needed). These classifications ensure that decisions are made from only known information. Anything unclear or assumed must be verified to the level of "known" to be used.	473
Figure 8-2	Operational conditions from normal showing the current primary structure in the control room. Most conventional operator monitoring relies on either alarms or serendipitously coming across the other problems before alarms. This chapter adds other tools to this.	474
Figure 8-3	Operational conditions from normal showing an extended awareness structure in the control room. In addition to alarms, the process of identifying and evaluating off-normal situations is added to the kit.	476
Figure 8-4	Alarms for abnormal situation identification are the primary responsibility of the process and equipment designers to build. All the other abnormal situations are left up to the operator to find, somehow.	477
Figure 8-5	Alarm design fundamentals. Every configured alarm must satisfy all four. A configured alarm that does not satisfy all four should not be an alarm (unless a special situation demands it and everyone accepts it as an exception). Anything that is not already an alarm, but satisfies all four fundamentals, should be an alarm.	478
Figure 8-6	Anatomy of a configured alarm. The normal range of flow is between 425 and 800 CFM. If this flow increases to 925 CFM (High Trouble Point), it causes production problems. To ensure that the operator has enough time to intervene to keep this from happening, an alarm will activate at 850 CFM.	479
Figure 8-7	Illustration of the method of flows to determine how to work through a plant for alarm rationalization. This method is best for significant flows where understanding what is happening before aids understanding of what follows.	483
Figure 8-8	Illustration of the method of elements to determine how to work through a plant for alarm rationalization. This method is best where understanding what is happening fully at each element is most efficient.	484
Figure 8-9	Example alarm response sheet. This is a fully described document that an operator will use each time an alarm activates. It includes all the information and steps to manage the specific alarm. Each alarm in the operator area will have its own alarm response sheet.	486
Figure 8-10	Example online alarm response sheet. This view depicts a formatted view of the information in the alarm response sheet in Figure 8-9. It is designed to be accessed by the operator on the HMI to guide the alarm response activity. This one includes access to the loop controller in the event the operator must make adjustments there. Otherwise, the proper response adjustments will be made elsewhere.	490

Figure 8-11	Operating regions showing proper locations for alarms. They are designed to help the operator see abnormal operation early enough to respond. Depending on how far from normal the problems arise, the response will involve increasingly more intervention. At the farthest distances, the operator is expected to move to a safe state instead of trying to keep operating.	491
Figure 8-12	Process safety time, fault tolerance time, and SUDA (see, understand, decide, and act) time illustrating the basic concept that alarm management is time management. Alarms are designed to activate early enough to provide the operator response time needed to successfully manage that specific abnormal situation.	492
Figure 8-13	Example alarm activation point calculations for two hypothetical temperature alarms: one that is easy for the operator to manage, but the temperature changes quickly; and the other that is harder to manage (due to perhaps a more complex process activity) but slower changing. The dynamics and difficulty are specific engineering and operational characteristics. The actual computations are quite simple.	493
Figure 8-14	Illustration of operator alarm loading for a hypothetical 1 hour. Shown is the base load (in blue) that is taken up by planned and regular duties (reports, regular monitoring, shift changes, etc.). Added to that is the time needed to respond to five low-priority alarms and one medium-priority alarm. All this activity leaves just a small amount of time left over (above the magenta pointer at the top).	502
Figure 8-15	Information and decision flow for shift handover. The leaving operator gets everything ready and then shares it with the arriving operator. The arriving operator must fully understand what is being communicated as well as test and verify that it is true.	504
Figure 9-1	Abnormal situations are divided into the portion that competent designers must identify during the equipment and operations design process, including the HAZOP and process hazard analyses and other applicable design and operational safeguards. For the former, alarms are designed. All the rest are left up to operators to find using effective monitoring during operation. HMIs, plant operation guides, protocols, and procedures facilitate this task.	518
Figure 9-2	Focus of operational conditions moving away from normal to not normal. Note that problems that cannot or did not get managed eventually may lead to faults.	525
Figure 9-3	Weak signals as off-normal operations showing the three possibilities. Each possibility is important, so the sorting of the weak signal into its proper category determines how it would be treated.	526
Figure 9-4	The situation management diagram detailing how the types of off-normal situations are sorted. Each situation would be handled using different methods and tools.	529
Figure 9-5	Situation management diagram showing operational regions (at left border edge) that identify what general activity is going on for each. The differences between abnormal situations (most likely leading to an eventual alarm activation) and the not-so-normal (off-normal) situations requiring special investigation to observe and consider managing are identified. Note that the operator does not ordinarily do the disaster management.	531
Figure 9-6	The vibrating cup of coffee from the movie *The China Syndrome*. These subtle vibrations (weak signals) were observed during an operational change. The true cause was faulty construction of the nuclear power plant that was covered up by	

	falsified welding inspections. The main subplot was the search for the meaning of the vibrations, which eventually led to the discovery of the fault.	538
Figure 9-7	Current operation display with icons for key operating parameters shown. The key variables are depicted by the graphic rectangles and ovals with embedded value bars. This encourages the viewer to readily see what appears to be off-normal or abnormal in context and in relationship to other key parameters.	544
Figure 9-8	Current operation display with icons for key operating parameters shown. This view is a comparison operation display highlighting all variables that have recently changed significantly (blue bars). The gray section shows all current values. Blue bars indicate 1-hour-ago values.	545
Figure 9-9	Plant with area selected (magenta dotted line) for calculation of a balance. All changes in the sum total of what is being examined must be in balance. Anything out of balance means that something is missing, suggesting a problem that needs further examination.	547
Figure 9-10	Mass balance gauge showing how it looks and its symbol key defining each visible characteristic. This provides a quick, at-a-glance overview including reference target, current status, and current direction of change. The symbol key is not a part of the operator display.	548
Figure 9-11	Mass balance gauge showing additional "right-click" information providing useful numerical values.	548
Figure 9-12	Plant with area denoted (dotted line) for depiction of the balance and the actual balance icon shown at the upper right. Note that there is nothing notable with the balance closure by the way the icon appears.	549
Figure 9-13	Mass balance gauge showing additional "right-click" information and a notice suggesting a possible leak (meaning that the mass balance calculation did not "close").	550
Figure 9-14	Energy balance gauge illustrating potential combustion problem identified by an energy balance calculation that is not "closing."	550
Figure 9-15.	Figure 9-14 with additional "right-click" information.	551
Figure 9-16	Plant showing identification of operational plausibility component boundaries (dashed lines). Calculations are made for the portion of the plant within each area to determine whether or not each seems to be operating properly.	552
Figure 9-17	Part of a plant (within a selected operational plausibility boundary) showing selected operational plausibility components that are used for observation and any calculations.	553
Figure 9-18	Operational plausibility model showing comparison values (larger purple bars at arrows) that single out ones different from current values (black bars).	554
Figure 9-19	Statistical process control gauge and its symbol key defining each visible characteristic. This provides a quick, at-a-glance overview including reference target, current status, and current direction of change.	556
Figure 9-20	Statistical process control gauge showing the process set point outside the desired operating range. This suggests an operating target was set in error either by an operator or by a cascade controller miscalculation.	556
Figure 9-21	Statistical process control gauge showing the current value is steady but outside of the desired operating range. This suggests that the process is unable to meet the desired value and has been there for some time.	557
Figure 9-22	Statistical process control gauge showing the current value at the edge of the desired range and getting worse.	557

Figure 9-23	Sparkline pop-up added to a comparison operating display to provide a mini history for the affected key variable. Each comparison variable would have a sparkline available for the operator to select if desired.	563
Figure 9-24	Weak signals suggested from the behavior of two trends. The observation is that the trends appear to be different in the beginning, begin following each other in the middle, and differ at the ends. Because both trends are expected to agree, this difference is now readily observable.	563
Figure 9-25	Nominations gauge illustrating a weak signal (annotated within the blue dotted circle that is actually not a part of the gauge but shown to identify the weak signal). If absolutely nothing changes, unlikely but possible over the next 15 hours and 43 minutes, the delivery will be under by 3.5 MMB.	564
Figure 9-26	Deviation diagram (repeated from Chapter 5, *"The Human Machine Interface"*) showing an overview of an entire operator area. Depicted are 36 critical operating values and their departure from expected value. The arrangement is the order of the significant movement of product from area entry (at the left) to area exit (at the right).	565
Figure 9-27	Weak-signal flowchart. This is a step-by-step work process. Follow the progression.	569
Figure 9-28	Depiction of weak signals 1, 2, and 3. The operator identified them by either looking for them or just coming across off-normal indications for each. This is the first step in weak signal management. Be careful not to get ahead of the process.	570
Figure 9-29	Arithmetic linear example of forward-extrapolation (solid black dots connected by a line to the open black dot at upper right) and backward-projection (dashed extension of black line to the open square at the lower left).	571
Figure 9-30	Forward-extrapolation illustration for a single weak signal (1) to the likely worst situation (B).	572
Figure 9-31	Illustration of backward-projection from a likely worst outcome for problem B to look for specific confirming evidence. While that evidence is shown in the figure (as δ, α, and β), let us wait until the next step to find them.	574
Figure 9-32	The backward-projections are done to look for confirming and disconfirming evidence of an abnormal situation existing and needing recognition. That evidence can be strong (clear and evident), sufficient (reasonably convincing), weak (maybe good enough, but probably not), or inconclusive (pretty much useless one way or another). Either way, it is on the table.	578
Figure 9-33	Illustration of two backward-projections. The first is from a likely worst outcome B. Projecting B backwards, we find specific confirming evidence 1 (the first original weak signal) and newly exposed α and β. Finding confirming evidence is sufficient to make a determination that (negative) outcome B is likely to be happening now. A second example is when we backward-project A, we find confirming evidence 2 (the other original weak signal) and newly exposed δ. Again, this is sufficient to make a determination that (negative) outcome 1 is also happening now. Because we found different confirming evidence for the two, it is unlikely that they are related.	579
Figure 9-34	The four types or "sizes" of signals observable by operators and how they are used to find abnormal situations needing attention. The first two, Not Visible and Barely Visible, are weak signals. The fourth, Evident and Distinct, is not. The weak ones are extrapolated and then projected to find evidence. The last one needs only to be projected to look for evidence. But is Visible but not Apparent weak or not weak? That classification will determine how we handle it.	589

Figure 9-35	The four "sizes" of signals observable by operators and how they are used to find abnormal situations needing attention. The third category, Visible but not Apparent, is now linked to the first two, Not Visible and Barely Visible, and are all handled as weak signals. The fourth category, Evident and Distinct, is not. The weak ones are extrapolated and then projected to find evidence. The last is only projected to look for evidence.	590
Figure 9-36	Texas City splitter pressure and level over time. Even at this level, things are not going as they should.	613
Figure 9-37	Texas City splitter process flow diagram depicting the specific process equipment in the operator area where the problem occurred. The excessive level was in the Raffinate Splitter (E-1101). The liquid that eventually exploded was expelled from the Blow Down Drum (F-20).	614
Figure 9-38	Texas City Splitter pressure and temperature over time. Refer to the list in Table 9-1 for the candidate weak signals that can be observed from this graph.	614
Figure 9-39	Relationship of abnormal situations and how alarms and weak signals are used to keep the operator informed. Alarms are configured for all known abnormal situations requiring operator action and identified at plant design. The rest of the abnormal situations may or may not require operator action but could not be identified at the design phase of the plant. These are left up to operator skill and diligence to find them; weak signals are a tool to help do that.	617
Figure 10-1.	Four-part situation management activity process that requires appropriate tools and technology to be able to find all potential problems, specifically confirm or deny their presence, and resolve them if there. Note the importance of factoring in any threats that are near, keeping the remediation relevant, and making sure that everything done is familiar (not invented on the fly).	645
Figure 10-2	Situation management success pathway illustrating both enablers and detractors for everything working together. We need to enhance the success factors and work hard to eliminate the things getting in the way (negations).	650
Figure 10-3	Traditional activities during a shift illustrating the large unstructured block(s) of time where the operator has to invent working processes within them. This can result in operators having their own methods or operators adopting unproven "common denominator" methods without proper enterprise involvement.	671
Figure 10-4	Recommended new situation management structure for the shift illustrating the clearly structured block(s) of time to manage operator responsibilities. This provides operators with approved enterprise methods for maintaining the proper balance between the various roles needed for good operation and for ensuring they keep on top of both the short-term regulatory performance and early developing problems that may impact things later.	672
Figure 10-5	Knowledge fork with the three "tines" describing how to qualify every bit of information available and intended to be used for problem identification and remediation (if needed). These classifications ensure that decisions are made from only known information. Anything unclear or assumed must be verified to the level of "known" to be used.	675
Figure 10-6	Air Florida Flight 90's 1982 crash into the Potomac River Bridge. The cause of the crash was too much ice on the plane. The accumulation was due to a series of pilot errors on the ground. The lack of knowledge of the problem was due to the pilot not listening to the copilot, who observed a problem during takeoff and attempted to notify the pilot.	677

Figure 10-7	Plant state versus operational modes and goals illustrating the clear change of operator goals. As the situation worsens, the operator changes the objective from trying to return to normal to bringing the plant to a safe state (perhaps shut down if that is the only appropriate state).	688
Figure 10-8	Regions of operation boundaries showing that the closer the plant is operating to the desired state, the more likely it is to remain there. As the plant begins to operate further away, the more likely the problem can get worse.	693
Figure 10-9	Permission to operate diagram depicted as a formal flowchart. Note the parallel "clock" that ensures that working on a problem resolution does not take too long without a change in tactics.	697
Figure 10-10	Escalation process framework illustrating the initiation events and approved transfer of responsibility that can take place to ensure that the enterprise brings to bear the right skills for the right operating situation.	704
Figure 10-11	Plant state versus operational modes and goals identifying the situation management escalation region (orange dotted box) in which the operator changes the objective from trying to return to normal to bringing the plant to a safe state (perhaps shut down if that is the only one appropriate). When sufficient time is available, a purpose-defined escalation team can best manage this mode of transition.	707
Figure 10-12	A single control room operating "chair" showing operator console and typical video displays.	720
Figure 10-13	Control room with four console operators and a supervisor illustrating an arrangement where each operator, in turn, manages a part of the plant before sending product downstream to the next operator.	721
Figure 10-14	Control room with four console operators and a supervisor depicting a serious upset at the second (in the flow of production) operator position.	722
Figure 10-15	Control room with four console operators and supervisor depicting a serious upset at the second (in the flow of production) operator position and showing the (temporary) assignment of his operating responsibility for his non-upset units to the first upstream operator.	722
Figure 10-16	The upset operating situation showing the assigning of responsibilities for the escalation team. The console operator is now only responsible for all hands-on changes and eyes-on monitoring of the upset unit. The managing operator is responsible for planning the proper operator responses and then asking the console operator to execute each step one at a time. The directing operator is responsible for filtering out all plant situations except for what is needed to understand the current operating risks.	723
Figure 10-17	Operational integrity measure "dial-type" trend icon that provides operators with an early notification of building abnormal operation (eventually indicating that the plant is not going to be adequately managed if kept operating).	729
Figure 10-18	US Airways Flight 1549 after a successful safe landing in the Hudson River with no loss of life.	733

List of Tables

Table 2-1	Operability integrity level (OiL) as an enterprise scorecard depicting the need to possess a higher degree of operational control as the inherent operational risks increase.	97
Table 2-2	FC-1-3-REACT-MIX-03452 showing two possible naming conventions used to construct a tag that incorporates a structured descriptive model.	116
Table 5-1	Work station general sizes for control room usage.	353
Table 8-1	Alarm priority for safety *consequences* and *severities* definitions only. The table is read row by row. There are additional consequences that will be identified and explained in a similar manner (not shown); see Table 8-2 for their labels.	495
Table 8-2	Alarm priority consequences and severities values. The rationalization team uses a table like this but will use the word *descriptions* (Table 8-1 etc.) for each cell to guide them to determine the priority of a given alarm. In this table, the relative weights of each consequence shown for each severity level are shown that are used in the numerical computations for a candidate priority assignment.	495
Table 8-3	Alarm urgency multipliers used to adjust the alarm priority numerical calculations to take into account how quickly an alarmed variable will actually get to the trouble point. This will slightly increase or decrease the priority for alarms that have a numerical value quite close to the deciding value between priorities (Table 8-4).	496
Table 8-4	Alarm priority from alarm score. Once the total score of the alarm is computed (Tables 8-2 and 8-3), this table is used to determine the recommended alarm priority. Notice that if the alarm score is very low, this is an indication that the candidate alarm is not important enough to actually be an alarm. Remember the four fundamental requirements for alarms?	496
Table 8-5	Maximum operator 8-hour alarm loading for a mix of priorities. Read each line across. An example would be that 33 low-priority, 2 medium-priority, and just 1 high-priority alarms activating during an 8-hour shift would fully occupy that operator in that shift. The plant in this example used only three alarm priority levels (not recommended).	501
Table 9-1	Retrospective weak signal analysis of 2005 Texas City disaster. Note that weak signals were not in use in 2005; therefore, this is all hypothetical and shown for after-the-fact illustration.	616

Foreword

Accurate situation awareness is critical to effective control room operations. Control room operators must monitor and understand a wide variety of information that can change rapidly, often across distributed complex systems. This can tax the ability of even experienced control room operators to stay on top of what is happening, and not just react, but project future trends so as to make proactive decisions that maintain the safety and efficiency of operations.

In *Situation Management for Process Control*, Doug Rothenberg addresses this need for situation awareness and provides tools for operators and managers of control rooms to actively manage the situation. This includes a consideration of both normal and emergency operations along with many real-world examples. Operator readiness, training and support tools are covered, along with key components of effective shift handovers. Rothenberg also covers the importance of effective control room design, with an emphasis on systems that are both user-centered and that conform to human-machine interface design standards. Overreliance on automation and alarm systems are shown to be particular hazards that Rothenberg focuses on.

This book is written for operators, engineers, and line managers of process control rooms who want to improve the safety and effectiveness of their operations. Rothenberg's focus on situation management emphasizes the need for the highest level of situation awareness—projection. It requires that operators be provided with early warning for proactive management of abnormal situations, plans for various contingencies, and the needed resources and authorities to resolve operational problems quickly and efficiently when they arise. The key tools for success can be found in not just relying on operator training, but on equipping operators with systems that support their real needs for situation awareness, coupled with organizational tools to support rapid decision-making and action when needed.

Mica R. Endsley, PhD
President, and CEO of SA Technologies

About the Author

Douglas H. Rothenberg is the president and principal consultant of D-RoTH, Inc., a technology consulting company that provides innovative technology and services for industry. D-RoTH, Inc. currently specializes in alarm management, control room management, fit-for-purpose product design, and innovation development for new products and services. Rothenberg is a leading authority in the field and provides professional consulting and services worldwide. He is the author of *Alarm Management for Process Control*, a defining reference book in the field.

Rothenberg was a faculty member in the Systems Engineering Department of Case Western Reserve University and spent over 20 years with Standard Oil, BP Oil, and BP Amoco where he was responsible for new state-of-the-art technology to support advanced manufacturing solutions.

Rothenberg has a PhD in philosophy, systems, and control engineering from Case Western Reserve University, an MS in electrical engineering from Case Institute of Technology, and a BS in electrical engineering from Virginia Polytechnic Institute. He has several patents in instrumentation and controls and alarm management. He is active in the International Society of Automation (ISA) and is a member of Sigma Xi. He is the recipient of the 2005 Educator-of-the-Year Award from the Cleveland Technical Societies Council in Cleveland, Ohio.

Note from the Author: Gender Neutrality

The book is intended to be gender neutral. The words *he, his,* and *him* are meant to be placeholders for individuals of all genders.

Doug H. Rothenberg

Acknowledgments

The author gratefully acknowledges the generous contributions of the following friends and colleagues:

To Frank Bradshaw, for his friendship and constructive support, and the many, many abstract and theoretical conversations we had, each leading to a better understanding and ultimately, clarity and purpose.

To Ian Nimmo, for getting me started in situation management and providing supporting materials to aid in the development of this book.

To Jack Pankoff, for his enthusiasm for the book, support, and provocative council and guidance along the way.

To Karen Smith, for her expertise and council in fit-for-purpose control room design and human factors.

To Steve Apple, Ardis Bartle, Walt Boyes, Irene Chang, Eric Cosman, Christina DeVoss, Don Dybowski, Richard Eastburn, Mica Endsley, Steve Elwart, Bridget Fitspatrick, Joseph Kaulfersch, Beverly Kuch, Zuzanna Kurowska, Dan Scarberry, Rick Schmotzer, Angela Summers, Russel Treat, Randi Shane Tollefson, Argerie Vasilakes, and Harold Wade, for their candid reviews and thoughtful evaluations that helped keep it real, accurate, and relevant.

To the remarkable ISA team of Susan Colwell for getting this started; Liegh Elrod for picking up the ball after Susan to provide competent shepherding, suggestions, and support; and Chloe Tuck for her dedicated editing and recommendations that improved clarity, impact, and form.

And to the others whose insights and ideas helped shape this book.

Part I
Operational Integrity

1
Getting Started

*If you think safety is expensive, try an accident. Accidents
cost a lot of money. And, not only in damage to plant and in claims
for injury, but also in the loss of the company's reputation.*

Trevor Kletz (Process Safety Expert)

*When we think about the future of the world, we always have in mind
its being where it would be if it continued to move as we see it moving now.
We do not realize that it moves not in a straight line ... and that its
direction changes constantly.*

Ludwig Wittgenstein (Philosopher)

Control room operators and operation center controllers (hereafter we will use the term *operator* for both) manage the real-time performance of a capital enterprise easily worth many hundreds of millions of dollars. We ask them to shoulder the burden of everything that goes wrong during their watch, all without any recognition when nothing does, and precious little (if not actual blame) when something goes wrong and they are just barely able to manage things. Within their area of responsibility and authority they must be able to view every control loop, most sensors, all the equipment, and much of the supporting utilities, and then adjust as appropriate. This is not an easy job. When something actually does go very wrong, the inability to maintain situational awareness is a major loss. Its loss directly contributes to almost every disaster event that was not the result of spontaneous complete surprise. No one wants an incident. But incidents and disasters happen. We now know to a high degree of certainty that they happen because those in charge of ensuring that they do not happen are not aware that they are. They fail to know the situation. They are unaware of what is really going on, what is likely to happen, or what is not happening that they think is.

There is a lot of material here. But it is not intended to be exhaustive. Rather, it is a very good overview. It has been written in a relaxed style. Most of all, it should make sense and give you a very good start at putting your arms around control room operational safety. This book is about how to design for, attain, and sustain a responsible level of operational safety. We will get more into safety in the next chapter. For now, let us get started. Each chapter begins with a framework of important points. These key concepts should be helpful to identify important thoughts about the material.

1.1 Key Concepts

Top Line	The only pathway to commercial success is to operate safely and responsibly. Safety and responsibility go hand in hand. All enterprises are capable of doing it that way. And the tools and skills are readily available and effective to use.
Enough Time	Having enough time is the single most essential ingredient for successful situation management. Notwithstanding everything else, if one has enough time, everything is possible. Without it, almost nothing is.
Manager's Role	It is management's duty to ensure that the enterprise is appropriate to task, the operations team is effective, and the operational requirements are set for safe, responsible, and effective operations.
Operator's Role	The operator is an essential part of the manufacturing team. Operator success will be limited by the extent that the enterprise is effectively designed, appropriately constructed, properly maintained, and responsibly managed (financially and administratively).
Situation Management	Situation management is a technology mature enough to provide a framework and methodology for an enterprise to significantly improve operators' abilities to effectively and successfully manage.
Cornerstones of Situation Management	The enterprise must provide: 1. Early enough warning to provide sufficient time for proactive and successful remediation to abnormal situations 2. Effective plans with appropriate contingencies 3. Sufficient resources at hand to be successful at managing, if success is possible 4. Conditional authority for all parties to act or continue to act to resolve operational problems so long as such action is appropriate

Accidents and Incidents	Accidents and incidents are not only growing in frequency, the extent of damage they cause is alarmingly increasing. If we are able to learn from the experiences of others, we might avoid many of the devastating impacts of firsthand knowledge.
	A failure-to-operate of any safety-related protection device, even if a second such device provided eventual full protection, is an incident and not a near miss.
Last Opportunity	Situation management provides the wherewithal and technology for the operator to attempt to successfully manage abnormal situations before existing infrastructure safe-operational safeguards are challenged.
Managers' Bottom Line	No reasonable amount of personnel selection, training, requirements enforcement, or anything else can surmount the reasonable ability of any individual person to perform. It is not responsible practice to place the majority of the burden on the hands-on operator beyond his appropriate capabilities.
	It is therefore a requirement for responsible enterprise operation that management provide appropriate tools, supervision, and assistance.
Operators' Bottom Line	It is the sum total of the decisions and actions the operator makes, within the capabilities of the enterprise, that determines whether or not the enterprise in his care operates safely and productively.

1.2 Introduction

This book is intended to give you options, not tell you what to do. But, it is not meant to be a "choose one from column 1 and one from column 2" and so on. Rather, the coverage is intentionally broad. It is designed to reinforce your background if you are already working on control room technology and operator support. The materials here are not the only ways to achieve the goal of effective situation management. On the other hand, if you are new to all of this, this book provides a supportive background and comprehensive introduction. Certainly, there are additional resources that go considerably deeper into the ergonomics of operator station design, control room architecture, detailed construction of human-machine interface (HMI) screens and controls, and more. The references at each chapter end provide a few selected resources. Consider those once you have a good framework of what to ask and how to approach things.

This book is intended to give you options, not tell you what to do.

Take the time to understand what is going on and what is needed to do the job well. *Situation Management for Process Control* unifies the understanding of how to deliver real value to control room operations. Properly understood and executed, it is a game changer for safe and effective real-time management of industrial plants and operations. It delivers a firm technical framework that ties together all the traditional individual aspects (procedures, HMI, control room design, alarm systems, etc.) into a technology to understand and design effective control room management operations for enterprises. This is a unified approach with explicit tools to deliver situation management to control room operators. An important new contribution is the concepts and technology of weak signals and their use to supplement the alarm system to cover the rest of the situations that alarms are not intended or able to manage.

Operators monitor, understand, and manage situations and events that need intervention. We use people because of the expense of fully automating proper management without them; the complexity of automatically managing that makes it very difficult to design appropriate technology; the inability to predict events or manage those events that might be predictable but are too "soft" to prearrange for their management; or, most often, a combination of these reasons. With the support of the operations team, operators ensure proper operation. The job to ensure proper operation must be accompanied with the ability to be successful. Doing this right provides the world's population with goods and services to make their lives more comfortable and enjoyable. From food and pharmaceuticals to the expansive host of consumer products, to the iron, steel, and engineered materials to make them, the process industry depends on operators to competently manage. They are on duty second-by-second, minute-by-minute, and hour-by-hour, day in and day out, attending to the making and delivering. To meet this responsibility they must be qualified, motivated, unimpaired, undistracted, and trained; and they must have all necessary tools.

Operators must actually be able to monitor what is in their charge, understand what that monitoring is showing them, and plan and actualize any needed changes. This begins with situation awareness. *Situation awareness* means knowing what is happening in the plant or operation and understanding what it means. The unfortunate reality is that there are few ready convenient indicators available to measure whether or not operators achieve good situation awareness, nor whether they even have the chance to achieve that awareness as a consequence of the design of their operator interface equipment. Missing situation awareness has been implicated in most of the incidents that have led to significant loss of life or plant damage and consequential environmental exposure. However, it is the combination of situation awareness, situation assessment, and the ability to successfully manage abnormal situations that we need—in other words, *situation management*. Proper situation management changes the game.

The undeniable and unavoidable facts that govern successful control room operator performance center directly on what is expected. *What is expected* is simple and profound: watching to make sure that the operations are properly progressing or not. When they are not, operators must understand what is going wrong and take corrective action to make it right. This requires two skills: the cognitive skill of seeing and understanding whether or not something is going wrong, and the skill for performing the direct intervention to make proper changes that work. The main thrust of situation management includes those cognitive skills as well as the resources for guiding intervention. The specifics of what intervention should be effected is left to experts in the process or enterprise as documented in the totality of operating procedures and protocols. This book knits those aspects together into an effective whole, supplementing missing structure and tools wherever possible.

According to a recent survey of process automation professionals conducted by *Control* magazine,

> Survey respondents acknowledged the potential for operators to significantly influence plant performance, as well as ongoing need to improve measures that would make them more effective in their jobs.
>
> When asked to what extent better prepared operator could positively influence key performance metrics, respondents placed significant accountability in the hands of the operators. Operators not only have a big impact on availability, equipment damage and personnel safety, but can play a big role in quality, environmental and economic performance as well….
>
> But an overwhelming majority of survey respondents also confirmed the increasing scope of board operator responsibilities, with more than three-fourths indicating a growing workload….[1]

Operational success rests squarely on operators' shoulders. Providing the tools, processes, and technology for them to be dependably successful has been a challenging task. There are many options for almost every situation. Most plants and operations have settled into similar tool kits and operation protocols. Yet production and operations continue to be fraught with an unacceptable level of incidents, accidents, and disasters. Personnel are deliberately hired and trained; operating schedules purposefully set and closely maintained; operating goals and requirements carefully laid out; most operating procedures designed, trained against, and largely enforced; and all manner of reporting and daily communications performed. But, when we strip everything extraneous away, operating was largely done by practicing what was

---

1 Keith Larson, ed., "Operator Interface—State of the Technology Report," *Control*, April 2015.

taught through on-the-job training, occasional qualifications testing, and other loosely structured ways. This is not to say that anything was careless or thoughtless. Just the opposite—a purposeful action plan was usually followed. Unfortunately, most of the action plans are weak in actual effective operational technology. The operator has a great deal of responsibility but few really good tools to meet it.

What we did not fully appreciate was that the control-room operating environment must provide all the necessary tools to allow the operator to meet that responsibility. Operating displays, instead of simply housing data, information, and control handles, must organize and convey the information in ways that operators readily, almost instinctively, use. This means information must be predesigned and appropriately located to be available so the operator does not have to graze and hunt for clues and causes. The alarm system must be designed to find all abnormal situations engineering and good operating experiences can identify that require intervention. Color is used only for information. Nested levels of display structure should be designed so that important problems and concerns are readily discernable and subtle problems and issues are easily noted. Icons, gauges, and dashboards are used to clearly expose early operational unusuals. Their design must explicitly ensure that operators understand the limits of their effectiveness. And we must provide powerful assistance to coordinate with additional personnel during serious events.

1.3 Cautions and Ground Rules

Design and Safety Notice

In addition to the broad and comprehensive approach to developing a working solution to the situation management problem, an important strength of this book is the wealth of examples, alternatives, and suggestions for your understanding and consideration. All control schemes, design suggestions, displays, diagrams, tables, figures, trend charts, lists, and the like described and illustrated in this book refer to materials that have been designed to amplify and explain concepts and practices intended to help get this material across to you. They are provided for training and understanding purposes only and are not intended as models for design or for implementation. The choice of what to design, which aspects to implement, and how that will be done must be retained wholly by qualified, authorized members of the enterprise staff, who must act with full knowledge of their specific plant configuration, process conditions, equipment, and applicable statutory practices and requirements.

No single work is capable of conveying the entire collective experience and important nuances necessary for success. After reading this book, if you have plans for implementing the suggestions presented, please seek additional specific guidance and experience from knowledgeable experts.

Conflicts with Established Protocols or Statutory Requirements

The express purpose of this book is to place a wealth of best practice information into the hands of the industrial and enterprise community for their understanding and consideration. It is the sole responsibility of the user community to decide their specific operational policies. No material contained herein is intended to circumvent that responsibility.

No claim in this book is meant to imply or otherwise suggest that any item, example, technology, tool, or illustration is or would be a sufficient solution for responsible operation. No included material is meant to suggest compliance with any policy, standard, or legal or other statutory requirement.

Exclusion of Special Systems and Responsibilities

This discussion mentions important aspects regarding the safety of personnel, the physical security of plants and enterprises, and related aspects that may in some way apply or relate to operators, control rooms, and operation centers. All of these topics require special knowledge and additional practices that are well beyond the scope and intention of this reference. No aspect of this coverage is intended to guide or inform as to their proper design or use, with the possible exception of alarm compatibility and control room human factors considerations. Even those must be carefully designed and implemented by appropriately qualified personnel.

How to Read This Book

You have opened a substantial reference book. Lots of pages. Lots of material. Lots of ideas and concepts; some you may already know, others are very new. The organization of the material has been carefully selected for a reader beginning at the start and carefully reading to the end. This helps orient you to relate what you read to what you already know. It allows each topic to progress within a framework and foundation so you can add each part as you come to it and choose to use it. And it provides a good way to tie it all together, rather than have you end up with many distinct ideas and tools. Sure, individually they are sharp and useful. But you already know that is not enough. For it all to make a difference for you, your operators, and your enterprise, it must fold back into your culture in a way that reinforces everything that is already there. That is what you will be doing. I provide the concepts, tools, and motivation.

This book is not an encyclopedia. Sadly, you are not going to read about everything you will need to bring situation management to your operators and into your control rooms. Other works go over detailed design, costs, project management, operator qualifications and training, procedures, and most of the rest you need to bring it into your enterprise. On the other hand, and it is a big "other hand," this book threads the needle for an integrated approach to operator success. Where needed, it introduces specific new tools or new ways of looking at older tools. It emphasizes where gaps exist and provides some of the ways they might be bridged.

To help make it easier for you to get into this, I have provided an extensive table of contents, numbered the chapter subheadings, provided a list of figures and tables, and provided an extensive index. There are many footnote references to materials that contributed to the coverage. Referring to them is not really necessary unless you have a desire to know more. In addition, the book is printed in full color. You can see how color works and catches your eye as you leaf through the pages. All of this is to suggest that if you usually dig into a book like this by looking around, have at it. Find what interests you and see what is here. If you want to follow a thread, there are cross-references to supporting and amplifying material in other chapters so you can do that. But please remember that I am counting on you finding enough ideas and effective ways to work them out. Getting everything working well requires understanding and perspective. Try and find ways to understand how the ideas are intended to work and be used when they are brought into an enterprise.

1.4 Audience

This book is written for managers, supervisors, operators, engineers, safety personnel, and technicians in industrial enterprises and operation centers. It is also written for regulators, specialists, engineers, system designers, and trainers at commercial firms (controls equipment manufacturers, architecture and engineering—A&E firms, and systems integrators) who provide monitoring and controls hardware, software, and technology to end users. You have a unique understanding of the needs and the influence and ability to address them for the control room. Without your care, innovation, and attention to purpose, operator situation management cannot be effective. You are the enablers, champions, providers, and deliverers of the technology.

Operators, Engineers, Technicians, and Support

Hello to you in operations and technology. Hopefully, your manager has provided this book and turned over to you the job of understanding what is needed in your operations and making it so. You should be able to find the explanations, reasons, and technology to deliver success to your control room operation. If your manager did not give this book to you, after you finish reading, please ask him to take an executive's tour of it. With your understanding and your manager's support, amazing progress is open to all.

1.5 Contribution and Importance

Studies of incidents, accidents, and failures in industrial plants, transportation systems, and dispatch centers are replete with examples of operators unable to understand what was happening, or understanding what was happening and yet unable to successfully manage. Traditionally, the activities of understanding and acting were approached separately. First, understand what is going on. Second, act on that understanding. More recently, these two activities have been understood as fundamentally linked. Rather than a handoff between them, there are powerful, effective methods of interactive collaboration of both roles. Identifying and covering additional important facilitators for this process strengthen the work.

This reference gives the reader an effective way to provide operators with tools by combining the technology and processes into a coherent functional approach. Here you will find out how to design appropriate mechanisms to provide operators with a unified and effective support technology framework and needed tools. This reference provides the unified approach, new technology, and explicit tools to deliver responsible situation management. Many of the individual tools and methodologies are currently in use in one industry or another. Their use is scattered and fragmented. Until now we did not fully understand how each might be used to bolster the other. This comprehensive treatment exposes a better basic understanding of each tool and methodology. But more important, you can understand how they fit together in ways that significantly improve the ability of operators to successfully manage. This synergy is both valuable and satisfying. It is satisfying to have a better understanding of how it really works. It is going to make a difference in your control room.

1.6 Situation Management

Situation management is the competency, ability, and willingness of the human operator to properly and successfully manage the enterprise or activity under his charge. It is the endgame role for all operators responsible for the effective real-time management of an enterprise or activity. Success requires the ability to recognize the current environment of operation, to develop appropriate and accurate assessment of that environment, to transform that assessment into needed action for proper management of abnormal situations, and to validate the effectiveness of the action. This is the message of the entire book. It is the sum total of the decisions and actions that the operator makes, within the capabilities of the enterprise, that determines whether or not the enterprise in his care operates safely and productively. It is the endgame role of all responsible operators.

> It is the sum total of the decisions and actions that the operator makes, within the capabilities of the enterprise, that determines whether or not the enterprise in his care operates safely and productively.

Success requires operators to excel at all the following critical activities:

- Attain awareness of the current environment of operation.
- Develop appropriate and accurate assessment of that environment.
- Transform that assessment into needed action for proper management of off-normal or abnormal situations.
- Properly implement those actions.
- Validate the effectiveness of those actions to ensure they deliver all required operational success.
- Modify actions and responses as necessary.

This book spotlights situation management as an essential work process in an industrial control room or operation center. It provides substantiation of its value to effectively frame the operator as a responsible component of successful enterprise performance. Situation management is not crisis management. It is not disaster management. For either of these, as soon as the operator initiates actions to move everything to a safe state and ensures that the actions to do so have been taken, but it is likely that a safe state of the plant is not attainable, his operational role ceases. From this point forward, the problem is shifted to crisis management responsibility or disaster management responsibility. Both are critically important. Both are covered in an extensive array of other resources. Both are beyond the scope and intention of this book, notwithstanding the possibility that the operator may play a role or perform duties for the furtherance of plant or enterprise safety.

Situation management responsibilities for an abnormal situation begin with the operator assuming charge of the situation (either formally or otherwise) over and above normal duties. These duties continue for each abnormal situation within the operator's sphere of responsibility until the abnormal situation either returns to sufficiently normal or fails to respond to the operator's best efforts and falls into a not-manageable category. This "falling into a not-manageable category" may be new to you. Please read carefully, for it is one of the more powerful operational safeguards there is. Call it *permission to operate*. You will find a very clear explanation of it and how it works in Chapter 10, "Situation Management." For now, simply think of permission to operate as a way to clearly and accurately decide when the operator is expected to stop trying to fix the problem and move everything to as safe a state as possible. It ends when the operator transfers charge to a relief operator or when the operation is prepared for a period of unmanned operation, shut down, or escalated to disaster management.

Situation Management Is Straightforward

This book has a lot of ideas and explanatory material. But situation management is basically rather simple. It is useful to you when it is kept simple. Yes, there might be many "moving parts," but those are later on. Understand how simple it is, and you can master it all. Every one of those moving parts is to empower you to get the simple part done and done well.

The first step is to determine what is not going well enough (see Figure 1-1).

This responsibility has an easy part, a good alarm system; and a hard part, an effective plant design, good operator training and procedures, and an effective operator interface to find out what must be found out before it is too late (see Figure 1-2).

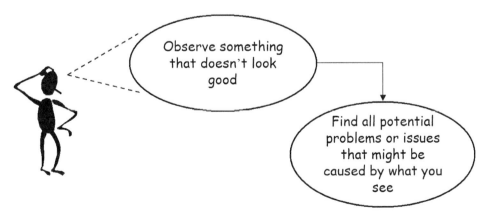

Figure 1-1. Our operator's first responsibility is to keep closely focused on the process to see everything unusual or abnormal. Everything observed will need to be followed far enough to either know that nothing important is going on or to correct things.

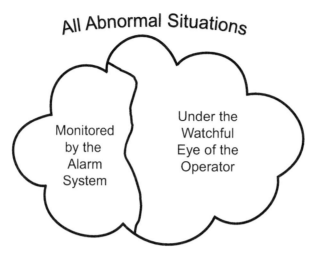

Figure 1-2. The operator has the important responsibility to find and remedy all abnormal situations that exist in his plant or operation. Unless he finds them before, the alarm system will announce those with alarms configured for them. He must find all the rest by whatever means that are available for him to use and that he knows how to use.

Industry knows what it takes to deploy a good alarm system. We will take a good look at that down the road in Chapter 8, "Awareness and Assessment Tools." So alarms are the easy part. Very important for sure! Let us assume that you have that covered—but make sure! See Figure 1-3.

What is left is all the other stuff; the hard part. Seeing all the rest forms the core of this book. Refer to Figure 1-4.

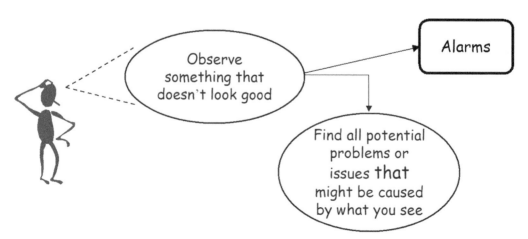

Figure 1-3. Alarms are the way process designers pass on to the operator for action all problems that should be identified by good engineering and a deep understanding of how the enterprise is supposed to operate. The rest must be found using something else.

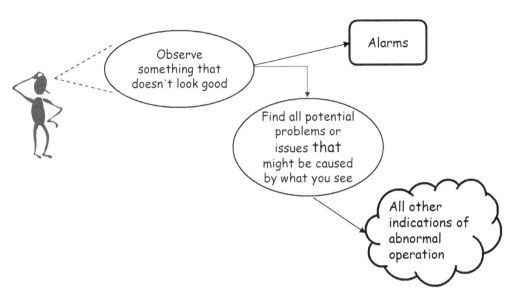

Figure 1-4. Once alarms are taken care of, the operator's responsibility boils down to how to see all the rest of the abnormal conditions and situations he must understand and plan for. Anything missed or misunderstood usually causes trouble sooner or later.

Successfully managing plants and operations is a critically important activity. This leads us to the second part where operators need to see problems. Figure 1-5 introduces this. Remember, this is all about keeping it simple.

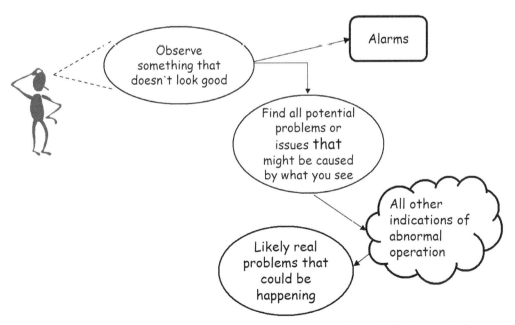

Figure 1-5. Operators will need to find all the "likely real problems that could be happening" suggested by the "indications of abnormal operation" that they must see. This frames the requirements for operator interface design, as well as the situation awareness and assessment tools. All need to work.

At this point, as shown in the previous figure, the operator must find all the other "likely real problems that could be happening" that have not been announced by alarms. Simple and to the point. The single step from knowing that there must be indications of abnormal operation, to ensuring that the operator both sees them and is able to detect the likely problem (but not the solution!), is the skill we require. Operators need knowledge of what they are in charge of, how it works, and how it malfunctions. This is deep knowledge, not the familiarity of just years of experience. The next step, deciding what to do with any problems needing attention and doing it, requires the same level of deep knowledge. See Figure 1-6.

Situation Management Fits Like a Glove

You are the operator. You are sitting in the control room operator's chair of a modest industrial plant that produces a chemical commodity product (e.g., ethylene, acrylonitrile, kerosene, or aspirin); or you are at an oil refinery, a metals refining plant, an electrical power generation station, or a natural gas distribution network, or you are making or distributing something else important. You keep everything running, safe, and profitable. And you do it in a way that makes the next shift even better at that job—you do your part to keep the operation going for the longer haul. You arrive rested and ready for work. While you have a life outside the control room, you put those worries and distractions away for the time being. You are ready to do the job. To

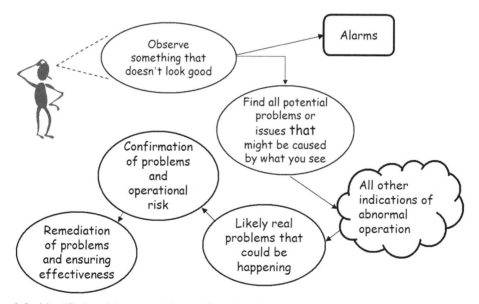

Figure 1-6. Identified problems must be confirmed to be true; not assumed or just thought to may be true. This requires specific additional investigation by the operator. Only once they are confirmed, can they be worked on. And included in the "working" is the necessity to ensure that the steps taken actually did remedy the problem.

do it, you need to know what your production plant is doing and how well it is doing it, *the entire time of your shift*. The design and condition of the physical control room plays an important part in keeping you comfortable, protected, and proud to be there. Its design manages distractions and permits all activities you need to do. If your control room is remote from the plant, you are at the operator interface equipment (the HMI). You have a productive way to interact with any outside operators at the site. Your HMI provides all the states, control handles, and other information you need to keep abreast of what is going on. Without that level of support, any awkwardness, missing information, information located in out-of-the-way places, and incorrect information gets in your way of a proper understanding. *All items missed or incorrect will get in the way of seeing, understanding, and managing.* These are not optional extras.

As you view and try to understand what your plant is up to, you rely on the quality and extent of your training into the production plant's design, equipment, and operational requirements. This is the fundamental training and knowledge that plant management and process designers specify and provide to you. Any gaps, misrepresentations, or confusion here can make the difference between seeing and managing problems, or coming face-to-face with disaster when you cannot. *This is not optional. All items missed or incorrect will get in the way of seeing, understanding, and managing.*

As you decide on courses of action to maintain or restore good operation, you rely heavily on your mastery of the needed competency skills and operational-task skills. Even the best-designed procedures will fall short of working if they are not fully understood or not properly followed. *Again, this is not optional. All items missed or incorrect will get in the way of seeing, understanding, and managing.*

As you intervene to take action, you rely on the proper design and working condition of the equipment and controls you use. Even the best understanding and following proper procedures cannot overcome design gaps and equipment problems that cannot be compensated for. *This is not optional. All items missed or incorrect will get in the way of seeing, understanding, and managing.*

As you evaluate the effectiveness of the intervention actions you are contemplating or taking, it is vital that you are able to do it objectively and honestly. If the operational conditions move you toward the "edges" of your skills or understanding, the processes of collaboration, escalation, and curtailment of operation must be in place and effectively utilized. Together they provide a vital safety net. *This is not optional. All items missed or incorrect will get in the way of seeing, understanding, and managing.*

To summarize: it is important to design an effective operator interface to the control and monitoring equipment. We know proper operator training is essential to their good performance. We get it that operators need carefully and completely designed procedures and good management processes that underlie and support them. We understand that all equipment must be properly fitted to task and kept in good working order. There is no debate among operators that the quality and propriety of raw materials have dramatic effects on operations and the quality of the manufactured products. Time and again, root causes of industrial manufacturing incidents and disasters can be traced in part to the physical and emotional state of the very personnel directly managing production. Proper situation management for hazardous operations then depends on all of these being first rate, everywhere, all the time. The material in this book can help you make the difference. It is the power you can have when you get it and use it.

1.7 Structure of Situation Management

You may be thinking that you have heard much of this before. Depending on your experience and exposure, that is entirely possible. Though exposure might not be enough. Exposure alone has a hard time delivering on operational success. In fact, it can stand squarely in the way. If you have heard a few things before, good. Some will have made sense. Some may not have. Here you will see how it was intended to function and how it fits together into a coherent whole. Situation management has power. It has a message that means something. It has a technology framework that puts the parts together in a way that makes sense, is doable, and will deliver an important added level of success to your enterprise.

Building effective situation management is like constructing any strong structure. It is done from the ground up. The bottom is effective control room management. The top is situation management. See Figure 1-7. It shows the heart of the technology. Understanding it will position you to be effective. Mastering it will be the game changer you are looking for. The explanation of a strong structure will begin at the bottom. Please start here. You need to build the bottom first.

Take the time to understand this foundation before moving on. It might be tempting to move directly to Chapter 8 for the awareness and assessment tools. There you would find a few that you have used before, and you could brush up on them. You would find a few new ones and figure out a way to get them in. You might get brave and delve into the newest technologies on weak signals, find out how they work, and see if some can be useful. Then you could try out some. You might consider trying

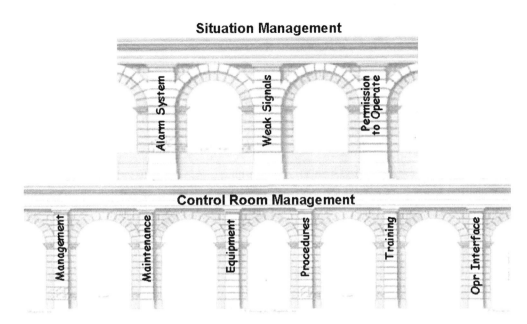

Figure 1-7. Framework for situation management requires effective control room management. Control room management is the competent foundation of management processes; everything in proper working order; proper equipment, processes, and training in place; and an operator interface that exposes the needed information and provides the necessary control handles. The situation management structure on top requires the alarm system, the ability to detect early concerns (via weak signals) before they become problems, and the safeguard of moving the plant to a safe state (via permission to operate) when good operation cannot be assured.

something similar with the tools and processes provided by the final chapter on situation management. Doing all or some of this might sound like a reasonable compromise to get by with less reading and more doing. Unfortunately, things will likely be on shaky grounds if you do. Without a firm *control room management* foundation, the tools for situation management cannot be made strong enough and effective enough to cover the lower-level gaps. Trust the process. Build the foundation you can rely on. Start at the bottom. Build.

Structure of Situation Management

You have a good idea of how useful and effective tools are formed. They are not cobbled together. They begin with a proven foundation and follow a construction pathway that incorporates a clear understanding with dependable functional components. Situation management is the same. Six critical pillars support control room management: management, maintenance, equipment, procedures, operator training, and the operator interface. Much is already written about each. Their best practices are clear and readily attainable. We review them to point out how they fit into a plan. Control room management provides the necessary foundation for situation management.

Without its technology and culture, a safe and reliable operational enterprise is impossible to achieve.

Let us take a closer look from the bottom.

Essential Components of Control Room Management

The six "pillars" that contribute to the building of an effective control room capability are reviewed next. Please note that all are essential. Without any one, or if one is ineffective, it will not be possible for your enterprise to be operated responsibly. They are a requirement for due diligence.

Sure, when equipment is new and everything is fresh and luck is on your side, operational goals might be relatively easy to achieve and consistently met. Unfortunately, as time progresses, production challenges change, maintenance slips, procedures become dated, and most operations become overly routine. Operational risk climbs almost unnoticed. It is not until something happens, or a chance audit, that the likely vulnerability is exposed. These principles and practices for control room management are designed for the long haul. They are important for sustainability.

Management (Pillar)
- Establish and maintain a clear and ever-present culture of safety and responsible operation.

- Put a fit-for-duty management program in place and enforce it (including fatigue, substance use, and emotional situational impairment) for every operator, for every shift.

- Ensure a robust shift handover policy is in place and used for every shift.

- Require full hazard analyses (hazard and operability studies—HAZOPs, PreOp safety reviews, etc.).

- Require and use an effective permit to work program.

- Run an effective near-miss program with all lessons learned incorporated back into the culture.

- Have and use formal and informal programs for performance audits and remediation.

- Ensure that line management is readily accessible to and cooperative with all levels of personnel at all reasonable times.

Maintenance (Pillar)
- Have an effective equipment monitoring program.
- Require fully trained and appropriately certified maintenance personnel.
- Have zero tolerance for broken or missing equipment without installed spares.
- Provide and require approved tools and technology for testing and repairing.
- Require appropriate access to qualified spare parts and replacement components.
- Follow a program of positive materials identification (PMI) for everything directly involved in the manufacturing or operation.

Equipment (Pillar)
- Require all manufacturing equipment to be appropriately designed and fit for specific purposes.
- Ensure that all equipment is operated within design boundaries and for designed purposes.

Procedures (Pillar)
- Ensure that all operational activities are covered by procedures of sufficient detail to be effective for new operational personnel as well as experienced personnel in normal times and times of stress and/or confusion.
  - *This means that procedures are in place that cover what to do during the unexpected or unusual.*
- Require that procedures be properly followed (and annotated as needed) for all operations, all the time, by everyone.
- Audit all procedures to ensure effectiveness and keep all documentation up to date.

Operator Training (Pillar)
- Ensure that operators are trained in operational competencies (not situation by situation).
- Require that all on-the-job training be led by the most experienced and qualified individuals who are certified to instruct.
- Ensure that operator qualifications and readiness are appropriately and periodically evaluated and all deficiencies are promptly remedied.

Operator Interface (HMI Pillar)
- Ensure that the operator interface is designed and maintained according to the best practices.
- Focus on the ability of the operator to observe overall operations as well as expose abnormal and unusual circumstances.

- Provide support for clear and effective navigation.
- Provide operators with clear and ready access to all infrastructure supporting information and documentation pertaining to effective operation.

Essential Components of Situation Management

Situation management is provided by the alarm system, weak signals (off-normal situations), and permission to operate (an operator protocol to manage unmanageable situations). The alarm system provides the first line of abnormal situation management and is a well-established technology.[2] New to you are weak signals and permission to operate and how they work to solidify the technology. Without any one, an operator can not be adequately responsible and appropriately effective.

Alarm System (Pillar)

Proper alarm system design and use has been extensively written about and forms the cornerstone for recognizing and managing abnormal situations for which the alarms are designed. See Chapter 8, "Awareness and Assessment Tools," for a full coverage. Alarms are

- designed and maintained according to the best practices (this is the first line of defense for managing operational risk that can be planned for in advance), and
- used operationally all the time.

Weak Signal Analysis (Pillar)

The concept of *weak signals* is very new. This book introduces it to the control of industrial operations scene. Weak signals are fully discussed in Chapter 9, "Weak Signals." Their use can transform situation awareness from a problem into a strong tool for operators.

- Incorporate the ability for the operator to see small but potentially important weak signals via the HMI and other sources of operational status.
- Include the analysis of weak signals in the operator's basic skill set.

Permission to Operate (Pillar)

Permission to operate may also be new to you. We borrow this technology from alarm management. It provides a strong safeguard for an operator to be better able to avoid

---

2 Douglas H. Rothenberg, *Alarm Management for Process Control—A Best Practice Guide for Design, Implementation, and Use of Industrial Alarm Systems*, 2nd ed. (New York: Momentum Press, 2018).

actions that might go well beyond the possibility of success. Refer to Chapter 10 for details.

- Design the technology to avoid unwarranted continued operation leading to substantial operational risk exposure.
- Ensure full management and peer support for its use at all times and for all situations.

Importance

These nine pillars form the essential message and technology in this book. Some of the material is covered in other works, of course. Here, we take care to ensure that the discussion and ingredients make sense by illustrating why and how the pillars are necessary and how they would be deployed in the context of effective operations. Each thread is woven in the right place for you to see it in action. For example, icons are being used as important information objects in operator interface graphics. Not only are they useful as informational capsules, in the right formulation and combination, they are also quite effective in conveying subtle but potentially important abnormal situations in the making—weak signals. And there are many other design aspects that, when taken in the context of building situation management capability, can both be crafted and understood as enabling.

Success Is Not a Trade-Off

> A plant or production enterprise can only be a commercial success if it also operates safely and responsibly. Safety and responsibility go hand in hand. All enterprises are capable of this. Good tools and proficient skills are readily available and effective to use.

There is an all too often "tug of war" in the minds of management between costs and everything else. To be direct about this, costs are important. If an enterprise cannot provide financial success, there is little reason for it. Put your time and treasure elsewhere. However, no investment in industrial production enterprises will go anywhere unless done for the long haul. That is the downside. The upside is that in the long haul, there are no such trade-offs.

Looking at this from inside the enterprise, success means everyone working there has the right to expect a work and business environment that is:

- **Safe** – Personal safety is expertly managed and everyone feels protected.
- **Rewarding** – Financial benefits are appropriate to maintain dignity.

- **Fulfilling** – Working there feels useful and important.
- **Environmentally secure** – The enterprise does not pass on any costs to society.

Looking at this from outside the enterprise, society has the right to expect the enterprise to do the following:

- **Provide** employment and social revenue (taxes and other benefits).
- **Produce** things or commodities of value (energy; chemicals; movement, transport, or distribution; etc.).
- **Protect** citizenship values (aid in society's advancement), the environment, economic and social values, and the safety of the community.

These pages offer experience and guidance (1) to ensure that trade-offs between financial, safety, and responsibility are not needed and (2) to show you how to claim the most benefit from your time, effort, and money. When your enterprise is safe, rewarding, fulfilling, and environmentally secure, and it provides, produces, and protects, then if it is managed with honest intentions and efforts, long-term success should follow. To say it another way, when any of these seven ingredients are missing or half-hearted, any short-term success you gain will have a very hard climb to reach sustainability.

1.8 Fundamental Tools for Situation Management

Plant safety and responsible operations are not going to get better just because everyone thinks they should. Situation management must be done carefully and deliberately. This means building an approach using the four activities for successful situation management: *identify, confirm, remediate,* and *confirm again*. When you build control room management protocols and processes around them, operators can more successfully guide their plants. See Figure 1-8.

Each fundamental activity represents a key component necessary to do a proper job with as much effectiveness as nature and responsible plant design allow. All are necessary. Let us look at each.

1. **Identify** – *Explicitly identify the current operational situation.* This means that "job one" is always to know what is going on in the entire plant or operation under operator responsibility. Not only must the operator know where things might be going amiss, but he must also know to a reasonable degree of certainty that the rest is operating properly. The absence of easily observable problems cannot be presumed to mean that any part is okay.

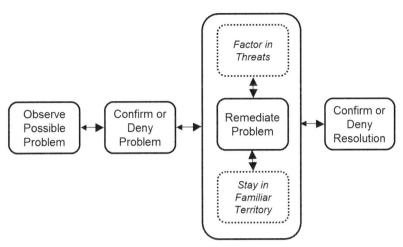

Figure 1-8. Basic four-activity situation management process depicting the progression from "observe" to "confirm" to "remediate" with some possibility to move back a step when evidence is not there or not enough to accept. Escalation and collaboration may take place at any step.

2. **Confirm** – *All information about the current operating situation will be confirmed.* As the operator makes his rounds to see how things are going, each conclusion about proper, off-normal, or abnormal operation must be explicitly confirmed by examining proper independent aspects. Independent aspects include laboratory reports, production values, mass and energy balances, and the like. As operators make their rounds (coming on shift, periodically during the shift, and preparing and doing the handover), observing must come hand-in-hand with confirming.

3. **Remediate** – *The operator is expected to remediate or resolve all identified problems and abnormal situations.* This is not a "shotgun" approach. Only the problems and situations that have been explicitly identified (and confirmed, of course) should be resolved. Operators doing something just to feel needed are out. Each intervention has a price. Do not pay for anything unnecessary. Just be sure to have it right. All exploratory intervention done while explicitly investigating problem resolution is proper. If a sticking valve were suspect, attempting to move the valve would be proper.

   a. *Factor in threats (identify and accommodate risk)* – *Expose and identify all risk to the enterprise coming from outside the immediate problem area.* These are the things that a proper resolution of an abnormal operation or situation requires or depends on but are outside of the control of the specific problem resolution to manage directly. They must be factored into the solution.

   b. *Stay familiar* – *Never let the process or plant take the operator to a place he has never been before.* This means exactly what you think it does. The operator should not ever be operating his plant in a state that he is not fully familiar with, in which he does not know exactly what is happening or what to do to keep it safe and in good operation. Sure, upsets count. So the operator is expected to do what he must to properly respond. But if his training, experience, guidance, and understanding cannot keep him familiar with where things are, he is not allowed to figure it out anew. Operators are not inventors; they are not innovators. Yes, they are talented and, of course, experienced but they should never be expected (nor permitted) to "figure it out on the fly."

4. **Confirm again** – *Now that the remediation action(s) have been initiated or concluded, the operator is expected to carefully monitor the results to ensure adequate remediation or resolution of the problem or abnormal situation.* This is not a "set it and forget it" activity. Lots of things might go wrong. To assume that the remediation actions were in the right direction and should have done the job is not enough. It might actually require a little more of this or a little less of that. On the other hand, if the needed remediation action were different than the one used, it should be quickly assessed and modified. There is a third aspect. That part is when the situation changes during the remediation process. What was the correct thing to do earlier now needs modification. All this means the operator is expected to remain vigilant and ensure the right things happen in response to his actions and activities.

Reminder about Management's Role

It is the nontransferable, nonassignable responsibility of enterprise management to ensure this happens. To meet this responsibility, management must require the following:

- The enterprise equipment must be designed and proper raw materials provided for safe and responsible operation.

- The enterprise production designers must build tools to expose all reasonable abnormal operating modes for discovery by the operators.

    - Alarms must be provided for all abnormal situations requiring intervention.

    - Operational integrity information (balances, operational plausibility, etc.) should be available to expose the normal and the improbable.

    - Clues and other reasonable leading indicators should be provided to expose unusual or abnormal operation.

- The operational leadership must fully train, properly shift, suitably supervise, and aid the sitting operator.

Getting to Good Situation Management Can Be a Big Step

A downside of you having to make all these design choices and alternatives is figuring out a way to decide what is important and what is just useful. First of all, recognize that there is a lot here. Everything should be useful. You might be using some of it now. You have heard about some of the other material before. Still, a lot might be new. There is so much to digest that it may appear to be either overkill or too huge to consider. As you get to know the pieces, they should make sense, and they can all fit together effectively.

Start with the basics:

- **Control rooms need operators.** We will see what a professional operator is all about; we will see what good control room design is and how important it is in the delivery of effective human work

- **Operators need supervisors.** We will see specifics about the ways in which top-level supervision must be provided.

- **HMI is a key component.** We will see how to structure the displays and screens, how to show information that will be critical to determining normal operation from all of the rest, and how mobility and communications play a vital role in effectiveness and versatility.

- **Rich enterprise culture is required.** We will see why long-term focus is the only sustainable one, and we will see how to organize the controls layout of plants to make operator management tasks work smoothly and effectively.

- **Situation awareness and assessment are tools.** We will learn how to see what is going on (and what is not) and what it means, and that all of this must be purpose designed and effectively tooled. A major new tool (weak signals) is introduced to supplement the alarm system for situations that must be recognized but are not far enough along or recognized as expected abnormal situations.

- **Understand human nature so it does not get in the way.** We will see that many of the ordinary ways we think and make decisions might get in the way of clear thinking and effective decision making. Use some of the suggested options to improve things.

- **Situation management is the goal.** We will see how to put it all together in a way that respects people and our limitations, and at the same time provide the sharp and powerful tools necessary for success.

If this seems daunting to you, please feel free to thumb through to find places that catch your interest. You will find examples, illustrations, encouragement, and know-how to understand and make it happen. See what is here. If it is interesting or useful, go as far as you are comfortable. Then go back and fill in the rest. Use this material to see how it might make your job easier, more understandable, and more appreciated. Included is a discussion of selected accidents and serious incidents from the transportation sector. Most are airline cockpit related. This book is not about airline safety or effectiveness, but air-related incidents are some of the best investigated and reported. So stories surrounding them are extensive. More to the point, the root causes and how the situations were identified (or missed, or miscategorized) are uncomfortably close to those in industrial control rooms or operation centers. The lessons they teach are valuable. We begin with just such a lesson.

TransAsia Crash

Just after a 10:53 morning takeoff from Taipei airport on Wednesday, February 4, 2015, an ATR-72-600 identified as TransAsia Flight 235 bound for the Chinese island of Kinmen crashed with both engines shut off into the Huandong Boulevard Bridge over the Keelung River (see Figure 1-9). More than 35 people were killed and 8 were missing. According to the *New York Times*:

> The plane's pilot had radioed the control tower shortly after its 10:53 takeoff to report an engine problem. "Mayday, mayday. Engine flameout," he said, according to a recording of the radio communication posted online. Local officials and aviation experts praised

Figure 1-9. TransAsia Flight 235 crashing into the Huandong Boulevard Bridge as a result of the pilot shutting off the only running engine, mistaking it for the other engine that had already shut down due to unknown causes (and therefore did not need attention during a takeoff). With no engines running, a crash was inevitable.

Source: Image from Twitter user @Missxoxo168.

the pilot, Liao Chien-tsung, for avoiding a greater disaster as the plane hurtled over a dense urban area, narrowly missing apartment buildings as it went down. "Yesterday that pilot, he tried his hardest," the mayor of Taipei, Ko Wen-je, said Thursday, his voice cracking and his eyes welling up as he spoke to reporters. Hugh Ritchie, chief executive of Aviation Consultants International, based in Australia, said the pilot did "An amazing job," and he added, "It looked like they were doing everything they could." Mr. Liao and the co-pilot, Liu Tse-chung, were among the 31 people confirmed killed in the crash. Fifteen people survived, and 12 others were unaccounted for on Thursday.[3]

Unfortunately, praise for the pilots was premature. According to *The News & Observer*:

> "One of the engines on TransAsia Airways Flight 235 went idle 37 seconds after takeoff, and the pilots apparently shut off the other before making a vain attempt to restart it," Taiwan's top aviation safety official said Friday. The details were presented at a news conference in Taipei by Aviation Safety Council Executive Director Thomas Wang as preliminary findings from the flight data recorder. Wednesday's crash into a river in Taipei minutes after takeoff killed at least 35 people and left eight missing. Fifteen people were rescued with injuries after the accident, which was captured in a dramatic dashboard camera video that showed the ATR 72 propjet banking steeply and scraping a highway overpass before it hurtled into the Keelung River. Wang said the plane's right engine triggered an alarm 37 seconds after takeoff. However, he said the data showed it had not shut down, or "flamed out" as the pilot told the control tower, but rather moved into idle mode, with no change in the oil pressure. Then, 46 seconds later, the left engine was shut down, apparently by one of the pilots, so that neither engine was producing any power. A restart was attempted, but the plane crashed just 72 seconds later.[4]

What happened in the cockpit after the starboard (right) engine unexpectedly, and seemingly without pilot control, transitioned to an idle mode, directly set in motion the events leading to the fatal crash. This plane is designed to fly if only one engine is operating. The loss of the starboard engine required no special action by the flight crew during the immediate process of takeoff and initial ascent except for the activity (not done) of transitioning their flight operation to account for the engine loss. There was no indication of fire or imminent loss of mechanical integrity (potential engine breakup). Nothing needed to be shut down. Even if the pilot ordered the engine to be to shut down, its shutdown would not have affected the flight integrity. We will never know whether the pilots erroneously thought the port (left) engine was in danger and needed to be shut down or they intended to shut down the starboard engine but in error shut down the port one.

---

3 Austin Ramzy, "TransAsia Airways Faces Scrutiny After 2nd Fatal Crash," *New York Times*, February 5, 2015, https://www.nytimes.com/2015/02/06/world/asia/transasia-airways-crash.html.
4 Christopher Bodeen and Ralph Jennings, "Both TransAsia plane engines lost power before Taiwan crash," *The News &* Observer, February 6, 2015, http://www.newsobserver.com/news/nation-world/world/article10259753.html.

From a situation management perspective, here is how things played out:

1. The pilot(s) failed to retain the necessary singular operational goal of being in takeoff and ascent mode.
   This mode calls for the singular focus of safely putting the aircraft in the air clear of all ground obstacles and nearby flight conflicts. No other concerns matter. A nonworking engine, unless on fire, needs no attention.

2. The pilot(s) failed to properly interpret the engine alarm that activated 37 seconds into the flight.
   The alarm should have identified the specific engine and should have provided sufficient information regarding the imminent mechanical integrity including the nonexistence of a fire. We rely on proper airframe certification to rule out that the alarm may have not been properly configured in the avionics or improperly messaged.

3. The pilot(s) erroneously took action on the wrong engine.
   Even though no action should have been taken, had the proper engine been forced to a shutdown state, no flight integrity would have been lost.

4. Once the pilot(s) realized that all engines were off, they were unable to successively restart any.
   The published record does not cover how the restart was attempted, so we are unaware if the pilot(s) finally knew to restart the port engine but it did not start, tried to restart the starboard engine but it did not start, or just ran out of time to do any restarts.

1.9 The Power of Time

If somehow, magically, one could stop time during an upset, all upsets could be successfully managed! If every disaster in the making could be paused, it would be possible to spend as much time as needed to understand what is going on, to assess the damage that would likely happen if it continues, and to fix it if we wanted to. Of course, none of us can reverse time. And the ability to stop time or even slow it down would be a wonder tool. Alas, such daydreams cannot come true. Yet exploring this fantasy helps focus on something vital and important: during an upset, the most valuable commodity an operator has is time. How much time is available and how it is used to manage the problem and reduce its unwanted impact is key to determining success or failure. Identify potential problems early. Work them properly. Modify those actions if needed. Situation management is time management.

See-Understand-Decide-Act and Process Safety Time

What the operator sees, understands, decides, and acts on, referred to as SUDA, is the important operational reality of successful production management. We will take a

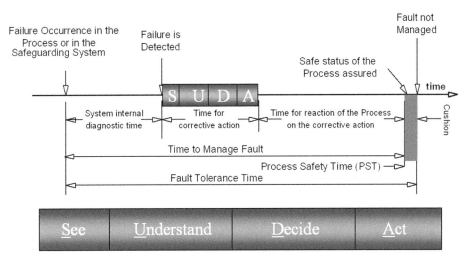

Figure 1-10. Process safety time and SUDA illustrating that everything takes time. (Illustration is not to scale.) It is only when the time required for proper operator management, "time to manage fault," is less than the time for the process to go into fault, "fault-tolerance time," that operator intervention is possible (and hopefully successful).

closer look at the effects of time on operator actions. Figure 1-10 illustrates an operator-centric view of the timeline of an event.

Examine this event. From the left of Figure 1-10, we see a failure occurrence. This starts both the time-to-manage-fault clock and the fault-tolerance time clock. (The time axis is not to scale to provide room for the labels.) If the time-to-manage clock ever meets or exceeds the fault-tolerance clock, the damage will be done. In the figure, we see a delay during which the raw field sensor detects the event, that detection is transmitted to the control system, the control system picks it up, and it is relayed to the calculation and display utilities available to the operator. At that point, we can assume that the fault has crossed the point where operator action is necessary. This is the operator's first notice that something is amiss. The operator might have already been concerned about the plant conditions or other aspects of operation; however, when the situation escalates, concern is confirmed. The operator now knows that something actually happened (*See*).

The operator must now find out what the process will do if a remedy is not found or not found in time (*Understand*). Next, the operator must determine what to do to remedy the situation and decide how to do it (*Decide*). Finally, the operator must actually make whatever changes and take other actions needed to implement the selected remedy (*Act*). Once these actions are complete, and not before, the process will have a chance to respond. If the actions are correct, the process can be on its way to recovery. The race has already started. If the process can respond to the operator's corrective actions before it enters the region of improper or dangerous operation, all is well. The objective is to ensure that the time to manage fault is always less than the fault-tolerance time!

This is all to ensure that the operator will be able to remedy the abnormal situation before the process escalates into an incident. One can see what is going on here with time, but how does seeing it manage it? Take a closer look at the timeline at the top of Figure 1-10. The system internal diagnostic time (or processing time for the signal) is controlled solely by the equipment in place. The time for the process to appropriately react to the operator's corrective action is governed by the fundamental nature of the process. The only thing left to do is to ensure that the time for corrective action is as long as necessary. This is the objective that alarm management uses to assign proper alarm activation points. The clock time for the operator to complete each of the four SUDA tasks is directly impacted by good situation management design. Training and procedures have their place. The underlying skill set operators have affects the effectiveness of their process management performance. Everything must work together.

1.10 Defensive Operating

Assume that all alarms have been taken care of or none are active at the moment. This means that the operator is no longer reacting to a situation. He can shift to being more proactive. Not getting into trouble is the best way of handling trouble. Defensive operating can be a good protocol for doing just that. Defensive operating is borrowed from the successful and well-known principles and tools of defensive driving.

> The Smith System for defensive driving employs five basic principles. Each principle is designed to reduce the risks involved in driving by teaching drivers to anticipate dangerous situations. By driving defensively, traffic-related injuries are reduced, even in adverse weather conditions. An important rule in defensive driving is anticipating other drivers' errors, mistakes in judgment and/or carelessness.[5]

Interestingly, with a few wording changes, the defensive driving principles also have a good deal to say about good control room practices. These can provide operators with useful policies and tools for *defensive operating*. Considering them that way is the "elevator message"[6] for situation management. Here is what defensive operating principles look like.

---

5 Andrine Redsteer, "5 Principles of Defensive Driving," accessed August 26, 2014, It Still Runs, http://www.ehow.com/info_8003043_list-five-principles-defensive-driving.html.
6 "Elevator pitch," Wikipedia, last modified April 25, 2018, https://en.wikipedia.org/wiki/Elevator_pitch.

Five Principles

These five principles form the practical guidance for how operators can deliver effectiveness to their work of managing things. Of course, operators' specific duties will have been spelled out in daily orders, shift handover instructions, general duty lists of activities and responsibilities, and other forms set down by the plant or enterprise. These duties, tasks, and responsibilities would be appropriately documented by existing plant requirements. The principles outlined below are complementary to the other good operating practices. They sharpen and focus the operators' competencies to do what they are tasked to do, but do it better. They are part of the professionalization of operators.

An effective operator will do the following:

1. **Aim high.** Know where you are in the current operating situation and then look ahead. Look ahead to see what your plant or operation will look like a few hours later. Is there a repair scheduled that you need to get ready for or to evaluate to determine if it must be delayed? Are feed tanks getting near full? Or near empty? How is the supply of chemical additives? Do you see anything down the way that you should think about to factor into what you are doing now?

2. **Get the big picture.** Check around all the operational limits and determine if there are any that you are close to, or quite distant from. Are there any production or other problems left over from previous shifts that you need to pay attention to or follow up on? Is maintenance work going on in your unit? Are storms coming your way? What do the production trends look like? Are lab values where they need to be, and do they appear to be solid? Are there other units around yours scheduled for significant changes in operation?

3. **Keep your eyes moving.** Observe current operations to the point of knowing where things are right and where they might need closer attention. Where are things going smoothly? Which things do you need to pay attention to now? Is there anything unusual going on in your unit? Are any other units around you experiencing problems (small fire, partial planned shutdown in progress, deliveries, maintenance activities, etc.)?

4. **Leave yourself an out.** Respond to operation that might be too near to operational limits by taking steps to reduce the chances of limit crossing. Do you need to change the operation of your overhead air coolers during the storm heading your way in the next few minutes? You are seeing pump efficiencies starting to drop. Should you consider cutting the rate to ease up on their load? The next operator position is taking 30% of your production but is experiencing a problem. Should you cut production in case he has to suddenly stop receiving?

5. **Make sure they see you.** Ensure that the rest of the enterprise knows what you are doing. If you are changing products or altering the operating situation, do the upstream and downstream operations know that? If you have an important maintenance operation scheduled soon, does the rest of the enterprise know? Are you sure?

The following additional points help clarify how to apply those principles to real tasks.

Supplementary Guidance

The five principles mentioned in the previous section are on top of an operator's regular task responsibilities. They will need to be continually kept in mind and changes made if needed. This is part of what an operator should do when he is monitoring things. The following principles, together with the five principles in the previous section, are powerful competencies for the operator's toolbox. They add substance and help the operator prepare.

- **Use extra caution when making operational changes.** Oversimplifying just a bit, operator activities break down into two broad classes:
    1. Maintaining current operation, including adjusting for load changes, managing disturbances and equipment malfunction, altering production rates, and the like.
    2. Changing operations like starting up, shutting down, diverting product, transiting from one product to another, and the like. When the operator is maintaining operation, he is engaged in activities that he has already been doing for a while. He is aware of what is going on because he has been managing it. This familiarity and operational history helps maintain a frame of what is normal and what might need attention. All of this changes when the operator will be making an operational change. Extra caution and a heightened sense of awareness are called for in this situation.
- **Leave some space.** Operating at a functional limit line may seem to be an efficient use of resources, but it will not be if a bit extra is needed to account for the unexpected. This is not to say that everything must be "saved in case it is needed." But there are resources that would be prudent to keep in reserve. Chemical production that operates near a transition that could lead to exothermic reactions must have sufficient cooling at the ready. Network trunks need to have capacity to handle unexpected yet vital communications.
- **Pay attention to current surroundings.** Operators need to be in the moment as they do their work in the control room. Having telephone conversations

or sending email to work out a disagreement or plan a vacation take emotional room away from the present control room task. Shifting back is not always seamless; intervening history is not so easy to see looking back as it is watching in real time. A task half complete is harder to pick up again without missteps.

- **Trust but verify.** Completed maintenance is not ready for service until you verify and prove that it is. For example, an outside delivery of a chemical additive is not proper until you check it to be sure; a customer saying he is ready for delivery of more product is not ready until you watch him taking it and it works. Never assume.

- **Always yield.** Teamwork and safe operation is all about working as a team and doing it safely. Yield first and find out why later. When coordinating operation with others, if your particular operation is in the lead but you see that maintaining that lead might cause a problem, back off, and then find out what is going on.

1.11 Situation Management Changes the Game

Operators of industrial plants and operations have a single primary task—keep it working well and prevent or fix any problems that get in the way. This is easy to say but difficult to do for every minute of operation. The operator roles for situation management are basic and quite simple to describe. We borrow from alarm management—it is being able to:

- **See** important and useful information that conveys the status of things and anything out of the ordinary that must be considered

- **Understand** the importance of the status that information might suggest

- **Decide** on appropriate actions to take

- **Act** (including inaction, waiting for more, looking for more, or active intervention)

Effective operators must know inside and out how their plants or operations work, how they might get into trouble, and how to prevent trouble from having a bad impact. If this were not enough, they must know how to see trouble coming in enough time to do their job in time. The heart of this book is the operating situation. Operations can be normal (where it is supposed to be), off-normal (not where it should be, but probably not too bad), abnormal (bad enough to need fixing), and really bad (a threat to surroundings). Let us put aside the really bad for now.

Figure 1-11 illustrates these regions. It is a very simple illustration to keep things clear and straightforward for now. The *normal* operating region is where we would like things to be and stay. Staying normal is the job of the operator. Knowing the process is normal and sensing any threats that might move it away from normal is what the operator does.

If operations are abnormal, we expect the alarm system to announce this to the operator. Figure 1-12 illustrates this. Alarms clearly announce to the operator that there is a problem. Operators select alarms and work to resolve the problem. The resolution steps first require the operator to actually verify that the underlying process or operation is really abnormal. If it is, the operator should work to rectify whatever problems exist. No problem should be worked until it is properly noticed and confirmed.

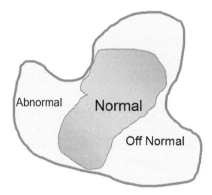

Figure 1-11. Normal, off-normal, and abnormal operating regions depicting the separation between normal (things are okay and need to be kept there), abnormal (something is wrong and should be visible), and off-normal (something might be headed toward a problem but is hard to see now).

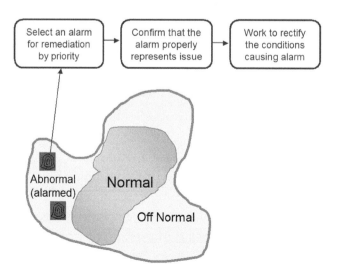

Figure 1-12. Operation in the abnormal region is announced by appropriate alarms.

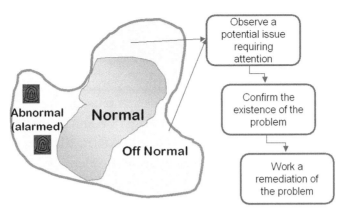

Figure 1-13. Operation in the off-normal operation region must be identified, confirmed, and remediated by operator activity not announced by alarms.

What about operation situations that may not be abnormal, but are not normal either? We term those *off-normal*. Figure 1-13 depicts this. The operator still must have a way to figure out that things are not normal. Somehow, if something off-normal seems to be there and the operator can see it, it is worked just like the alarms are: confirm that an off-normal situation exists and work to make it normal again. This approach is the same as for alarms, but this time it is going to be harder to identify.

Situation management is simple. It is being able to effectively perform the four activities: identify, confirm, remediate, and confirm again. This book is designed to give operators good protocols and tools to be effective at all four activities.

Situation Management

Situation management is the ability to identify possible and potential threats to good operation; confirm the validity, significance, and extent of any threats; undertake appropriate response to those threats, and remediate or, if remediation is not possible, limit the extent of damage; and evaluate the results of those efforts.

Situation Management for Process Control is going to build your understanding of control room management. Situation management provides understandable and highly effective processes for you to put into practice. When you do, it will make a fundamental difference in the way your operators perform the job of managing things effectively. No longer will they need to struggle with that nagging concern of "How in the world can I avoid making that huge mistake that looms ever present over every shift?" You now are in a position to provide an infrastructure, proper tools, and effective working protocols. You will know what they are. You will understand how to design them for your company or enterprise. Operators can use them.

This book is also targeted to top-level managers of industrial enterprises and their direct reports to understand what it takes for the control room to be effective and responsible. Surprised that a technical reference book is aimed squarely at management? Please do not be. Managers set the rules. Managers set the expectations and culture. Managers provide the resources. Managers provide the leadership. This book is designed to provide the reasons why, the energy for, and guidance to do. This technology, these practices, and this know-how can make the difference for safe, responsible, and profitable operation of an enterprise. With them, we can understand and deliver what managers need. The only way to do that is for management to ask and fund. This way we get the resources and the buy-in, and can set the objectives to do it well. All of this starts and ends with managers because ultimately they are responsible for how things are done and the successful outcomes. "Process safety begins in the board room."[7] People cannot work beyond managers' expectations, cannot exceed their requirements, and cannot use anything above what the managers explicitly provide for as tools and resources.

Intuition and insight together with initiative and ideas that "seem to work" have been shown over and over again to be risky and unreliable operational support. We now better understand how to select what is needed for an effective control room and how to get it in and working. It is the reason for this book. Its power is how it weaves the myriad of individual components of control room design, operator interface design, operational protocols, and operator support technology into a coherent and usable methodology. The underlying concept is to make tools and processes explicit where all too often they are either implicit or just missing from the control room operator tool kit. *Implicit* means that something might be there, or not, but it is neither formalized nor included in the operating culture (qualifications, procedures, training, and the like). *Explicit* means that it clearly and specifically identifies what the operator is responsible for and how this can be provided to the operator to assist in delivering successful operation. It is all about what infrastructure is needed to manage the burden on the operator. This coverage provides options to provide a fuller understanding and develop good tools to do the job effectively and efficiently. It is all aligned to help enterprises tip the balance strongly in favor of operational success.

Much of the technology, practices, and design guidelines have been available for a while but in bits and pieces. Here it is gathered together, explained, and put into perspective so it can be appreciated and understood. We dig into *why* it is so important

---

[7] Graeme Ellis, "Process Safety Begins in the Board Room," *Chemical Processing*, March 21, 2013, https://www.chemicalprocessing.com/articles/2013/process-safety-begins-in-the-board-room/.

and *how* to choose what to use and get it into place. Also missing before was a key process for identifying small problems early. This has the power to change the operations success equation for the better. This material will expand your understanding of the entire operational setting.

1.12 The Overall Operational Setting

We return to the operating setting by revisiting the operating regions. Good operating is all about "driving" production around the potholes and barriers, and missing guardrails so that it stays properly working. You might think that good driving is the "taken for granted part." But it needs the right tools and playing field. Situation management is a big part of that. Figure 1-14 illustrates diagrammatically the entire operator arena using a two-dimensional depiction of the preferred operating target (small magenta circle) within the desirable operating area (the unlabeled green area around the operating target). The operator is tasked with keeping things at the target if he can, but otherwise, for sure within the normal operating region (blue). The operator's job is clear. What is not clear is what tools the operator has to do that job properly. There are some rudimentary ways an operator can see how far from the operating target he is. Most plants have an entire suite of controller set points and measurements that the operator can refer to. Most of those set-point values and measurement values come with "experience" ranges of deviation that, for the most part, are sort of known to be okay. There are a lot of them to watch. There are alarms to announce problems. There are other indicators of problems that the operator watches for. When problems arise, there is manual intervention the operator can do to make needed adjustments. Let us take a look at the technology and tools to enhance the operator's ability to keep production or operation away from trouble.

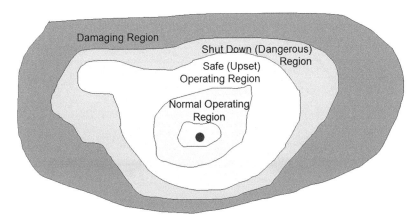

Figure 1-14. Operating regions identifying the "nested" nature of the differing regions as operation deteriorates from being right on target through upset and potentially damaging.

Alarms (Something Is Wrong for Sure)

An important line of defense for operators managing things is the alarm system. We use alarms to make sure that the operator does not miss something important. Alarms are shown in Figure 1-15.

There are two important things to point out here. First, alarms are near but not at the border between safe operating (upset) and shutdown (dangerous) regions. You will see later on that alarms are used only to notify the operator that his action is needed to prevent an upset from leading to something bad. Bad things are to be avoided. Alarm activations ensure that the operator has notice of them. Here is where things get important. It is the clear that the combined responsibility of the plant designer, the experienced operating team, and the technology support team is to identify *all reasonable* abnormal situations requiring operator intervention and provide alarms to notice them. This is a built-in responsibility. It defines the scope and intent of what the alarm system must provide.

The second thing to note is that there is a part of the upset region (building upon it first depicted in Figure 1-10) that is not protected by alarms (the region beyond the dashed line to the left and outside the magenta oval). This is not because of sloppy work. It is the nature of plants and other operations that not all things that can go wrong can be pre-identified. Alarms only cover the parts that can be. The rest we leave to the operator to find in time. This reality must give us pause. The operator has a duty to notice and respond, but there is no good mechanism for notifying. This point should have your real attention. The technology and process of situation management is tasked with doing as much as possible to bridge that gap.

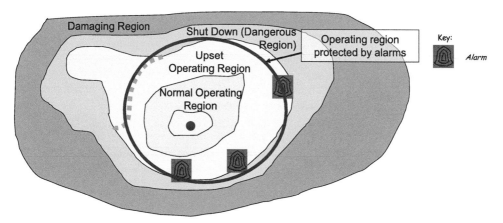

Figure 1-15. Operating region showing alarms announcing (protecting) likely transition out of the upset region toward worse operation. Operator intervention will be required in order to improve things.

Strong Signals (Something Is Wrong but Not Sure What)

Strong signals is a new term. These signals have always been present, but now we need to give them a name. Every incident post audit turns up several that were clearly present and should not have been missed, but were. Figure 1-16 depicts them on our operating region diagram. Strong signals are easy to identify. They are missing or broken parts of the tools and infrastructure that the operator must rely on or use. They are clear indications of operations gone amiss. And they are clear evidence of an essential situation management need or requirement. Basically, a strong signal is a clear indication of a problem. Once you see one, you know that something is wrong and needs attention. They appear as a result of a failure of an operational procedure, a personnel staffing or availability failure, or any other faulty or missing enterprise responsibility directly affecting operations. They can result in alarm activations. All must be tested to ensure that what they seem to point to actually is what is wrong and needs attention. In other words, signals might not necessarily indicate what you initially might think they do. Verify. Chapter 6, "Situation Awareness and Assessment," will cover them.

Weak Signals (Clues Something Might Be Wrong, Find Out)

Weak signals is another new concept. This book introduces it. They provide a tool for operators to detect early or subtle problems in the making. Each weak signal is what we might call a small indicator of something that does not appear quite right. Weak signals offer an important methodology operators can use to understand them and decide what they mean. Chapter 9, "Weak Signals," provides a full discussion and coverage of this topic. For now, know that their important value is that using

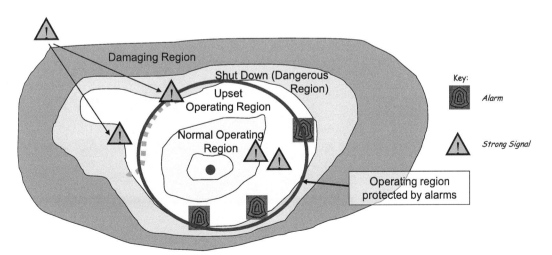

Figure 1-16. Operating region showing strong signals announcing problems that were likely protected by alarms, but normal alarm response does not seem to be doing the job.

them will be a planned activity operators can use to see if something odd might bear fruit if looked into more carefully. Figure 1-17 shows weak signals (the little white clouds) placed about the operating region. They are in all regions except the damaging region. If they can be discovered, processing them will lead to valuable clues and then confirmation of something going amiss. The technology is an important way to "fill in the cracks" of your operator's tool kit. It is all new; but do not let that get in the way of seeing how it works.

The objective of weak signal analysis is to be able to categorize something that "just looks a bit out of the ordinary" and decide if it is something that really needs attention (turning the observation into a strong signal), something to keep an eye on, or an insignificant variation that suggests nothing at all. The takeaway here, be sure you do not miss it, is that alarms combined with weak signals provide a robust awareness and notification capability for operators. Together they cover a lot of ground. When coupled with the effective decision support and situation management methodology, the operator will have the best chance to deliver success.

1.13 Standards, Guidelines, and Practices

Standards, guidelines, and best practices provide invaluable benefit by informing us of what has been understood, accepted, and tried out by those before. They are useful because it is generally not possible to evaluate the full extent of a design until it is fully built and tried. Understanding the experience of others enables us to move our efforts

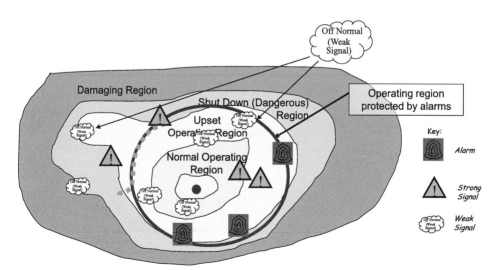

Figure 1-17. Operating region showing weak signals that must be found and processed by the operator, suggesting problems that were not considered serious enough to be protected by alarms but could eventually develop into problems if not recognized and handled.

from "trial and error" to the arena of "very likely to work and be of real benefit." These resources form the cornerstone of good engineering design. They may be prescriptive, outcome based, or a combination of both.

Outcome based means that they describe an outcome or requirement but without providing specific ways or procedures for producing the required results. *Prescriptive* means that they describe a specific procedure or steps to follow, and they must be followed no matter what. In cases where there may be a reasonable number of ways to achieve an outcome, and those ways are best selected by the plant or enterprise, the outcome-based approach works extremely well. In other cases where it is either difficult to select the correct approach, or there exist undue influences on that selection that usually prejudice proper selection, the prescriptive approach is preferred.

Note that when it comes down to the specific level of company or corporate policy, almost everything that is set out internally by that company is done prescriptively.

Standard

A *standard* is described as any specification for physical characteristics, configuration, material, performance, personnel, or procedure, the uniform application of which is recognized as necessary for the safety or regularity of a covered activity. A standard is an established criterion or set of rules that have been set by an approved entity (in a legal or governmental jurisdictional sense, or in a technical or professional organizational sense) governing acceptable practices. The main purposes of standards are to promote safety, reliability, productivity, and efficiency. Such practices may involve design, operation, construction, or other aspects. For standards that have been approved by governmental entities, within the applicable jurisdiction, all behavior subject to the standard is expected to comply. Standards that are not adopted by statute or law are considered applicable only to the extent that a duty of due diligence is expected.

Examples of standards bodies are the American National Standards Institute (ANSI), the Canadian Standards Association (CSA), the International Standards Organization (ISO), and the International Electrotechnical Commission (IEC), among others.

It is important to observe that a standard that is required within an applicable legal jurisdiction can be extremely useful for all other jurisdictions where it is not a legal requirement. For those situations, consider it as a recommended practice. This benefit should not be underestimated.

Recommended Practice

A *recommended practice* is a way of doing things (for which the practice applies) that comes directly from the available pool of resources (standards, guidelines, longtime experience, etc.) and is considered the appropriate way to do things. In general, a recommended practice is more long-standing than a best practice and usually lags it in time.

Best Practice

A *best practice* is a way of doing things (for which the practice applies) that may come from the available pool of resources (standards, guidelines, recommended practices, longtime experience, etc.) and is considered the most appropriate one to use by those with sufficient experience and expertise to make such a judgment (see Figure 1-18). Unless the practice conflicts with applicable jurisdictional law, it would be the most appropriate one to use.

Best practices describe methodologies, processes, or techniques that have been demonstrated to consistently produce results superior to other alternative designs or approaches. Additionally, by being identified as best practices, they become the focus for improvement as knowledge and experiences add to the best understandings. While they generally do not carry the weight or responsibility that a mandatory requirement may, oftentimes they evolve, in the absence of such requirements, as the standard of care that responsible organizations strive to employ. Importantly, the Health and Safety Executive (HSE) in the United Kingdom generally relies on industry best practice as a statutory indicator of due diligence.

Guideline

A *guideline* is a standard or principle by which to make a judgment or determine a policy or course of action. It is generally found that there may be many guidelines

Figure 1-18. Dilbert weighs in on best practices. They are the recognized way to do things by industry. In the United Kingdom they generally have the interpretation of due diligence.

Source: DILBERT © 2008 Scott Adams. Used By permission of ANDREWS MCMEEL SYNDICATION. All rights reserved.

available for consideration. These guidelines may come as encompassing sets (e.g., for shift handover or electronic logging) or expressed individually (e.g., if 8-hour shifts are used, be sure to change shift times to a later time slot rather than an earlier one). They offer a wealth of things to consider. Many independent authors of guidelines exist, and their work is as useful as it may be incomplete. There is a common wisdom that where guidelines are used, it is best to adopt a consistent approach rather than pick and choose different parts from different authors.

Recognized and Generally Accepted Good Engineering Practice

Recognized and generally accepted good engineering practice (RAGAGEP) is the category we could likely find the most useful. *Recognized* means that a sufficient number of those in the direct field of application have looked at it in a suitably substantial way. *Generally accepted* means that there is consensus among those who did the recognizing that they agreed with what they saw. And *good engineering practice* means that what they saw and agreed to is the right way to do the work. The strength of this category is that RAGAGEP carries the dual benefit of *effectiveness* and appropriate exercise of *due diligence*. In the United States, there is a category in the Occupational Safety and Health Administration (OSHA) regulations that includes RAGAGEP requirements in certain disciplines (e.g., alarm management) and carries the weight of law. A brief glimpse into the regulation is included to provide a general framework for understanding the intent of the regulation. The text has been preselected, so not all of it is reproduced here.

OSHA 29 CFR 1910.119

1910.119 Process safety management of highly hazardous chemicals.
   Purpose. This section contains requirements for preventing or minimizing the consequences of catastrophic releases of toxic, reactive, flammable, or explosive chemicals. These releases may result in toxic, fire or explosion hazards.

   (a) *Application.* (1) This section applies to the following:

   (i) A process which involves a chemical at or above the specified threshold quantities listed in appendix A to this section;

   (ii) A process which involves a Category 1 flammable gas (as defined in 1910.1200(c)) or a flammable liquid with a flashpoint below 100°F (37.8°C) on site in one location, in a quantity of 10,000 lb (4535.9 kg) or more except for:

   (A) Hydrocarbon fuels used solely for workplace consumption as a fuel (e.g., propane used for comfort heating, gasoline for vehicle refueling), if such fuels are not a part of a process containing another highly hazardous chemical covered by this standard;

   (B) Flammable liquids with a flashpoint below 100°F (37.8°C) stored in atmospheric tanks or transferred which are kept below their normal boiling point without benefit of chilling or refrigeration. [Remainder of this section omitted.]

(c) *Employee participation.* (1) Employers shall develop a written plan of action regarding the implementation of the employee participation required by this paragraph.

(2) Employers shall consult with employees and their representatives on the conduct and development of process hazards analyses and on the development of the other elements of process safety management in this standard.

(3) Employers shall provide to employees and their representatives access to process hazard analyses and to all other information required to be developed under this standard.

(d) *Process safety information.* In accordance with the schedule set forth in paragraph (e)(1) of this section, the employer shall complete a compilation of written process safety information before conducting any process hazard analysis required by the standard. The compilation of written process safety information is to enable the employer and the employees involved in operating the process to identify and understand the hazards posed by those processes involving highly hazardous chemicals. This process safety information shall include information pertaining to the hazards of the highly hazardous chemicals used or produced by the process, information pertaining to the technology of the process, and information pertaining to the equipment in the process. [Remainder of this section omitted.][8]

Please understand that while it is possible for every plant in every industry to develop their own conforming approaches and safeguards to ensure proper process operations, the ones laid out here (by OSHA) have the benefit of being peer verified and field proven. In some cases, adhering to OSHA 29 CFR 1910.119 is a legal requirement for doing business. It would not be good to learn that your "plant-grown" ideas and methods did not work well enough resulting in people and communities paying a high price for your inexperience.

Best Available Technology Not Entailing Excessive Cost

Best available technology not entailing excessive cost (BATNEEC) is most often used to decide the limitations of considering an alternative remedy or technology to control environmental exposure. Yes, it is a bit open to interpretation in the area of "not entailing excessive cost." However, this is not as discretionary as it might first appear. That is, the "excessive" part must be the result of objective consideration. For manufacturing, the determination of excessive is not at all related to the availability of funds for implementation or use, but rather to the extent of the operational risk (to safety or environment) and the general costs and profits around the enterprise operation. A manufacturing operation that uses, say, $20,000 a day in raw materials and capital equipment costing tens of millions of dollars cannot suggest that an expense for

---

[8] 29 CFR 1910.119, *Process Safety Management of Highly Hazardous Chemicals* (Washington, DC: OSHA [Occupational Safety and Health Administration]).

needed technology is excessive when it may cost $100,000 (even if the enterprise is operating at a loss) when the likely risk of serious issues can result in life lost, extensive environmental damage, or property losses.

Enterprises are encouraged to consider whether the costs to include appropriate technology (including procedure changes, training changes, equipment changes or additions, etc.) to satisfy RAGAGEP qualify as BATNEEC. If so, prudence would suggest that it should be done. If not, ensure that your due diligence is recorded.

Duty of Care

Duty of care is

> … a requirement that a person [or enterprise] act toward others and the public with watchfulness, attention, caution and prudence that a reasonable person in the circumstances would. If a person's actions do not meet this standard of care, then the acts are considered negligent, and any damages resulting may be claimed in a lawsuit for negligence.[9]

This will be your guiding light. Please do this not merely to avoid litigation. It will help do that, of course. Do it to avoid action or occurrence in the first place that caused harm to others that should not have happened had you given proper consideration.

Discussion

This book covers how to provide effective situation management in the control room. Many of its components are already known in the technical community. To summarize, some of the individual components carry the force of law, for example, control room management in the US Department of Transportation (DOT) pipeline management regulations promoted by the Pipeline Hazardous Materials Safety Administration (PHMSA). Some components are well covered in standards for HMI published by the International Society of Automation (ISA) and the American Petroleum Institute (API), to name a few. The purpose of this book is to provide a cohesive and understandable approach for implementing and evaluating a wide number of appropriate practices for safe and effective operations. It is in your hands.

1.14 Management's Role

Management sets the entire stage for operations. First, management must provide the infrastructure to ensure early enough warning to all appropriate personnel to provide sufficient time for proactive and successful remediation. Management must

---

9 "Duty of care," Collins Dictionary of Law, accessed May 8, 2016, http://legal-dictionary.thefreedictionary.com/duty+of+care.

provide sufficient resources at hand for all to be successful at managing, if success is possible. Management must ensure that there are proven and practiced effective plans with appropriate contingencies. Management must provide the policy for appropriate authority for all parties to act or continue to act to resolve operational problems. And, finally, when action is not likely to be successful, management must provide responsible safeguards to contain the situation. The purpose of this book is to put it all together for you to understand how situation management works and empower you to build effective policy, technology, and infrastructure for your enterprise.

This could really start with you. Can your engineers set such a policy? Can your operations foremen develop operation procedures, policies, and rules to ensure that this happens? Can anyone else buy the resources needed to make it so? Can anyone else but senior management bring this to life and sustain it? Of course they cannot. But is that not exactly the *why* of management? Is it not the job of management to ensure that what is necessary for success at any level the enterprise requires, gets there? Success is set to your values, not mine. Getting it is your task, not mine. Giving you the knowledge, the tools, and the best practices is my job and goal. That is the story behind why this book came to be. Welcome. And when you have finished reading what you need, please pass it on to the people who will help bring it into your enterprise. With your power behind them, they will be positioned to deliver.

1.15 The Human Operator

When we place a human operator in the control room, we are asking him to monitor, understand, and manage situations and events to consistently maintain good operation and avoid all harm. The unfortunate reality is that there are few convenient indicators that can measure whether or not operators achieve good situation awareness, or whether they even have the chance to achieve that awareness as a consequence of the design of their operator interface equipment. The lack of situation awareness can be costly. However, situation awareness is not an end goal. The needed result must be the operator's ability to successfully manage abnormal situations. Operators must exploit their understanding of the current status to understand the operation and execute successful decisions to keep the enterprise on target.

The progress for improved control room practices through the development of infrastructure and deployment of tools capable of really improving the operator's ability to manage is slow. There are many vendors, often with complete "solutions" to sell. On the other hand, many enterprises look for a more incremental approach. In the belief that it can save money, some use *trial and error*. Plants are already able to make do, so why do more? Limited budgets can stifle any plan to evaluate real risk. So what to do? To be effective at the role of managing, the operator must be able to actually monitor what is in his charge and understand the meaning of what is provided by that monitoring. You will need purposeful design and careful plans. Your operator then needs the wherewithal to put into action what must be done to properly respond.

1.16 Frontline Supervisor's Role

When we examine the contributing causes for serious operational incidents and disasters where the control room operations either caused the issue or failed to lead to appropriate resolution (where resolution should have been possible), we discover clear supervision failures. These failures fall generally into five categories. Supervisors failed to do the following:

1. Adequately ensure the effectiveness of operator training.

2. Appropriately evaluate shift operators' fitness for duty.

3. Be ready and consistently available for consultation and proper escalation decisions.

4. Call for additional assistance in a timely manner. (Poor behavior almost always gets in the way because senior management fails to recognize that it is better to provide assistance even if it is eventually determined to be unnecessary than to not have the proper resources at the ready to use when needed.)

5. Establish a broad requirement of situation awareness to train and empower the operator to be able to delay "jumping to conclusions" or avoid "continuing on a solution approach" that reasonably would be neither appropriate nor adequate.

We must not place any burden on the hands-on operator beyond his reasonable and appropriate capabilities. This is why you have control room supervision. Who is best suited to do that supervision? How should proper supervision be designed? These are proper and important questions. The answers involve understanding the nature of supervision in the control room. This impacts whether or not the control room operator is able to rely on supervision being readily at hand and up to the task. Control room operators and supervising managers must team up to work hand-in-hand operationally.

We explore supervision in three categories: administration, development, and operations.

Administrative Supervision

Administration is the framework by which an enterprise ensures that its functional structure is followed and maintained. Here is where performance standards are ensured. For example, it includes scheduling personnel and ensuring that the scheduled personnel actually show up. Administration will maintain training needs and proficiencies records. It will ensure that proper equipment (tools, procedures, etc.) is available and used as needed and intended. It will ensure appropriate production record keeping and utilization. It will set and ensure appropriate product/production standards are in place and followed effectively.

Development Supervision

Development is the life-cycle process for delivering proper and sufficient training to operations personnel. It will deliver safety awareness and training. It will deliver operations skills assessment, coaching, and training. In order to do so, supervisors employ mentoring, programmed sequence of critical skill building, and performance feedback. All this is done in a coaching and team-building atmosphere that is absent of blame and understandingly tolerant of honest missteps.

Operational Supervision

Operations management is the active manager participation with the control room operations team to ensure the suitable availability of experience collaboration. It will ensure suitable personnel resources are present and functional during operational needs that require it. Such resources may include additional hands-on personnel, decision guidance, and management coordination for policy implications. These resources may be physically co-located or available via effective tele-collaboration.

Takeaway Message

No reasonable amount of personnel selection, training, enforcement of performance requirements, or anything else, can make up for the lack of ability of any individual person to perform. It should therefore be considered a requirement of responsible enterprise operation for management to provide appropriate hands-on duty selection, supervision, and adequate assistance. By doing so, the end result will be a clear and continuous partnership between operators and supervisors. Most operations failures can be traced to supervision failures. Neither is necessary.

1.17 How Situation Management Really Works

Remember, situation management is the competency, ability, and willingness of the human operator to properly and successfully manage the enterprise or activity under his charge. It is the end game in every control room. Be they managing large chemical manufacturing plants, extensive natural gas distribution systems, a lonely offshore oil platform, or a complex electrical power generating station, these control rooms all rely on the human operator. Situation management brings to the operations table a highly effective and understandable methodology to integrate all that we know about best-practice operations into one coherent, effective whole. It organizes everything in a way that you can make your own and deploy. Situation management preparation includes the following tasks.

- Explain why the entire manufacturing infrastructure must be appropriate and in place before any enterprise can be properly operational.
- Provide explicit guidance into what is needed for the content and design of the HMI to effectively convey current operating conditions.
- Demonstrate the criticality for best practice alarm system design.
- Provide the technology of weak signals to look for, find, and evaluate ways that good operations might be at risk (or provide an alternate method to the operator).
- Lay out a process for ensuring appropriate personnel collaboration during abnormal situations.
- Provide safeguards to ensure operations incidents drive toward proper resolution.
- Incorporate the ability to place escalation directly and clearly in the operations management path.

We ask operators, both experienced and novice alike, to sequester themselves in a control room or operation center for as long as their duty period lasts and keep everything running as needed. They have regular duties like following scheduled events, coordinating third-party activities (sampling, maintenance, and answering management inquiries), preparing for duty change, and the like. They have another duty—keep this expensive operation away from harm. It is ever-present. It is a relentless responsibility. A good day at the last duty provides no protection for what might happen on this one. In fact, more often than we admit, a long run of success mentally seems to poise one to come crashing down because good luck is being stretched too far. We know that operators understand that a lot of money is at stake: thousands, millions, or more, sometimes much more. They know that people's well-being and lives are at risk if grievous error is made. And they know that the quality of life for near and far communities can be at risk.

But, when it comes down to what they must do minute by minute, much is left to their own devices. Operators take some comfort in cycling through displays to look and see. There is the value of variety as they take time to issue a permit to work or take a phone call. There is a bit of reassurance that comes from taking readings or looking them over, or getting ready for duty change. But the bottom line is that there are far too few good tools to identify what is wrong. Situation management provides those tools. They include the following:

- Establish and follow a full duty period (shift, if you like) plan and schedule.
  - Set up and enforce "entry to duty" and "exit from duty" protocols and responsibilities.
  - Ensure production schedules are understood, followed, and adjusted as needed (including sampling, material routing, and scheduling or rescheduling).
  - Manage, request, and/or coordinate maintenance and other mechanical integrity activities.
  - Understand and manage observable production abnormalities as they appear.
  - Provide and make available all needed production documentation.
  - Engage in expected self-directed training, orientation, and skill-building activities.
  - Utilize appropriate tools and processes to remain duty-ready throughout the entire duty period.
  - (Add to this list as appropriate for your enterprise or operation.)
- Rely on the alarm system to clearly identify and guide abnormal situations requiring timely operator intervention.
- At regular, preplanned intervals, conduct a weak signal sweep and process all that arise.

We know that when the operator is not following this list, he will be just looking around. And "looking around" might be the same thing he did before. And that is fine as far as it goes. What is different, and what is really important, is that this casual looking around is not his only chance or his expected chance to find problems. Sure, he likely will find a few. You need him to find them all! This means that now you can have a reliable set of new tools that should change the operator's entire feeling of exposure and vulnerability. You can design for success.

1.18 How Real Is All of This?

This reference can be your guide to the heart of understanding operations and operator needs. This material can help move your enterprise's efforts into focus, and put you on a pathway to make the difference. It can guide you to understand what the problems are. It will assist you in grasping the technology available in a way that opens up necessary possibilities. And it will show you how to get it. Not step-by-step; that is too dependent on what you have and where you want to end up. But concept by concept, so that you can walk through the entire geography of what it is all about and how to get it. And you will know that you are not walking alone.

The major controls vendors are already to the point that they want to help you get there. And remember, their work will benefit from the materials you are discovering in this book. While the vendor descriptions are really short, please understand that each provider has undertaken serious efforts and has useful offerings. Here is a brief peek. The vendors are listed alphabetically.

ABB

ABB is full swing into a program for improving operator effectiveness. Its team is marketing it as a way to understand how their various platform products and support products and services built a capability to do just that. ABB recognizes the need for improved operator support. It defines four pillars for operator effectiveness.[10]

1. **Integrated operations** – Bring the information and control of the operator's entire sphere of responsibility to the operator in a way that is accessible and has a united form and feel.

2. **High-performance HMI** – Use the best practices in user interface design; make sure that your alarm system is standards compliant.

3. **Ergonomic design** – Ensure that the control room, including furniture, arrangement, and other features, is in full support of user-centered design practices.

4. **Operator competencies** – Provide the operator with the tools, training, and experiences to succeed at the tasks required.

---

10 Hongyu Pei Breivold, Martin Olausson, Susanne Timsjö, Magnus Larsson, and Roy Tanner, "The Effective Operator" (white paper, November 21, 2011), https://www.automation.com/library/white-papers/the-effective-operator.

Emerson

Emerson is working to firm up the HMI through style guides and the ability to use "themes" for operator display. The objective is to enhance salience (the ability for the operator to see what must be seen) for screens. This is close to a best practice. Another control room enhancement is the iOps command center concept. The term *iOps* refers to integrated operations. Its purpose is to provide when-needed collaboration with the operators in the control room and the rest of the team. It is presently not considered for crisis or other serious and immediate operating problems.

> iOps is "The integration of people, organizations, work processes and information technology to make smarter decisions. It is enabled by global access to real-time information; collaborative technology; and integration of multiple experts across disciplines, organizations and geographical locations."[11]

Emerson's efforts include close collaboration with the Center for Operator Performance[12] located at Wright State University. Specific topics include operator workspace design, large [display] survey, and "data into information."

Honeywell

Honeywell has launched an Operator of the Future initiative that has focused on the workstation configuration leading to the replacement of the mix of work displays with a single very large display to create a single seamless workspace.[13] The operator creates combined views on the fly to coordinate with current tasks and operations. The tools include use of pan-and-zoom navigation to permit quick and intuitive access to needed information and control handles.

Rockwell Automation

Rockwell's newer systems employ HMI screen design elements that are modular and easy to deploy. They are designed to include appropriate operating ranges, desired performance objectives, and relevant historical data and trends. The objective is to provide reliable information to operators in context and with guidance where appropriate.

---

11 Jim Cahill, "iOps Command Centers Connect Experts Together," Emerson Automation Experts, May 28, 2014, http://www.emersonprocessxperts.com/2014/05/iops-command-centers-connect-experts-together/.
12 "Welcome to the Center for Operator Performance," Center for Operator Performance, accessed August 15, 2015, http://operatorperformance.org/.
13 Larson, "Operator Interface—State of the Technology Report."

Schneider Electric (Foxboro, Invensys)

Schneider's approaches improved operator support from two vantage points:

1. Provide enhanced situational awareness tools and features.

    a. HMI tools including screen hierarchy; enhanced contextual screen information; and improved use of sparklines, trends, and dashboards

    b. Alarm system design and implementation

2. Provide operator-level business measurements to drive continuously improving business benefit by operational decisions.

    a. Business value models

    b. Goal-orientated designs

The overall drive is to enhance operators' ability to remain situationally aware, understand the information in context, move in the direction of proactive operations, and achieve real-time business management.

Yokogawa

> Yokogawa provides various kinds of Advanced Operation Assistance Solutions to enable plant operators in large process industry companies to improve their operations skills on a daily basis and to consistently achieve ultimate daily operations.[14]

Yokogawa places high value on both the cultural understanding of the need for integrating operator support into the controls and enterprise management picture, and the provision of a wide suite of product tools to assist users. The initiative VigilantPlant is composed of the following component items: production management, data integration, alarm management, safety management, energy management, and operator effectiveness. Each component is supported by specific tools and integrated capabilities within its platform. The operator effectiveness part includes

- improving operators' skills—training simulators and knowledge management;
- advanced decision support—alarm management, advanced operating graphics, and predictive controls; and
- procedural automation.

---

14 Yasunori Kobayashi, "Advanced Operation Assistance Solutions for Operator Enhancement and Optimization" (Yokogawa Technical Report, English Edition, No. 43, 2007).

1.19 Contribution and Importance

Situation management is just a tool. Yes, a sharp one, but just a tool. Like all tools, it is what is done with it that makes the difference. Only when it is properly used does the benefit get delivered. That benefit is enterprise sustainability. Effective situation management is necessary but not sufficient for realizing the long-term benefit called *sustainability*. You need it. You cannot achieve proper sustainability without effective operational situation management. But it alone is not enough. This book is about how to get that part. You will use other resources for the rest. Sustainability delivers the following:

- Better operational control to the point that fewer operational mistakes are made and abnormal excursions are less intense and less often result in reduced operational stress on equipment

- Increased pride of operation with diminished personnel stress due to better control of the enterprise production

- Reduced exposure to the public, including less danger to people, less exposure to environmental challenges, and better chances for the enterprise to continue to provide jobs and other revenue

Audience

This book is intended for a broad audience. Plant and operation center managers will find the coverage spot-on and relevant. Safety and environmental personnel will be able to recognize both the need for responsible operation and the clear ability to achieve it through appropriate plans and implementation. Operations teams will readily appreciate the breadth of coverage and the accessibility of the presentation. They will be encouraged to seek and adopt strong tools and processes for doing their job. Engineers and technicians will find the technology within their grasp: easy to understand, practical, and useful. They will be able to conceptualize, design, build, evaluate, and maintain effective operations tools.

The entire enterprise team will find that purpose, technology, and terminology build a common ground to work together for effective operations management.

Suggestions for Reading

You might think that me telling you how to read is absurd. It is. So that is not where this is going. This is not about reading, per se. It is about what you might be thinking about as you read. Let us keep this simple. There are three reasons to read. You read something new to:

1. Check off what you are already know and confirm that you know it.
2. Find all the things you are doing wrong you need to know about.
3. Actually learn about something new.

What you do when you read is entirely up to you, of course. While you are thinking about how you are going to use all of this material, please let me suggest a few things. First, many topics in this book are already out there. So you have heard about them before. Second, there are other topics or material here that you are not going to find anywhere else, at least not soon. So the power of this book is that it provides a better understanding of why and how these familiar topics or materials fit together and work. You likely already know that a good HMI is important and necessary. And there are many ways to build good ones. Here you find out how those many parts of a good interface can be built so when they are actually incorporated into the operator interface, it works effectively. Moreover, and just as important, there is new material that provides operators with important tools that are either missing now or there but not well understood. The more we understand, the better we can make things for the operator to use. However you chose to read, I am glad you are.

1.20 Things to Keep in Mind

The material here is categorized into two broad areas:

- Ideas and technology that should be used and properly integrated into the appropriate enterprise infrastructure (specifications and design, implementation, management of change—MOC, training, and all the rest). *Identify which might be useful; determine if it can be brought in.*

- Ideas and tools that are consistent with existing enterprise infrastructure and that you could use better to make things more effective. *Understand how it works, practice, and use as needed.*

It should be relatively easy to know which category is which. Anything that seems useful and helpful and is both consistent with the enterprise guidelines and practices and supplementary to them would likely be available for you to use. Anything else should wait until it is studied and approved (or disallowed). You are encouraged to discuss everything with peers and supervisors as you go along. As you read and explore this, I offer a few special "words" of encouragement to take to heart.

To the Reader

I am delighted that you have opened your book. It is the first of a kind. As you continue to read, please think about what this book really might be about. You will see clues that some of this could be right-on to your experiences. On the other hand, some parts will be completely new. It is hoped that you will find the concept and content of this book intriguing and worthwhile enough to bring into your company or enterprise. When you do, your operators will find their jobs easier and more effective, and it will engender more pride in their work. And your management will soon experience a visible improvement in overall operation.

It is also possible that after reading here and there and jumping around to find topics that seem interesting, you decide not to do more—you have read enough. Plus, you already have more real work than you have hours for. What you have now seems to be working well enough, thank you very much. Not to say all works perfectly. But who has time for perfect? If this is you, I ask you to give the material and practices a try. You will gain important insight and tools. This book can tip things in a dramatic way. Take advantage of all of this. Take the look.

To Operators

You know that this is all about providing tools and confidence for you to feel and be effective and comfortable during your shift and afterward. There are things here that you have seen before and likely used. You should find more reasons for why they are there and how better to use them. There are things here that you have heard about but were not sure what they meant or how they might help. For these, you might find much that can be useful; give them a try, but carefully and thoughtfully. And there are things here that are going to be entirely new. A few might seem either scary or too "off the wall" for you to consider. Others might be worth a try. Discuss them with others, including operators, engineers, and supervisors. When you find enough value, give it a try.

The bottom line is that all of this is supplementary to your existing enterprise operating policy. No tool or idea or anything else you read here should be utilized when it appears to violate any existing procedure or directive. If you think it might be useful, get permission first. If it works, refine it and fold it back into your infrastructure.

To Engineers and Technologists

Operators depend on you to provide the complete operational set of tools for the controls infrastructure and operator interface. You are their resource. There is quite a lot here for that purpose. Understand, evaluate, experiment, design, and implement. When you identify ways of improving your enterprise, work to include them in the procedures, protocols, and culture. There is a lot to consider here.

To Supervisors

All the procedures and training are your areas of responsibility, as are fitness-for-use evaluation and approval. The material in this book is valuable for understanding what tools might be effective for operators, how they work, and what it might take to get in place the ones you would like. Please take time to understand how all this might work and what parts you would encourage operators to understand and incorporate into their competencies of knowledge. Additional parts could be useful but need your input to be designed and formally incorporated. Start the "ball rolling."

To Senior Management

Your senior staff, line operations and support managers, and plant and operation center managers will see that the coverage is right-on. Your safety and environmental personnel will be able to recognize both the need for responsible operation and the clear ability to achieve it through appropriate plans and implementation. Your operations teams will readily appreciate the breadth of coverage and the accessibility of the presentations. They will be encouraged to seek and adopt strong tools and processes for doing their job. Your engineers and technicians will find the technology within their grasp easy to understand, practical, useful, and implementable. They will be able to conceptualize, design, build, evaluate, and maintain effective operations tools. That happens best with your leadership, encouragement, and, of course, facilitation.

What gets in your way? Well, the first barrier is that you might not be learning the lessons of history. You might not be taking the time to find out what those lessons are. If you sort of know what the lessons are, you may not believe they are telling a lesson. When you do believe the lesson is real, you truly believe that it only happens to others. "It will not happen to me." It will. Unless you go out of your way to understand your operational risks and keep up with the problems encountered by others, you are mostly betting on luck. In nature the house always wins. Change luck to good fortune by learning wholesale from others; please do not pay retail by having to learn from your own misfortune. When you pay retail, it often costs a great deal. Not only that, the bills come all at once. Your earlier financial acumen at understanding costs and setting prices did not set aside a nest egg sizeable enough to cover this.

Managers can have a hard time understanding that it is just fine to use money to make decisions, *but* they forget to use the right accounting method. The only one that works is long term. The cheapest daily way to pay for maintenance is "run equipment to failure." You do not need to spend for predictive or preventative maintenance. If someone before you bought quality equipment and it was installed properly, it usually runs for quite a while without attention. You can save a bundle day to day. The cheapest way to pay for operators is to put them out there and let them "operate until there is a fault." You do not have to pay for procedure writing, training, and proficiency testing. You are saving every day, and nothing bad happens. But when it comes, and it will, the cost can often threaten the entire framework of the enterprise. Please do not "bet the enterprise" against short-term gain and luck.

Every incident of note—from the ones costing $10,000 to the ones costing well over $10,000,000—has one or more primary management failures as a key contributor. Certainly your hand was not on the valve that was left open when it should have been closed. But it was not too far away. The failure to fully set safe operations policy, the failure to effectively walk the safety walk, and the failure to appreciate and then provide proper resources and checks and balances to safe operation requirements were the real contributors. Fortunately, those bad things, for the most part, happen rarely. "Rarely" is a chance event, not a result of policy. Safe operation should not be a wager. Please do not bet on it. Your employees depend on you both for the opportunity to earn a livelihood and the expectation that they will be able to do it safely and with pride. Your stakeholders depend on your enterprise producing the needed product and revenue. The public depends on your good citizenship and shepherding their resources and community. Steward your enterprise so that everyone can look back on it years later with gratitude and respect.

Rosa Antonia and Neil Samuels set this stage with their work on safety conversations:

> During the authors' many years of work to improve safety, they have interviewed hundreds of employees from all levels in many organizations. Yet, one interaction in particular stands out. After a fatality at a Georgia plant, a large group of employees and supervisors gathered to provide insight into the incident's root causes. Despite being pushed to identify other possibilities, they remained adamant that the root cause was a lack of trust and open communication. The group reported that it had long tried to bring the potential dangers of that situation to management, but "they just didn't listen."

> The authors have spent most of their careers helping people listen to each other because they have come to agree with those plant employees. Often, information to prevent a failure is available, but management does not understand it or employees do not discuss it because they are afraid or they feel it will not make a difference. Unless

managers can conduct and encourage the right conversations across all organization levels, preventable fatalities and incidents will continue.[15]

Hear the message from gaps that contributed to recent major accidents. Here are the six essential leadership principles from Ellis (2013) that must be in every safe enterprise. When any are missing or inadequately provided, the plant or enterprise suffers.

1. Ensuring senior management actively supports process safety through its investment strategy and focus on the safety culture of the organization
2. Reinforcing the importance of safety by personal example
3. Thoroughly understanding major accident hazards and key risk-control systems
4. Investigating process safety incidents and near misses to find the underlying causes
5. Developing world-class safety management systems
6. Identifying [and remedying] weaknesses in these systems using targeted performance indicators[16]

The best operators in the world cannot invent and acquire the tools they need to do their job well. The most talented technologists cannot free up enough time from an overloaded schedule to discover what they need to provide operators. Nor can they locate the revenue if it is not already budgeted to purchase the necessary ingredients. The most conscientious supervisor cannot train and prepare operators without proper operating tools. Here and now, on this page, and on behalf of all of your team, you are asked to see that your operations team gets the wherewithal they require to perform the way you and they would want. The present situation is not good enough. They and you have been getting by, but nature and reality are relentless—together they will eventually demand their due. The only real way to slow those forces down is through deliberate and honest effort. This book can show how.

Can your engineers set such a policy? Can your operations foremen develop operation procedures, policies, and rules to ensure that this happens? Can anyone else but you authorize the purchase of the needed resources to make it so? Can anyone else but senior management bring this to life and sustain it? You ensure what is necessary for success actually gets there. Success is set to your values. Achieving it all

---

15 Rosa Antonia Carillo and Neil Samuels, "Safety Conversations—Catching Drift and Weak Signals," in *Professional Safety* (Park Ridge, IL: American Society of Safety Engineers, 2015), 22–32.
16 Graeme Ellis, "Process Safety Begins in the Board Room," *Chemical Processing* 3, no. 57 (2013), accessed April 8, 2013, http://www.chemicalprocessing.com/articles/2013/process-safety-begins-in-the-board-room/.

is your task; no one else's. Giving you the awareness, knowledge, tools, and the best practices is my job and goal. That is why this book came to be. When you have finished reading it, please pass it on to your team. Ask them to bring it to life for your enterprise. With you behind them, they will be able to deliver and want to.

To Regulators and Inspectors

You have vital and important roles in all of this. A great deal of this body of work both reinforces your duty to ensure safe and reliable operations and validates that it is doable. The public is all too familiar with your "red pencils" and "enforcement actions." And that is part of your collective responsibility; we get that. However, you have another important role that you can choose to play—one that can be dramatically effective and useful. That role is education and encouragement. Please take every opportunity to cross-fertilize these good practices across your audiences.

Dual Responsibilities

Operators come in all shapes and sizes. They are newly minted (in their early 20s), well into their career, established in one career and shifted into this one, or seasoned and learned (late 50s and older). Their experiences, values, norms, and expectations vary as widely as the population from which they come. It is the burden of the enterprise to ensure that the plant and equipment design, operational requirements, and performance and duties of operators take these differences into proper account. *Taken into account* means that these are responsibly managed to be in conformance with good engineering practices and established industry norms. It is the responsibility of the operator to accept, understand, and accommodate his personal preferences and points-of-view so that he does not compromise operational requirements and responsibilities.

1.21 Review of Book

This book is about how to deliver that success. It opens with the foundation and practices for operators of industrial, commercial, and governmental enterprises and the management infrastructure they use. This forms the operational integrity portion of the picture. It builds on the understanding of situation assessment by exploring the physiology of enablers and inhibitors to achieve appropriate situation awareness. You will see how successful designs of operator interfaces, control rooms, and the tools to assist in the successful management of abnormal situations fit together. The tools to see problems are followed by the tools to manage them. This is the situation management part of the picture—the successful objective of it all. The integration of situation awareness and decision support into a cohesive structure produces a robust and powerful operations capability to deliver situation management.

Plants are managed by human operators from a remote (not necessarily distant) control room or by operators in direct local contact with the machinery and apparatus. Direct local contact provides a great deal of sensory, tactile, and observational information at the potential expense of access to historical, analytical, and computational information on the operational situation. Remote control provides the broader frame and depth of operations at the expense of the direct feel of the heartbeat of things. A well-designed plant or enterprise will provide the comprehensive ability to control from the remote control room integrated with the local direct controls function provided by specific on-site sensing equipment and local hands-on personnel. So to manage well, all the rest of the enterprise must be proper. We spend time to understand what it takes to put together a sustainable enterprise (Chapter 2, "The Enterprise"). You learn what goes into empowering and preparing good operators (Chapter 3, "Operators"). You discover the importance of housing operators in control rooms that work (Chapter 4, "High-Performance Control Rooms and Operation Centers"). And you work out how and why operator interfaces are designed the way they are (Chapter 5, "The Human Machine Interface"). Operators will use the carefully prepared tools to identify what might be going wrong (Chapter 6, "Situation Awareness and Assessment") early enough to make the difference. Along the way, we take a careful look at the inherent "baggage" that all of us tend to bring along (Chapter 7, "Awareness and Assessment Pitfalls"). Then, we bring out the tools for managing operations (Chapter 8, "Awareness and Assessment Tools"). A sharp tool of weak signals is introduced and fully covered (Chapter 9, "Weak Signals"). Finally, we wrap this all up by tying things carefully together (Chapter 10, "Situation Management").

Part I: Operational Integrity

"Operational Integrity" introduces the end game—how to ensure the best possible operational results for an enterprise. Planners and entrepreneurs envision an enterprise to provide a service or a produce product. Designers (engineers, scientists, and the like) prepare the plan to fully create the physical enterprise. Entrepreneurs provide the needed resources to bring the plan into reality. Operational integrity is the ability to operate the enterprise to achieve the intended benefit. Operators manage the real enterprise to deliver that benefit.

The chapters start with grounding in what makes your operation actually operable. Just because you have a control room and put an operator in it and tell him to do his job, does not mean for a moment that he can and will. And we are not talking about skills and motivation or other personal attributes. Of course, those are needed. The other part we are talking about is whether or not this plant or enterprise you ask the operator to manage is really manageable. We are asking if you have provided the proper tools and support structure.

Part II: Situation Awareness and Assessment

"Situation Awareness and Assessment" includes an introduction to the concepts of situation awareness: the ability to fully appreciate what the enterprise is up to at any given moment. Its first duty is to expose what the operator must know to enable him to understand what he sees. It covers the psychology of situation awareness and operator readiness, and how both depend on situation assessment. Situation assessment is what the operator does to understand the current situation being announced by all of the situation awareness information. This is a careful examination of the equipment to provide the information we give operators to work with. It includes the HMI, how to design the displays and arrange them spatially in the control room, how to provide information on each display, and how to integrate audio and video information into the mix. It also includes the technology of decision support mechanisms and graphical interface display design. It is likely that you have met situation awareness and assessment before. Operational literature is heavy with excellent work that is inviting. But most of the hundreds of articles and books are for aircraft pilots and cockpits. Your operators are not in an aircraft cockpit. Our goal here is to recast this information to make it useful for control room operators.

Here, we take the essence of cockpit experiences, add quite a bit of new understanding and relevance to operators, and frame it for your use. It will leave you with the tools and process for ensuring that operators in the control room will be ready to take on what they see and use it to their advantage. They will be ready to use the kit. A significant tool you will see is the methodology for weak signals. It is one of the most significant advances in operations support methodology in recent years. It provides the tools to do the *seeing of the unusual* and *investigating its potential impact*. The additional concepts covering cultural psychology and operator readiness closes out this part. Your operator should now know what is wrong. Next, we will work with that.

Part III: Situation Management

Our technology pathway now comes to the last way station: "Situation Management." Here we dive into ways to enable your operator to identify the operational threat and effect an action to reduce or eliminate it. This is where the full force of procedures, training, simulation advice, escalation (if needed), and safely backing off (if success is not achievable) is designed, deployed, and used. It enables you to establish protocols, expectations, and training designed to ensure that it all fits snugly together. Situation management covers the requirements to successfully merge these concepts. The operator is then able to manage the enterprise and achieve the intended result of providing services and goods for the benefit of the owners, while meeting the responsibility needs of the society within which it operates. It brings together the insights and tools so you can formulate and begin your plans.

Putting It All Together

This work closes with how to make sense of all of this. Some concepts you already knew. Some you had thought about before. Yet many might be new. This is less about hardware and software and more about *thinkware* and *peopleware*, to coin a few new but useful terms. It is less about procedures and policies and more about culture and protocols. Its purpose is to move managers, senior and junior, to feel comfortable about feeling uncomfortable. Please understand that the technology for situation management is real, accessible, and effective—but only if it is put to purpose in your enterprise.

1.22 Suggestions for Using This Material

You will discover material, ideas, and technology in this book to understand and consider using. A large part has been out there for years. What is important for you is that most of the pieces and parts that were "out there" are brought "in here." They are explained from a point of view that operators and supervisors can appreciate. The material is organized in a way that makes sense and would fit into your way of safe and effective operations. It is all on the table. Is it for picking and choosing? Or is most of it really needed to meet your responsibilities? The answer is that most of it is needed, either as laid out in this book, or by your choosing an alternate method that works better for you. And there is new material to help you resolve long-standing concerns and real gaps that heretofore had too few useful options.

Good Engineering Practice

All enterprises are expected to operate in as safe and effective manner as befits their responsibility to owners and society. This includes the design for and active (1) prevention of injury to personnel both in plant and out, (2) conservation of materials and energy to the minimum needed, (3) protection of the environment from harm and remediation stresses, and (4) respect for the necessary financial objectives of both the enterprise and the greater society. You place operators at the helm and charge them with doing what they need to do to fully and competently achieve those objectives.

Good engineering practices are dedicated to assist in realizing infrastructure for the operator to achieve these objectives and more. Practices do not make decisions between what should be used and what should not. People do. You do. This book will help you understand, accept, and use these practices. Let us take a look at some of the end objectives and capabilities.

Roles and Responsibilities

Operators have a special job to do. Yet they cannot be successful unless the enterprise sets out what they must do, and how, and with what. Thus management must explicitly set out what the roles and responsibilities of their operator must be during normal, abnormal, and emergency operating conditions. The expectation is that there will be both clear general guidelines and operations requirements and clear, detailed, and specific instructions for each important task and responsibility. The general guidelines would include precisely how the operator is to resolve any mismatch between conflicting duties and how to prioritize duties that both require very timely response as well as compete for available time to complete. One of the cornerstones of these requirements is the need to carefully lay out the functional requirements and specific actions that govern the operator being able to pursue the goal of trying to right an abnormal situation versus the recognition that righting the situation is no longer reasonable so the operation must be curtailed or otherwise rendered safe. Operators following these requirements will receive the full support of the enterprise when they determine that continuing to operate places undue burden on safe and responsible operation.

Provide Adequate Information

Enterprise management must provide their operators with the necessary information, tools and equipment, processes, and protocols and procedures necessary for operators to successfully carry out the required roles and responsibilities (set out above). This requires the ready availability of documentation including written procedures, checklists, appropriate drawings, and other materials that comprise enterprise information the operator will need and should use. Effective shift handover is but one important example.

Operators on the Job

Enterprise management must lay down requirements to ensure that when the operator is "in the chair" he is fit for duty. This includes programs for educating and monitoring for fatigue, chemical influences, emotional influences, and illness as it might impact operating. Evaluations would be integrated into shift handover as well as supervision monitoring.

Alarm Management

The most evident tool for identifying and handling abnormal operating situations is an effective alarm system. There must be one that is designed and utilized according to appropriate established industry guidelines. We will note from the outset that a proper alarm system is one that has been carefully designed and built from the ground up. It requires a comprehensive written plan that identifies all aspects of alarm management, including specific operator considerations and responses for each and every alarm.

Operating Experience

Enterprise management must ensure that all lessons learned from both their own operating experience as well as similar ones from related industries are fully understood and adequately incorporated back into their entire control room operating environment (including training and evaluations). Near-miss programs also provide a vital vehicle for this, though certainly not the only one.

Training

Careful planning, adequate documentation, carefully designed equipment, and all the rest can be of little success unless the operator makes effective and appropriate use of them. Experience is second to training in importance. We now know that it is unreasonable to the point of being impossible to train operators to recognize and handle every specific required intervention needed for safe and effective operation. Moreover, and this is the important message here, because bad things almost never happen the same way the second (or third) time, training for them specifically has little value. So rather than train for skills, we train for competencies. These include how to start up things, how to shut down things, how to locate problems with operating things, and the like. And of course, all training requires assessment and refreshment.

Maintenance

Improper or lacking maintenance is a leading indicator for operational failures. Continuous and conscientiousness maintenance of all aspects of an enterprise is vital if everything is to be kept safe and operational. With the exception of installed spare equipment, no enterprise is designed with the ability to function properly and safely with broken or weakly performing equipment, missing or inaccurate documentation, ineffective training and auditing, or anything else missing that the designers built into it. It is a diabolical twist of a really good design that when maintenance is deferred or improper, the bad effects usually take more time to show up. This lack of immediacy is always interpreted as reinforcing the decision to do a poor job of maintenance. Any missing equipment or delayed maintenance must be immediately categorized and accounted for as an operating hazard.

Conservative estimates of the extent of this problem reveal that about 20% (15%–25%) of equipment is in a readiness or performance state below design. The list of candidates is as long as it is disruptive:

- Sensors or transmitters
    - Inappropriately designed for the application
    - Improperly ranged for zero and/or span

- Damaged
- Intermittent or faulty operation
- Missing due to removal; or planned but not yet installed
- No longer being used but still commissioned and/or installed

- Control valves
    - Inappropriately designed for the application
    - Improperly ranged for zero and/or span
    - Damaged
    - Intermittent or faulty operation
    - Missing due to removal; or planned but not yet installed
    - No longer being used but still commissioned and/or installed

- Control loops
    - Poorly tuned
    - In manual
    - Cascade and ratio controls in automatic (instead of cascade or ratio)
    - Advanced controls not in "advanced" mode

- Required plant maintenance
    - Delayed or tabled
    - Poorly scheduled so as to interfere or curtail operations

- Operating procedures and protocols
    - Missing
    - Imbedded into other nonrelated activities so as to require too much modification or interpretation to be used properly
    - Poorly designed or incomplete

And there is more in addition to this list. They require more operator attention just to manage the usual. They threaten good operation during abnormal situations.

And they expose the plant to excessive damage, escalated personnel injury, and significant environmental challenge during upsets. *Lest this message get lost in the list of problems and bad impacts, please understand this: not only is the equipment not going to function as the operator needs it to, the very procedures necessary to manage things cannot be useful or followed.* No one designs procedures to account for all the possible equipment problems and maintenance lapses—they would be too long and it would be too hard to find out what to do! So set a policy to timely fix everything that is broken. If operation must continue, specifically lay out the procedures for doing so safely and successfully. It is likely that it is far easier, less costly, and less risky just to fix it now.

Evaluate and Improve Everything

Most problems cannot be solved or resolved completely. Sure, we try. But doing so seems almost always a bit beyond reach. "If money were no object" often starts the discussion, yet while the costs are important and there is no denying that, the hard problems really are hard and usually do not give up their secrets too easily. What we do is solve or resolve problems to a sufficient degree that there is a comfortable balance between the cost and what it delivers. Forget cost-benefit analysis; it is too biased on the costs and often is used to excuse lack of proper responsibility. The balance we really look for is the one that most of us feel comfortable with even after something not so nice happens.

How do we ensure that this balance thing is not going to get unmanageable? We do it through monitoring, evaluating, and fixing. Alarm systems are watched. Operating experiences are monitored. Training is tested. Incidents are evaluated and lessons learned incorporated back into the technology and culture. Near misses are a powerful tool. The critical message is that for every program, protocol, or operation, make sure that evaluation and continuous improvement are also designed and followed. And make sure that they are effective.

1.23 Actually Getting Started

You might have a feeling that your site has a pretty good plan for effective control room operation. It seems to work well. So, you may be reading to see if this book at least figured out what you already know. Then you are likely thinking, "Okay, they get it and I've already gotten it. I'm getting back to my long list of other things I really need to get to." On the other hand, if not, or you are not completely sure, consider looking a bit more into this. Either way, please hit the "pause" button. Just because we have some overlap does not mean that there is nothing more for you. There is more—likely a great deal more. This discussion lays an honest foundation for your genuine interest and likely some new understandings.

Suits and Coveralls

Let us be overly simplistic to make an important point. Plants and enterprises generally are staffed by operators, technicians, supervisors, and frontline support personnel (we might affectionately term *coveralls*). Executives, managers, and entrepreneurs (at times termed *suits*) provide most of the financial, business, and political support. Both are essential for even a modest measure of success. The enterprise thrives where this cooperation extends into strong collaboration and teamwork. The suits expect the coveralls to produce effectively, responsibly, and economically so the enterprise will be a financial success. The coveralls are at the operating front lines for every second of operation. They depend on the extent and quality of procedures, policies, equipment, training, and availability of enough support to do that job every second of every minute. The coveralls expect and need the suits to get it all into place, with cooperation and collaboration, of course.

Situation management is about how to provide successful practices and supporting equipment and technology to make success happen. In short, once the framework and infrastructure for proper situation management is in place and responsibly followed, the coveralls must be able to depend on the suits to both expect and respect that incidents will happen. And when they do, unless negligence is at the root, the coveralls did their job. If that is not good enough, the infrastructure must change.

Key Ingredients for Operational Success

This book brings together the following key ingredients for operational success in enterprises that rely on human operators for effective situation management.

- Ensure the enterprise infrastructure is up to its intended design, and that design is effective.

- Provide early enough warning to give sufficient time to respond.

- Have proven, effective plans on hand with appropriate contingencies.

- Have sufficient resources at hand to be successful at managing.

- Specify and enforce authority for all parties to act or continue to act to resolve operational problems with limits to prevent overreach (*as a safeguard for things that go wrong enough to outreach reasonable management*).

This reference book covers the needed principles and methodology for you to provide these ingredients. This depends on the enterprise (plant or other facility) being designed, constructed, operated, and maintained appropriately for the purpose required. No amount of operator diligence and skill would be sufficient to overcome such deficiencies. Be careful enough to not interpret your lack of serious incidents or other significant abnormal operational events as valid evidence of a proper plant. Nature does not gamble. Rely not on luck.

This book is also a call to consciousness of the close link between financial failure and operational failure. Clearly, when an enterprise gets into financial limitations (short-term cash flow, overly invasive or overly pandering demands from shareholders, or just inadequate or poor business decisions), there is a shortage of money. Its effect on the business is not often immediate and visible. Its effects on operations are partly immediate and partly long-term. The better the manufacturing parts are beforehand, the less immediate and more hidden the effects. It takes time for neglect and degradation to show up sufficiently for all to see. When it does show up, we often do not attribute it to policy—bad luck and wear and tear are the "red herrings" of blame. The truth is that most, if not all, of it was caused by management neglect. Disasters are "planned for," not by deliberation, but by that neglect. It is imperative for senior management to be just as involved, just as invested, and just as proud of operational success as they are of financial success.

You have just been introduced to a technology: situation management. Situation management provides a framework with tools for creating operational capability for successful and responsible production. Situation management is now part of the operations management dictionary. It provides value to effectively frame the operator as a key responsible component of successful enterprise performance. It brings together the important and useful control room and operations center requirements. Before this book, the many discussions and approaches were fragmented. While each aspect might be quite insightful and independently competent, it may have been difficult for you to see how to combine the various ideas and approaches into a workable enterprise culture. By integrating the needs and the solution process, situation management provides unified understandings, approaches, and solutions.

Some Causes of Poor Operation

Accidents and incidents are not only increasing in frequency, they are increasing in the extent of the damage they cause. The reasons for this are straightforward enough:

- Lack of appreciation by management of the extent of actual exposure to operational risk
    - Inadequate plant design
    - Lacking or faulty maintenance

- Overly sensitive production requirements
- Highly complex production operations
- Excessively dangerous production process or ingredients
- Failure to recognize early warnings
- Failure to provide appropriate operations management tools
  - Inadequate or ineffective control room design
  - Inadequate or ineffective operator interface equipment and tools
  - Failure to provide proper skills and effective technology to ensure reliable operator management of operational abnormalities
  - Operator overload
  - Lack of an effective learning environment where weak or missing skills will be recognized and remedied in a constructive, respectful, and productive manner
- Conflicting or otherwise incompatible management requirements to be the lowest cost producer and maximize profits, versus responsible manufacturing operation
- Failure to provide operations management direction
  - Weak or absent effective operational procedures
  - Confusing or overly vague operational directives
  - Inadequate or ineffective situation management tools and protocols
  - Ineffective control room management with regard to working conditions, operator overload, operator readiness, and the like
  - Inappropriate or ineffective operator selection
  - Weak or ineffective operator training
  - Avoidance of remedy for anticipated or recognized operational risks
- Weak or absent operator readiness
  - Weak or inadequate operator fatigue assessment and management
  - Weak or inadequate substance impairment assessment and management
  - Weak or inadequate emotional affects assessment and management

Each of these contributes to operational problems. They are real. Bringing them to your "front burner" by examining how control room design, operational tools, and operator actions affect all of them is the reason for this book.

As author, I am pleased that you have opened this book. I hope you will come to appreciate that the ideas, the concepts, and most recommendations have merit for you. They may encourage a critical examination of your situation management capabilities. Here are some of the tools and insights to assist in your examination as well as formulate effective responses. Moving forward, the ball is in your court. The results can improve operational confidence and enhance operational success.

Knowing What Is Right

How might you learn that this material is worth reading and understanding? And if you get that far, how might you figure out that it could work for your enterprise? The first question has an easy answer: contained in this book are principles and best practices that work. You should find answers to questions that you might not have ever asked, but having seen them asked, can now see that the answers may have a great deal of use for you. Without knowing the approaches suggested here, your operations might be unduly exposed to difficulty and unwanted impact. This has a lot to say about how open you might be to rethinking your operations. If you are, here are the insight, discussion, tools, and technology to put you on the path to success for a very long and productive time.

The three following methods can help you keep an open mind to judge if what you find here would make a difference to your enterprise. Please keep in mind that they are thought points. They are intended to guide approaches and plans, not as design requirements in and of themselves.

1. Simplicity versus Correctness

How can anyone identify what is the truth from what is not? How can we know whether a particular solution is the way to go? These questions have vexed thinkers and decision makers for as long as there have been thought and important things to be decided. These are easy questions to ask. As you might imagine, definitive tests do not exist. However, there is the guidance of simplicity. If there are two competing theories or explanations for something important, we believe that the simpler one, that does not require suspension of intellect, of course, is the more correct one.[17] This test is known as *Occam's razor*.[18]

---

[17] Richard Swinburne, *Simplicity as Evidence of Truth* (Milwaukee: Marquette University Press, 1997); Thomas Aquinas (Italy: 1225–1274 ACE).
[18] "Occam's razor," Wikipedia, accessed December 4, 2016, https://en.wikipedia.org/wiki/Occam%27s_razor.

For those of you who might be familiar with the competing theories for the orbits of planets before Kepler, Newton, and Copernicus, there was an attempt to force a circular orbit of the Sun around the Earth to work. The Earth was the center of the universe. Circles were well understood. They were regular. Solutions should be regular. But circles alone did not work very well. To help them work, the early astronomers made modifications of a circular orbit with the addition of an imbedded additional circle called an *epicycle*. An epicycle is a circle within another circle. Think of it as if you had a coin rolling around inside of the circumference of a circle. The pathway that a mark at the edge of the coin makes as it is rolled around inside of the circle is the construction they were hoping would work. When one epicycle did not do the job satisfactorily, more were added. Eventually, even with a very large number of them, the orbits could not be adequately described. Once the Sun was made the center and the Earth the satellite, the problem was solved. Yes, the ellipse is slightly more complicated than a circle. That the single ellipse did the job perfectly, better than sets of orbits comprised of circles, is the very definition of simplicity.

Lest this thought be carried too far, we all might agree that genuine solutions can be multidimensional. That is, good solutions should be understood to perhaps require several "moving parts." But the end result should fit together simply, naturally, and efficiently.

2. Concept of Scale

The *concept of scale* means that if one is on the right track to solve a problem, then more of the same solution actions should solve the problem even better. If using "more of the same" quickly gets out of hand and becomes an equal burden or worse of a burden than the original problem, it strongly suggests that you were using a wrong approach. It probably means that that solution approach was more about only resolving symptoms. It failed to work on the root cause. Look elsewhere for a proper solution.

3. Not by Subtraction Alone

Proper resolution of real problems almost always involves a combination of reduction of unwanted aspects with an addition of aspects that are missing. Initially, those problems show themselves as too much of this or too much of that. It is tempting to approach the resolution by working hard to reduce the "too much" into "just right." That is okay except, more often than not, the "too much" came about because there was a real problem. The "too much" was arrived at by trying to solve a problem. Cutting back properly will certainly reduce the "too much." But then it often leaves the rest of the problem, the real problem, unresolved. Therefore, as you approach a problem with too much of something, such as too much data on a display, try to understand the real need and add it to the cutting-back solution. Reduction of too much data on a given display must be accompanied by effective increase in maintenance, better ways of recognizing abnormal plant operations and conditions, and more effective options for managing plant upsets and disaster avoidance. Subtraction would be reducing unneeded data; addition would be more maintenance, better recognition of abnormal situations, and better ways to handle problems. Proper resolution also needs deciding what needs adding back.

1.24 Limitations of Situation Management

This book can help you set your sights on a course to provide your operations personnel with a purpose-built methodology and infrastructure for a more effective way of doing business. Safe operation is not always a reachable outcome. Unfortunately, even the best-conceived and competently operated enterprise has no immunity from trouble. Using the processes and methodology and concepts here can help a lot but they are not a guarantee. You will want to build on what you can do for the control room by revisiting your crisis and disaster management plans. Work out and implement a way for them to seamlessly integrate into what you provided for situation management to move incident avoidance, mitigation, or remediation into a functional and effective master plan.

1.25 Close

This book is groundbreaking. The approaches, materials, and insights here can provide you with the ability to make the paradigm shift to successful operations. The entire concept is integrative in nature. Integrative in that each component builds on the parts before and locks together in a way that provides a strong foundation for the parts that follow. Because we see this "locking together" thing happening, it should reinforce your comfort in the material having real value in the industrial world. It can literally be a game changer for you. But only if you understand and use it. From this moment, situation management can be an inseparable part of your operations. Situation management brings new industrial insights into situation awareness, situation assessment, human machine interface design, and control room management. Situation management develops a cohesive whole in a way that is understandable, readily accessible, and therefore doable; makes clear sense; and when properly applied, will deliver value well beyond investment. Understanding the principles and methodology will inform and empower a wide range of solution pathways for you to gain positive, broad-ranging, and lasting benefit.

There are many ways to learn from experience. Almost no serious failures happen "accidentally." An accident is really that. It is an unexpected event usually resulting in damage. The power of this simple definition lies in understanding the word *unexpected*. Read unexpected to mean "not expected after all of the responsible work done to ensure that all reasonable and potentially likely things that might happen have been planned for." These are things that would have been expected by exercising due diligence of operation. This entire book is aimed at providing you with the wherewithal to be able to firmly stand behind how you operate your enterprise and to equip you to provide the substance for asserting responsible operation.

All can be lost unless you accept that there is a level of operational responsibility that you must not require of operators unless you provide the proper tools and kit for them to succeed. Every day they do without is a day of undue exposure to unacceptable risk. If you are a technologist, after you read this book, please pass it on to your senior management. If you are in management, please pass a few extra copies to engineering and technology and ask them to follow your lead and plan and implement change. This book can be a call to action for you. Answer the call.

I wish you a safe journey.

1.26 Further Reading

CCPS (Center for Chemical Process Safety). *Recognizing Catastrophic Incident Warning Signs in the Process Industries.* New York: John Wiley, 2012.

Endsley, Mica R., Betty Bolté, and Debra G. Jones. *Designing for Situation Awareness: An Approach to User-Centered Design.* Boca Raton, FL: Taylor and Francis, 2003.

Graham, Bob. "Deep Water—The Gulf Oil Disaster and the Future of Offshore Drilling." Report to the President of the United States. Washington, DC: National Commission on the BP Deepwater Horizon Oil Spill and Offshore Drilling, 2011.

US Chemical Safety and Hazard Investigation Board. "US Chemical Safety and Hazard Investigation Board—Urgent Recommendation (BP Texas City Explosion and Fire, March 2004)," news release, August 17, 2005.

2
The Enterprise

The fact is that management can not learn by experience alone what they must do to improve.

The job of management is inseparable from the welfare of the company.

W. Edwards Deming (Quality Expert)

Don't be tempted to equate transient dominance with either intrinsic superiority or prospects for extended survival.

Stephen Jay Gould (Evolutionary Biologist)

The enterprise or plant working as needed is the entire reason for its being. So it is important to know what is useful to be there so that everything that is needed is there and works to deliver. This way, the operator will be able to do his part. Safe and responsible operation is the only way to stay in business. You cannot set a safe level of unsafe that agrees with a budget or allows you to manufacture your products at a selling price that returns a decent profit and expect that any safety level will be okay. So how does management decide what to require and what to "take chances with"? The answer will determine whether or not the plant or enterprise will be sustainable. Safe operation is a threshold. This is the minimum. Everything below it will not work. Above that level is not guaranteed to be safe, but it should meet the generally accepted criteria for due diligence. The need is to achieve "best practices." If they do not seem needed, appear too complicated for your culture, or cost more than you are prepared to spend, perhaps you might want to place your time and treasure into another investment—this one is going to let you down when you can least afford it.

The *situation management* premise, and core message of this book, is that for an enterprise to expect an operator to properly manage, the enterprise and operator must possess a complete and proper infrastructure: equipment, maintenance, processes and procedures, training, and pride of participation from all. Collectively, these constitute best practices. A few items may be entirely new to you. Others you may know. You can find them in many industrial management resources. We cannot completely cover the topics here. Rather, the intent is to introduce them into a larger frame of situation management. When you are ready for more, there is a wealth of available information out there. Situation management provides the integration pathway that puts it together in ways that both make sense and work. It is the reason why and the glue for getting the job done. Without a strong infrastructure in place and fully operational, no amount of attention and care during operations can be strong enough to overcome missing parts.

2.1 Key Concepts

The Importance of Safety	If you intend to stay in business, safe and responsible operation are not parametric. Safety is not anything until it is everything.
Quality of Operation	Management culture and style is the most important determinator of long-term operational success and operator effectiveness.
Short-Term versus Long-Term	Any responsible long-term objective when compared to a similar short-term one will *always be worse in the short term*. Therefore, all attempts to compare them in the short term are a deliberate effort to be unfair. All plans and work must be long term.
Enterprise Design	Enterprise design determines the upper limit of operational ability. No amount of operator attention and proper operational decisions can overcome the inherent weaknesses in equipment design and maintenance, ineffective procedures and protocols, and lack of proper training and supervision.
Perfect Process Understanding	A perfect understanding of an entire plant or enterprise is not essential for good and proper operation. However, every piece used must be reasonably well understood and specifically accounted for in the operation procedures, protocols, and training.
Essential Decomposition	The ability to decompose a plant or enterprise into well-defined, understandable pieces is essential to being able to understand and manage it.
Perishable Skills	All skills tend to degrade if not refreshed and then properly evaluated and reinforced.

2.2 Introduction

This chapter begins with the basics of enterprises and the inherent responsibilities of their senior management. The culture and constitution of an enterprise are the most reliable predictors for long-term operational success. The chapter concludes with important technology for carving out small enough parts of the enterprise to have a good understanding of how it works and where it can benefit from your attention. The ability to decompose a plant or enterprise into well-defined, understandable pieces is essential to being able to manage it.

The Enterprise

An enterprise (plant or operation) is often an extensive organization with many moving parts and complexity. It is the organized action of making of goods and services for sale; enterprises can be any type of organization, including businesses, nonprofits, and government agencies.[1] The term *commercial enterprise* combines the meanings of the words *commerce* and *enterprise*. Therefore, a commercial enterprise is a business that engages in buying, producing, and selling activities for the purposes of making a profit.[2]

Enterprise is one of the broadest terms used to describe an organization. In general, an enterprise is an organized collection of people and systems working toward shared goals. A more specific interpretation of enterprise expects that all departments and employees within the organization have synergistic responsibilities to achieve goals.[3]

It is All About Culture

A poignant experience with the then-manager of BP's Toledo Refinery (Toledo, Ohio, United States) reveals the mark of a leader. This story goes back a few years.

> Soon after taking over the leadership, he noticed that the equipment painting in the refinery was in dire need of refreshing. He tasked the maintenance manager to review the situation and produce an estimate for doing the work. The estimate was done and duly delivered. After carefully looking it over, the refinery manager asked whether in fact the entire refinery could be properly repainted for the estimated amount. There was a rather long and silent pause. At the end, the maintenance manager admitted that it could not. When asked, he replied that he did not think the actual estimate would be accepted since it was a considerable expense. Whereupon the refinery manager said that it was the maintenance manager's responsibility to produce appropriate, truthful, and accurate reports. It was the refinery manager's responsibility to decide the appropriate action. In this case he noted that if the estimate were too much for one year's budget, he'd stretch it over to two. The maintenance manager got the message. The refinery got properly repainted.[4]

---

1 "Enterprise," Wikipedia, last modified May 17, 2018, http://en.wikipedia.org/wiki/Enterprise.
2 Neil Kokemuller, "What is a commercial enterprise?" Bizfluent, last modified September 26, 2017, http://www.ehow.com/facts_7207187_commercial-enterprise_.html.
3 Kokemuller, "What is a commercial enterprise?"
4 Jim Shaeffer, personal communication with the author, ca. 1993.

Unless your plant or other enterprise has its "ducks in a row," it is unlikely that it can perform. It does not happen without purposeful design that includes the physical equipment (including maintenance), operational procedures and protocols, a willing and able professional staff, and a rich culture that melds it all together into a reliable and safe entity. Let us visit some important parts.

Walk the Walk

Leadership requires capable leaders (knowledgeable, honest, and communicative) with a message that captures the hearts and hands of followers, and methods that are fair and effective. Responsible leadership requires that the desired goals be reachable and sustainable. Effective leadership requires that the leaders share the workload and the prizes. Sure, different hands do different work. But work is done. Managers must do the hard work and share equitably the spoils of success.

Some of this hard work is actual work. For example, three important uses of emergency and crisis management teams (Chapter 10, "Situation Management") are to conduct readiness assessments, provide coaching, and perform training evaluations. While most of the detailed design and implementation of an enterprise is done by subject matter experts, senior personnel should do the evaluations of performance. Delegation must not get in the way of responsibility and control.

Walking the Walk

Leaders walk the walk. This is a not-so-subtle expression that it captures all—leaders must "get down into the trenches" and see what their followers see; feel what their followers feel; and give to the enterprise what they ask their followers to contribute. Sure, the workforce (middle managers, supervisors, engineers, technicians and journeymen, operators and attendants, and all manner of other support personnel) does not expect that managers wear the same shoes as they wear. But they do expect those fancy shoes to get dirty when they should. But dirty shoes cannot be the only ticket to leadership. Workers expect to see hard decisions made carefully—just as leaders expect careful attention by all their personnel to their duties and responsibilities. Leaders are expected to lead in integrity, responsibility, honesty, and success. Here are examples:

> When Hewlett-Packard (HPQ) faced a recession in 1970, co-founder Bill Hewlett took the same 10% pay cut as the rest of his employees.
>
> During the early years at Charles Schwab (SCHW), whenever the customer-service phone lines got really busy, founder "Chuck" Schwab dropped everything and answered calls along with everyone else at the company who had a stock broker's license.
>
> Whenever Wal-Mart founder (WMT) Sam Walton traveled on business, he rented the same compact economy cars and stayed in the same inexpensive hotels as his employees.

For the first nine years that his Union Square Café was in business, owner Danny Meyer was there in person every day—clearing tables and mopping spills along with his staffers—as together they made it a top-ranked restaurant in New York City.

Ray Kroc picked up the wastepaper in the parking lot whenever he visited a McDonald's (MCD) to show cleanliness was a continual job for everyone—even the CEO.[5]

These brief illustrations get to the heart of the message. Please walk that walk.

Talking the Talk

Communication is a great tool. Leaders need to use it effectively, consistently, and often enough. Well, that is the easy part. Easy because leaders talk "top down." Their talk is broadcast using the very fabric of the enterprise infrastructure: newsletters, memoranda, town hall meetings, and the like. But "talk" is not communication. Effective communication begins with listening. Responsible organizations develop powerful mechanisms for all stakeholders to communicate—to talk and to listen.

Understanding the Plant or Enterprise

A perfect understanding of an entire plant or enterprise is not essential for good and proper operation. However, every piece used must be reasonably well understood and specifically accounted for in the operation procedures and training. This goes way beyond familiarity. We are talking about a level of expertise that permits those in control of operations to recognize proper operation, understand the nuisances of operation going astray, and professionally manage situations that require intervention. This understanding must be embedded into the fabric of the enterprise to ensure continuous improvement and continuity between stewards.

The Mindful Organization

A *mindful organization*[6] may seem like something made up to sound important and provide good filler for a politically correct chapter in a book. This is not the case here. A mindful organization or, as you may be more familiar with it, a *high-reliability organization* is one that has achieved a sustained level of successful operation.[7] Much of the success is attributable to the inherent fabric of its design. It is so important that no organization can aspire to achieve such success without it. To the point of this book,

---

5   Alan Deutschman, "How authentic leaders 'Walk the Walk,'" *Bloomberg Business Week*, September 18, 2009, https://www.bloomberg.com/news/articles/2009-09-18/how-authentic-leaders-walk-the-walk.
6   Karl E. Weick and Kathleen M. Sutcliffe, *Managing the Unexpected: Resilient Performance in an Age of Uncertainty* (San Francisco: Jossey-Bass, 2007).
7   Todd La Porte and Paula Consolini, "Theoretical and operational challenges of 'high reliability organisations': Air traffic control and aircraft carriers," *International Journal of Public Administration* 21 (6–8): 847–852.

yes, there are going to be a large number of useful tools and ideas presented. At the end of the day, they are going to be largely a waste of time unless your organization itself is healthy and fit—highly reliable. There is more later in the chapter.

2.3 Silos

This topic is important for an effective enterprise. Probably everything you have heard about silos in organizations was negative: make sure you do not have any silos; if you have them, get rid of them. That is because of the "silo mentality." Silos here are a very different story. Most silos are really beneficial. Having them well developed and competently used is going to be an enterprise best practice. First we split silos into two discussions. There are information silos; those are the bad guys. And there are functional silos; the ones that produce expertise and competence for well-functioning component parts to an organization. Let us begin with what a silo is and go from there.

A *silo* is usually thought of as a tower used to store bulk material like grains. Most silos are in the form of a cylinder and bring to mind an idea of something large with a lot of internal content but not much connected to anything else. On a farm they would look as shown in Figure 2-1.

Our silo is a conceptual term used to describe a collection of stuff (people, equipment, documents, collective memory, etc.). Most of what is inside goes together. That going together can be a nice organized fit or just a mishmash. Our silos are mostly organized around the organization and function of plants and industrial enterprises. Figure 2-2 depicts one representing "mechanical" from the set of engineering silos. Inside is a trained group of mechanical and related engineers. They have an appropriate range of experience. They are required to maintain a certain acceptable level of

Figure 2-1. Agricultural silos illustrating their individual significant size but showing no visible connection between them.

Figure 2-2. Pictorial depiction of a mechanical engineering silo.

competence. This may include certification and possibly registration. They are encouraged to participate in professional enrichment programs. They have access to technical resource materials in their discipline. There is a well-established protocol for consulting outside resources for guidance, assistance, or handoff. Summarizing, within the mechanical silo are all the necessary resources for it to be relied on by the enterprise for mechanical engineering services.

Figure 2-3 depicts a larger organization composed of silos for engineering operations, maintenance, and sales and marketing. Management and support, legal, and environmental silos are also needed but are not shown in this figure. Notice there are silos within silos. The silos within silos depict the way individual centers or functions

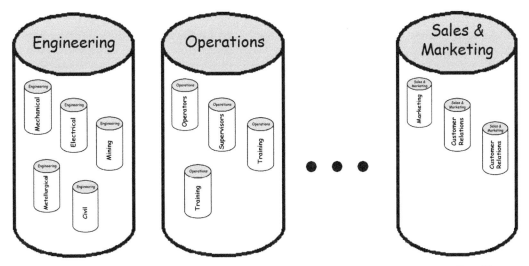

Figure 2-3. Depiction of an enterprise with component organizational parts shown as silos. Three are illustrated. Of course, there are more. Within each is a complete set of organized components (let us also refer to them as subsilos, but each is a silo as well) that together make up a well-functioning component. Each silo and each subsilo has the essential properties within itself to properly function.

(silos) are grouped and then combined together to cover the plant or enterprise. Within the engineering silo are the protocols for collaboration; resources for project management; standards and expectations for ensuring project quality and efficiency; and programs for apprenticeship, recruiting, retention, and remediation. The engineering silo is also responsible for the proper coordination with the enterprise to ensure that it delivers on its charge and responsibility.

This discussion continues with the substantial internal content as well as connectivity.

Information Silos

An *information silo* is not actually a silo. The term is used to suggest that whatever information is within an individual silo is not shared outside the silo. Information silo suggests the powerful visual metaphor that information does not pass between one silo and any other. No matter how well built each might be and no matter how important the stuff inside is, nothing much passes between them. That is the problem. We need them to communicate. We call this phenomenon the *silo mentality*. We need ways for them to do their work within and communicate effectively between each other. More on that later.

Functional Silos

This is the part about silos that is beneficial. To get things started, let us list some of the functional silos an organization might have. There are broad categories. Within each category, depending on the organization's specific responsibilities and ways of doing business, we see different ones. For some components, a silo organization can be a compact and robust way to achieve proficiency and completeness. A silo provides an internal framework for a critical mass of

- knowledge and expertise,
- unique and shared tools to do the job,
- experiential memory so lessons learned stay in place,
- efficient and effective management protocols and personnel, and
- internal quality controls and requirements for maintaining skills and proficiency.

There is sufficient infrastructure within a silo that is tailored to its overall mission and internal functionality. For example, we saw a silo before for an engineering enterprise known as *mechanical engineering*. Mechanical engineers are tasked with a broad area of expertise and application. For example, look at a silo that supports a mining operation. These engineers would be experienced in heavy equipment; conveyor systems and other bulk transport mechanisms; crushing and separation systems; ventilation, moisture, and dust control; operations scheduling (shared with mining engineers and sales); and more. This silo would be responsible for selecting qualified personnel, keeping them current in their fields, ensuring the quality and professionalism of their work process and results, offering pathways for advancement, and more. Of course, it is possible to build weak or incomplete silos. Please do not go that way.

Management of Silos
An enterprise normally needs more than one silo. Each silo is expected to carry the full weight of its entrusted responsibility without prejudice or compromise to the overall enterprise. While the methods, tools, and expertise may differ quite a bit from one to the other, their coordination must be shared. External to the silo and a part of the organization, there will be requirements for control of the individual silo, including its function and management effectiveness and stability, keeping its capabilities and role aligned with the overall requirements of the enterprise that it supports, ensuring efficient and effective assignment of technical and managerial responsibility and coordination between silos and their clients, and providing proper resources to it to do the expected job. In all respects, each individual silo would be subordinate to the enterprise and its mission and function. This "tail" must not "wag the dog." Note that this structure of organization and effective management would be present at each level of silo. Again, refer to Figure 2-3 where we saw that there could be an overall engineering coordination activity involving all the engineering silos where each of the disciplines would have a coordination activity, and certainly a coordination responsibility. Similarly, so would each of the other structural silos in the plant or enterprise.

Communications between Silos
In an effective organization, what goes on in one part has an effect on other parts. Sometimes it is minor, sometimes not. The default position of determining the level of importance must be left up to the information-receiving part, not the information-sending part. In short, effective organizations expect that each part (silo, if that be the case) puts out sufficient information to the others so that each of the others can know enough about the rest to properly function. If additional information is required, then the part that needs it knows enough to ask.

A matrix organization for a project team is one way of both staffing a specific project and ensuring effective communication between the various parts. In a matrix organization, individual participants are chosen from all relevant and necessary resources (think of the silos in Figure 2-3) to form the project team. The project team does the work to meet their deliverables. So their task might be to design a new plant. It might be to find a new market. It might be to scale down production. Each of the team members knows what is going on in his discipline (silo). As a normal part of work, they will share that with other team members (from their different silos). Coordination of information becomes a natural result. And it works both ways. The individual team members still belong to their silo. They bring ideas, technology, and experience from each project they work on back to other members of their silo.

There are other ways to form project-specific teams. When they are used, effective communication must be ensured.

2.4 Enterprise Capabilities

Plants and other enterprises are conceived by entrepreneurs, designed by engineers and architects, and constructed by builders and fabricators. Everyone intends to do a proper job, and for the most part they do. We have no right to assume that they got it right from the start or that they fixed it up later and it is good to go. Despite best efforts, or as a response to irresponsible actions, things go wrong. Responsible organizations accept this and strive to identify and resolve problems and issues before they escalate.

International Association of Oil and Gas Producers on Process Safety

The International Association of Oil and Gas Producers (OGP) mission is to collate and distribute valuable experiential and evaluative knowledge to industry as valuable good practice guidelines. The association's stated goal is to "ensure a consistent approach to training, management and [safety] best practices throughout the world."[8]

> OGP's Human Factors Sub-committee believes that improved understanding and management of the cognitive issues that underpin the assessment of risk and safety-critical decision-making could make a significant contribution to further reducing the potential for the occurrence of incidents.
>
> A focus on engineering issues alone may not be sufficient to prevent future incidents. The role of people in the operation and their support of safety-critical systems require significant attention in parallel with engineering solutions. A better understanding of the psychological basis of human performance is critical to future improvement.
>
> [Enterprise management] should ... work on the operational and management practices that should be in place to ensure operators are able to perform these tasks reliably. That means, for example: avoidance of distractions; ensuring alertness (lack of fatigue); design to support performance of critical tasks in terms of use of automation,

---

8 "Vision, mission, objectives," International Association of Oil and Gas Producers, accessed August 9, 2015, https://www.iogp.org/about-us/#vision.

user interface design and equipment layout; increasing sensitivity to weak signals and providing a culture that regards mindfulness when performing any safety critical activity.[9]

This OGP study is a very readable and persuasive look at what went wrong and how better to approach operations. The study concluded that a significant number of serious incidents were caused by operators failing to monitor or detect information that was available to them. This speaks strongly against more data; it speaks directly to the need for effective tools to understand the data that comes their way. The two most important tools are the understanding of weak signals and the ability to manage the untoward effects of confirmation bias. To say it another way: look for what might be really going on, but look without deciding until enough is known.

Institutional Failure: PG&E San Bruno Pipeline

At 6 p.m. on Monday 9 September 2010, a 30-inch-diameter natural gas transmission pipeline in San Bruno, California, ruptured.[10] The resulting explosion could be seen as a 1000-foot-high fireball in the midst of a highly populated residential area. Eight people were killed; many others were injured. Thirty-eight homes were destroyed and another 70 damaged. The root cause was old and improperly installed natural gas pipe. The pipe condition was known at the time. There were other causes for the failure. The management at PG&E failed to institute an appropriate construction protocol for pipe in a high-consequence area, failed to install appropriate remote shut-off valves that *controllers* (the term for *operators* in a control room of a pipeline) could use to properly isolate a leak, and failed to set control equipment design and operation practice for loss of control room power. However, the most telling message to everyone about the corporate attitude toward safe and responsible operation was management's decision to divert funds away from pipeline replacement to pay for executive compensation.

Process Safety Begins in the Boardroom

Process safety really does begin in the boardroom. We introduced this in the previous chapter. Let us get into it more now. Without the clear direction, active leadership, and provision of adequate resources by senior management, organizations will fail to incorporate sufficient safeguards to ensure safe and reliable operations. Leadership

---

9 International Association of Oil and Gas Producers (OGP), "Human Factors Engineering in Projects," (Report No. 454, August 2011), https://www.scribd.com/document/80232761/OGP-Human-Factors-Engineering-for-Projects.
10 National Transportation Safety Board (NTSB), "Pacific Gas and Electric Company Natural Gas Transmission Pipeline Rupture and Fire, San Bruno, California, September 9, 2010," Accident Report NTSB/PAR-11/01 PB2011-916501 (Washington, DC, August 2011).

failure leads to operational failure. Getting to operational integrity requires the following:

- Ensuring senior management actively supports process safety through its investment strategy and focus on the safety culture of the organization
- Reinforcing the importance of safety by personal example
- Thoroughly understanding major accident hazards and key risk control systems
- Investigating process safety incidents and near misses to find the underlying causes
- Developing world class safety management systems
- Identifying weaknesses in these systems using targeted performance indicators [11]

Enterprises must be purposefully designed and operated for safety.

High-Reliability Organizations

This is a good time to take a closer look at high-reliability organizations (HROs).[12] All manufacturing enterprises that we depend on need be operated reliably and, of course, safely. In this context, *safely* refers to people safety and environmental safety. We depend on them to provide important products, such as chemicals, power, and communications, that can only continue as long as they are financially viable. Design for HROs should be the goal. Their organization really matters.

> [Many manufacturing] organizations have a number of similarities. First, they operate in unforgiving social and political environments. Second, their technologies are risky and present the potential for error. Third, the scale of possible consequences from errors or mistakes precludes learning through experimentation. Finally, to avoid failures these organizations use complex processes to manage complex technologies and complex work.[13]

All the players have roles. Managers must require that their enterprises be HROs. Engineers must anticipate problems and failures by using proper design. Operators must identify operational challenges, react, and cope. Doing so will provide the organization (plant or operation) with the structural capability to find problems, concerns,

---

11 Graeme Ellis, "Process Safety Begins in the Boardroom," Sustainable Plant, April 5, 2013, http://www.sustainableplant.com/2013/04/process-safety-begins-in-the-boardroom/.
12 Chrysanthi Lekka, *High Reliability Organisations: A Review of the Literature*, Health and Safety Laboratory for the Health and Safety Executive, Research Report RR899 (Sudbury, Suffolk, UK: HSE Books, 2011).
13 Paul R. Schulman, "General Attributes of Safe Organizations," *Quality and Safety in Health Care* 13 (2004): Supplement II, ii39–ii44.

and abnormal situations and manage them. The organization must be built to be robust and capably operated. The characteristics built in include the following:

- **Preoccupation with understanding and preventing failure** – Work hard to detect emerging small failures; anticipate future weaknesses and risk exposures.

- **Reluctance to simplify** – Identify and understand first; resist all categorization and all labeling by keeping it unique until it is clear that more is needed.

- **Sensitivity to operations** – Know the operations inside and out.

- **Commitment to resilience** – Ensure that first, second, and third lines of defense are independent, capable, and in place.

- **Deference to expertise** – Do not guess; seek out those who know and listen to them.

The ability to successfully deal with crisis situations is dependent on structures that have been developed and in place well before the crisis began.

2.5 Short-Term versus Long-Term

We all know that long term is better than short term. Remember the parable of the race between the tortoise and the hare.[14] The first one to cross the finish line is the winner. Both begin at the same starting point, leaving at the same time. The hare leaves in a flash, gets far along the course, and then stops, exhausted. Believing that he has traveled too far and too fast to be overtaken, he takes a nap. Meanwhile the tortoise adopts a pace that, while slow and prodding, is steady without missing a beat. He passes the sleeping hare and wins the race. The sustainable long-term plan delivers. Sure, parables are interesting and maybe we learn a thing or two. But let us leave this story and build one for ourselves. This one will get to the heart of the concept. It is the story of two (hypothetical) enterprises.

Story of Two Enterprises

We begin with two identical companies (Company A and Company B) with a single top executive. Each company has its own boss who reports to the top executive. The companies start out making the same products using identical machinery and similarly prepared sales forces and marketing material; they have the same suppliers with

---

14 Laura Gibbs, trans., *Aesop's Fables* (Oxford: Oxford University Press, 2002).

the same cost structure and profit structure. They are separate but identical companies. Their only overlap is that they share one national marketing region.

Each starts off with the same structure and assets but with very different performance goals. The top executive instructs Company A's boss that he will be judged on a strict yearly basis. Each year, to keep his job, he must demonstrate that he operated his company the best it could be operated during the previous year. No excuses, no delayed benefits—each year must be the best it can be, period. Company B gets different instructions. Company B will only be judged at the end of the first 5-year period and every 5 years thereafter. If at the end of each 5-year period, Company B is not doing its best, so long boss.

How does it look at the end of the first year? Figure 2-4 shows the progress.

Looking at Figure 2-4, we see that Company A (orange line) has increased in value more than 75% by the end of the first year. This is excellent. We see something very different for Company B (blue line). It has dropped about 50% in value. If this keeps up at the same pace, there will not be a company at the end of the second year. Company A gets a gold star. Company B gets a pass for now because the boss upstairs is not looking.

Let us pause this competition briefly to ask what might be going on to produce these results? Company A's results at the end of year 1 are understandable. This boss is going all out. He got in there, got cooperation from his entire team to go for the maximum, pushed the production machinery to its highest rates, priced his product to sell as much as possible, and got results to match. Company B operated very differently. It invested in production, sales, and marketing systems. It sent

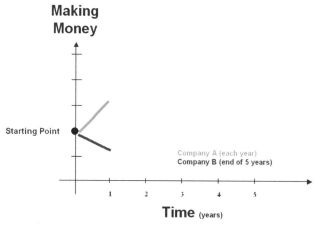

Figure 2-4. Two competing companies at year 1 showing a clear gain of pursuing short-term goals over long-term goals in the short term.

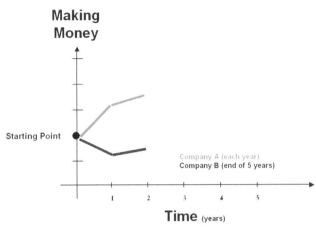

Figure 2-5. Two competing companies at year 2 still showing a gain of pursuing short-term goals over long-term goals but much less than that of the first year.

some people back for more training, spent considerable effort starting a comprehensive market research program, and worked on production optimization plans. Quite predictably, Company B's sales team had less time to sell and its equipment received much attention and more than a few tweaks, but at the expense of manufacturing quantity and sales.

We move to the second year shown by Figure 2-5. Company A has increased in value about 30% over the first year, certainly a very respectable progress. We do see improvement for Company B but nothing noteworthy. Yes, the decreasing trend is reversed, but its current position is below what it was at the start. Dig a bit deeper. Company A achieved less the second year than the first likely because its sales team needed to back off a bit to catch their breath; the machinery needed a bit more maintenance to get over working overtime the year before. Pushing at maximum works only for a while. On the other hand, Company B started seeing better results from the sales team using new approaches and skills. Their improved supplier agreements got some lower costs and better delivery, and they were able to get their machinery working more reliably. You can see how this is shaping up.

Fast forward to the end of year 5. Figure 2-6 depicts these results.

At the end of 5 years the difference in policy is telling. Not only has the long-term company (Company B) dramatically exceeded the results of the short-term one (Company A), but also the short-term one appears to be headed for collapse. It should be clear that it had likely spent considerable, if not all, of its reserves in earlier years—holding back then would have meant less achieved during any year it was held back. Long-term benefit was forfeited.

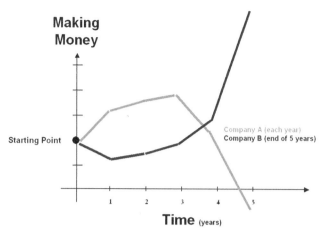

Figure 2-6. Two competing companies at year 5 showing a clear and dramatic loss by pursuing short-term goals over long-term goals in the long term.

Valuable Message

We are now ready for the message: any long-term best plan or activity will always perform less than any short-term plan in the short term, oftentimes much less. Look at this the other way around: to do better in the long run, it is always necessary to do less than better in the short run.

> All long-term strategies will lose in the game of competition if judgments are made in the short term. Always! Every time!

If you are a manager, please ask for long term every time. If you are reporting to the manager, please remind him of this message if ever he forgets. This is the bottom line of sustainability. Without it, all the ideas, technology, and good intentions will not have an honest chance. The work you do deserves that honest chance.

2.6 Delegation of Responsibility

Delegation of responsibility is a powerful operational tool. If management were to learn only one thing from this book that might make a difference in safe and reliable operations, it is probably the concept model here for delegation of responsibility, also called *delegation of authority*. What you are about to read differs in only one small respect from what is common practice. Yet this difference makes all the difference.

Delegation is used when the primary responsible individual is unavailable or unable to perform the duties or actions he is responsible for. A dictionary definition for delegate is: "*noun*—a person designated to act for or represent another or others;

deputy; representative.... *verb*—to commit (powers, functions, etc.) to another as agent or deputy."[15] Delegation allows an organization to continue activities without undue delay caused by an unavailable responsible person. We are conditioned to view delegation as being done *down* the chain of organization. If a team leader were on vacation, he would usually choose a senior individual working under him on the team to stand in for him. But the act of delegation does not at all suggest such a direction. It is the choice of direction that determines everything! We continue with a true story.

There was an Ohio oil company named The Standard Oil Company (a residual of the massive Rockefeller Standard Oil that retained the original name after 1911) that marketed under the Sohio brand (and others) before the merger with BP in 1978. Its established policy for delegation was to delegate upward. For example, if the manager of civil engineering were on an extended overseas trip, the engineering manager to whom the manager of civil engineering reported would manage the civil engineering team during that time. It was a simple change of delegation direction—with a profound operational effect. Two important organizational benefits resulted. First, in order to be able to step in and be responsible for teams under, the above manager must maintain a clear pulse on the activities going on under. Regular reports and happenings now had a function and an informed audience. No longer were they just paper or meeting activities—they needed to communicate adequately and in a timely manner. Second, any decisions or actions taken by the delegate could be relied on. After all, he had the experience and authority. Sure, the returning manager might revisit what was decided and modify as needed, but that would be an expected role for his job. No one below him would be offended or undermined.

You may wish to refer to the "Collaboration" and "Escalation" sections in Chapter 10, "Situation Management," for a discussion of a communication framework for this activity.

2.7 Operational Integrity

Every plant must be designed for operability. This is quite different from the traditional design for functionality. Yes, every plant must be functionally designed to do the thing it needs to do: produce a chemical, refine an ore, manufacture a substance, clean and prepare a material, and more. Doing this properly relies on the technology of fields such as engineering, basic and applied sciences, materials, and construction. Of all the choices for good functional design, designers must select only those that are sufficiently operable so that the operation poses an acceptable risk to the enterprise

---

15 "Delegate," Dictionary.com, accessed July 5, 2018, http://dictionary.reference.com/browse/delegate.

and the larger society. Successful design requires the designer to fully anticipate and manage all operational risks through the following:

- Paying attention to safe operating regimes to ensure the process operational envelope remains preferentially safe

- Identifying needed information required by operators for proper understanding of the risk and effectiveness of current and likely operational states including abnormal ones

- Providing the operator access and controls necessary to maintain proper operation

- Providing the operator access and controls necessary to restore proper operation during threats to it

- Providing operator access and controls necessary to safely end operation or move operation to an allowable risk level, if continuing to operate "as is" should not be a proper option to select

This turns the table, so to speak, because we now need to ensure that a given plant is operational as built. Measuring operational integrity helps in understanding this. Operational integrity is the inherent capability of an enterprise to be responsive to proper operation. Responsive to proper operation is a collective attribute of whether or not an enterprise is capable of being appropriately managed by automatic controls and other instrumented mechanisms appropriately supplemented by human individuals to monitor and intervene where necessary to accommodate abnormal operations.

An enterprise's capability for good operation is a complex matter. Operability integrity is a methodology for examining the individual aspects of the enterprise for purposeful design and its ability to be effective, productive, and safe during operation. By developing and applying an objective measure of the operability of a plant, an enterprise will be able to assess its ability to safely and effectively operate its production facility. The measure must be simple enough to readily identify sites at risk. The measure must be rich enough to delineate strengths and weaknesses, identify risk exposure, and point the direction toward a prioritized list of operability improvement activities. The measure must be sufficiently objective to permit general, widespread usage to achieve similar results. Additionally, the measure must provide tools for the various providers of operations improvement technology to more effectively communicate their message, thereby increasing their ability to provide quality products and expert technical services.

Operational Integrity Levels

Analogous to the measure of the safety integrity level (SIL) as an unambiguous way to qualify the effectiveness of safety equipment and processes and their suitability for delivering a sufficient level of protection, operational integrity level (OiL) serves as a comprehensive indicator of a production facility's ability to properly meet its operability requirements. Low OiL identifies a plant at risk. A high level indicates a facility that meets the expected requirements for safe and effective operation. It should be noted that a high result from using this measure will not assure safety, nor will a low result presage imminent problems. We discuss it here just to reinforce the situation management message: it is not possible for an operator to properly manage a deficient plant or enterprise.

The OiL level would be composed of two general components: awareness and effectiveness. Each component has many ingredients. The *awareness* component measures the site's awareness of where it stands on the OiL measurement scale. The *effectiveness* component measures where they actually need to be. Observe that a site with a high effectiveness score but a low awareness score would not rate a high overall OiL score. The reason: because the awareness score was low, it is likely that the high effectiveness score has been obtained either by lucky coincidence or by design from an infrastructure no longer present. Either way, it is unlikely that under changing conditions, the high effectiveness score can persist. A high OiL score, therefore, represents a high degree of current operability as well as acts as a predictor, albeit not guaranteed, of future success.

Plant Operability Components

Let us take a look at the breakdown of the key aspects of plant operability.

Management Competency

Is the management team competent in the skills of enterprise management including the ability to motivate people; effectively communicate and listen; understand and manage technology; understand and manage risk; delegate responsibility; and manage an enterprise for overall, long-term success? Is the management team sufficiently knowledgeable about the entire spectrum of operational aspects for the enterprise? Is the management team sufficiently informed of the necessary regulatory requirements (financial, enterprise operation, health and safety, environmental, etc.) applicable to the enterprise?

Management Effectiveness

Has management put in place the proper responsibility chain? Has management provided sufficiently clear and complete direction so that those responsible for plant operation know their responsibilities and authority? Has management put in place a system for monitoring the enterprise achievement such that it is clear what is working effectively and what is not? Has management instilled in the plant staff sufficient motivation and resources for them to deliver on their respective responsibilities?

Plant Complexity and Inherent Stability/Safety
Does the complexity of the production facility or other enterprise provide an appropriate match to the difficulty of the requirements and the inherent safety of the enterprise? Is the design appropriate to good practices?

Equipment Suitability
Is the physical production equipment (pipes, vessels, roadways, transmission lines, control and information infrastructure) appropriate to good practices? Is everything properly integrated? Is the documentation and approval infrastructure up to good practice and current?

Equipment Readiness
Is the physical production equipment in good working order? Is everything properly maintained? Are all operations approvals and practices current? Is there a comprehensive program of continuous monitoring to ensure readiness? Is all needed documentation up-to-date and readily available?

Staffing Quality
Do the background, capabilities, and experience of the operational, engineering, and maintenance personnel match the requirements for a successful enterprise? Is there an ongoing program to monitor their well-being?

Staffing Competency
Do all personnel possess the appropriate competencies to perform at a level consistent with good practices? Is there an ongoing program to monitor their performance? Is there an ongoing program to improve their performance as well as provide career coaching and improved job satisfaction?

Staffing Readiness
Are the personnel physically and emotionally ready to assume their needed duties?

The Record
What is the operational history of the facility? Is the product of high quality? Is the rate of production suitable to the investment and responsive to the business environment? What upsets have been experienced? What are their causes, what are their costs, how well were they managed, what was done to prevent future occurrences, and so on?

	Inherent Safety/Complexity		
Actual OiL Level	*Low*	*Normal*	*High*
High	**Prudent**	**Prudent**	**Responsible**
Medium	**Prudent**	**Responsible**	**Dangerous**
Low	**Inadequate**	**Dangerous**	**Negligent**

Table 2-1. Operability integrity level (OiL) as an enterprise scorecard depicting the need to possess a higher degree of operational control as the inherent operational risks increase.

OiL as a Measurement Evaluation

Let us take a look at how a plant's OiL rating might be used to assess its risk status. Above is a matrix that scales the inherent safety/complexity of the plant against the actual measured OiL level.

Referring to Table 2-1, if after an audit the OiL assessment level rating for a plant is *medium* and its inherent safety/complexity level is *high*, the plant is in a *dangerous* operational state with regard to its ability to manage the production. In like manner, if the OiL is *high* and the inherent safety/complexity is *normal*, then the production is being managed *prudently*. Observe that the highest OiL status for an enterprise that has a high inherent safety/complexity level is *responsible*. This reality reflects the inherent difficulty to render dangerous operations completely safe and reliable.

The bottom line here is that plants or operations that cannot achieve the overall level of *responsible* or *prudent* should be winnowed out.

Definitions
Actual OiL Level

Low – The current enterprise operation is unreliable; incidents, accidents, and personnel safety are an ever-present and significant risk.

Medium – The current enterprise operation is managed, but there are important gaps in infrastructure and protocols that unnecessarily expose the operation to danger and personnel to injury.

High – The current enterprise operation is fully managed up to industry best practice.

Inherent Safety/Complexity

Low – The operation of the plant or enterprise poses little risk to safety or community; the design of the equipment is traditional (to the point of being a commodity) and well understood by all.

Normal – The proper operation of the plant requires the exercise of "ordinary care."[16] However, it can place personnel and community at limited risk; the design of equipment and operations requires experience.

High – The operation of the plant requires special care to manage properly as it places personnel and community at extensive risk; the design of equipment and operations requires significant expertise and specialized experience.

Enterprise Scale Points

Negligent – Enterprise poses a clear and present danger and should completely and immediately cease all operation.

Dangerous – Enterprise has existing safety and community risk management programs in place, but programs consistently fall short of minimal requirements; operation should cease or be appropriately curtailed until defects are remedied.

Inadequate – Enterprise has existing safety and community risk management programs in place, but programs have important gaps that require timely remediation.

Responsible – Enterprise adequately manages risk appropriate to achieving responsible safe operation and respect to the community.

Prudent – Enterprise fully managed risk to personnel safety and community welfare up to the best available standards and practices.

Discussion

It is tempting to say that the case where the OiL is *high* and the inherent safety/complexity is *low* is an overexpenditure or overemphasis. This is not likely true. Whatever the complexity or safety exposure of managing an enterprise might be, there are always useful production efficiency benefits yet to capture. The higher the OiL, the better the plant's ability to realize those benefits. Measuring a site's OiL and comparing it to industry best practices can be a powerful motivational tool for an enterprise to improve its competitive position. Moreover, if measures are developed to track

---

16 "Standard of Care," Wikipedia, last modified June 2, 2018, http://en.wikipedia.org/wiki/Standard_of_care.

changes online, useful operational risk information can be made available to operators and supervisors.

2.8 Safety

"Safety is the state of being 'safe,' the condition of being protected against physical, social, . . . financial, . . . emotional . . . occupational . . . or other types or consequences of failure, damage, error, accidents, harm or any other event which could be considered nondesirable. Safety can also be defined to be the control of recognized hazards to achieve an acceptable level of risk."[17] *Safety* is "… the quality or condition of being safe; freedom from danger, injury, or damage; security."[18] At its most basic level, safety is the state of being secure and out of danger. This book is all about safety—improving it to the point that we can count on responsible industrial process operations is why this book is in your hands. Everyone knows about safety. The word safety is part of our culture. Most nations have legal requirements for it and have agencies dedicated to its care and advancement. However, you will not find an in depth coverage of safety here. This material is intended to supplement or explain your existing programs. The intent is to provide information that you should find useful and important to: (1) motivate you to want to improve your operations, (2) have a good chance to understand *what* and *why* the material is here, (3) learn what would be useful to you, and (4) provide an ability to bring what you choose into reality. So let us briefly talk about safety.

> Situation management's role is to bridge that gap between *safety being available* and *safety delivered*.

Safety is so confusing on the one hand and so simple on the other that it is easy to say you have it and discover in actuality that you do not have very much. Let us get basic. If an enterprise fails to plan for and deliver the full extent of safety, it cannot be safe. History is overwhelmed with examples of failure. The BP Texas City disaster was enabled (almost made an inevitability) by a culture that placed personal safety as the primary safety program to the extent that all other levels were largely ignored.[19] BP

---

17 "Safety," Wikipedia, last modified May 15, 2018, http://en.wikipedia.org/wiki/Safety.
18 David Bernard Guralnik, ed., "Safety," in *New World Dictionary of the American Language*, 2nd ed. (New York: Simon & Schuster, 1980).
19 BP US Refineries Independent Safety Review Panel, "The Report of The BP US Refineries Independent Safety Review Panel" (Washington, DC, January 2007).

failed to adequately ensure infrastructure safety and operational safety. Piper Alpha had a program for infrastructure safety but failed to ensure its quality, to enforce it, and to adequately maintain it.[20]

Safety is not the job of a safety department. It is not the responsibility of a safety manager. It is the job of management. It is the job of senior management.[21] And it is a job with the responsibility to deliver the requirements to demand what is needed to render safety, the actions to meet the requirements, and the commitment to gain lasting compliance to ensure that what is needed is evaluated, supplemented where needed, and fully maintained. Having said that, it is everyone's job to act and operate safely. It must be part of the DNA of the organization.

Characteristics of Safety

If industrial plant safety is required, it must be obtained first and foremost by the inherent design and build of the plant. Atop this, safety is completely dependent on how the people and equipment function together to meet the operational requirements necessary for safe operations. It is not gained by slogans, by tracking safety records, by contests, or by any other secondary achievement. Rather, safety is gained by having sufficient technology and operational excellence. It requires a robust engineering community constantly tasked to participate in the evaluation for fitness, the installation for robustness, and operation as designed. It must be supplemented by the constant and consistent practices that ensure the enterprise is run properly. And it must be critically supported by decision-making leadership to ensure the highest needed standards are met.

In your enterprise, it must be readily possible to assure the constant presence of sufficient experienced, trained, and confident engineering and technology personnel and effective operations leadership. Their numbers and their responsibility cannot be less than required to ensure a strong atmosphere of honest involvement in the robust and safe operation of the enterprise. It must be done with continuous evaluation of the competency of the equipment to operate properly and responsibly in all production campaigns. Personnel must have the ability to identify abnormal operation and to clearly and responsibly manage during abnormal operation to ensure that safe and responsible steps are taken to return to normal operation; or, if return to normal is not a reasonable option, safely discontinue operations.

---

20 Lord W. D. Cullen, *The Public Inquiry into the Piper Alpha Disaster*, Volumes 1–2 (London: Her Majesty's Stationery Office, 1990).
21 Cullen, *The Public Inquiry into the Piper Alpha Disaster*.

Further, it must be possible to identify a complete culture, from the highest manager to the lowest-level employee, that understands safety, accepts safety as the only appropriate operational parameter, and is provided with all necessary means to achieve the individual responsibilities to provide the full measure of safe and responsible operation.

Components of Safety

Safety means different things to different people. Unfortunately, being safe does not depend on an organization's definition of the word *safety*. Safety cannot be defined into being, and danger cannot be kept at bay by what someone thinks safety requires. Safety can only be improved by actually working hard to be safe. Being safe has four distinct aspects:

1. Personal safety
2. Safe conduct of personnel
3. Safe equipment and operational design
4. Safe performance of activities

Safety depends on the unrelenting adherence to all these aspects, all the time. No one aspect is more important than any other. None is optional.

Personal Safety
Personal safety is about safety glasses, hard hats, Nomex work clothes, gloves, dust masks, earplugs, steel toed shoes, nonslip soles, and a lot more. Individuals are expected to use the proper personal protective equipment (PPE) all the time it is needed. Each activity and each enterprise will understand the risks and exposure and decide what to use and when. It is the first line to directly protect people from ordinary harm. This is only one part of safety. It is the beginning, not the comprehensive requirement. It is the easiest to understand and enforce. While parts of it might not be the most comfortable or stylish, most of us know that we personally benefit in real and important ways.

Safe Conduct of Personnel
This means that people going about their work do so in a safe and responsible manner. This is about how individuals work to be safe. It means using safe work practices like wearing the proper PPE, notifying the proper personnel when working in a specific area, using proper tools and equipment, climbing ladders properly, entering confined spaces properly, and all the rest. There are no shortcuts here.

Safe Equipment and Operational Design

This means that the process and equipment being operated has been safely designed and contains all the engineering and operational safeguards that are reasonably needed to keep things safe even when someone makes a mistake, something breaks, or something gets nearly out of control. This includes infrastructure safety. Infrastructure safety is all about HAZOPs, training, procedures, permit to work systems, ladder cages, railings, CPR training, accurate information, near miss processes, safety audits, effective safety budgets, operator qualifications including readiness, personnel impairment testing, near end of life equipment tracking, management of change, and a lot more. Operating interlocks are one example. Overtemperature, overpressure, high- or low level shutdowns, dikes to contain spills, and vessels designed to withstand fire are additional methods. Taken together, they cover the framework for ensuring everything needed for safe operations is present and working effectively. There are many others.

Safe Performance of Activities

Now we come to the place where it all comes together. Even though your personnel are wearing and using the right equipment, they are doing their work correctly, and the plant has the requisite safe operations safeguards, people must conduct activities that are designed to be safe and effective.

Operational safety is all about following procedures, effective tactical operations management, delivery of essential operating information, effective situation management, and seamless integration of crisis and disaster management operations, when needed. When operational safety is not present, even though there is a good infrastructure safety, operations are in danger because:

> ... the information needed to safely operate the facility are present in the procedures and practices of the facility or know by facility personnel. Yet, ... well intentioned, well trained workers committed grievous errors.[22]

Following operating procedures is a part. Not inventing ways to handle situations for which no appropriate procedure exists or existing procedures do not appear to be working is a part. Effective training and practice is a part. Situation management's role is to bridge that gap between safety *being available* and safety *delivered*. Its role is to foster a clear understanding of and provide resources for safe control room process operations.

---

22 CCPS (Center for Chemical Process Safety), *Conduct of Operations and Operational Discipline for Improving Process Safety in Industry* (New York: Wiley, 2011).

Delivering Safety

There must be a central technology and operations team charged with maintaining the collective responsible infrastructure. This team must be fully qualified in their individual technologies. They must be aware of the full nature of recognized and generally accepted good engineering and practices (RAGAGEP). And they must be included in robust roles of all parts of the enterprise. There is an unfortunate trend to do away with centralized resources in favor of lean, compact local teams. Having too much responsibility to local customs and expediency disadvantages these local teams. They are disadvantaged by having too little experience with anything outside the daily tasks and operations to which they have become accustomed and are responsible. They are disadvantaged by being too few in number such that their line tasks leave nothing left over for anything out of the routine. This situation leaves them without sufficient time and resources to do even primary tasks. Making matters worse, the individuals of these local teams are, by and large, without benefit of proper professional supervision and nurturing mentoring from senior, experienced practitioners. Work involving customary situations is often done reflexively, without sufficient care, time, or knowledge to fully understand the appropriateness and effectiveness of the job. New situations are either press-fit into older molds or handled by stumbling through as best as one can.

Nor is subordinate (or junior) management relieved from the press of too little time, too little resources, and too little responsibility for the long-term performance of the enterprise entrusted to their leadership and care. Rather than being trained and mentored from above, they are all too often thrust into position by proxy or convenience. Thusly placed, they are reluctant to perform the critical evaluations needed for proper situation assessment. Lacking comprehension of the full nature of what is going on, they find refuge in the misconception that doing little to change is the safer route to avoid future blame for any failure. They are rarely able to rise above their inherited situations and cultures. Tragically they fail to appreciate the inability of words and slogans to change what actually requires resources, infrastructure, and honest commitment.

2.9 Responsible Engineering Authority (REA)

Plants and operations that have a responsible engineering authority (REA) are better positioned for successful and safe production and operation. The position of an REA is a best practice to reduce the chances of important items falling through the cracks. An REA is an experienced and respected individual who is the main contact point for collective operating experience and technical knowledge. He is the knowledgeable resource of rules and procedures, is a champion for hazard evaluation protocols, and functions as a general watchdog for safe and reliable activity. He is responsible to ensure that all relevant safe operation aspects are designed, in place, operational, and integral to the daily life of the enterprise.

Careful attention to providing a strong and effective production infrastructure cannot by itself guarantee success. No amount of propping or operational finesse can adequately replace what is missing. There are many moving parts in any industrial plant or operation. Yes, there are the actual physical moving parts like machinery, flows, vehicles, and such. But the moving parts we visit here are infrastructure ones, like procedures, permits, training, compliance, HAZOPs, and the like. Each represents an important aspect necessary for safe operation, efficient production, or regulatory compliance. If any are missing or inadequate, problems and failures usually result—not necessarily immediately, but sooner than anyone would like. Managing all of them can be a daunting responsibility. However, the concern is keeping all these "parts" running right. The problem is that "being concerned about them" can be so spread out among different individuals, departments, or offices that it is hard to know who does what, whether it was done, and if it was done, how well.

If all of this sounds too much for a single individual, you are partly right. One person is not doing it all. The REA is charged with being the go-to resource, at times the first, but not the only. He will draw heavily on and seek advice from others in your enterprise. His duty is to ensure that the whole of an operation does not get forgotten during the regular course of business. Here is a part of the REA's coordination and facilitation list of responsibilities. The REA is a collective conscience to ensure the following:

- Plant design and equipment specification are monitored to ensure fit for purpose and appropriate for safe and effective operation; this is also done for modifications and extensive repairs or retrofit.

- Adequate and effective HAZOPs are done and up to date.

- The Management of Change (MOC) program is effective and fully utilized.

- Safe operating limits (SOLs) are established for all areas of the plant or operation that can cause unacceptable injury, financial loss, and environmental damage. Adequate reviews are performed to keep them appropriate.

- The enterprise is 100% compliant with the positive materials identification (PMI) program, including predictive and actual maintenance inspections and monitoring.

- The enterprise reviews 100% of "lessons learned" from similar experiences in relevant industries.

- The enterprise monitors 100% of near misses, documents them, and folds the relevant experiences back into the enterprise.

- Operational hazard isolation and mitigation systems are designed, placed, and properly used, including identification and proper operation of systems to protect the SOLs, such as corrosion or excessive wear, electrical protection, over- and undertemperature and pressure-relief systems, and all other aspects of stored energy and exposure to operational risks.

It is not the REA's role to do all of this. His responsibility is to ensure that all of this is "kept in mind" and not lost or diluted during the press of daily operations or reluctant resources.

2.10 The Magic of a Control Loop

How is it that a single feedback control loop, properly applied, can produce such a remarkable taming of the wild nature of Nature? What is it about such a loop with a simple construction that enables it to provide effective operations over a wide range of operating conditions? All this amazing efficiency has little to do with the details of the algorithms used or the clever methodology of implementation. The magic is fundamental. And once you see it, it should forever change how you understand and use a feedback control loop. Ready?

A feedback control loop works its magic by moving unwanted disturbances from where they cause harm or other unwanted effects to another place where they are not much in the way! The loop moves the trouble from where we do not want it to another place where it does not matter what it does. Okay, your imagination and interest are piqued here. But this probably still does not make enough sense for you to say, "Ah, I understand." So let us take a look at some simple examples.

Temperature Control Example

Consider a temperature control loop. Say we have a large drum partially filled with a liquid that we need to keep at an elevated temperature. Liquid is constantly flowing into and out of the drum. The drum is internally heated by steam that flows through a submerged coil. A control valve manages the steam flow. A controller sensing the temperature of the liquid in the drum manages the control valve. What can go wrong? The liquid temperature is "disturbed" by changes in inflowing liquid, changes in outflowing liquid, changes in the pressure of the drum, changes in the weather outside the drum, or changes in the heating value of the steam itself. The place where we do not want the disturbances or variations is the temperature of the liquid within the drum. So to where does the temperature control loop move them?

The control loop manipulates the steam flow in order to respond to temperature variations. Regardless of what might be the cause of the temperature variations (remember that list we just went over), the steam flow is manipulated to provide additional heat if the liquid in our drum is too cool or less heat if the liquid is too warm. The act of control is moving the temperature variations in the drum liquid into flow variations in the steam system. The steam system is designed to handle those variations.

Flow Control Example

The steady management of materials flow is important to the uniformity of many manufacturing processes. The way this is usually done is to measure the flow and then use a flow-modulating device (usually a valve for liquids) to increase or decrease the flow until it is in agreement with the desired flow value. On the surface, this seems like the practical way for this to work. Below the surface, where the actual activity takes place, we have an entirely different story.

All efforts to modulate a flow in one part of a series process will upset the flow both upstream (at another processing station that has the same type of flow controller that we have) and downstream (again at another processing station where the steady management of flow is important and might be a different value than our upstream one). We have a paradox. Engineers have long recognized this. To deal with it properly, the equipment design includes places for the flowing material to accumulate between the flow control loops. They call those places *surge*. Surge usually takes the form of a separate surge drum, a tank, an accumulator, or simply a space at the lower end of processing or separation vessels for the liquid to ebb and flow.

In this manner, plant designers are able to move the flow-affecting disturbances, which get in the way of good production, into level variations, where it is presumed they will not bother the process very much. This is a good plan. Unfortunately, sometimes the designers of the level control systems are not privy to the master plan. Most control designers value level management in the same way the other control loops are valued—they want them to control perfectly. As a direct consequence, levels are controlled to be as close to a desired value as they can be. But that desired value almost always has no process importance. In the end, we see tight level controls interacting with other tight flow controls. These loops fight each other by trading disturbances back and forth. Each one works very hard on its own problem. None can work together as needed.

Let us end this illustration with the suggestion that plants examine all of their level control loops for proper construction. Where strict level control is needed, the flow must be available to be modulated at will. Where level is used for a disturbance catcher, then its control structure must be changed to enable managed swings that permit good flow control without resulting in level movements outside the physical limitations of the surge volume. Standard nonlinear level controls work well here.

Useful Message about Moving Disturbances

The important thing about all of this is the value of examining control loops everywhere to ensure that they are moving the disturbances to where they should be. They are sometimes passed around to other places where they are still not okay instead of away from the plant production process. There is a rich reward to understanding this concept, locating all the places where the disturbances are ineffectively or improperly moved, and moving them to other proper and better places. Good process control will ensure that the controls design does this properly.

2.11 Selective Automation

Plants use a number of partial or selective automation methods. They use them to improve the production operation and reduce the manual control and process monitoring load of the operator. Appropriate automation can make a significant difference. Automation methods are best implemented within the structure and design of the basic process control system (BPCS). Doing it this way significantly increases the reliability of the plant's operation (no special code or programming required) and makes its design completely visible to all having access to the control system. This means that operator graphics can include all the plant's information, including switches from automatic to manual, from on to off ratio, and from cascade to off cascade. All alarms needed are fully integrated by the BPCS and therefore require only conventional configuration and documentation.

Nonetheless, operators will require sufficient knowledge of how alarms work and what to do when they seem to be having problems or need to be turned off due to issues with a sensor or valve or the plant becoming upset so that its operation is outside the design limits for those controls. Training is needed on how to observe potential problems and how to manage them when they present. Proper documentation must be available to permit manual adjustment when practical. All of this requires periodic practice and monitoring. Most of the suggestions provided in the previous section would be useful.

Advanced Basic Control

The steps up from basic regulatory control can take many directions. The most accepted and widely used is advanced regulatory control. It usually involves cascade control loops, where one simple control loop, say a flow, has its set point determined by another simple control loop, say a level. If the flow is an inflow from the level controlled area, as the level increases, the flow set point will decrease to keep the level from becoming too high in spite of a separate control on the level, and the other way around as the level drops. The advantage of having both variables under basic control is that both can be regulated by the feedback mechanism (somewhat) independently until they need coordination. They coordinate through the cascade controller.

Another common type of advanced regulatory control is ratio control. It is used where the desired values (controller set point) of two or more independent control loops need to be adjusted together. For example, if a mixing operation requires the manual adjustment of the primary mix component, all the other mix components would then be automatically adjusted (their set points determined by a ratio from the primary one) so that the overall mix ratios remain the same. Certainly, each of the other mix components could have their set points manually adjusted at the same time as the primary one. However, if the ratios of the mixes are different, or change based on the primary value, there is much room for operator error (wrong ratio, waiting too long, missing one, etc.). Here again, a little bit of control design would provide a substantial amount of production accuracy and reliability.

Advanced Control

Advanced control is a broad but important tool for reliable operation of complex and highly integrated operations. This control is quite different from the advanced basic controls just discussed. Advanced control utilizes special engineering and process design information to coordinate various basic controls in a customized way. A few common examples are the final blending of automotive gasoline into a delivery pipeline, controlling a chemical reaction that requires careful intervention during the various stages, and network controls to accommodate node outages or wide changes in loading. These controls work in such a coordinated way that it would not be straightforward for operators to observe or intervene when necessary. Operator protocols to ensure their ability to manage them will generally follow the options suggested in Section 2.12, "The Automated Plant," later in the chapter.

Models

Plants and their associated major pieces of equipment can be complicated. This complication can arise due to the equipment's size, the nature of its pieces, or simply because

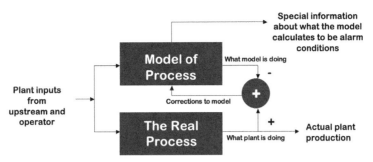

Figure 2-7. Model reference design diagram illustrating how information from the real process is used to "build" a model that then permits a "look inside" to see what could not be seen in the real process. In this case, information about the alarm conditions inside.

what happens inside it is hard to see, or hard to understand, even if it could be seen. Suppose there was a way to see and understand everything. Suppose we could build a model, a replica, of that plant or piece of equipment according to our own rules of understanding. While there are many ways to build models, suppose ours is built of mathematical equations. And suppose this model is "tuned" so that it behaved closely enough to the real thing. Because we constructed this model, it is built of pieces we understand. It is easy to look inside it to see what could not be seen in the original. Moreover, we can also use our model and make it do all sorts of things, even things that might damage actual equipment. It permits us to see what might be going on inside the actual thing. Knowing what is going on inside permits us to draw conclusions and make decisions. We have described the process of model-based reasoning. Figure 2-7 illustrates this design.

The lower box depicts the actual process. The model is the upper box. The real plant inputs (streams, energy, etc.) go into the process. We also tell the model mathematically exactly what inputs go into the plant. The plant produces results (products). The model calculates these results as data. Both results are compared, and the model is adjusted to bring the model results as close as we can to what the plant is actually doing (instrument and sensor readings, lab reports, etc.). Once that is done, our model matches the plant. If we were to look inside for data (e.g., for reaction products or internal temperatures), we can know them from the model even if we cannot measure them in the actual plant. This information can be used to advise the operator. Such advice can be used to prevent upsets or better manage those that occur.

Model Reference Adaptive Control

A specific arrangement of models used indirectly for adjustment of controls is model reference adaptive control (MRAC). MRAC uses the same model building technique described in the previous topic to adjust the internal parameters of the process control algorithms to better match with the process needs of the moment. Figure 2-8 depicts

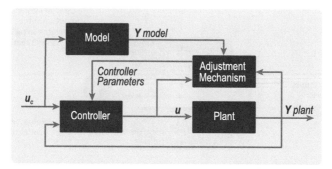

Figure 2-8. Model reference adaptive control diagram where a model of the real process is used to constantly update a controller for the real process to ensure it works as needed despite significant changes in operation or situation.

how this fits together.[23] In this case, the model is already built and is used to assist the "adjustment mechanism" to compute the proper changes needed for the controller to work as a better match to what the process needs at the moment. Other forms of this advanced control arrangement include multivariable process control (MPC) and inferential control.

This situation brings up an important issue for operators. This type of process management is quite different from the conventional control loops operators are accustomed to managing. If this specialized controller stops keeping the plant in good control, outputs not in their proper place will be quite visible. Not so the causes for that trouble. Unless these advanced types of controls include specific diagnostic information about their inner workings and provide remediation suggestions,[24] the operator is usually left with two awkward alternatives: (1) turn the whole control scheme off (sometimes bypassing, sometimes shutting it down) or (2) live with the problem until an expert comes and fixes it.

Procedural Automation

Most industrial plants have two general modes of operation: regulation and transition. In regulation mode, the plant is operating where it is desired. The control system and operator work together to keep it there. Variations in controlled variables would be automatically accommodated for by the feedback and other controls. When the accommodation is too slow or inadequate, the operator will make adjustments to help things along. Transition mode is where the current operation itself needs adjustment; say to accommodate different raw materials, to change over from one energy source to

---

23 Drexel University, "Model Reference Adaptive Control (MRAC)," blog post, accessed July 19, 2015, http://www.pages.drexel.edu/~kws23/tutorials/MRAC/MRAC.html.
24 Steven Obermann, "Measuring Multivariable Controller Performance—How to Tell if Your Multivariable Controller is Doing a Good Job," *InTech* 62, no. 3 (2015): 16–20.

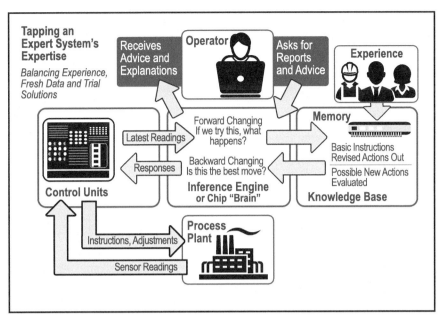

Figure 2-9. Concept of operator as coordinator—no hands.

another, or to switch operating modes from peak to off-peak; and so on. To make these sorts of changes, the operator must make many complicated adjustments. He will utilize complex or involved procedures to make the necessary manual adjustments over a potentially long period of time. All of this opens up operations to a lot of extra attention and exposure to error. Figure 2-9 suggests how an operator might conceptually employ knowledge-based information.

The procedures the operator uses will have all been carefully designed and spelled out in detail. Where sufficient instrumentation and remote-control devices exist, those procedures can be automated.[25] An excellent example of how this works and its effectiveness can be found in an *InTech* article by Maurice Wilkins. A portion of the article follows.

> An example helps illustrate the concept [of procedural automation]: An oil refinery in Japan underwent crude oil feedstock switches two or three times a week. The efficiency of the operation depended on the experience and skill of the board operator running the distillation unit.
>
> With a skilled operator, the time to reach normal steady-state operations was typically five hours, but it could take more than eight hours with a junior operator.

---

25 ISA-106 Series, *Procedural Automaton for Continuous Process Operations* (Research Triangle Park, NC: ISA [International Society of Automation]).

An automation supplier was called in; they interviewed the board operators from different shifts and were able to uncover and document best practices.

For instance, when ramping up feed temperatures, junior operators would typically ramp feed temperatures at a linear rate throughout the temperature zones in the column, but veteran operators had the operational experience to ramp temperatures at different rates depending on the zone.

It was also discovered operators typically had to make more than 100 adjustments to the process through the distributed control system during the switchover.

Implementing procedural automation enabled the refiner to make significant improvements in the switchover time to a predictable four-and-a-half hours, regardless of which operator was on shift. The operators' workload was also reduced significantly, with over 100 control system adjustments reduced to 10, and more than 2,000 process alarms cut substantially by configuring the system to be operationally aware of process conditions. Finally, there was increased process knowledge sharing, a significant reduction in operator errors, and reduced required operator training.[26]

As is the case for other forms of automation for operators, the design will need to include any specialized operator refresher training and procedures for proper manual operation should that need arise.

2.12 The Automated Plant

For well over the past 70 years, industrial plants have been operated under increasing levels of measurement instrumentation and controls automation. Part of this can be attributed to the increased availability of more reliable instruments to measure conventional process conditions like temperature, pressure, level, and flow. Their use has been enhanced by specialized sensors that were more accurate, more tolerant of wider process conditions, easier to install and maintain, and more affordable. Additionally, as separable components of a plant became more integrated due to production efficiencies, energy integration, and sophisticated production procedures, additional instrumentation and significantly more controls devices were required. It is not too much of a stretch to imagine the degree of controls required to increase to the point of near full automation.

Even at present levels of automation, it is not unusual to operate for long periods of time without significant operator intervention needed. Certainly there are the usual snags and snafus due to uneven automation or faulty equipment or other operational annoyances. Most are quickly managed and result in more nuisance than operational threat. Oh yes, let us not forget the odd disaster when everything that can go wrong

---

26 Maurice Wilkins, "Improving Operations in Continuous Processes through Procedural Automation," *InTech* 58, no. 6 (2011).

does go wrong in the worst possible way. These situations are important. They provide the necessity and framework for design and implementation of safety integrity systems and emergency response protocols. We put them aside for this discussion. One of the unintended consequences of increased automation and the resulting operations variability reduction is that operators have less to do and, by doing less, are less able to test their skills. While they might be effectively trained to do a wide variety of tasks and operations, those particular skills are perishable without frequent exercise and validation. Plants now face increasingly complex operations with operators less able to step in and take charge when something goes amiss. If we take the leap to a nearly fully automated plant, the exposure to stale operators now becomes vastly more important.

There are four general approaches to this situation: resist automation, selective automation, force periodic operation without the automation, and cease operations during upsets. All have their proponents. The latter three are discussed below.

Selective Automation

There are a share of plants and plant management and plant operators for whom any significant degree of automation is distasteful. They quickly suggest that it is too costly, needs too much maintenance, is too hard to understand, and is much more than they feel is actually needed. For them, "hands-on" is a personal enrichment. Unfortunately, this feeling is more an expression of wishful thinking than operational benefit. Operators in these situations find themselves burdened by the necessity to always be very close to all operations and remain overly diligent about watching for things amiss. They become highly susceptible to the monotony of repetition and boredom. In end, there is really less operator satisfaction and even less operational benefit. The bottom line is that it is important to use automation as a tool: provide what is needed and no more or less, and provide only what will always work.

Operate Periodically with Automation Off

An important way to keep the skill levels up is to exercise them on a regular basis. This can be done quite effectively using high-quality simulators. Simulators permit skills testing through the point of failure without any risk except a little time and perhaps a degree of pride. They allow repetitions of scenarios to hone skills. They also facilitate the ability to qualify required skill levels and compare relative mastery of individuals. All in all, when simulators sufficiently match the real thing, they are extremely valuable. Nor do all simulators need to be a physical mimic. Simulations, sometimes called *tabletop simulations*, have a proven value.

The other way of operating without automation is to do just that: operate the plant with as much of the automation that can be properly supplemented by manual operations turned off. Not so fast. There is more to this than there seems. The plant must be in a production mode that is not overly sensitive to operational variability more than usual. There must be adequate procedures in place for this. There must be additional personnel present to monitor and assist with extra hands and additional coaching. All of this has the important benefit of developing operator skills as well as testing procedures, the alarm system, collaboration, and escalation. Carefully monitor the events to make the appropriate recommendations for improvement. Then audit the improvements for next time.

Before we leave this topic it is important to identify a principle: if a plant or other enterprise cannot be operated without full automation in place, then it must not be. If the failure of any automation element cannot be satisfactorily managed by manual action, then either (1) the plant must be designed so that any such failure cannot occur, either by automated safeguards or replication, or automation to move to a safe state; or (2) the plant cannot be operated at all. There is no safe middle ground.

Cease Operations during Significant Upsets

There is another alternative. When the process becomes upset to the point that it should be evident to the operator that the automation is not going to properly manage, shut it down or move it to a safe operating situation. We will look at this carefully later in Chapter 10, "Situation Management." Let us see how this might work. The benefits are minimal risk of abnormal operation escalating into anything significant, no need to have extra personnel present just to manage upsets, much less chance for equipment to be operationally stressed, and much less chance for off-specification product.

Right about now, you might be thinking that this option is turning you off. Please stay with this a bit longer. Without proven and reliable procedures and the rest of it in place for your operator to safely and efficiently move the production process to a safe operating state (or shut down), your plant is really not safe. It is always better to avoid damage beforehand than repair it afterward. You will need all sorts of spare parts you likely do not have. The expertise to repair may have to be brought in from outside, or at the very least taken away from other scheduled and necessary work. And the outage will almost always come at a time when you would not have wanted it.

Summarizing

The proper level of automation should be set by production requirements with due consideration to operating costs. Once set, it is imperative to ensure that operators properly utilize it (studies have shown that 20% to 30% of all instrumentation and controls are not working at any one time) and that they maintain the needed skills to keep the plant safe and effective. The lower the level of instrumentation and controls, the higher the minute-by-minute requirements on operator presence, on very close attention to operating conditions, and on making all the ready hands-on adjustments needed to keep things going properly. Having the operator under this load prevents his use for other purposes. Having the operator operate under this daily chore load is wearing on motivation and job satisfaction in the long run. The few who take a sort of primal pride in being essential are far outweighed by the many for whom each shift is a relentless grind with constant exposure to unnecessary untoward events. The bottom line is that having a proper level of automation more than pays for itself in the long term.

We now turn our attention to how to understand a large enterprise by being able to subdivide it into manageable smaller parts.

2.13 Plant (or Enterprise) Area Model

This chapter now addresses the subject of organizing your plant's infrastructure, nomenclature, and naming conventions in a way that the various related parts can be identified and referred to by collections. This will enable you to avoid fully enumerating each part of each subcomponent or component to tell what it is composed of. Some of the tools and procedures may be new to you. All can be most helpful.

Get ready for an important subject. It is so important that if you get it wrong, you will not even know anything is amiss until you are well into writing procedures and training operators to locate related equipment. It is truly one of those "pay me now or pay me later" situations. Only the later "payments" might be so costly that they will not be affordable. And then, just as your careful design is poised to reap real benefit to the operation, you will not be able to deliver. The reason for that failure has a lot to do with how your plant's component labels are constructed in the process control system (PCS). If the labels are constructed to incorporate clear information as to exactly where in the plant the component is located and with which equipment it is associated, then the operator's ability to quickly navigate as the plant operates can be easily and clearly managed. Otherwise, all of your management must be done from large lists of explicit equipment and situations. We call that process "management by enumeration." It is not a good way to do things.

A plant area model is all about geography and labels. Refer to Figure 2-10 for an illustration of the concept. The figure depicts a single plant. In this typical plant, there are four to six operator areas. Within each typical operator area, there may be three to five or so individual, significant process units. Each process unit might contain 10 to even 50 individual pieces of equipment (pumps, heat exchangers, towers, etc.). Within

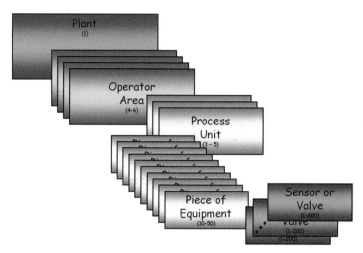

Figure 2-10. Conceptual plant area model showing the various levels of equipment and how it is "attached" in a diagram to the levels above and below. When such a structure is used for naming conventions, it is much easier to use automated or semiautomated procedures to modify operating conditions or alarm settings and the like.

each piece of equipment, there might be between one and several hundred individual sensors, valves, and so on.

But the plant area model is really not about numbers. It is about location and relationship to other locations and pieces of equipment. (Note: A tag is a label given to each unique entity appearing in the PCS that is labeled and later used for control or computation or view.) For example, if the tag contains the area model, it will be easy to manage alarms. Consider the following example flow controller tag identified as FC-1-3-REACT-MIX-03452.

Table 2-2 shows the information used for the construction of a tag. There are likely many tags for the reactor mixer, all with a unique label and number. Plants need not use letters to identify geography. A tag can also be constructed of only numbers, after a few letters for a prefix. For situations with a fixed format, the dashes or spaces are not needed. Therefore, our example tag for the flow control loop identified in Table 2-2 might just as well read FC131241503452.

Plant ID	Operator ID	Process Unit ID	Piece of Equipment ID	Sensor, Valve, or Element ID
1	3	REACT	MIX	03452
1	3	124	15	03452

Table 2-2. FC-1-3-REACT-MIX-03452 showing two possible naming conventions used to construct a tag that incorporates a structured descriptive model.

We are now ready to make use of this capability. Suppose that the mixer in the reactor was shut down. It is a simple matter to identify most related equipment with tag FC1312415 XXXXX where the five Xs refer to all such tags in the database with the FC1312415 prefix. Using this prefix as an automated search filter (say in a script), we find all related equipment and shut them down. Alarms could also be managed this way. Otherwise, in order to do the same thing without this ability, you would either need a list of items that are part of FC1312415 and use some sort of automated way of using the list, or read the list and manually shut down the individual required parts.

2.14 Decomposition

Decomposition looks like a complicated word. And thinking about it so early in this book might seem like too much work. Sure, the word is complicated, and this topic is just a bit technical. But it is going to pay off. The payoff is that you will use this tool to take a large enterprise or plant and snip it into smaller pieces. The smaller pieces are going to be much easier to understand, much easier to analyze and manage, and much more efficient to design.

Things are taken apart in order to work on the smaller pieces in an easier and more complete way than possible when whole. It is really easy to take anything apart. Any scheme will work. Anyone can do it. On the other hand, it is extremely difficult to take things apart in a way that permits them to be effectively put back together afterward. How often have you heard the story of an item taken apart and then reassembled only to find three screws left over? Or even worse, it cannot be put back together at all. The art and science of taking things apart in a way that permits analysis of the individual parts and then reassembling into a whole, with the work done on the parts being fully useful to the whole, is called decomposition.

Most plants are composed of interconnected pieces. Conventional manufacturing plants are, for the most part, built up with pieces that do specific jobs. These pieces may be individually designed purposefully to minimize cost by avoiding any unnecessary complication or size excess. However, they most often conform to a conventional norm of design and construction. There are many norms out there for each. Then the plant is designed by connecting all the needed parts into a workable whole configuration.

In most situations it is relatively easy and convenient to lump pieces together. What lumps together to create a subsystem is quite straightforward. Yet there are plants that appear to have no useful easy way to define a subsystem boundary. In these cases, transformational analysis (discussed later) would be a useful tool to decide where breaks would be most effective.

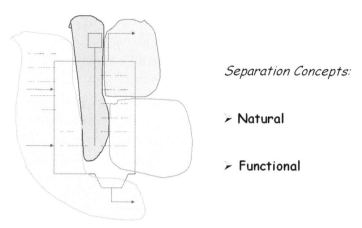

Figure 2-11. Illustration of typical decomposition boundaries using natural (physical) relationships or functional relationships as decomposition boundaries. Shown are different colored parts of four separate functional components, such as a separator, a heater, a mixer, and an analytical processing component. Natural boundaries might be first stage, second stage, third stage, and finishing stage.

Decomposition Basics

Decomposition means to take apart. To do our work, we will need effective ways to take a large plant and break it down to decompose it into manageably small parts. If it were just a matter of breaking it apart, then any forceful way would do. We could be as haphazard as we wanted. But breaking it apart has a purpose. We will want to be able to understand what is happening in each of these pieces in a way that will make sense. Random pieces do not lend themselves to planned processing. Once the work of understanding each separate piece is complete, we put things back together into a functioning whole.

An effective way to break something large into smaller pieces is to take advantage of boundaries. Boundaries can be natural; for example, rivers, lakes, mountains, forests, and the like divide land. Boundaries can be functional like political boundaries. Figure 2-11 illustrates hypothetical boundaries.

Notice that this example has four identified regions or subcomponents. We were able to identify the boundaries between the regions by finding places within which most of something was, and that same something was not much elsewhere. Or, we found the boundaries by looking for places where the adjacent piece(s) were less connected than elsewhere. Fortunately, there are some good techniques we can use to help identify places where separation can be best made.

Subsystem Boundary Attributes

You have been introduced to two separation concepts to decompose a rather large system into manageable pieces. And there are more, of course. You may have a few

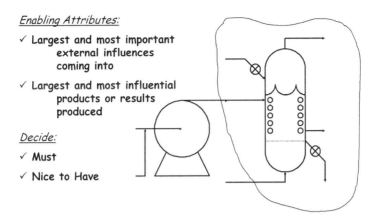

Enabling Attributes:
- ✓ Largest and most important external influences coming into
- ✓ Largest and most influential products or results produced

Decide:
- ✓ Must
- ✓ Nice to Have

Figure 2-12. Input and output classification questions to accompany a decomposed element to choose the connectivity for each decomposed element with each other and the other components of the plant.

favorites of your own. Once the dividing up is done, it is time to look at what you have. With our process decomposed into workable subsystems (or pieces as we sometimes refer to them), we can turn our attention to the task of finding the key influential aspects between the pieces. We first will find the significant influences at the boundaries between subsystems. Then we will find the significant influences within each subsystem. Refer to Figure 2-12.

Our example boundary has been drawn around a separation tower. The tower and two coupled heat exchangers are inside; the pump and everything else are outside. The first thing to do is to list all the items (e.g., mass flows, energy flows, and information flows) that cross our subsystem's boundary. Make a list of the largest and most influential items that come into our subsystem. Make a similar list of the largest and most influential products or outflows leaving our subsystem. Categorize these influences as (1) vital to know or (2) important but not essential. These influences provide a way to start building a candidate *situation awareness* list. We continue with this in the next topic.

Subsystem Internal Attributes

We now turn our attention to the interior of subsystems. Identify all key variables that have a fundamental role in the operation of the subsystem. There should only be a very few key, fundamental variables that indicate its health. Find the key ones that are inputs, the key internal ones, and the key output ones.

Figure 2-13 puts it all together. Using the list of key variables in, key variables out, and key variables within, determine the smallest subset that can provide the operator with an understanding of the process. We want this list complete enough for us to

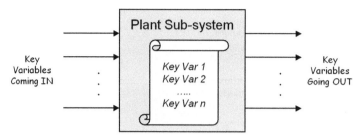

Figure 2-13. Generalization of how a decomposed element relates to other elements and the rest of the plant. This provides an effective but simplified model to use for alarming, human-machine interface (HMI) design, training, and more.

identify abnormal operation and to be able to understand what affects what, how, and how much. For most industrial systems this will work out quite nicely.

Now we will take a concrete look at how this usually works for real industrial plants. They are a bit more complex, but the same idea applies nicely. Also, the nice thing about industrial plants is that their parts are mostly conventional and easily recognized. We will use this benefit to good advantage next.

2.15 Decomposition Underlying Situation Management

Large systems can be difficult to understand and work with if they must be dealt with as a whole. If at all possible, we will seek to decompose them into clearly recognized, understandable, and analyzable parts. It is usual to use several levels of decomposition. This means that we might start with a large system (the word *system* is used for a plant or other enterprise). The first step (or level) is usually decomposing into major subsystems. Each subsystem in turn may be decomposed into smaller subsystems. While it is certainly possible to continue to decompose into ever-smaller subsystems, three or possibly four levels are most common.

Structure of Decomposition

Let us see how this might work for a typical chemical plant. Rather than start with the overall large system, it would be easier to build up our understanding if we start at the bottom of the levels. We know that we want to use this for situation management purposes.

Key Repeated Elements

Start with the basic building blocks of the plants that operators manage. We term them *key repeated elements*. They are "repeated" in the sense that most large systems usually contain many of these. They are "key" because our understanding of each will help us understand how the larger system works. There is no magic here. It is just that heat

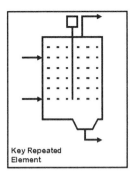 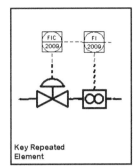 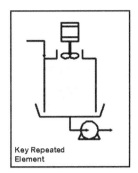

Figure 2-14. A few examples of key repeated elements. Each key repeated element represents a production module that is compact (clear boundaries, inputs, and outputs) and appears often in a plant. Managing them this way is efficient and promotes uniformity. Depicted are a vertical separation tower at the left, a significant flow loop in the center, and a stirred-tank reactor at the right.

exchangers, reactors, separation towers, mixers, and the like all go into the makeup of many chemical plants. Where there are several similar heat exchangers, understanding one should provide most of what we need to understand the rest. Other types of plants or operations have their own key repeated elements. Figure 2-14 shows a few chemical plant examples.

From the left, there is a vertical separation tower, an important valve/flow station, and a reactor. These are but a few of the key repeated elements that are used for plants. For example, the job of identifying appropriate alarms (including alarm priority and the alarm response sheet) can be done for a typical key repeated element. Each one can be reused everywhere the key repeated element appears in the enterprise, taking care to ensure that all items match the specific situation for each occurrence by making adjustments as needed.

Key Subsystems
The next level up in our "composed" plant would contain many of the key repeated elements (as needed to produce the intended product), but now they are combined into larger entities. What we are doing is looking around the plant and finding larger "lumps or hunks" that form subsystems. Each one is referred to as a *key subsystem*. As before, we use the word *key* to denote that if we understand each of them, in particular, then everywhere else we find them we are well on our way to understand them as well. Figure 2-15 depicts one *key repeated subsystem*. Notice that a key repeated subsystem would likely itself be composed of some of the key repeated elements we just discussed. To identify alarms, as appropriate, each key repeated subsystem will use the alarms identified for each of the key repeated elements determined earlier (including alarm priority and the alarm response sheet). And in like thinking, a large plant might have a few of the key subsystems used more than once. Treat them as we did

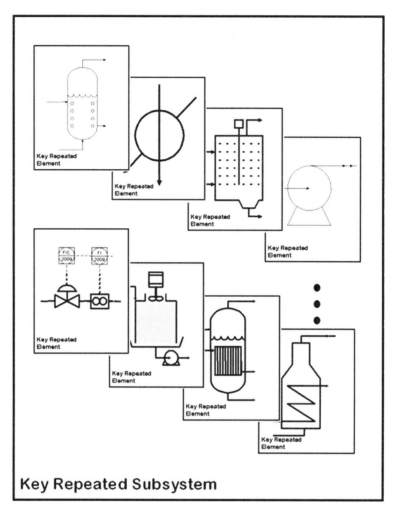

Figure 2-15. Example of a key repeated subsystem that is constructed with key repeated elements defined earlier that fit the specific subsystem being described. Other repeated subsystems use other repeated elements as needed to adequately describe each. Managing them this way is efficient and promotes uniformity.

the key repeated elements, using each as a template starting point for understanding, alarming, and such.

Notice that Figure 2-15 uses eight of our previously identified key repeated elements inside of it. Because we already understand what each key repeated element is and how each works and misbehaves, when they are put together into this subsystem, we should not have a difficult time understanding how they form the subsystem. Please note that the example subsystem may have more than one of each of the key repeated elements. They would have what is needed to do their intended job. As before, once we understand how the *key repeated system* functions, we are going to reuse it wherever appropriate.

The System

The next level up is actually the top. It is the whole system. See Figure 2-16.

As you can see, the overall plant will contain many key repeated subsystems (but some different ones and a few very similar). And there will be many odd key repeated elements here and there that are not associated with any larger subsystem. They are not visible in the figure, but they are there.

Looking for Abnormal Situations in Key Repeated Elements

Let us bring back an example key repeated element, say, a simple condenser. Figure 2-17 illustrates one, including a pictorial representation of its operating regions.

Each specific operating condition, like a flow, temperature, and such, is located somewhere in the operating area. The normal operating area for the heat exchanger is shown in the gray area marked *normal*. To be in the normal area means that it is operating in such a way that no intervention is needed or sought, so things go on as they are. Adjacent to the normal area at each side are *abnormal* areas. The brown one at the left represents all the abnormal operating conditions that would be reasonable to fully understand and would be alarmed through application of the alarm management plan. This means that when the condenser operation enters that area, an alarm

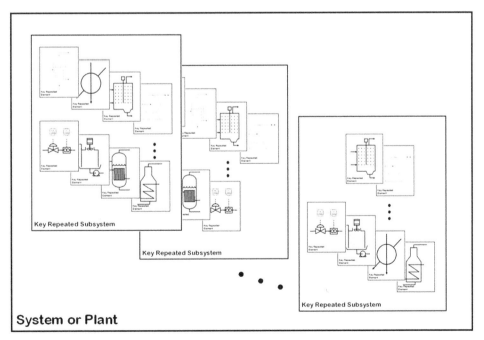

Figure 2-16. Entire plant or enterprise constructed of the building blocks defined earlier (key repeated elements and key repeated subsystems) and other pieces as needed to complete the description. Managing it this way is efficient and promotes uniformity.

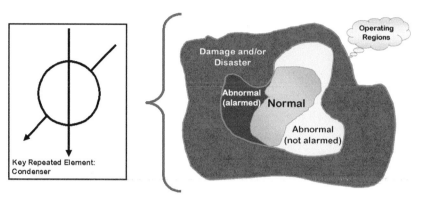

Figure 2-17. Condenser and its operating regions. Each key repeated element is operated, alarmed, and understood in relationship to its own specific operating regions and associated risks.

will activate for the operator to take action to manage the problem. The yellow area to the right represents the rest of the abnormal situations. Some might be known, but their occurrence would be unreasonable to alarm. Most are unknown. Either way, none in the yellow area are alarmed. The operator is still required to ensure that operation does not enter the red *damage and/or disaster* area. The situation management responsibility for the operator is to adequately and properly identify and appropriately respond to these using the rest of the tools and policies at his disposal. These tools and procedures are discussed in Chapter 8, "Awareness and Assessment Tools." To summarize, everywhere a condenser appears in the plant, the methodology for ensuring good operation will be based on the one used in the template, but taking care to make adjustments and modifications to match each individual situation.

We do the same thing for each of the other key repeated elements. Figure 2-18 depicts a pump. Please note that while the operating regions for it may appear to be

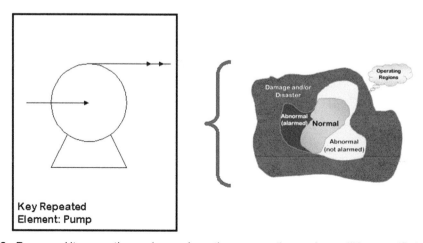

Figure 2-18. Pump and its operating regions, where those operating regions will be specific to the pump.

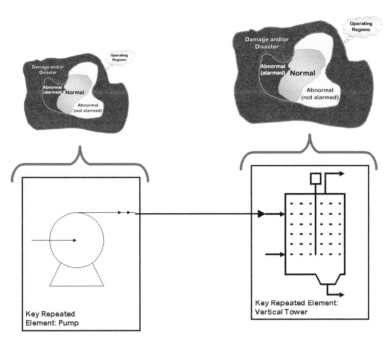

Figure 2-19. Pump and tower connected by starting from a basic pump and vertical tower and customizing each to precisely fit the specific pump and tower here. Most of the basic element will remain unchanged to the degree that it had been carefully defined as a template.

pictorially similar to those for the condenser, in reality they will be quite different. They are identified using the same methodology as before.

In this manner, each key repeated element is examined and appropriate operating areas identified. Examining each key repeated element by itself generally outlines operating areas. This is a starting point. In actual use, these objects are interconnected. For Pump 3402 that is directly connected to Tower 3422, the specific appropriate operating areas are determined for this exact configuration. For example, if the working pressure for the tower is below the maximum output pressure that the pump can produce, the tower working pressure is used to define the appropriate normal and abnormal regions for the pump. Figure 2-19 depicts the two connected items.

With a different pump, and/or a different tower, or had a condenser been there instead of a tower, the operation regions would be different. They would be determined to match the equipment present in the way they are interconnected.

Looking for Abnormal Situations in Key Repeated Subsystems

Finding all key repeated elements inside and adding any one-off items that are present, build the subsystems. Each one-off item would be examined and accounted for exactly as if it were a key repeated element, but in these cases, the items are not reused

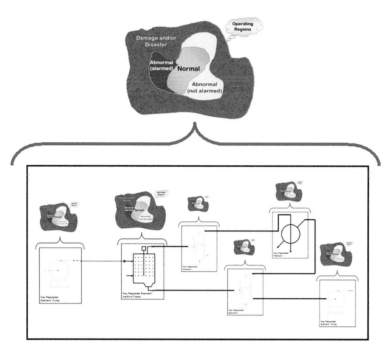

Figure 2-20. The operating regions for a key repeated subsystem do somewhat reflect those of the individual elements from which they are built; however, by virtue of interconnections and operations, the regions are a new overall situation.

because there is only one of each. Rather than begrudge the one-offs, let us be thankful for the repeated ones! Once you have each key repeated subsystem, you can use them as starting point templates for all other places that the subsystem is found. Observe that there is tailoring of the subsystem in the same way we tailored the repeated element to the specific situation. As always, one is required to understand the specifics of each subsystem, but by starting from the template for it, extra work is unnecessary and a great deal of consistency is realized. Figure 2-20 depicts the "construction" of an example subsystem.

Notice that not only are all the internal operating regions there (from each actual key repeated element that is part of the subsystem), but we have an additional overall operating region for the entire subsystem (as shown above the figure box). The specifics of the overall operating region may be different from those of the key repeated elements and other key repeated subsystems. Any differences reflect the actual differences due to the equipment being different or being operated for a different purpose. Of course, if any equipment is a clone of another, these operating regions are likely to be the same or very close. If this is all a bit esoteric or confusing, perhaps think about each subsystem as if it were a "packaged system" bought somewhat "off the shelf." Each might be nearly identical as delivered. But if each were operated to produce different products, then their individual operating regions would likely be different.

They might get into operational difficulty differently. This should pose little problem for us here, because you understand each of building blocks. And that is the objective: carefully analyze and design, and repurpose everywhere applicable.

Summarizing

Let us review where we are here. We started with the entire system representing a complete operator's area of responsibility for management. The first thing we did was to find all the small pieces that seemed to recur over and over again in that system. Each separate type of small piece was analyzed and understood. An operational template was constructed for each. Using the templates as starting points for everywhere else the piece was used, we adjusted the template to match that individual specific use. Then we noticed that these smaller pieces also seemed to be grouped into larger subsystems. As we did for the small pieces, each separate type of subsystem was analyzed and understood, and an operational template constructed. Using the template as a starting point for everywhere else the same subsystem was encountered, we adjusted the template to match that individual specific use.

Decomposition is the tool used to divide that large entity (a plant, a pipeline system, or a network routing center) into useful smaller parts. This allows each part to be understood and designed for. It is the sum total of the larger subsystems plus a few missing odds and ends that combine to form the overall system. We keep track of the overall system by finding ways to keep track of the individual building blocks we used when we decomposed it. When the parts are connected together, the specifics of connection determine the design for the parts. All the while, the approach preserves the underlying understanding of each part so the reuse of the design is efficient and productive.

2.16 Transformational Analysis

Transformational analysis was first developed and used by Imperial Chemical Industries (ICI), a British chemical company now owned by AkzoNobel in the Netherlands. Unfortunately, it is an undocumented process. So we introduce it here. Nonetheless, it is easy to understand and use. See what you think.

How It Works

This is a convenient and effective method of evaluating a process or operation to identify operational risk in a consistent way. The basic framework for building the diagram is to follow the flow of major raw materials from entry into the enterprise until they are transformed into finished products (and then shipped or transferred out). The plant

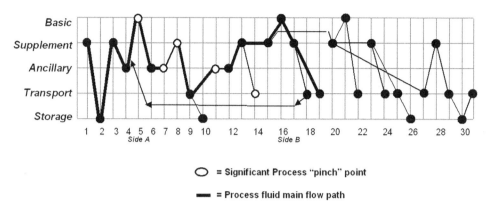

Figure 2-21. Transformational analysis diagram depicting the (left-to-right) flow of operations in a plant where each operation step is identified as to complexity (with simplest at the bottom and most complex at the top).

acts upon those raw materials in a series of steps. Each step involves a specific action. Action categories range from simply moving the product along via a truck, conveyor, or pipe; heating it; or mixing it with something else, all the way up to operations that change it into something very different (reactions, welding, etc.). Figure 2-21 depicts an example *transformational analysis* diagram. Let us take a closer look.

The diagram depicts an operator's entire process, from beginning to end. On the horizontal axis, at each block along the way are single elements (or steps in operation) of the process. At the left enter the "raw" materials, either true basic materials or the end results of other upstream process areas that come to our operator for additional processing. At the right are the exiting "finished" products or the end result of our processing that is passed on to the next part of the plant for additional processing or shipped to customers. Our particular example has 31 steps or blocks. This means that there are 31 process-type actions in this particular operator's processing area.

The vertical axis of our diagram contains five separate levels. Each level represents a different degree of sophistication and/or complication of a process action. The most simple is located at the bottom, the most complex at the top. Starting at the bottom, the Storage level is just that, a place that material is being stored. That could be a tank or a warehouse. The next level up is Transport. Transport is also exactly that, a process component whose only job is to move product from one part of the unit to another. That could be a pipeline, an over-the-road vehicle, or a belt conveyor system. The next level up is Ancillary. At this level, simple processing is done to our material. Examples are heating, cooling, pressurizing, aerating, stirring, and so on. The important distinction is that some of the gross properties of the material are being altered in a very simple and basic way. The next higher level is Supplement. Here is where our material is altered in a way that changes its intrinsic makeup. The mixing of several materials together or the separation of a mixture into components, say by distillation or centrifuge, are example processes at this level. While perhaps costly, it is generally possible to reverse what is done at this level. At the top level, we find Basic. Here fundamental changes in the material are made. A good example is a reaction. The product at this level does not resemble any of its constituent materials. It is not generally possible to reverse what is done at this level. So what is done here cannot easily be undone as, say, something that is too cool might simply be heated a bit more.

The level of processing for each of the 31 steps (or actions) is indicated on the diagram by either solid or outline "dots." We will assign the appropriate level for each operational step. Outline dots are used to identify steps that are known (by those who understand this process well) to be places of unusual difficulty to manage. Solid lines connect the dots to permit our eyes to follow the diagram. Heavy lines are for main material progression. Light lines are for minor (sometimes feedback) amounts of material flow. Our example process shows the material entering at progression Step 1 and proceeding to progression Step 31. Notice that most of the material stops at progression Step 19. This is not at all unusual for complex processing operations. Starting with Step 1, we see that the first thing our process does is a Supplement. In the next step, the product goes to Storage. Next, it comes out of storage and goes directly to another Supplement step. Then, at Step 4, it receives an Ancillary type of processing. Immediately after that, it goes into a Basic processing, and so on.

Using Transformational Analysis for Risk Assessment

Let us test our hypothesis of whether it is more difficult to manage the higher levels of processing (Ancillary, Supplement, and Basic) as opposed to the lower ones (Storage and Transport). The test we use is to go to the actual operating records for this process as it has performed historically to determine where the incidents and accidents happened, to see if they correlate with the indicated level of difficulty. Figure 2-22 shows incident statistics overlaying the transformational analysis diagram.

The incidents shown in Figure 2-22 were compiled over 6 months. For this process unit, there were 54 incidents recorded by the site. The most, 29, were at the most complicated Basic Step 5. Total incidents for all Supplement levels were 16. Total incidents for Ancillary levels were 8. Incidents for Transport amounted to 1. There were no Storage incidents. A cursory examination indicates that incidents fall off by a factor of two at each descending level. It is also interesting to note the breakdown of incidents into their contributing factors. Shown are the contributing factors of people (in blue, from the 12 o'clock position clockwise), next equipment (in deep magenta), and then process (in yellow). Contributing factors were determined by investigation and represent the

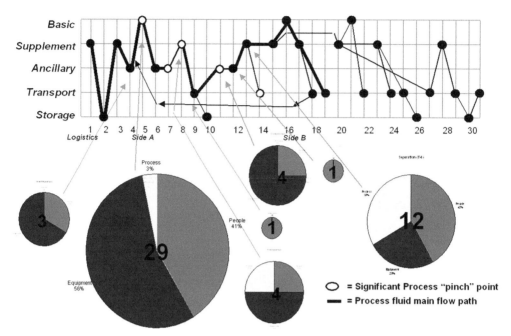

Figure 2-22. Transformational analysis diagram showing the degree of agreement between the number of important incidents and the degree of complexity with more incidents at the more complex step.

one that was the most likely cause of the incident. At Step 5, the breakdown is as follows: people, 12; process, 1; and equipment, 16, for a total of 29 incidents. This all seems to correlate well with our intuition.

Using Transformational Analysis for Decomposition

Now that we have a good idea how the diagram works, it is time to see how it might help determine where good places exist for decomposing the process into smaller parts. By now, it must be occurring to you that the steps that place lowest in the diagram are probably the better places. This is best illustrated by first thinking about what might result if we selected a step at the top. Suppose we decided to decompose this process at Step 5, a Basic step location. First, remember what it means to break up the process. Breaking it apart means that we will want to consider the part before the break as one part and the part after the break as a separate part. The first part will be Steps 1–4. The second part will be, say, Steps 5–9. But we already know that Step 5 is extremely complicated. To do it correctly, Step 5 must be able to know and manage as much of the inflowing energy and raw materials as possible. But that part is mostly Step 4 and earlier. And those steps are in another part. Therefore, Step 5 must do without this information and management of what is coming in, require extra effort by the part containing Steps 1–4 to do its job correctly, or require extra information from the part containing Steps 1–4 so Step 5 might try and compensate for inadequacies at Step 4 as best as it can.

On the other hand, if we were to break the process apart at Steps 2 or 9, we will already know most if not all the properties of its inflow material, because all that happens before is either Storage, in the case of Step 3, or Transport, in the case of Steps 10 and 11. We are left with a strong suggestion that transformational analysis can be useful in process unit decomposition.

Use for Operational Area Division of Responsibility

On the one hand, dividing up a relatively large plant or operating enterprise into manageable pieces to assign to individual operators can be really straightforward. This is done most often where there is far too much operator load for one but quite a bit less for two operators. So from a loading point, there is too much load flexibility. That reduces any special needs to have to avoid very clear or natural breaks between areas of responsibility because loading is pretty much not an issue. As comfortable as this might appear at the beginning, there is a natural tendency for plants and enterprises to grow over time. As they do, operator load may become an issue. When it does, the assignment of individual boundaries between areas becomes important. Transformation analysis can expose places to divide the areas to minimize cross-area interactions.

Another situation is where a relatively large plant is either experiencing a need to reduce operators or is finding one or two operators with excessive load and others with available load. The number of operators is clear. Having a risk-based way of identifying potential places to divide operator responsibility would be useful. Once again, transformational analysis can be a useful tool.

Identification of Structural Operational Issues

It is difficult to see how a transformational diagram can be fooled. The specific steps (horizontal axis) are clear and evident. The assignment of complexity level (vertical axis) is also remarkably straightforward. Apply the definitions at each step and the level is cleanly determined. If any specific step has more than one simultaneous level of complexity, the higher is used.

We now suppose that the transformational analysis diagram has some special results.

Not Enough Low-Risk Steps

A given operator area of a plant is examined using transformational analysis, and the results indicate there are few low-risk steps. This means that few of the steps have levels of *storage* or *transport*. This suggests that many of the things going on can have abnormal operating situations that spill over from one part to another. Let us further indicate that the particular operator area we have in mind has an operator that is not overloaded but has a relatively high load. The abnormal situations' tendency to propagate makes it much more difficult to both analyze them and manage them. Our operator could quickly experience a transition to overload, even for abnormal situations themselves that might not be very dramatic. Running out of operator resources would be dramatic.

A careful analysis would expose this situation and suggest a need for additional operational resources: improved training, shorter shifts, better controls, and so on. The picture here is that you have a tool at your disposal that with relatively little investment can provide an important insight into places that might have operational risk exposure that were not so evident before.

Too Many High-Risk Steps

A given operator area of a plant is examined using transformational analysis, and the results indicate there are quite a few high-risk steps. This means that more than a few of the steps have levels of Supplement or Basic. This suggests that many of the things going on can have abnormal operating situations that will have a substantial negative impact on operational integrity. There are a lot of places where things can go wrong, and when they do, they are going to be impactful. For this situation, operator loading would have much less of an effect. Even for a relatively low normal operating load on the operator, having lots of places that require careful watch and ready course adjustments will quickly absorb all load and then some. Extra resources here are likely not going to be able to rise to the level that the risk exposure demands. This operation must be structurally redesigned to provide solid infrastructure handling of production rather than simply rely on an operator to stretch to the situation.

Of course, if the operator area does have a place where it can be split into two operator areas, and the resulting split is not too risky, then splitting and providing two operators might be an answer. At the end of the day, this type of analysis should be done to ensure exposure of certain hidden operational risks that might not have been known otherwise.

2.17 Near Misses, Incidents, Accidents, and Disasters

If nothing ever went wrong and everything went according to plan, we probably would not need operators. When things do go wrong, operators are the front line to manage things. If they cannot or do not manage things, it can go very wrong in ways that affect the operation itself as well as the surrounding communities. These losses can be slight and unremarkable. Or they can expose an unacceptable risk to the future existence of the enterprise through severe injury and loss of life, as well as expose surrounding communities to physical and environmental damage.

The incident literature is replete with serious incidents and disasters. Reading about them can be upsetting. We do it because of the important benefits that come from understanding. Obviously, a major benefit would come from the incorporation of direct knowledge into related operations at your enterprise. If Company X did not provide reliable instrumentation that was capable of providing needed HMI data to the operator during a major upset, and your company had manufacturing operations similar to Company X, you would want to check on what types of instrumentation you had. But this type of situation is the least important. One reason is that such a match is not the norm. Another reason is that most dramatic incidents do not often repeat themselves. Therefore, careful study and effective means for managing their incidents would fail to provide an effective deterrent against your future incidents. Instead of this narrowly focused understanding, we look for a broad appreciation of what went wrong and use that appreciation to develop root causes. From these root causes, we evaluate appropriate remediation at an enterprise operational risk management level to ensure that the lessons are actually incorporated into the enterprise's kit.

Pay Retail or Pay Wholesale

There are two types of experiences to learn from: your own and others'. Firsthand knowledge comes from things happening to you. It came on your watch. It happened to your company. People you know were involved. It cost you. You paid retail for this experience. You paid full price. You will learn. If you are able, you will make changes so that the incident should not happen again. But you paid in full. This is a hard learn. Maybe the other way would work better.

We can learn from the experiences of others. This is easy to say, but not so easy to do. To do it well, one must give up a dearly held belief: bad things will not happen to me; they only happen to the other guy. In reality, unless you are incredibly careful and exceptionally lucky, it has not happened to you yet. Unless we learn, make changes, and have a bit of luck remaining, it will happen to us. To reduce the chances of it happening to you, do the following:

- Investigate *all* your near misses and make appropriate changes to ensure you are really covered.

- Keep fully abreast of what is going on in your industry, plants, and operations.

- Read and understand relevant and related reports from OSHA, the Chemical Safety Board (CSB), the National Transportation Safety Board (NTSB), the HSE, the Dutch Safety Board, the European Agency for Safety and Health at Work (EU-OSHA), and the rest that apply.

- Evaluate all of those experiences for both proximate cause and root cause.
- Evaluate your vulnerability to the root cause effects.
- Make appropriate changes to ensure you are covered.

If you do all of this, you can buy your experience wholesale. That is the best price anyone can find. Pay it and move on.

What follows is a discussion of how we classify things going wrong. Reviewing each category will help you understand how they go wrong. Knowing this, you will better understand why this book is concerned about them. Please do not spend too much time on the definitions. Just get a feel for each one. There will be no quiz.

Hazard

The US Federal Aviation Administration (FAA) was one of the earliest examiners of complex accidents. A *hazard* is defined by FAA Order 8040.4 as a "condition, event, or circumstance that could lead to or contribute to an unplanned or undesirable event."[27] As the result of the broad experiences gained from the difficult task of determining root cause, the FAA learned a very difficult lesson. The vital lesson is this: seldom does a single hazard cause an accident. More often, an accident occurs as the result of a combination or sequence of causes. It is this lesson that can profit us well down the road. We should choose to heed it.

Abnormal Situation

An *abnormal situation* is any unexpected or unintended situation that differs from what is expected. A plant is abnormal when it is in a state, or status, different from what is expected. A plant is expected to produce the required product in the required manner at the required rate and quality. When it does not, the situation is abnormal. Even when a plant is shut down, it may be in an abnormal state if the plant has a mismatch between its actual state and its expected state.

Where people are involved, abnormal situations have two related but very different aspects. These two situations are very different, but both present an unfortunate opportunity for things to go wrong.

---

27 FAA (Federal Aviation Administration), Appendix G to *System Safety Handbook* (Washington, DC: Government Printing Office, 2008).

Aspect 1: Plant Actually Abnormal

Here the plant or facility is actually abnormal. Your operator may or may not know that it is abnormal. If he knows, this is good—steps can be undertaken to manage.

If he does not know, he must know. Either something in the situation awareness resources kit is not working as planned or the problem is deeply hidden. Either way, we have a double problem.

Aspect 2: Plant Actually Normal

Here the plant or facility is actually normal. Your operator may or may not know that it is normal. If he knows, good—no extra attention is needed. But if he suspects something is amiss, there is a problem. The plant is normal, so something in the situation awareness tool kit is not working the way it should. Interesting. What has just happened is that you have uncovered a second important requirement for situation awareness: not only is situation awareness expected to ferret out what might be going wrong; by not finding things wrong, that result must be strong enough to be believed.

You want the operator to sniff about and find early problems so he can deal with them. You do not want the operator to think he has problems that are not there and spend lots of time looking for them; time that is better spent on other necessary tasks. By the way, you do know that often enough, if we look hard enough for a (nonexistent) problem, we find something. It is not broken, but because we think it is, we do things—usually improper things.

Near Miss

First, let us be clear that every plant and enterprise should be actively capturing and processing near hits. Near hits (the usual expression is near misses) are lucky events. But the term *near miss* really does not capture the situation. Nothing nearly missed; it did miss. There was no hit. The *near* part was that it nearly was an actual hit. Bad things heading your way somehow missed; so they did not cause harm or other damage. What about next time? Investigating near hits, identifying the causes, and working to make sure that all the causes that can be managed are managed, are powerful safeguards. A near hit (near miss) results from the outright failure of one or more protection barriers designed for good operation. A near hit can also result because an incident was just barely avoided by the good fortune of not failing a protection barrier only because the stress on it was not quite enough to fail it. Had the stress been a bit more, it would have failed to protect. The incident was only avoided because one or precious few of the protection barriers managed to hold. This close is too close. Coming this close exposes the enterprise to unnecessary and avoidable risk. Managing this is necessary to stay out of eventual harm.

A near hit can come from anywhere. It can come from an abnormal situation where no alarm activated (but should have) and the abnormal situation corrected itself just shy of any consequence. It can come from an abnormal situation for which the alarm activated and the operational response was not done properly, or soon enough, or at all, but the situation corrected itself just shy of any material consequence. It can come from an off-normal situation that was serious and recognized, but too late for proper working; not recognized at all; or recognized and worked incorrectly yet the situation managed to correct itself just enough to avoid material consequences.

Accident

An *accident* is:

> Some sudden and unexpected event taking place without expectation, upon the instant, rather than something that continues, progresses, or develops; something happening by chance; something unforeseen, unexpected, unusual, extraordinary, or phenomenal, taking place not according to the usual course of things or events, out of the range of ordinary calculations; that which exists or occurs abnormally, or an uncommon occurrence.[28]

> An unforeseeable and unexpected turn of events that causes loss in value, injury, and increased liabilities. The event is not deliberately caused and is not inevitable.[29]

Not very much in the way of disasters are accidents.

An accident must be accidental. To truly be an accident, it must be earned. It will be earned by an enterprise having met the full delivery of due diligence. This means an effective policy exists and was followed that provides safe operation: proper design, diligent maintenance, capable operating procedures and guides, responsible training, and all the rest that goes into responsible stewardship of an enterprise that could injure people, threaten the environment, and cause undue financial loss. What is commonly called an accident is properly an incident (or worse).

---

28 "Accident," The Free Dictionary by Farlex, accessed June 15, 2015, http://legal-dictionary.thefreedictionary.com/Accident.
29 "What Is Accident?" Black's Law Dictionary, accessed June 18, 2015, http://thelawdictionary.org/accident/.

Incident

An *incident* is:

> [a]n occurrence, other than an accident, associated with the operation ... that affects or could affect the safety of operations.[30]

No enterprise needs incidents. It should be the primary objective of senior management to put the entire proper infrastructure in place to ensure that the enterprise can operate responsibly, and to make sure it does just that.

A failure to operate of any safety-related protection device, even if a second such device provided eventual full protection, is an incident and not a near miss.

To be sure we understand the significance of incidents, any challenge of a safe operation safety-related physical component (overpressure relief valve, high-temperature shutdown switch, low-level pressure shutdown switch, loss-of-pilot interlock, etc.) that should have caused the device to activate, but it did not, is an incident. Even if a second, independent safety-related physical component (e.g., a dual overpressure relief valve) then properly functioned and the safe condition was preserved, the first failure is an incident. It should be investigated just as any other incident would be. It should be followed up on just like any other incident would be.

Disaster

A *disaster* is:

> [a] sudden calamitous event bringing great damage, loss, or destruction.[31]

Pretty clear. We want to avoid them all. Let us not have any. Let us hope not to read about any either.

2.18 Critical Failures

Almost no serious incident or disaster happens due to a single point of failure. This is partially by design. You plan for operations that can potentially cause damage or harm.

---

30 "Definition of Key Terms," AirSafe, last modified July 2, 2017, http://www.airsafe.com/events/define.htm; "NTSB Accident and Incident Reporting Requirements," NBAA (National Business Aviation Administration), accessed June 18, 2015, http://www.nbaa.org/ops/safety/ntsb/.
31 "Disaster," Merriam Webster, accessed June 8, 2015, http://www.merriam-webster.com/dictionary/disaster.

One way to plan is to conduct specific hazard reviews during design. Knowledgeable personnel get together and review what might go wrong during operation and decide which incidents pose sufficient exposure to harm. For those, specific safeguards are set in place in an attempt to ensure that the planned-for thing does not go wrong. For example, if over temperature can cause a fire, a secure high-temperature shutdown is used. If over pressure can cause an explosive rupture of a pipeline, a secure high-pressure relief is installed. If a tank overflow will cause the release of environmentally damaging agents, a high-level shutdown or inflow shutoff valve is installed.

It is most unfortunate that managing a single point of failure does not yield sufficient safety or protection. We need more.

Chains

Let us talk about chains.

You have heard the expression that "a chain is only as strong as its weakest link." Nice phrase. It is intended to suggest that there is something at the end of a chain that you want. You want to be able to pull on the chain until that thing is in hand. You want the chain to hold. You would make sure that each link was strong enough. You want what is at the end (see Figure 2-23).

This "chain is only as strong as its weakest link" wisdom is often used to motivate us to do the right thing: to do our share of the work. If we were the one whose failure caused that chain to break, we would naturally be the bad guy. When the chain breaks, the prize at the end is lost. We do not want the chain to break!

Figure 2-23. A chain is only as strong as its weakest link. When it stays intact, you can have whatever is attached to it.

Figure 2-24. Disaster chain represents an intended safeguard where each link that *does not break* represents a critical safety factor that *did not work*. If none of the links break, you pull the disaster right into your lap. Most serious industrial incidents have disaster chains more than three links long. This means that all three safeguards did not work as designed (and required) to ensure safe operation!

Disaster Chain

But chains work both ways. Yes, we want the chain to pull what we would like to have. But if the chain is attached to something that we do not want to have, we now want it to break. We do not want what is at the end. We would rather not, thank you just the same, have the chain pull a disaster to us. We really want this chain to break! Same chain, but now it has a vastly different connotation. A single broken or missing link will provide the chance to avoid the problem. An intact link might stand for improper training (for the link to break, training needed to be fully up to the task), another for faulty alarms, others for inadequate maintenance, a faulty HAZOP policy, a failed lockout/tagout policy, and so on. If any single one of these safeguards were to work, the corresponding link in the disaster chain would break, leaving the disaster behind (See Figure 2-24). We would be safe.

You may have heard this message from another concept: the Swiss cheese model of incident causation (see Figure 2-25).

> The Swiss cheese model of accident causation is a model used in risk analysis and risk management, including aviation, engineering, healthcare, and as the principle behind layered security, as used in computer security and defense in depth. It likens human systems to multiple slices of Swiss cheese, stacked side by side, in which the risk of a threat becoming a reality is mitigated by the differing layers and types of defenses that are "layered" behind each other. Therefore, in theory, lapses and weaknesses in one defense do not allow a risk to materialize, since other defenses also exist, to prevent a single point of weakness. [In actuality, if the slices of cheese move (due to lapses in attention to the integrity of each layer of protection, slice) then

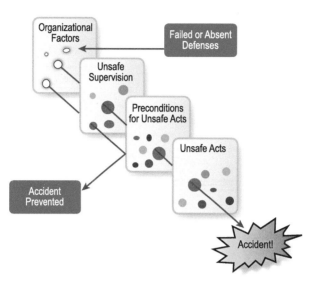

Figure 2-25. "Swiss cheese" model of how an incident (accident?) can occur despite several presumed safeguards in place. If each safeguard has significant weaknesses or faults, it is just possible for a single event to find a fault for each safeguard and therefore cause the sum total of all safeguards to fail. Therefore, it is the quality of each safeguard that is important, not just the number.

it would be possible for a single event to pass through all layers since the holes will line up. An incident results despite the apparent protection hoped for by the many layers.][32]

It looks like Figure 2-25. Each slice stands for an infrastructure safeguard that is missing or done so poorly that it becomes ineffective. Even if there are a lot of them, if all the ones responsible for the protection of an unwanted production safe operation condition are not fully working, the protection is not there.

Disasters in Review

We now take a closer look at disaster chains for a few notable incidents. They clearly expose the critical importance of everything—people, equipment, procedures, training, and a whole lot more.

Milford Haven

This incident happened at the Texaco refinery in Pembroke, United Kingdom, in July 1994 (see Figure 2-26). Lightning struck the refinery during an early morning thunderstorm. The ensuing damage caused a power outage at four of the production units: vacuum, alkylation, butamer, and fluid catalytic converter (FCC). Five hours later, around 20 tonnes [sic] of hydrocarbon was released downstream of a fluid catalytic

---

32 "Swiss cheese model," Wikipedia, accessed November 27, 2016, https://en.wikipedia.org/wiki/Swiss_cheese_model.

Figure 2-26. Fire and explosion at Milford Haven (1994) that was the "last straw" in serious industrial incidents in the United Kingdom. Following it, the HSE commissioned new regulations including the current best practices in alarm management.
Source: Health and Safety Executive

converter unit (FCCU) flare knockout drum that had overfilled. Ignition followed and an explosion ensued. Twenty-six people were injured, none seriously. Damage to the refinery was in the neighborhood of $80 million.

The disaster chain:

1. Plant actually shuts down but the shutdown activity is not properly monitored
2. Faulty display design
3. Process operational fault not recognized
4. Process modification not properly assessed
5. Failure to maintain situation awareness
6. Faulty alarm system design
7. Faulty operational goals and policies

Piper Alpha

At the Piper Alpha oil and gas platform in the North Sea on July 6, 1988, 167 men died as a result of a series of fires and an eventual gas explosion (see Figure 2-27). This set off a major inquiry leading to the discovery of a culture of lip service to safety, resulting in an extreme exposure to untoward incidents from operational safety indifference.

The disaster chain:

1. Faulty maintenance isolation procedures
2. Faulty permit-to-work system
3. Faulty/missing return-to-service procedures
4. Ineffectual managerial inspections
5. Defective fire control systems
6. Inadequate attention to explosion risks during the design stage

Figure 2-27. Piper Alpha platform in the North Sea (1988) engulfed in a gas explosion and fire resulting in large loss of life caused by numerous operational and safety lapses.

Source: Reproduced with permission from Associated Press.

7. Inoperable fire control equipment
8. No emergency response policy to ensure safety and life
9. No plans or training for interplant emergencies

BP Texas City

BP was attempting to start up an isomerization unit (see Figure 2-28).

> During the startup of the Isomerization Unit on Wednesday, March 23, 2005, explosions and fires occurred, killing fifteen and harming over 170 persons in the Texas City Refinery, operated by BP Products North America, Inc.[33]

The disaster chain:

1. Violating safe start-up policy
2. Faulty installation of a critical level transmitter
3. Alarm activation failures
4. Failure to follow procedures
5. Failure to maintain situation awareness

Figure 2-28. Aftermath of the BP Texas City, Texas, refinery fire (2005) that resulted in large loss of life and was caused by numerous significant safety and operational lapses.
Source: US Chemical Safety and Hazard Investigation Board.

---

[33] BP, "Fatal Accident Investigation Report: Isomerization Unit Explosion—Final Report" (BP Internal Report, December 9, 2005).

6. Faulty relief valve operation

7. Failure to understand basic process operations

8. Failure to manage exposure to personnel danger within hazardous process limits

Any one but the last would have probably prevented the release and subsequent explosion and fire. Had the last been properly managed, all deaths would have been avoided.

Deepwater Horizon

On the evening of April 20, 2010, a well control event allowed hydrocarbons to escape from the Macondo well onto Transocean's Deepwater Horizon, resulting in explosions and fire on the rig. Eleven people lost their lives, and 17 others were injured. The fire, which was fed by hydrocarbons from the well, continued for 36 hours until the rig sank. Hydrocarbons continued to flow from the reservoir through the wellbore and the blowout preventer (BOP) for 87 days, causing a spill of national significance.[34] [See Figure 2-29.]

The disaster chain (management failures were the root cause of all other eventual failures listed as follows):

1. Management failures

    a. Ineffective crisis leadership.

    b. Ineffective communication and siloing of information.

    c. Failure to provide timely procedures.

    d. Poor training and poor general supervision of employees.

    e. Poor management and supervision of contractors (and other third parties).

    f. Inadequate use of technology.

    g. Failure to adequately examine and prepare for the risk likely faced during operations.

2. Cement slurry failure. (The cement to seal the well was of insufficient design, ineffectually installed, and inadequately tested.)

---

34 BP, "Deepwater Horizon Accident Investigation Report" (BP Internal Report, September 8, 2010).

Figure 2-29. Deepwater Horizon drilling rig before and during the disaster (2010) caused by ignoring safety issues to keep drilling on schedule.
Source: National Commission on the BP Deepwater Horizon Oil Spill and Offshore Drilling (left) and BP.

3. Faulty temporary well "abandonment." (Procedures were ad hoc and faulty.)

4. Failure to adequately interpret well and sealing integrity tests (which had clearly suggested faulty sealing).

5. Inexperienced site technical leadership.

    a. Failure to provide leadership responsibility.

    b. Failure to provide sufficient requirements detail and operational guidance.

    c. Failure to place prudence, caution, and safe operations above costs.

6. Failure to train employees in emergency procedures.

7. Failure to include lessons learned from similar near misses into the operating culture.

8. Failed, disabled, and ignored alarms on the drilling rig.

9. Failed blowout preventer. (This failure was the last line of defense; its proper operation could have controlled the extent of the leak but not prevented the proximate deaths.)

10. Lack of appropriate upward communications to technical and management leadership positions who could have made a difference in the outcome.

11. Failure of appropriate governmental regulatory bodies to adequately permit, inspect, and monitor operations (before incident).

Figure 2-30. Olympic Pipeline disaster in Bellingham, WA, caused by a culture of neglect for pipeline monitoring and failed follow-up of problems and near misses.
Source: Kiro 7.

Olympic Pipeline

On June 10, 1999, at approximately 3:28 in the afternoon, a 16-inch gasoline-filled pipeline ruptured in a residential area of Bellingham, Washington. Three deaths, eight injuries, and $45 million in damages resulted (see Figure 2-30).

The disaster chain:

1. Construction damage to the underground pipeline during plant modifications
2. Inadequate evaluation of the incident and a failure to properly inspect thereafter
3. Failure to appropriately test essential safety equipment prior to placing pipeline in operation
4. Failure to investigate repeat incidents (which had the identical root cause)
5. Failure to ensure appropriate operational integrity of the control system

2.19 Process Hazard Management

Proper situation management is not going to happen just because it is a category of design that we work out and try to implement. It cannot just be something added on to existing enterprise goals and objectives. Rather, it depends on many aspects of design being in place that are necessary to underpin good operation. Process hazard management is an important one. Process hazard management is an accepted and well-understood discipline. There are a wealth of experiences and tools that have been demonstrated to be reasonable and effective. While all hazards cannot be totally eliminated, appropriate management can be approached through a careful assessment of hazards and an effective program to control them. Very few actual incidents and disasters repeat themselves (although Milford Haven and Texas City do appear to come awfully close). Thus the simple study of history would be sobering and perhaps enlightening, the way to manage is best done by understanding the root causes of failure. Attacking and mastering them is one of the best deterrents. We list the root causes that appear again and again. Please read with an eye toward really understanding them. Refrain from using the list as a check-off for your program. While it is certain that many causes are covered, many are likely not. Each cause that is faulty or is weak represents a serious opportunity for failure. A passing score must be 100%. Each not fully mastered is poised with the power and deliberation to do harm.

- Know the process fully.
- Plants and equipment must be designed by experienced experts with a full understanding of process vulnerabilities and the foibles of people.
- All equipment must be fully maintained to the original design standards or better; end-of-life limits must be carefully managed.
- Processes must be operated entirely within design safe operation limits.
- All personnel must report for duty unimpaired and remain unimpaired throughout their period of responsibility.
- Safe and responsible operation is placed before profit or other gain or benefit.
- All personnel are appropriately trained, and such training is kept up-to-date with both personnel skills as well as process changes or modifications.
- All operations must be proceduralized, and these procedures must be incorporated in the training and certification process.
- All incidents and near incidents are investigated, lessons learned compiled, and appropriate infrastructure modifications made.
- All documentation must be up to date and readily accessible to all personnel who can benefit from it.
- An effective anonymous or no-fault reporting program must be implemented at all levels of personnel, including contractors and visitors.
- An effective program for discovering and reporting all suspicious or unsafe situations and near misses must be implemented, used, and continuously improved.

- When the supporting business entity is under financial stress or is attempting to significantly curtail responsible investment in operations, cease operations or operate with renewed care and diligence at all times.

You have probably seen most items on this list before. It is here to help you understand that the items are not optional. No protection from hazard can result if any single one is absent or faulty. Every one is necessary. The message is that safe and reliable operations must be part of the culture and planned for and worked for all the time by everyone. Safety is no accident. Any incident is not enough safety.

2.20 Root and Other Causes

It is important to go over "causes" of incidents and accidents (and, of course near misses). You do so in order to develop ways of prevention. We do so here in order to provide a useful lesson about the difference between blame and responsibility. Situation management is delivering on your responsibility to ensure safe and reliable operations. It goes to great lengths to identify reasons why improvement may be needed and how to do much of the "improving." Root cause identification is the thing that you will want to get into tip-top shape to reduce the chances of nasty things happening again (even if they happened to the "other guy," you will want to stand in the *pay it wholesale* line).

Definition of Root Cause

Paradies (2005) has defined *root cause* as follows: "The most basic cause (or causes) that can reasonably be identified that management has control to fix and, when fixed, will prevent (or significantly reduce the likelihood of) the problem's recurrence."[35] The disaster chain has links for each cause. Each can and should be fixed. An unbroken disaster chain leads to an incident or disaster. If a broken disaster chain still leads to an incident or disaster, it means that you are missing important safety principles!

Explanatory Case for Root Cause

IDCON (a plant reliability management consulting firm; North America) is convinced that the process of analyzing the root cause of failures and acting to eliminate these causes is one of the most powerful tools in improving plant reliability and performance. But what is a "root cause"? One definition is "The cause of a problem which, if adequately addressed, will prevent a recurrence of that problem." Let us look at an example.

Imagine that a bearing has failed, and that an investigation shows that it had not been lubricated. Asking the question "why had it not been lubricated?" may lead to the

---

[35] "Root Cause," Wikipedia, last modified March 17, 2018, http://en.wikipedia.org/wiki/Root_cause.

discovery that the grease point for the bearing had been missed during a lubrication survey and it was not on the lubrication mechanic's route sheet. Using the above definition of "root cause," this problem can be prevented from happening again by simply adding this grease point to the lubrication route sheet.

But if the definition of "root cause" is changed slightly to "The cause of a problem which, if adequately addressed, will prevent recurrence of that problem or similar problems," this raises the bar to a higher level. The next question then is "Why had the grease point been missed from the lubrication route?" The answer may be "that lubrication routes were set up by a single person, with no checks or confirmation that the routes were complete." This new understanding should cause a change to the procedure for the development of lubrication routes, which will ensure that there are no other missed lubrication points in the plant, nor will there be in the future.

By asking the question "why" a few more times, the root cause of a problem is often identified as a procedural, or management, shortcoming. Addressing these root causes often requires a change of thinking and some pain and effort, but the results will be much longer-lasting and higher-value than correcting individual failures.[36]

Proximate Cause

Proximate sounds fancy. It has an important legal definition. It is used to identify the at fault party in tort law (torts are where people sue each other over damages done). So when *someone* has allegedly done *someone else* harm, they look for the cause of the harm. If the *someone* caused it, he is liable for fixing things for the *someone else*. The general test is the "but for" test. The traditional example is *someone* running a red light and hitting *someone else*. But for *someone* running the red light, *someone else* would not have been hit.

"But for" the incorrect mounting instructions of the Flagelre Pump (fictional name, of course), it would not have broken loose and set fire to the plant. So, yes we can blame the Flagelre Pump Company for faulty documentation. But another cause (and here we invent for illustration a few possible scenarios) was that the plant's purchasing department did not require general instructions to be marked up so that they only apply to the equipment for which the instructions provided. The correct instructions were there, but it was difficult to make sure they were selected. At night, in the rain, the mechanics missed it. Actually, the purchasing department had made the instruction marking a requirement beginning last year; unfortunately, the pump was purchased just before. It should also be noted that had the mechanics not been so rushed, they would have easily noticed that there were other installation aspects that were not reasonable for the actual pump at hand. Why were the mechanics rushed to the point that

---

36 "What's a Root Cause?" IDCON, accessed January 30, 2014, http://www.idcon.com/resource-library/articles/reliability-tips/833-whats-root-cause.html.

they neglected due care? Why did operations wait to replace the pump until it failed? For the past 2 days, it had been acting poorly and the maintenance department recommended it be stopped for inspection. Operations did not stop it! It will be left to you to suggest root cause here.

2.21 Layers of Protection

Layers of protection are safeguards against bad things getting worse. They are identified and specifically designed, or exist due to the inherent design capability to safeguard enterprises that can cause damage to personnel or the environment if misoperation occurs. None are ad hoc. All are intentionally designed.

Layer of Protection

LP is the term for layer of protection.

> Protection layers typically involve special process designs, processing equipment, administrative procedures, the basic process control system (BPCS), and/or planned responses to imminent adverse process conditions; and these responses may be either automated or initiated by human actions.[37]

Independent Protection Layer

IPL is the term for independent protection layer.

> A system or subsystem specifically designed to reduce the likelihood or severity of the impact of an identified hazardous event by a large factor, i.e., at least by a 100 fold reduction in likelihood. An IPL must be independent of other protection layers associated with the identified hazardous event, as well as dependable, and auditable.[38]

Layers of Protection and Situation Management

The first thing to clear up is that layers of protection have nothing to do directly with situation management. Yet both are important. Both are ways that enterprises count on to keep bad things from getting worse. Good situation management should greatly reduce any challenges to the enterprise where a layer of protection would be called into action. And having effective layers of protection is important in case operators lose situation management. This book will assume that your enterprise has appropriately designed and implemented layers of protection required by good engineering and operating practices. That design is outside the scope of this work.

---

37 CCPS (Center for Chemical Process Safety), *Safe Automation of Chemical Processes* (New York: Wiley, 1993).
38 CCPS, *Safe Automation of Chemical Processes*.

One thing to keep in mind is whether or not operators should be considered a layer of protection that would be counted for safe operations. There are those who honestly think that they are, and those who honestly think they are not. For the purposes of informing the reader, this author states that operators, as important and effective as they might be, should not be considered a layer of protection in the formal sense. The argument is straightforward: people being people, they cannot be counted on to be perfect enough to be right and effective enough of the time. Of course, good operators protect lives, fortunes, and the environment every day—that is their job. However, modern industrial society knows to backstop any operator error with reasonable protective defenses. Enterprises must work to implement best practices in situation management without any short cuts.

2.22 Close

We began this story in normal. The process was normal, the operator was on the job, and the general expectation is that things will go right. The operator's job is to "maintain his finger on the pulse." He is on the job to nurture operation: smooth out small irregularities, and eek out improvement as opportunity presents or exploration exposes. He is there to modify operations as production calls for regime change or modification. He is there to restore good operation when production strays. He is there to exercise his considerable skill at restoration or curtailment when serious operational challenges demand so. He is, in effect, the first opportunity for hands-on adjustment and the last opportunity for human intervention before the safe operation infrastructure is challenged. Without operators, there would be no need for anything in this book. You should find a treasure trove within these pages. Explore. Examine. Take what you will for your own.

2.23 Further Reading

Deming, W. Edwards. *Out of the Crisis*. Cambridge, MA: Massachusetts Institute of Technology Press, Center for Advanced Engineering Study, 1982.

3
Operators

Everyone doing his best is not the answer. It is first necessary that people know what to do.

W. Edwards Deming (Quality Expert)

It should not be necessary for each generation to rediscover principles of process safety which the generation before discovered. We must learn from the experience of others rather than learn the hard way. We must pass on to the next generation a record of what we have learned

Jesse C. Ducommun[1]

Recalling from Chapter 1, *Operators* form the front lines of safe and effective manufacturing, transportation, and distribution. Operators are a backbone for responsible activities to provide the world's population with the goods and services to make their lives more comfortable and enjoyable. From food and pharmaceuticals to the expansive host of consumer products, to the iron, steel, and engineered materials to make them, the process industry depends on operators to competently manage. They are on duty second-by-second, minute-by-minute, and hour-by-hour, day in and day out, attending to the making and delivering. To do this right requires proper tools and competencies. *Situation management* is about providing sufficiently capable, properly tooled, and available human operation to a control room or operation center.

---

1 BP US Refineries Independent Safety Review Panel, "The Report of The BP US Refineries Independent Safety Review Panel" (Washington, DC, January 2007).

The hard work of operating is the operator's (or controller's) responsibility. The mastering of sufficient skills and capabilities is his responsibility. The arrival for duty and continuation of attention during the full course of duty is his responsibility. Skillfully performing the needed activities is his responsibility. But all the rest is the enterprise's responsibility. The enterprise must establish proper personnel requirements and exercise appropriate measures to ensure they are fully met. The enterprise must establish necessary skill and capability requirements and exercise appropriate measures to ensure that they are mastered and kept up to date. The enterprise must set forth appropriate interpersonal cooperation and respect expectations, and take all necessary steps to ensure that they exist and are effective. The full list is longer, but you get the idea. Unless the enterprise delivers on its part, no amount of rulemaking or strong enforcement will make the operator effective enough. Here is what the enterprise will need to consider for the operator to have a responsible chance to do his job well.

3.1 Key Concepts

Operators As Professionals	Develop and support the full professional expectation and policy for all operators.
Basic Needs	In order to be appropriately functional, all operators need to (1) be in a comfortable environment free of unnecessary distractions and encumbrances, (2) feel physically and emotionally safe, and (3) feel accepted.
Impairment	There is no such thing as an okay impairment. All functional and/or performance limitations must be explicitly and effectively managed for everyone, all the time.
Readiness Responsibility	It is the responsibility of operations management to ensure that all operators on duty are ready (free of impairment including overload). Ready access to appropriate management support is an integral part of operator readiness.
Training Responsibility	If training is needed to prepare any individual for his assigned job, it is the responsibility of senior management to ensure it is done, adequate, and effective.
Training Objectives	Competency training followed by critical skill development is the only effective protocol.
Conducting the Training	Training should only be done by the most experienced and expert personnel. Training for task and on-the-job training are to be avoided.
Shift Handover	Any time a shift ends or begins, even if the same operator resumes work after a gap (overnight or over weekend), an adequate shift handover is required.
Maintenance	Maintenance personnel must be granted temporary ownership of equipment, and the operator must formally take back ownership at the end.
"Long-Arm" Operations	Where remote site personnel are assisting the operator at distant equipment locations, the operator retains full equipment ownership at all times.

3.2 Operators' Creed

The production *operator* of an industrial plant or other operator-managed enterprise activity is the responsible individual entrusted with the second-by-second, minute-by-minute, hour-by-hour, day-in and day-out successful operational outcomes. From the moment he assumes operational control until the moment it is either ended or passed on to another, we expect him to do all in his power to ensure safe, effective, reliable, and responsible operation. He is an essential member of the enterprise team. He is a professional.[2]

A creed in these modern times is understood to be a deeply held expression of dedication to a belief. It is taken to be an expression of professionalism. Among the better-known technical creeds are the Engineers' Creed (United States) and the Obligation (Canada). In recognition of the responsibility society places on the shoulders of operators and as an expression of and respect for this level of professionalism, a draft Operator's Creed is offered here for the consideration of professional operators. This is new.

Operator's Creed

I accept responsibility for the safe, reliable, effective, and responsible operation of the facility under my control. I arrive fit for duty. I take responsible charge. I properly utilize the resources and tools at my disposal. I place safe and responsible operation above all else. When duty is done, I effectively close operations or pass on responsibility to my relief. I leave duty in a responsible manner.

This expression of responsibility is offered for operators to consider as an expression of their professionalism and acceptance of responsibility, as they may find useful.

3.3 Operators and Operations

Definition of an Operator

1. "The person who operates equipment for its intended purpose. Note: The operator should have received training appropriate for this purpose."[3]

2. "The person who initiates and monitors the operation of a process."[4]

---

2 "Professionalization," Wikipedia, last modified April 1, 2018, http://en.wikipedia.org/wiki/Professionalization.
3 International Society of Automation (ISA), *The Automation, Systems, and Instrumentation Dictionary*, 4th ed. (Research Triangle Park, NC: ISA, 2003).
4 ISA 82.02.01-1999, *Safety Standard for Electrical and Electronic Test, Measuring, Controlling, and Related Equipment—General Requirements* (Research Triangle Park, NC: ISA [International Society of Automation]).

Peopleware

Peopleware is an invented word. You will find it useful to separate it from the hardware and software operators use. The word is intended to be a collective term for the tools and identifiable resources operators use to get the job done. Peopleware includes the "soft stuff," meaning things such as skills, training, and performance protocols (e.g., call out and escalate before working on hard problems). It is the appropriate combination of peopleware, hardware, and software that leads to control operating room success. Peopleware includes an important designer component: the plant or operation designer is obligated to pay sufficient attention to the operability of his creation. This is over and above the attention devoted to the plant's actual concept, design, and implementation. Those are important dual requirements. We are now ready to ask the question: what does it take to properly equip an individual operator to do his job in the control room effectively with neither inappropriate nor excessive demands? And having identified those needs, we must immediately provide for them.

Each plant or operation is complex and unique. As you might imagine, it is well beyond the scope of this chapter to be able to fully identify those needs for you; rather, it suggests that they are necessary.

Plants and Operations

Throughout this book you will find the dual term of *plants* and *operations*. Operators operate both. A *plant* is the collection of physical equipment, energy, and resources usually used to manufacture or produce something. Things like petroleum refineries, chemical plants, electrical power generation, pipelines, electric transmission lines, food processing, and raw materials processing, come to mind. In general, they could not operate without supervision. An *operation* is more often an existing entity that is placed under an operator's management because of its complexity or requirements of public safety or critical public resources. Things like highway transportation networks, electronic data networks, and air traffic and security monitoring systems come to mind. In general, they might operate without supervision, but supervision brings an important level of performance not attainable without it.

The scale of plants and operations is of little concern unless they are very small or very large. The very small ones are for special purposes; those purposes generally require special operational management. The very large can be so impactful to a community or a locale that special operations practices and regulations govern.

It is because of the benefits of including human operators into this setting that both very small and very large plants are covered in this book. Real things are going on that require personal monitoring and responses as opposed to just mechanical (or computer managed) watchfulness and control. Notwithstanding the frequency of operator intervention, an operator's role is not "babysitting." Operators are expected to fully understand the plant or operation. They must keep fully abreast of what is going on through active monitoring and testing. They are asked to initiate and follow through with any and all corrective actions required to keep things properly operating. And they are required to seek outside assistance for all situations that challenge their resources to properly manage.

3.4 Boundaries and Responsibilities

Operator-responsible operational boundaries and responsibilities and how they are adversely challenged is an important aspect of control room management. We take for granted that the operator is charged with ensuring effective operations. After all, that is why we use operators. If everything could be fully and completely instrumented and automated to manage things as good as or better than manned operations, and it were cost-effective, no plant or operation would need operators. Alas, such a plan is presently too expensive to consider. Eventually, that may not continue to be the case. When we arrive at that place, everyone can revisit. For now, let us agree to take the presence of an operator for necessity.

An enterprise (or modest to large plant) will usually have several operating areas. An operational area is the physical extent of equipment that a single operator is responsible for. Figure 3-1 schematically depicts this situation. There is a small interior portion

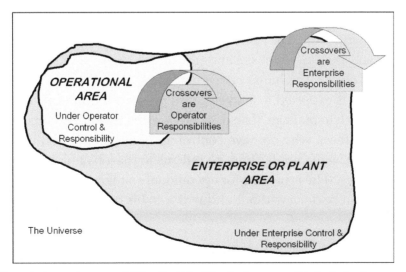

Figure 3-1. Boundaries and responsibilities that identify who is responsible for trying to prevent operational excursions, from what region of operation, and who is in charge once an excursion has occurred.

noted as the operational area. The *operational area* is contained within an *enterprise or plant area*. The enterprise area is contained within the *universe*, which is shorthand for everything else outside of the *enterprise* (e.g., the community).

The operational area is expected to be fully under the control of the operator who has the responsibility for it. There may be other operational areas, each with their own operator (not shown in the figure). Generally, all the operating portions of the enterprise will be within one or another *operating area*. When things go wrong inside an operational area, we expect the relevant operator to manage them. So long as the effects or issues about that operational area remain within it, that plan will work. Once the issue or abnormal situation threatens to cross a boundary into another operational area, into the enterprise in general, or outside the enterprise, enterprise resources must be brought to bear. Now the issue or abnormal situation transfers to the enterprise personnel (perhaps shared with a relevant operational area whose boundary has been crossed). Any operational details of this transfer process will be addressed in escalation protocols.

The following discussion is extremely important. At first glance, it may appear to be more about definitions and categorization. It is far from this. As we will see, when we understand how responsibility and control works, everyone must understand how abnormal situations that get beyond local hands to manage are handled. First, let us set the proper stage. The operator is expected to properly understand and manage any area where he has both *control* and *responsibility*. Now consider situations where both are not necessarily concurrently present.

Responsibility

Consider responsibility first. The term *responsibility* is used to clearly identify who must bear the operational burden—whose hands are on the controls. Within his operational area the operator must take charge of the outcomes. If he also has control (see next discussion) then he is expected to actively, with his own actions, make the necessary adjustments and such to manage. If he does not have control yet retains responsibility, then he must alert those who do have control and have responsibility for the situation. Once he has done the notification, he returns to the job at hand of managing his own operational area. If he is needed for operational changes to assist with the alerted activity, and that can be done without undue risk to his operations, he would do so.

Control

Having control means that the responsible individual, the operator for us, has at his disposal and use all the existing "handles" for changing and modifying operations.

He can start up conveyors, change the speed limit on the eastbound bridge lanes, and reduce the distillate product flow from a separator. If he also has responsibility, he makes these changes at his own discretion. If he does not have responsibility, he will make changes and modifications only at the direction of others with responsibility.

Enterprises must carefully clarify the delegations for control and for responsibility.

Crossovers

A *crossover* is when the impacts of operation cross the boundary between responsibility from one operator to another, from an operator to the enterprise, or from the enterprise to the rest of the world (the universe). It is in these situations that the usual operating procedures and protocols are at the most risk of failure. This is because of the sheer complexity of alternate scenarios of failure and the inherent difficulty in providing comprehensive plans for all. It is compounded by the fact that different individuals are involved—with different responsibilities, possibly different cultures, and often different managers. The approach is to recognize when a crossover is either likely, occurring, or has occurred, and adjust operational activity to take it into proper account.

Crossover Management

Basic Procedure Framework

Operating procedures and training should be explicitly designed to handle crossover situations. This includes the following:

- What defines a crossover (beginning and ending)
- What must be coordinated
- Who will do the coordinating
- Who will be lead
- How the coordination is to be carried out (protocols, procedures, decisions of how to deal with the conflicting aspects)
- What, if any, additional within-area operational changes are needed

Crossing from One Operational Area to Another Operational Area

Abnormal situations (really, the effects from one) that cross from one operational area to another are the most common crossover. Most crossovers are directional. That is, the problem from one affects the other, but not necessarily the other way around. What has likely happened is that the *affecting operator* (where the problem comes from) has tried to manage and/or contain the abnormal situation but has failed. The *affected operator* (getting the problem) now has the control and responsibility to adjust his operation to manage the new problem. Each operator would be diligently working to manage things within his separate area. This is how it should work. However, because a crossover is involved, there is the additional duty of coordination. Now each operator must explicitly consider both the local effects as well as the crossover effects of his actions. Before crossover, coordination was either unnecessary or implicit: unnecessary where the normal interactions were minimal; implicit where the normal interactions were well-understood and according to usual material, energy flow, or processing schedules. In this case, so long as the interactions were within requirements, nothing more needed to be done. This may all need to change after an actual crossover. Explicit crossover plans and protocols are needed.

Piper Alpha Lesson
This disaster was introduced in Chapter 1, "Getting Started." Perhaps the most diabolical error made during the Piper Alpha fire and resulting devastating explosion was a complete failure to coordinate problems between operational areas.[5] In brief, due to a series of operational failures, a fire broke out on an offshore oil and gas platform in the North Sea. Communications to an adjacent platform were disrupted, but that platform knew that Piper was on fire and that the situation was more than trivial. However, at no time did the adjacent platform stop pumping oil into Piper, thus actually feeding the fire on Piper. That fire lasted long enough to rupture two natural gas risers that then caused the complete destruction of the platform and the resulting large loss of life. The public inquiry into the disaster discovered that there were no inter-platform protocols or training for handling such emergencies.

Crossing from an Operational Area to the Enterprise

In this situation, the crossover has occurred between an operational area to the enterprise, but outside another operational area. In this case, there is only one formal operator. Therefore, it must be predetermined who will perform the pseudo role of *enterprise operator* in the basic procedure framework process.

Crossing from an Operational Area or Enterprise to the Universe

It is always serious whenever the deleterious impact of the operations of an enterprise crosses over the boundary to the outside (universe). These situations place the public at risk. The public usually does not have the information at hand to fully understand the

---

5 Lord W. D. Cullen, *The Public Inquiry into the Piper Alpha Disaster,* Volumes 1–2 (London: Her Majesty's Stationery Office, 1990).

risks and properly respond, does not generally have suitable protective equipment and suitable alternative resources (such as safe drinking water) available, and does not possess the resources and skills for remediation. This situation is explicitly recognized by the mandated requirements for municipalities and governments to have disaster management plans ready. Readers of this book are in a position to understand why all of this is vital. You appreciate how necessary it is for the public to appropriately respond for their own safety and comfort. On the other hand, most municipalities thankfully will never have to execute any such plan. Unfortunately, this low probability engenders a lack of real need for them to do the work and pay the price to have plans ready. This is a prime reason why a large share of the emergency response burden is placed on enterprises. Please consider the enormous public service benefits from an informed public properly engaged. Far from scaring the public, such planning and discussion can have the effect of real empowerment in a spin-off way often unappreciated. Good general disaster management plans first derived for some industrial thing gone wrong are almost never used for that eventuality can often form the backbone for natural disaster management.

Control Room Coordination with Its Operational Area Field

Every control room or operations center has a "field." The field is the physical (or virtual) location(s) that contain the equipment, other components, materials, and energy that are managed partially or totally by operators in a control room. The field may begin at the doorstep of the control room or be at a far distance. When things go wrong in the field and understanding and/or remedy is beyond the timely physical reach of the control room operator, the work in the field is done by others. The term *outside operator* describes a field-resident person (or one designated to visit the field) who performs much of the routine work there. Of course, there are others in the field including maintenance personnel, inspection personnel, delivery personnel, visitors, and supervisors.

With rare exception, the control room operator is the single-point individual in charge. Here are some general considerations to keep in mind.

Control Room Operator
- Focuses on coordination with the schedule and other operations or work
- Is in charge before, now, and afterward
- Retains overall responsibility for all operations
    - *No handover of tasks or duties*

Field Operator
- Focuses on the immediate time frame
- May perform general tasks (such as being eyes and ears) but is usually devoted to a specific plant component or localized situation
- Possesses responsibility for competently observing, reporting, and recommending
    - *Hands-on actions must proceed only with the agreement of and under the strategic control of the control room operator.*

Field Maintenance
See Section 3.9 for specifics on operator coordination with field maintenance.

Maintaining Responsibility

It is important to maintain a chain of responsibility. Kendra Cherry (2016) defines an important impediment to doing that in a group setting:

> Diffusion of responsibility is a psychological phenomenon in which people are less likely to take action or feel a sense of responsibility in the presence of a large group of people. Essentially, in a large group of people, people may feel that individual responsibility to intervene is lessened because it is shared by all of the onlookers.[6]

Looking ahead to Chapter 7, "Awareness and Assessment Pitfalls," we will see that in a control room setting, when a problem arises in a group situation that all observe, each individual believes that someone else is going to take responsibility for it. Even where there should be a clear chain of responsibility, when individuals of senior organizational authority are present together with those of junior or equal responsibility, there is a tendency for the operator to wait, yet at the same time the senior individual defers. Where time to act is important, action is excessively delayed. In Chapter 10, "Situation Management," we lay out the tools for managing this problem. Those tools form the backbone of the requirements for a qualified operator (formal operator qualifications that are a prerequisite for sitting in the operator chair). It is one of the key reasons for the formality of delegation and escalation procedures. Escalation (and also delegation) of authority will include specific mention of the situation being escalated (or delegated). The discussion also has a clear fit with *crew resource management*, discussed in Chapter 8, "Awareness and Assessment Tools."

---

6 Kendra Cherry, "Diffusion of Responsibility," Verywell, last modified March 11, 2018, https://www.verywell.com/what-is-diffusion-of-responsibility-2795095.

Here we offer additional insight into how to manage the problem associated with who is responsible, despite who, if anyone, is actually supposed to be in charge. John Darely and Bibb Latané (1968) proposed a course of action. They suggested that regardless of who else is in the control room in addition to the operator, it is the operator who must follow these activities:

1. The first step involves actually noticing a problem.

2. Next, the individual must decide if what they are witnessing is actually an emergency.

3. Next is perhaps the most critical decision in this process—deciding to take personal responsibility to act.

4. Then the individual has to decide what needs to be done.

5. Finally, the [individual] must actually take action.[7]

Where the operator has formally shifted responsibility to another, then the preceding steps suggest that the "action" would be to notify the individual in charge of the observed situation (a.k.a. crew resource management). Where the operator is in charge but senior, even if very senior individuals are present, courtesy would suggest that the operator announce that the situation has been noticed and he will manage it. All current established protocols for managing major situations, escalation, permission to operate, sterile control room, restricted control room, and others may be used when the problem is judged serious enough.

Just to be sure this is clear, at all times when the operator is present and on shift (not relieved), he is responsible for making sure that someone is in charge of his chair. Even if he has escalated operation to another qualified person, and that person has accepted operating responsibility, the original operator must retain the explicit responsibility for ensuring someone is in charge. Without individual enterprise requirements to the contrary, if the operator initiated escalation and that escalation was accepted, the operator must retain the responsibility to ensure the escalated individual remains in charge until control is returned, and to order it back if the situation requires it. Even if someone else initiated the escalation, the operator must be responsible for ensuring someone is always in charge.

This detail, if one chooses to call it so, is extremely important. Because of his regular schedule of work in the control room, the operator is familiar with the entire range of the job. On the other hand, the person escalated to is likely not. Certainly that

---

7 J. M. Darely and B. Latané, "Bystander Intervention in Emergencies: Diffusion of Responsibility," *Journal of Personality and Social Psychology* 8 (1968): 377–83, doi:10.1037/hoo25589.

person has special skills and would be fully qualified to operate. However, maintaining operational control for long stretches of time (a shift) would not be so familiar. If the situation were to stabilize or even be resolved, there is a natural tendency to (mentally) shift away from the "in charge" part and step back. This presents a serious continuity problem. It might be clear to one party who should be in charge, yet it might not be the same for all. This ambiguity cannot exist. Therefore, someone must be tasked with ensuring that this does not happen. Someone must always be in charge. The sitting operator must make sure.

3.5 Operator Readiness

From the moment your operator begins his journey to his shift site until the moment he arrives at his end-of-shift intended destination, his employer is at least partially responsible for his safe and effective actions. The trip to work and from work includes travel time as part of the operator's total fatigue load as it affects adequate time off for normal recovery activities like personal hygiene, eating, sleeping, and relaxation. The trip-to-end-of-shift destination responsibility is broader and extends to ensuring that the operator is sufficiently alert and otherwise not emotionally limited due to activities on shift or otherwise physically limited in his travel to the end-of-shift destination. The entire scope of on-shift readiness must be managed.

The three categories of *fatigue*, *impairment*, and *overload* are discussed next. Each must be understood and managed with assured effectiveness. No plant or enterprise that utilizes operators for management can be complacent unless the human operator is on the job. Even with a perfectly designed control system, with the most carefully deployed safeguards, and state-of-the-art plant equipment in perfect condition, every plant or enterprise with no need for operator intervention will go awry if the operator messes with it inappropriately.

Understanding Fatigue

Being too tired is something each of us can surely relate to. You might be somewhat near that right now—as you are reading this chapter! Yet all fatigue is not the same. Sure, getting away from the immediacy of the moment and getting rest is usually restorative. Yet fatigue is really a complex situation. Understanding its causes is important for recognizing the potential for fatigue and setting out plans to manage both its onset and the damaging effects. According to Martin Moore-Ede of Circadian, "excessive sleepiness" is perhaps the most useful definition for fatigue.[8] It is widely accepted

---

8 Martin Moore-Ede, "The Definition of Human Fatigue" (white paper, Stoneham, MA: Circadian, 2009).

in medical diagnosis, regulatory agency approval of prescription drugs, public policy and work–rest hours regulation, and criminal law. "Thus fatigue and its major symptom excessive sleepiness, is well recognized in law and regulation as a significant body impairment, malfunction and source of ill-health and of accident and injury risk."[9] Fatigue management is quickly moving to the top of the list of enterprise safe operation activities. It must be managed in the control room.

Operator fatigue is a double-edged sword. On the one side, because alertness is significantly dulled, it is likely that subtle changes are missed, and thoughtful regular monitoring will be much reduced in effectiveness. This reduces the chances that small problems and issues will be discovered in time to properly manage them. On the other side, when the operator does take action, either in response to identified abnormal situations or misperceived situations, it is likely that that action will be flawed. Procedure steps will be missed or done without due caution. Escalating situations will be missed entirely, or either overstated or understated. Either way, the risk to proper operation is much higher than it need be. If all of this were not enough, chronic fatigue itself is a contributor to ill health.

Circadian Rhythms

The human body uses a daily cycle that greatly helps us to do the things we do. This cycle has been woven into our genetic structure. It is called *circadian rhythm* and is illustrated in Figure 3-2.

This cycle should be easy to relate to. When we are "in sync" with it, our body has a natural way of helping us move through our day. It prepares us to awaken in the morning, it aids our work or other activities during the day, and it gently moves us into the restful state conducive to sleep at night. This cycle is reset every day by sunlight. Research has shown that, left to its own, this cycle is a bit longer than 24 hours. Our earth day is exactly 24 hours. Why the mismatch? The answer is that because our internal clock is longer than the actual cycle, the daily reset will only happen once. If our internal cycle were shorter, it would often reset on its own (say after 23.68 hours) and again with the sun, resulting in havoc.

Fatigue and Incidents

There is an almost straight-line link between operator fatigue and serious incidents and accidents. The human body has a strongly programmed daily cycle that will admit a small bit of tinkering but in the end is intolerant of most movement efforts. It

---

9   Moore-Ede, "The Definition of Human Fatigue."

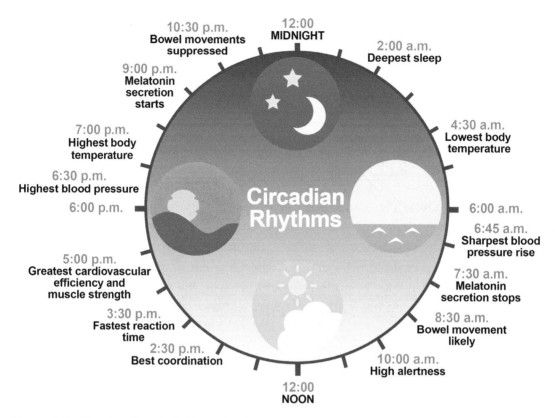

Figure 3-2. Our circadian clock illustrating how our natural body cycle of alertness and functioning follows a daily rhythm. If our internal clock is much different from the actual time clock, we are not at our best.

has been said that the human body never adequately adapts to shift work. Shift workers themselves will admit to that truth.

This internal clock is quite difficult to adjust. It adjusts better (but not by much) when we move our bodies to a later time rather than earlier. World travelers have pretty much accepted that it takes about a full day to move their internal clock by one hour. However, this is not a hard-and fast-rule—most of us take longer. And the transition is not smooth. For a time change of 6 or more hours, there is usually a "brick wall" at day three. And a lingering sense of being not fully there lasts for almost as many days as the number of time zones that were crossed. All of this must appear abstract to those of you who change shifts on a weekly basis. Abstract or not, it is real. It is inescapable—we never really get used to such changes. And our performance and well-being show it.

One of the indicators that operators never quite manage shifting their internal clocks is the strong relationship between circadian time and major incidents.

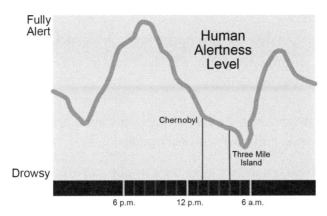

Figure 3-3. Major incidents occur at our least-alert times. Notice that both Chernobyl and Three-Mile Island initiated near the "drowsy" level of the rhythm.

Figure 3-3 shows where Chernobyl[10] and Three Mile Island[11] line up. The time of cycle of the onset of these incidents is not uncommon.

Causes of Fatigue

According to Moore-Ede[12] there are six primary causes of fatigue:

1. **Sleep deprivation** – Too little sleep or sleep of poor quality, including the effects of shift work.

2. **Sleep disorders** – Sleep apnea or other related pathology.

3. **Illness or disease** – A common cold, other physical conditions inducing pain or discomfort, or chronic sleep deprivation.

4. **Therapeutic side effect** – Side effects from use of prescription or over-the-counter drugs.

5. **Heavy stressful physical or mental exertion** – Extended efforts at heavy physical labor (manual road building), intensive mental activity (studying for or taking important examinations), and post-traumatic stress. It should be noted that these effects could be delayed or, on the other hand, recur as a rebound phenomenon.

6. **Stimulant drug use** – Rebound or aftereffects from stimulating street drugs or legitimate pharmaceutical prescriptions.

---

10 "Chernobyl Accident 1986," World Nuclear Association, last modified November 2016, http://www.world-nuclear.org/info/Safety-and-Security/Safety-of-Plants/Chernobyl-Accident/1
11 "Three Mile Island Accident," World Nuclear Association, last modified January 2012, http://www.world-nuclear.org/info/Safety-and-Security/Safety-of-Plants/Three-Mile-Island-accident/.
12 Moore-Ede, "The Definition of Human Fatigue."

Managing Fatigue

It is generally understood that not all fatigue effects due to these causes are fully reversible once the causes are properly managed. Therefore, an effective fatigue management program will include the following:

- Conducting awareness training for operators and supervisors on the nature of fatigue, its causes, and how it is best managed.

- Setting out expectations for operators to obtain sufficient time away from the job to allow for reasonable outside activity between operating shifts. This is specified by law for some professions (e.g., airline pilots, pipeline controllers, and truck drivers).

- Developing effective protocols for assessing operators' degree of fatigue sufficient to deny work to those who are fatigued. An effective aspect of a program is the no-fault ability of an operator to self-declare being too fatigued to work, either before reporting, during reporting, during a shift, or when leaving work.

- Designing careful work schedules that take into account the length of each shift, the number of consecutive shifts, provisions of specialized home sleeping accommodations for day sleeping, and careful management of extra-job employment to ensure lack of conflicts.

- Allowing careful and moderate use of caffeine and other properly consumed alertness enhancers.

- Providing on-shift work variations and limited recovery options (e.g., temporary relief for a power nap, exercise, and light therapy).

- Ensuring proper medical and emotional health through encouragement and access to appropriate medical resources (and some form of monitoring to identify problems and issues).

Enterprises should undergo fatigue awareness training, set up a program, put it into place, and make sure it is working.

Understanding Impairment

Impairment of operators can result in a wide variety of effects. Its effects on good operation run the spectrum from reduced benefits by operating too conservatively to endangering the enterprise and surrounding public. Make no mistake: fatigue is one of the more ubiquitous causes of impairment but not the only one. The entire list falls into these general categories:

- **Fatigue** – Being overtired to the point of significantly reduced sensory awareness, diminished motor skills, inability to focus and concentrate, and manifestation of situational detachment

- **Stress** – Being in a state where one feels unable to cope with current conditions as they appear to exceed available resources and appear to endanger one's own sense of well-being or safety

- **Chemical impairment** – Having dulled senses due to illicit drugs or side effects of prescribed medications; addiction

- **Emotional or psychological impairment** – Experiencing pronounced distraction from duties and current happenings due to life events (such as a very sick child or spouse, loss of a dear one, falling in love, or unmanaged anger)

- **Pathologic mental impairment** – Succumbing to underlying mental disease, distortion, or impairment

- **Nonpharmacological addiction** – Being addicted to gambling, pornography, or excessive risk taking

- **Illness or undisclosed physical limitations** – Suffering from the physical effects of cancer or response to cancer treatments, diminution or loss of senses (such as hearing, sight, or touch), movement impairment, life-cycle events, or dementia

Managing Impairment

The effective enterprise will take steps to understand that these impairments can be present yet masked (intentionally or unintentionally) and therefore need to be monitored and compassionately but effectively addressed. When sufficient sponsored resources are available to help operators to understand and manage these, incidents decrease. Shift handover is a good time for monitoring. Remember that both oncoming and outgoing operators have a monitoring role.

Enterprises also need to understand that the above effects may be temporary, based on a particular situation or point in time and coming and going, or chronic. While shift monitoring is tasked with checking for the presence and degree of impairment, it is not equipped to deal with any chronic aspects. Certainly, the immediacy is evaluated and necessary steps taken to ensure that the operator is fit for duty or leaving duty. But the recurring aspects must be looked for and managed by the enterprise.

Understanding Overload

Setting all bravado and exuberance of corporate allegiance aside, operators, like the rest of us, have limits. And like most other systems with feedback, performance will not vary proportionally as limits are approached. So there are not likely to be traditional warnings of overload. Things may be fine until that uncomfortable moment when they are not. And when "they are not" may not be a graceful event. To be sure, there will be signs of overload early on. Morale may show it. Absenteeism may be an early result. An increase in near misses or small accidents almost always is present. And there is the ubiquitous increase in fatigue.

One or more of the following can cause or contribute to operator overload:

- The actual load exceeds available operator resources.
- The mix of situations leads to excessive cognitive workload.
- A lack of adequate supervision leads to either operational reluctance or excessive risk taking.

Workload

Obviously, there is the physical part that comes from the sheer magnitude of the tasks to be done and the time and effort to do them. Even without unusual complexity or extra effort to understand what to do, if the list of things to do is too long, the operator will run out of time to do them all. Running out of time automatically adds another weight to the load: what to do about the things that cannot be done? Not only what to do, but there is often an added behavior burden of feeling both overwhelmed and inadequate. "After all, wasn't I hired because they thought I could do the work?"

For these reasons and more, it is important for management to understand that they—and not the operator—have the primary responsibility to prevent overload when it is possible and to manage it when it is not. Set the culture for operators to let you know when things are getting tight. Set the expectation that you will affirmatively manage things when they do. This means that in addition to assessing operator workload, effort must be directed to uncovering and managing individual stress to a responsible level.

Work Complexity

When operations become operationally or situationally complex, your operator's understanding and capabilities may move outside his ability to properly manage. This is one of the primary reasons for the concept of *permission to operate* set out in Chapter 10, "Situation Management." Crew resource management, discussed in Chapter 8,

"Awareness and Assessment Tools," also aids in managing these complex situations. Both technologies and tools are intended to keep your operator's performance matched to situations the operator is facing at the moment. It will be up to first-line supervisors and trainers to assess the effects of complexity on operators and take steps to ensure appropriate operation. This may mean that procedures need to be reexamined for clarity and directness. It may mean that the present process design has inherent inadequacies or excessive operational difficulties. In any event, situations that lead to excessive operator attention or any sizable number of abnormal operating events should be reviewed for overcomplexity, inadequate instrumentation and controls, poor maintenance, ineffective training, and inadequate procedures and protocols.

Supervision
All operators should be more than adequately prepared and able to manage things when they go reasonably as planned. Few operators have the wealth of experience, full set of tools, and wisdom to enable them to handle everything that comes their way. And even for those few, if the situation threatens to exceed their boundary of operation or authority, others are required for guidance, assistance, and permissions. In all situations where the requirements for operation may be adversely affected if *permission to operate* were the only controlling guidance, access to supervision is required.

It must be made clear that even where supervision is directly in the loop, the requirements of *permission to operate* should never be bypassed.

Managing Operator Load
Operator loading will need to be explicitly and adequately assessed. This is also a formal requirement of proper alarm management.[13] On the one hand, loading assessments are relatively straightforward to accomplish. On the other hand, once assessed, management seems ever-pressed to increase the average load without due regard to either preserving an operating reserve to handle the unexpected, or the long-term effects of driving people "full steam ahead" for long periods of time. There is a common misconception that operator loading is set to be a balance between normal operations and abnormal operations. Studies have shown that neither is the actual case. During normal operations, most operators are loaded around 30% doing general monitoring, preparing reports, preparing and conducting shift handovers, coordinating maintenance, and such. This allows the remaining 70% to be used for a bit of training,

---

13 Douglas H. Rothenberg, *Alarm Management for Process Control—A Best Practice Guide for Design, Implementation, and Use of Industrial Alarm Systems,* 2nd ed. (New York: Momentum Press, 2018).

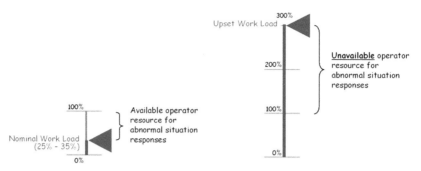

Figure 3-4. As operator loading changes from normal to upset situations, the amount of available time to attend to the problem not only disappears but without aid will overload.

responding to alarms, following through on weak signals (Chapter 9, "Weak Signals"), and other operational uncertainties.

Along comes a significant abnormal situation, and your operator is immediately transformed to have way too little time resources to understand what might be happening, decide what to do about it, and to act to remediate matters. Studies from significant planned abnormal events (major start-up or shutdown) have repeatedly shown the need for one to two extra pairs of expert hands. See Figure 3-4 for a comparative illustration.

At this point it should be noted that any attempt to set any sort of "balance" in operator loading between these two situations could not possibly work. Rather, the approach is best done from the vantage point of ensuring adequate time available to handle the manageable upsets that generally come along. When the extra reserve is not needed at the moment, use the time for tasks such as training, special reporting, or writing work orders for optional improvements. People function best over the long haul when their load is variable and reasonably far from full steam ahead. For all larger upsets, some other mechanism must be in place. Chapter 10, "Situation Management," suggests options and approaches to consider.

Operator Alertness

Twelve hours on duty is a long time. Even eight hours on shift can tax an operator's ability to remain fully alert and attentive to the myriad of responsibilities and tasks. Therefore, enterprises must find effective ways to ensure operators remain alert. To some extent, each individual operator will discover and use his own personal "tricks" and task variations. Usually they are not enough. Shift design will need to incorporate appropriate task schedules and actual physical changes. The ability to work sitting or standing can help. The provision for and ability for operators to cook their own meals is another. Close-at-hand training aids and modest exercise equipment help. Supervisor visits are helpful for both formal activities and informal coaching or just a brief kibitz about last night's ball game.

The coverage here is only meant to suggest the important need to understand, monitor, and manage operator alertness. Other resources are more suited to advise you of how all of this might be done. There is one aspect, however, that must be brought to your attention. It concerns the practice of practical jokes or teasing or baiting that sometimes finds its way into the situation. Practical jokes and teasing might appear effective but are damaging in the control room. Please institute a policy to eliminate these practices and substitute those that work.

Improving Operator Performance

The bottom line is that "[e]nhancing operator performance means tailoring the right combination of situation-aware displays, rationalized alarms, ergonomic consoles, and field-capable interfaces."[14] This is supplemented by the judicious deployment of better control applications, better operator performance measurements, active reinforcement of the need for and tools to do situation awareness, better and more relevant training, extensive use of ergonomics for operator station design, and a widespread use of now-familiar tools from other smart devices.

3.6 Operator Training

The responsibility for ensuring that all personnel are qualified, trained, and fit for their assigned job rests solely with management, specifically senior management. Senior management ensures that everyone gets all the needed training and other skills to occupy an assigned post. Although senior management may delegate portions of this task to others, including subordinates, they retain the responsibility for continued monitoring and ensuring adequacy. No person should be assigned a task or role for which he is not fully qualified. Individuals participate in a skills and training assessment. However, management has the ultimate responsibility to assess and render them fully qualified. In case it was not clear, no one in the organizational chart below the senior management is expected to self-assess, self-train, or make do with any missing skills that are necessary to appropriately perform the job for which they are assigned and responsible.

---

14 Jim Montague, "Perfect Fit: Operator Performance," ControlGlobal.com, last modified March 14, 2014, http://www.controlglobal.com/articles/2014/perfect-fit-operator-performance/?show=all.

Training and Skills

If something is rather simple in its construction, most of its aspects are easily gleaned through the exercise of ordinary skills employed using ordinary care. Things that are not simple, including the highly complex, are not so readily understood. Moreover, when we must understand them, doing so often requires specialized skills and capabilities. Industrial manufacturing enterprises, like the ones we address here for operational improvement, are generally manned by individuals from one of four plant organizational areas: management, operations, technical support, and administrative support. For operations, we usually find two groups: operators and supervisors/managers.

Most operating positions are hourly (often supported by union representation). They are recruited largely from the general labor pool in the community. At times, they are staffed by other enterprise members transferring into the operations entity. They receive hiring qualification testing, some level of individualized and/or group formal or semiformal foundation training, and a sizable amount of on-the-job situational-response training. Their job is to make sure that the production facility produces the required amount and quality of product at an acceptable material and energy consumption for those aspects under their control.

Most supervisor/manager positions are promotions of the most qualified and experienced operators. The position is usually salary. There may be promotion qualifying testing but usually not. There is little if no executive or manager training, although some enterprises do offer a general suite of salary-level training that is also offered to the general salary staff of all departments. The job of supervisors/managers is to ensure staffing of the team, resolve scheduling and other daily requirements and differences, assist/advise during emergencies, improve the general performance levels of the team, and satisfy all other administrative requirements.

The overall operations team, the frontline of the production enterprise, functions with its workers mostly reacting to the production situation of the moment, and its managers making due with an implied model of success (based on experiences as an operator, supplemented by what could be recalled from working with previous supervisors, and observed from current peers). There is very little formal continuing training and experience enhancements except what might be gleaned from daily experiences. Finally, and perhaps just as telling, throughout their careers, few real evaluative and strength-probing evaluations are done and shared with the individuals. What little is done tends to be highly variable and is generally not followed up with careful coaching and development.

Contrast the above situation with another "purpose-designed" model described next. Military entities, for all their historical use and perhaps even misuse and abuse, have very structured personnel categories. Again, we have the two functional groups: officers (managers/leaders) and soldiers (workers/technicians). Soldiers are prepared for duty through an extensive regimen. First, they are fully screened (physically, emotionally, and intellectually). Then they are carefully indoctrinated with their purpose, mission, and functional model. Following that, they are trained by experts in a dual-role configuration: first as fighting men and women to analyze first-level tactics, to perform against that role as individuals and a team, and to assume limited roles of authority as required in the heat of crisis; and second as specialists with specialized fighting skills or technical skills. Then they are assigned to a daily working team to further develop their performance skills and training. Their training and development will permit them to be reassigned to any similar unit as well as substituted for other positions within such units. Periodically, they receive additional specialized training from formal experts.

Officers are prepared for duty by the two-phase process of careful selection and then extensive training. Their selection comes not from the general pool of soldiers but from a pool of more highly educated and older citizens. They are asked to meet the same minimum physical, emotional, and intellectual standards of soldiers as well. Their training is also done by experts. First, they are carefully indoctrinated with their purpose, mission, and functional model. Interestingly enough, part of this training is done by senior soldiers. This has two effects: it provides clear skills taught by those who know it works because they have done it, and the officers are left with a new respect for the capabilities and roles of the soldiers who will serve under them when they assume command. Second, to advance in command responsibilities (rank), officers are required to undertake periodic formal advanced training. At the upper levels, this training will be as extensive as any advanced, postgraduate degree in the non-military workplace. Within their large sector of operation (including infantry, air, and armor) they are fully interchangeable with other officers and often rotate units as they advance significantly.

This discussion does not suggest that industry immediately adopt the military organizational and training model. But, by introducing these aspects to the industrial manufacturing thought, perhaps important ideas for change might receive consideration. Moreover, the understanding of training strategies and the appreciation of plant complexity are vital to the establishment of effective rules, procedures, and training. Now that the responsibility part is clear, let us move on to the actual training part. There are two general forms of training: skills training and competency training.

Skills Training

Skills training is the "dexterity or coordination especially in the execution of learned . . . tasks." A *skill* is "[t]he ability to use one's knowledge effectively and readily in execution or performance."[15] Skills are acquired by review and practice of specific tasks in order to master them. For example, when network routing centers experience an overload, the procedures will be referenced and carried out as designed with care to accommodate the current state of equipment and available alternate routes. Another example would be the process of shifting a major product flow from a main pump to a backup one without unduly disturbing the flow characteristics. The advantage of acquiring skills is the certain knowledge that the specific task or tasks trained for will be executed competently. And so long as the procedures are accurate and robust, this would likely be the case.

On the other hand, if the operator is a bit "rusty" on the procedure or something unexpected pops up that is not covered, there is the likelihood for missteps. It is a bit like using a map to navigate. Good map-reading skills and keeping an eye on where one is will enable navigating to the intended destination. However, if there is a map error or one somehow moves out of the map area into a region where road names change every few blocks, all bets are off. Getting back on the map might be the only recourse for this training, but getting back would depend on an entirely different capability. An ability to use ground features, the sun, stars, and moss on trees becomes essential. The ability to use dead reckoning will be vital. These latter tools belong to a class we term *competencies*.

Competency Training

Competency is "[a] valid, observable, and measurable representation of the knowledge, skills, and attitude demonstrated through behavior in a specific job, which underlies and drives optimum job performance."[16] Competency is foundational training that melds core skills and a deep understanding of underlying concepts and processes. It is training that is core and portable so that it carries over from mission to mission and situation to situation. The following partial list of examples should help you appreciate how this works.

Competency is the ability to

- follow a procedure and recognize and properly understand and react to differences in degree of specificity and generality and accommodate missing or misleading aspects;

---

15 "Skill," Merriam Webster, accessed July 14, 2018, https://www.merriam-webster.com/dictionary/skill.
16 Technology Training Systems, "Abnormal Situation Management Training for Process Operators—A Competency-based Approach to Sustained Performance Improvement" (student handout notes, Denver, CO: Technology Training Systems, 1998).

- recognize when an activity or situation is not behaving as expected or appropriately;

- understand and manage complex situations in a way that manages effective risk for operational safety;

- understand and practice the ability to remain situationally aware during the entire shift without undue reliance on alarms;

- set up and manage process control information on active display screens to ensure appropriate visibility of needed information and controls;

- understand and effectively practice the process of collaboration;

- understand and effectively utilize the processes for escalation; and

- manage the effects of fatigue or other impairments to alert operation; when this is not possible, properly and promptly alert relief resources.

The above examples illustrate the idea.

"Personal" Tools

Little of what we do in everyday life follows "the book," so to speak. Some of us carefully read instructions for a new appliance or a newly acquired software program. But most of us do not. We just jump in and try working our way through the basics. Most times we seem to get to where we want to be, if the product is adequately designed and allows for intuitive action. But once in a while, someone shows us how to do what we were doing intuitively, the way the instructions would have suggested. Wow, this method is so much easier and faster! Before, we were using our personal interpretation of how to do it. For the most part, it worked. But remember how there was something you think you should have been able to do but could not? Or another thing that you tried to do but it never turned out the same each time you tried? We were using "personal tools." Personal tools might be just fine personally, but they are unacceptable in the control room.

Management must challenge all "private" and "personal" tools developed by operators. While they may seem useful and appear important, they must be demonstrated to be in concert with the established operations technology, procedures, and good protocols for utilizing them. This is not to say that all personal tools must be disallowed; it is that each must first be formally evaluated. The few that are found to be beneficial may be rolled into the infrastructure through the established management of change process. The rest are to be quickly removed and proper ones explained and reestablished.

This "challenge of personal tools" is one of the core safeguards for training. Failing to address this challenge is a large contributing factor to the failure of on-the-job training (OJT). Management should insist that no OJT be used unless done by the most experienced and qualified trainers. When the operator who has been sitting in the job for a year or so attempts to provide OJT for others, he passes on all the errors and misunderstandings from his earlier training as well as some new mistakes (and personal tools) of his own.

A good illustration of how OJT can begin with good intentions and eventually end up in a very unacceptable place is the children's game of Telephone. In Telephone, everyone sits in a large circle. The first child starts off with a "telephone call" (whispering in the ear) to his neighbor with a clear and explicit message. It may be a single sentence. The neighbor "calls" his neighbor and repeats the message as accurately as possible. This continues until all in the circle have been called. By the time the original caller is called by the last caller, the message received is almost never recognizable as coming from the one he originated. The reason for the change is entirely understandable: people hear messages not as they are said but through the "filter" of their experience and expectations. For example, the phrase "in hot water" might be heard as "in trouble." But "in hot water" was meant to be literally in hot water. The listener missed just a bit of the part that went before and understood the phrase "in hot water," but he did not use it because his family always used the more direct "in trouble." Because the listener had to remember the message, it was much easier to remember "in trouble."

The children's game of Telephone and the "in hot water" example are far afield from real control rooms. Yet their message is clear. Control room management must be comprehensive. Small aspects have considerably more importance than they seem to have at first glance.

Process Understanding

An operator must have a proficient and clear understanding of the plant or process he is in charge of operating. A "passing understanding" is definitely not where this is going. Sure, it might appear that the chemistry or mechanics behind the plant or production activity is too complicated to fully master. From a process design point of view, that is right; that level of understanding is not needed for operators. Operators need to know how the plant or operation they are managing reacts to the full range of normal and abnormal operation. They need to understand the full range of operational faults that may result from an inability to maintain safe and responsible operational conditions.

3.7 Qualified Operator

A *qualified operator* is an operator or controller or other individual placed in charge of the operation of a plant or other entity who possesses the necessary required knowledge and skill to operate safely (safely includes without injury and environmental exposures). While it is up to the management of each plant or enterprise to set the qualifications for their operators, there are important requirements to meet. A latitude is intended to allow individual enterprises the ability to tailor the requirements to their specific needs; it does not permit avoidance of the full measure of responsible design. The best working definition of what constitutes *qualification* can be found in the Pipeline and Hazardous Materials Safety Administration (PHMSA) regulations for pipeline operations. In the following excerpt, I have substituted the term *managing owner* for the equivalent PHMSA term *operator*. No other regulatory or compliance entity has provided alternate definitions. For our use these regulations are the best available. Reliance on them would be responsible.

> [The PHMSA rule] requires pipeline managing owners to document that certain employees have been adequately trained to recognize and react to abnormal operating conditions that may occur while performing specific tasks.
>
> [Note: To be qualified,] an individual must be able to properly perform assigned covered task(s) and be able to recognize and react appropriately to any AOC [abnormal operating condition] that may (reasonably be expected to) be encountered while performing the covered task—whether the condition arises as a direct result of his/her work performance (e.g., be specific to the covered task being performed) or not (e.g., be generic in nature, but still observable because the individual is present on site).[17]

The basis of the requirement is to ensure that pipeline controllers (*operators* in a general context here) are properly trained to ensure safe and reliable pipeline operation under their watch. They must possess adequate skills. The skills include the items listed below. The reference to "specific tasks" is understood to encompass all activities and operational states that are "safety related," which are broad and important.

Brief Glossary

Here are a few terms that are carefully defined. They are taken from the PHMSA guidance mentioned earlier. We do this defining because, while they each have an intuitive meaning, it is important for us all to agree on exactly what they mean. This clarifies what we really mean when we use each term throughout this book. We expect the operator to be fully *qualified* by possessing the skills to operate effectively by being able to *observe* and correct for all abnormal operation including *safety-related* ones using the available *resources*. Only qualified operators should be entrusted with process or plant operation.

Resources

> [This requirement is] intended to assure that [management] would provide [operators] with accurate information and the training, tools, procedures, management support,

---

17 49 CFR 195.450-452, *Definitions, Pipeline Integrity Management in High Consequence Areas* (Washington, DC: PHMSA [Pipeline and Hazardous Materials Safety Administration]).

and operating environment where [an operator's] actions can help prevent accidents and minimize commodity losses.[18]

Observe

The act of watching; to watch or perceive. For purposes of conducting qualification evaluations using on the job (OTJ) performance, observations must include the interaction of the evaluator and qualification candidate to ensure that the candidate's knowledge of the specific task requirements and procedures (and the reasons for key task steps) is adequate to ensure the continued safe performance of the task.[19]

Operate

Starting, stopping and/or monitoring a device or system.[20]

Safety Related

PHMSA considers safety-related to mean any operational factor that is necessary to maintain pipeline [plant] integrity or that could lead to the recognition of a condition that could impact the integrity of the pipeline [plant], or a developing abnormal or emergency situation.[21]

Skill

A demonstrable competency to perform a given task well, arising from talent, training or practice.[22]

Qualified

An individual has been evaluated and can (a) perform assigned covered tasks and (b) recognize and react to abnormal operating conditions.

. . . a written qualification program that includes provisions to identify covered tasks and the intervals at which reevaluation of the individual's qualifications are needed.

Requiring additional [operator] qualifications to measure or verify a [operator's] performance, including the prompt detection of, and appropriate response to, abnormal and emergency conditions likely to occur.

The intent . . . was to ensure that [operators] would have the necessary knowledge, skills, abilities, and qualifications to help prevent accidents.

Requiring additional [operator] qualifications to measure or verify [an operator's] performance, including the prompt detection of, and appropriate response to, abnormal and emergency conditions likely to occur.[23]

---

18 49 CFR 195.450-452, *Definitions, Pipeline Integrity Management.*
19 49 CFR 195.450-452, *Definitions, Pipeline Integrity Management.*
20 49 CFR 195.450-452, *Definitions, Pipeline Integrity Management.*
21 Byron Coy, "Control Room Management," presented to the Southern Gas Association, Jacksonville, FL, July 2011.
22 49 CFR 195.450-452, *Definitions, Pipeline Integrity Management.*
23 49 CFR 195.450-452, *Definitions, Pipeline Integrity Management.*

Roles and Responsibilities

Each [enterprise] must define the roles and responsibilities of [an operator] during normal, abnormal, and emergency operating conditions. To provide for [an operator's] prompt and appropriate response to operating conditions, [management] must define each of the following:

(1) [An operator's] authority and responsibility to make decisions and take actions during normal operations;

(2) [An operator's] role when an abnormal operating condition is detected, even if the [operator] is not the first to detect the condition, including the [operator's] responsibility to take specific actions and to communicate with others;

(3) [An operator's] role during an emergency, even if the [operator] is not the first to detect the emergency, including the [operator's] responsibility to take specific actions and to communicate with others; and

(4) A method of recording [operator] shift-changes and any hand-over of responsibility between [operators].[24]

Message of Operator Qualification

The bottom-line message is that management must ensure that only qualified operators manage control room operations. No one, not even supervisors or managers, should attempt to *require* a sitting operator to conduct an action or engage in a control or management activity unless they themselves have met the requirements for being a qualified operator. Otherwise, their guidance is limited to just that, guidance. While the details of what constitutes adequate qualification are left to management to determine, the requirements for an effectively qualified operator are quite clear. This section has attempted to lay out those essential needs.

3.8 Operator Tools

Operator training, basic skill set, and tools go hand-in-hand. In this section we discuss the basic tool set. All of these tools should be quite familiar to you, so this is not going to be a review. Rather, the purpose of this discussion is to nail down the advantages and disadvantages of each. We do this to help you appreciate when one tool might work better than another. It should also aid you in shaping the specific uses to better suit your intended purposes. Of course, there are more tools than the ones mentioned here. The general format for discussion is to lay out the following:

- What the tool is all about
- Why the tool is useful/important

---

24 49 CFR 195.450-452, *Definitions, Pipeline Integrity Management.*

- How to utilize the information in the tool
- Benefits and limitations of the tool
- How the tool fits into the job of situation awareness, situation assessment, and situation management

Checklists

A checklist is a specialized tool that is usually a part of a larger procedure or protocol. It consists of a specific list of required items to check, the order of the checking, and the condition to be observed or confirmed against. The main purpose is to ensure that every item that needs attention will be attended to. A checklist is useful for routine tasks that might be performed constantly. It is useful for tasks that might only be performed once. It is extremely useful for high-stress situations where the likelihood of something being missed or done out of order is important to control. A checklist is compact tool that combines what to do, the specifics on how it must be done, and documentation that it was done, all in one place. The immediate ticks (√) and any accompanying annotation document the progress to ensure tracking of what has been done and what has not.

One of the dangers of using checklists is that they will be completed perfunctorily without reviewing the full item and without doing a full check to see what's what. Training and insistence on proper use helps. For one or two vital steps or items, an additional user entry (not a simple check) can be requested at the end of the checklist item line. Missing entries there indicate an incomplete or perfunctory check.

The checklist must meet these structural requirements:

- **Clear** – Every item must be written with the conventional terminology, phrasing, and expressions that are unambiguous and understandable to the expected user. Any bilingual aspects are to be handled by enterprise policy or according to applicable statute.
- **Complete** – Every item that is required to be checked or performed must be explicitly on the list. Even if an item or task is minor or is included as a part of many other items, it must be listed explicitly. If order is important, a strict order matching the requirements must be specified.
- **Identified** – Where a checklist is used, the process must be documented as to when it is being used, on what is it being used, by whom, and for what purpose.
- **Fully optioned** – There must be space by each item on the list to check or documented to match the found conditions. For example, if an item's status cannot be verified or is normally important but today is not, there must be space

to document this situation on the checklist. If an explanation is required to appropriately denote the meaning of the specific "check," it must be present and adequate.

- **Explained** – If special instructions are needed to check any line item on the checklist, those instructions must be present, close at hand, or fully referenced and a note added to document that those instructions were referred to as this item was evaluated or checked.

- **Contemporaneous** – The checklist must be immediately at hand so each item evaluated or checked can be entered on the checklist without remembering or jotting down notes or other reminders for later entry on the list.

- **Operationalized** – Where certain conditions must be present in order for the list to be used, those conditions must either be built into the list (e.g., by having a different form of the list for each condition) or be able to be added during the check-off process (e.g., by tracking date and time where the checking process might be paused or interrupted).

- **Archival** – Every aspect of the checklist must be sufficiently expressed, documented, or annotated so that anyone, at any time later will be able to fully understand what was checked and the condition for the check has been entered.

Protocols

A *protocol* is a description of a general process for doing a specific class of operations. For example, there might be a general protocol for starting up equipment. It would apply to all equipment. It would contain items like locating the controls, checking with lock-out/tag-out to determine if one is required and properly in place, checking work orders to ensure that it is not on one, calling for additional assistance if such is usually required, and checking other aspects common to all or nearly all equipment start-ups. When procedures are to be used, the protocol will indicate what specific procedures are to be used and where they might be found and verified current. The protocol will specify whether there are any global operating situations that affect the task.

The usefulness of protocols is that there are less of them and they are broadly applicable. Consider that they provide the upper-level framework within which procedures are properly used.

Procedures

A *procedure* is a fully explained guidance on how to conduct a specific operation or task. It is generally written in a narrative form. This means it would be read and then followed. It may be designed for a step-by-step use, either fully or partially. Different from checklists discussed on the previous page, each procedure may reference the applicable protocol rather than include it within the document.

Procedures include the following attributes:

- **Applicability** – Under what conditions or situations the procedure is to be used.
- **Intent** – Exactly what the procedure is for.
- **Explicit** – Exactly what to do, how to do it, when to do it, and who should do what. Documentation and reporting requirements are listed and explained or referenced.
- **Expected results** – What to look for when things are correctly done; what to look for as signs of problems; any special confirmations needed to move to the next step or pause (indicate why and how long) and then resume; what to look for as indicators that the step should be skipped or the entire procedure stopped or ended. If the step or procedure is stopped or ended, there should be specific instructions on how to stop or end it.
- **Limits of operation** – What are the safe limits of operation that must be respected regardless of the process or activity under way; what are the preferred limits of operation that should be used as indications of normal?
- **Complete** – Designed so that faithfully following it should produce the expected results (unless obtaining the expected results is not possible due to existing problems or outside effects).

Simulators

Simulators are predominantly thought of as training aids. And if their quality and fidelity are good, they are excellent for training. They are also well suited for engineering and other design and testing uses. For example, before a new alarm management design is implemented, simulation can provide a check on the design as well as be used to produce training aids for both the new alarm system and the redesigned procedures and protocols.

However, simulators can also be extraordinarily useful during actual operations. For example, if an operator is planning to make an operational change that might have broader implications than usual, say due to unusually high loading or a tight schedule, simulating various ways to make these changes would readily provide a great deal of feedback as to which alternatives might be more appropriate. All of this happens without disturbing the actual process or operation.

You might have been expecting this discussion to lead off with using simulators to help handle abnormal situations. They can handle such situations, but it would be quite unlikely that the operator would have enough time to make use of them. If time permits, they could be useful to search for causes of the abnormal conditions. Usually, this means "guessing" a problem and using the simulator to determine if the expected abnormal situation develops for it. Also, if the problem is known, various remedies might be simulated to determine which ones work, and which ones might work and give better results. All in all, the time constraints of managing upsets and abnormal situations are usually too demanding for using simulations in real time. There is one notable exception: if other personnel are available and they have access to simulators and are trained to use them, they might use them while the operator works the problem so long as the expected available time for operator success permits. Yes, this is a long shot but it is another potential use for simulators.

Reports

The first thing that comes to mind when we think of a *report* is probably not an operator tool. Reports are for the managers, or the government, or for the customer, right? Aren't reports the necessary evil of working that go to other people? Well, yes. But reports are really communications. They are intended to capture information in an organized way (the report form or format) and make it available to others. The operable part of that is (1) capture, and (2) available to others. Properly designed and appropriately used, reports ensure that the "stuff" that was important during production or operation is captured and available. Now everyone can rely less on remembering (we rarely remember what really happened), and the report can be shared with the writer later on, with others, and searched, if in electronic format.

3.9 Shift Handover

The shift is changed for a single reason—to provide a replacement operator to perform operational duties when the existing operator will not continue in that role. Handover is the process for doing that replacement in a controlled way. Depending on shift lengths, for 24/7 operation, shift handover will occur two or three times each day. Thus, handover is one of the most—if not the most—important shift communications events.[25] For success, it requires the *leaving operator* to fully assess what happened on shift, what is likely to greet the *arriving operator* coming on duty, and suggestions for handling those issues. The handover must be carefully designed and competently

---

25 Center for Chemical Process Safety (CCPS), *Conduct of Operations and Operational Discipline: For Improving Process Safety in Industry* (New York: Wiley, 2011), http://www.aiche.org/ccps/publications/books/conduct-operations-and-operational-discipline-improving-process-safety.

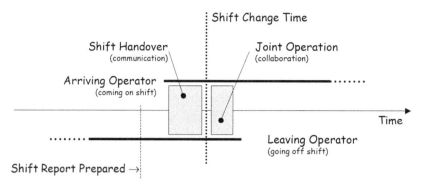

Figure 3-5. Shift handover timing activities showing first communication to transfer knowledge followed by collaboration to transfer ownership.

executed to ensure the arriving operator has an opportunity to orient himself and come up to full speed. This must happen every time.

Figure 3-5 shows the time overlap periods for control room operators with responsibilities identified. Note that the *communication* part before handover transfers operating responsibility and the *collaboration* part after transfer ensures a smooth transfer of ownership. During this collaboration period the leaving operator assumes the role of coach and "second pair of eyes" briefly after transfer before going off shift. No times or durations are shown on this or any other timing diagrams in this chapter. There are general conventions, of course. Before-shift handover overlap might be 15 minutes to a half hour. The after-shift handover joint operation overlap might range from 5 (a bare minimum) to 15 or more minutes. Of course, if there are operational situations or events to manage, such as a start-up, shutdown, or abnormal situation, these times should be adjusted "on the fly" according to established procedure. This section will cover all three types of handovers:

1. Control room operators
2. Field (or outside) operators
3. Operations supervisors

While the personnel differ, of course, the handover act and actions are remarkably similar. Thoughtfully coordinating them will significantly improve their effectiveness.

Reasons for Shift Changes

Relief is required when the existing individual is to be relieved due to

- sufficient hours of duty completed (end of planned shift);
- "temporary" shift change due to operational or personal need to leave the control board for a short time, where manning of the control board is required at all times;

- inability to continue duty due to inability to remain vigilant, illness, injury, personal emergency, or unauthorized absence; and

- inability to remain in the designated physical area due to weather, fire, physical security breach, personal needs, and other circumstances.

Except for field operators, not all shift changes need to occur between individuals located at the same physical location. Operators and supervisors may begin duty at other locations prearranged and designed for them. One example is where day operators use one control room and night operators use another. A second example is where remote central control rooms are located around the globe in a "follow-the-sun" configuration. Here each control room can be manned during the normal day shift for its time zone. Another typical example is emergency operation due to weather or other situation. The complete contents of the reporting and reports prepared for these situations are left up to the specific site style and standards.[26] However, the requirement for providing a full shift assessment must be met. This section will discuss how shift change assessment fits into the overall picture of situation management. As you read through this section, you will notice a lot of similarity between your activities and those proposed here. That is as it should be for a well-thought-out practice. You will also note differences that exist but may not have been expressly required or performed. Two will be pointed out: fitness for duty assessment and overlapping joint operation. These are to be conducted at all handover events.

There are important requirements for doing readiness assessments of fitness for duty and for travel home. For example, in the transportation industry (those regulated in the United States by the federal Department of Transportation), there is a specific requirement to assess any fatigue or other potential impairment of the controller (what we are calling the control room operator) coming onto duty as well as of the one leaving to make it home safely. For large control centers, supervisory personnel would usually conduct this task. Where supervisory personnel are not available, the individual operators must do this. This section presumes that supervisory personnel are not necessarily available.

There is also a strongly suggested practice for a limited period of joint operation. This means that after the arriving operator takes active control, the leaving operator remains in the control room to assist. The reason for this is the mounting evidence[27] that it takes

---

26 API. RP-1168, *Pipeline Control Room Management* (Washington, DC: API [American Petroleum Institute], 2008).
27 Mica R. Endsley, Betty Bolté, and Debra G. Jones, *Designing for Situation Awareness: An Approach to User-Centered Design* (Boca Raton, FL: Taylor & Francis, 2003).

some time for the arriving operator to fully incorporate everything being managed. Yes, the shift handover communicates what it must. But taking over is much more extensive. It takes a short while before everything is "put into mental place," so to speak. Both operators should perform a "think through" that carefully reviews all specific tasks planned for the next shift. Here is where they mentally simulate the activity and check on their plan. For demanding operations or those in a bit of an upset, more than ordinary diligence is needed. The assistance of someone fully up to speed can make all the difference. Figure 3-6 shows a shift timeline identifying the various handover activities and when they might best occur. Notice that this diagram is constructed so that time starts at the top and increases to the bottom. We begin with control room operators.

The timeline starts at the top and flows downward. The leaving operator finalizes the handover report before the arriving operator arrives. Information, notation, and

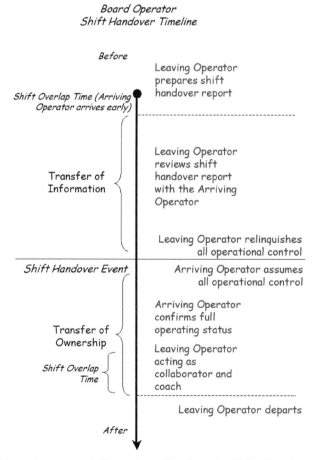

Figure 3-6. Detailed board operator shift handover timeline identifying leaving operator preparing the handover report, sharing the report with the arriving operator, transferring the operational control, and sharing joint operation and collaboration leading to the arriving operator assuming full ownership of shift.

observation activities occur throughout the shift. These activities can be significantly aided by any of the many shift handover reporting tools commercially available. Operators best remember the critical details if they are captured as near in time to the event as possible. The report and all other needed information is shared with the arriving operator in a give-and-take atmosphere. Operational control passes to the arriving operator at shift change time. This is followed by a period of operation with the arriving operator as lead and the departing operator as collaborator and coach. The arriving operator assumes full ownership of the shift at the end of this overlapping period. The departing operator leaves unless the plant is upset and extra hands are needed.

One must understand where the operator's role begins and ends. Once it begins, the sitting operator is in charge and is expected to exercise that responsibility according to enterprise requirements and established procedures and protocols, using responsible and prudent stewardship for the duration of the shift. Once it ends, all operational activities and responsibilities for this operator cease for the shift.

Beginning of the Operator's Shift Role

The operator's role begins either formally or (rarely) informally:

- Formally, through a shift handover and subsequently taking charge from the current operator.

- Informally, by showing up and beginning duties for which he assumes responsibility. This type of shift change is unusual. If used, it must be expressly permitted by the plant or enterprise and have some type of functional connection to key personnel with operational responsibility. Otherwise, there should be no informal shifts. It is not a preferred or even good practice.

- Casually, by just being there and observing or not, acting or not, assuming responsibility or not; all with the ability to do so whether or not he has the training or authority. This type of shift change is unusual. It must be expressly permitted by the plant or enterprise. Otherwise, there should be no casual shifts. It is not a preferred or even good practice.

The usual way we think about this is at the start of a shift. For operations that are 24/7, it happens in such a routine and matter-of-fact way that we hardly notice it as an event. It is like getting dressed in the morning; you just do it. You make some decisions in the moment (e.g., what to wear depending on the weather) but all is quickly placed in the bin of "unimportant and unremarkable." It gets much more interesting for 24/5 operations or 8/5 operations. Here we have shifts where the operator is not relieved or comes to duty when no one is presently there, either because the operation is shut down or simply left (purposefully or not) unmanned. See the topic "Noncontiguous Handover" later in this chapter.

A practical illustration of what it might feel like to an operator to come on duty without adequate briefing is something we all come across almost daily. We tune into a news program on television or our smart mobile device about a minute after it has started. We can see that something is going on; we get a general nature of it from the way the announcer is reporting it or the drama being played out on the video clip in view. But we do not know for sure what is going on until we either watch long enough to piece things together or the program itself recaps things. Once we do know the frame for the piece, often the pieces start to fit together in a different way. How much more important having that frame must be for an operator of a vital, dangerous, or valuable operation.

It turns out that something that seems as simple on the surface as shift handover is actually extremely important. Handovers require focus and effort. Both the acceptance of operational responsibility (coming on duty) and the relinquishment of operational responsibility (leaving duty) are events that must be carefully constructed and faithfully executed. This is because of the following:

- Enterprise operations have a history (what happened in the past). Anyone taking charge of an enterprise must learn that history and operate with its knowledge. Anyone leaving charge of it to others (either proximately or at a later time) must add his notes to the history.

- Enterprise operations have context (the current operating state). Anyone taking charge must know the context and operate with that knowledge.

- Enterprise operations that require operators to be present must be able to count on operators actually being present, functionally ready, and capable.

Shift handover has evolved to be one of the critical operational responsibilities for managers and operators alike. A shift handover that meets all requirements laid down for that process must include the final formal transfer of responsibility. The arriving operator must declare to the leaving operator words to the effect, "I am now operating this unit."

Ending of the Operator's Shift Role

The operator's role normally ends formally, through a shift handover and subsequently relinquishing charge to the new operator and the new operator accepting responsibility. This is the expected way. Established procedures, protocols, and training would be designed to handle this. However, there are other ways for the shift to end that need understanding and management. Here are a few:

- Formally, through a shift handover.

- Situationally, through the individual becoming sufficiently impaired so as to not be able to perform responsibly (including sudden illness, accident, or emotional shock).

- Functionally, by an abnormal situation escalating into an incident or worse (thus transiting into a direct responsibility of the incident management or other responsible process).

- Informally, by completion of limited duties for which the individual is properly qualified and not barred from doing, but the duty was assumed informally by being in the right place at the right time. In any event, this activity must include notification to appropriate personnel before leaving the area.

Plants and operations centers need to have protocols in place to handle all the aforementioned end-of-shift possibilities. If able, the leaving operator must declare to the arriving operator words to the effect, "I am no longer operating this unit." Until both operators accept the change, it has not been done and the arriving operator remains in full responsibility. If the leaving operator is unable to actively participate, the arriving operator must validate assuming responsibility through established escalation policy.

This chapter will focus on normal shift handovers.

Functional Components

Providing checklists and writing specific procedures and protocols are beyond the scope of this work. This discussion develops requirements. Starting with the requirements provides valuable insights into what is really needed to do that handover effectively. Understanding why these requirements exist will be critical to your ability to interpret what they mean for your enterprise. Then, you can develop the process and write the procedures with a renewed assurance that your operators will realize the full benefit.

Start this with the core value and power of the shift change process. Continue by taking for granted that all operators are fully qualified and trained to do their assigned jobs effectively and efficiently. If they are not, fix it. Next, a successful handover depends on the plant data being accurate and available. If the data are not, fix that. Finally, understand that the single reason for changing operators is physical—replace the current operator with another. But there is more to this. And that more is where the synergy starts. Here you have a real opportunity to approach this from a different direction, and by doing so, you gain valuable opportunities to bring to your enterprise. This activity of exchanging operators goes beyond a mere change of physical persons with a good briefing and fond hopes for the best.

The report-building process must be deliberate and diligent, not tedious. Much of the information will come from a shift log created by the operator pretty much as things happen during the shift. It generally consists of a run-of-time list of entries of everything notable that occurred during the shift. However, the shift log is not used as is for the handover report. Instead, it provides the raw material for the report. There are four basic components: report the full status, show up ready, understand and own the shift, and participate in short joint operation. Each component will become the basis for understanding and developing efficient and effective handovers. As you read, please keep in mind that they may start along familiar lines, which is of course comforting. However, each contains valuable elements that go beyond current practices; in a few cases, well beyond.

First Component: Prepare Full Status Report
Toward the end of the shift, the to-be-relieved (leaving) operator sets out to consolidate the documentation (paper or electronic or a combination) of the shift for the relief. The shift handover report at first seems like a cut-and-dried activity, yet it is one of the most challenging tasks. The operator must capture a complete history, identify those items that are important, and place it all in proper operational context. Most sites have an itemized list of what to write down. Many use a form to make the job easier. Here we set the requirements for those lists and forms.

> The sum total of all information in the shift handover report must be such that an arriving operator will be able to assume informed consent ownership of the operator position upon taking over.

This means that the arriving operator should be able to be at full speed from the start of independent operation. It does not mean that we expect the arriving operator to *begin* situation assessment after sitting down and spending the first half-hour checking this and checking that to understand and orient. One of the best ways to really understand what this component means is to reframe the objective in our mind. Reframed, it asks, what does it take for the current operator to build a coherent evaluation of where his plant or operation is that exposes as much as is known about the operational status? What might be any unmet objectives (projecting objectively to end of shift, if that is appropriate, or end of campaign, if that is what is going on)? What might be any suspected or actual threats to good operation in progress or coming down the road? And what enterprise coordination must be kept in mind? Once all are understood, build procedures, protocols, lists, and competencies.

Before we leave this topic, there is another quite useful benefit. Start thinking about this benefit by remembering how we developed the components in the first place. We asked the question, "what does the operator need to fully understand where, at this moment, his operation or plant is?" And we used the answer to develop a comprehensive plant or operation status for shift handover. So, if it is that comprehensive, what about using the same framework for check-pointing operation during a shift? What if we asked the operator to do it, say midway in the shift? What if we asked the operator to do it after an upset, when everything was leaving *abnormal* on the way to *normal* again? A review here would ensure that the upset had actually recovered and was not just looking better on the way to something amiss or worse. This could provide another useful structural tool for the operator to be able to move through a shift and effectively keep an eye on things in a carefully planned way. By also incorporating this comprehensive check-point into the shift work process, it should increase the operator's confidence to keep track of things that matter. It would be another tool for more predicting and less reacting.

Second Component: Physical Presence
The arriving operator must be present. Wherever the leaving operator is functionally located, the arriving operator must also be in the self-same place. The exception is where a shift handover is authorized and will take place with one or more operators that are not colocated. Examples include shifting from a main control room to a backup one, or shifting from a control room to another operator located elsewhere. For a shared physical control room, the arriving operator will come to it and the leaving operator will leave it. This is about as straightforward as it gets. Yet, there is more. An operator is of little value if he is present yet unprepared or unfit for duty. Therefore, all shift changes must include a proper fitness-for-duty evaluation. That evaluation must be sufficient to assure that the operator is unimpaired and ready. Moreover, the leaving operator must also be evaluated to ensure the ability to responsibly arrive at a chosen off-duty location (normally, the residence or other quarters). Refer to the earlier Section 3.5, "Operator Readiness," where this was introduced.

> Both the leaving and arriving operator must be present during the handover and both must be unimpaired for the duties they are responsible for and conducting.

Before leaving this component, there is a not usual possibility for control room location. It is entirely possible, given the current or near-future state of technology, for the control room to be in different physical locations for some or all operators. Each physical control room can be very differently sited. The differences can be for the convenience of individual operators who will work close to where they live. The different sites can be for the safety and security of the operational team and equipment, for example, where major weather disruptions can affect both the ability to man a facility as well as the potential physical and functional security of the facility. The different sites may be far-flung in a "follow-the-sun" design so that every operator works a day shift. This, of course, means sites scattered around the world. Regardless, principles of shift handover apply to all.

Third Component: Understand the Status, Conduct a Second-Sight Review, Take Operational Control and Ownership

Either before arriving at the control room (unusual, but possible) or on arrival (the norm), the arriving operator will be fully briefed on all information required by plant or enterprise procedures, including the shift handover report and all that is provided personally by the leaving operator. Discussion regarding what is happening and potential threats should focus on worst-case potentials, not best-case or likely situations. This part is complete when the arriving operator is familiar with everything. However, the familiarity part has a mirror part. After becoming familiar with everything, the arriving operator is expected to reassess everything to ensure that what is believed to be the status of the plant or operation is actually the case. Together the arriving operator and leaving operator think through the plan for the shift, walking it through in their minds. This should include ensuring that the arriving operator's mental model of what is happening aligns with the leaving operator's. The arriving operator is encouraged to "ask questions and rephrase material"[28] to expose any differences in understanding between the operators. This overall second-sight portion is essential for the arriving operator to be able to take proper operational control and ownership. A full review of alarms would be done here.

> The leaving operator is responsible for transferring all shift handover information to the arriving operator. Together they review the specifics for the next shift. The arriving operator is responsible for taking informed ownership of the operating chair.

---

28 B. Parke, and A. Mishkin, "Best Practices in Shift Handover Communication: Mars Exploration Rover Surface Operations," in *Proceedings of the International Association for the Advancements of Space Safety Conference*, Nice, France, October 25–27, 2003, http://human-factors.arc.nasa.gov/publications/Parke_MER_SurfaceOps_Handovers_05.pdf.

Fourth Component: Overlapped Operation

It has been observed that operator error rates are highest during the first few minutes after a shift handover.[29] Errors do not always lead to incidents, but errors are the major contributor to them. No other correlated error-prone periods are observed at any other time of the shift—certainly not at the end. That they occur during the very first part of a shift is sufficient to raise concerns. The difference between the very first part of a shift and the rest of a shift is clear: the arriving operator needs some time to come up to speed. Most objective studies indicate that this error-prone period lasts about 5 minutes. Five minutes is enough time for a mistake to have serious consequences, yet it is not a long time.

A simple remedy is close at hand. It is time to take advantage of the fact that at shift handover both the leaving operator and the arriving operator are present. During this overlap, the leaving operator is still at the job; he remains responsible for all operations until the moment of relief. He is fully aware. Let the shift handover execute. The arriving operator will take charge. But let the leaving operator remain as an observer, coach, and consultant for a brief but defined period afterward. Let the arriving operator be a second set of eyes. When both operators are sitting together is an excellent time for the leaving operator to see cues and other reminders of points or observations that were either inadvertently left out of the shift change reporting or appeared to be unimportant then but may have relevance now. Those could be readily shared. This is the perfect opportunity to identify and work weak signals (see Chapter 9, "Weak Signals"). As soon as the shift change transition has time to settle out (a few minutes unless problems appear), the sitting operator can go it alone.

> Leverage the fact that both operators are present at shift change to reduce the exposure to beginning-of-shift errors by briefly overlapping operators at the board after handover completes.

If all of this seems to be too much, consider how this fits into operator professionalization. First and foremost, please leave trust out of this. Not for a moment is anyone distrusting the arriving operator. His competence and ability to come to the board and take full and appropriate charge is not at question. This is about developing and using an important and beneficial activity. Let us take a brief look at the general skills needed for operators. If the situation required, he would be

---

29 Endsley, Bolté, and Jones, *Designing for Situation Awareness.*

part of the extra operational support on planned events such as unit start-ups or shutdowns. He would support other operators if called in during major incidents or accidents. He would be part of the training process for new operators. In short, the skill set for operation includes the ability to observe others in a supporting role and coach and consult as part of that support. A natural way to keep these skills sharp is practicing them at every shift handover as part of overlapped operations. It fits into an enterprise's goal of operational responsibility. Exploit this for your benefit.

Noncontiguous Shifts

A noncontiguous shift is when the end of one shift does not mean that the next shift will begin immediately. This type of shift is quite common for industries such as manufacturing. These industries have a need for continuity, responsibility, and effectiveness that is quite similar to that for contiguous shifts (24/7). There are two general situations: (1) the same operator resumes duty for the next shift as was on duty for this shift, or (2) another operator resumes duty for the next shift. The actual handover operation is functionally quite similar.

Upon returning for duty, the operator reviews the last end-of-shift report completely. It goes without saying that each end-of-shift report must be sufficiently complete to be fully utilized by another operator different from the one preparing it, even if the same operator is expected to staff the next shift. The resumption of the shift will verify present operational status against the expected status. All anomalies and differences will be examined, explained, accounted for by remediation if needed, and appropriately documented. The operational plan for the shift will be reviewed if it was previously determined or set if it was not already. If all is ready, operations begin.

Logs and Reports

Logs

A log has two main characteristics. First, a log provides a frame to capture all that happened with situational notes and any control system aspects. Second, it organizes the information in a structure that readily breaks into categories. A list of categories might be (1) run-of-time items (items arranged in sequence by the time they occurred) that need to be archived; (2) events and/or items that require follow-up by others, not operators; (3) events and/or items that require follow-up by the next operator; and (4) events and/or items that specifically frame the operating situation for both the current operator and follow-up operators in a way that assists in interpreting current or future events or modifying current or future plans.

Reports

The contents of the reports will rely on specific site style and standards to deliver the necessary full assessment. The more these reports are integrated into the regular flow of operations, the easier they are to do and the more accurate they can be. Sites have already begun to use online operator logs, with online access to laboratory analyses of production, electronic (the current word for computer-based) maintenance management, electronic scheduling, and other directly related activities. Properly designed, these records can be assembled in fit-for-purpose reports. One of these can be a shift report. Figure 3-7[30] illustrates an online log entry for pipeline use where the controller (pipeline term for the operator) is able to both enter contemporaneous information and flag that information for later use in shift reports. Remember that this is not the report itself but rather provides a mechanism that assists in building it.

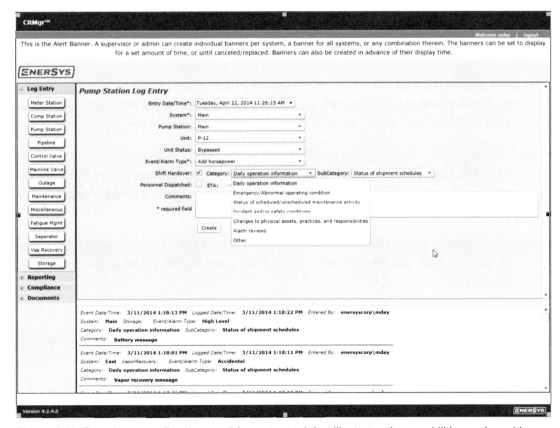

Figure 3-7. Example controller (operator) log entry tool that illustrates the capabilities and provides an example of a structure that is easy to use and promotes completeness and understanding.

Source: Reproduced with permission from EnerSys.

---

30 Russel Treat, "Discussion Document—Abnormal Situation Management—HMI," Version 1, April 22, 2014, personal communication.

This example contains a few interesting capabilities:

- **Operating notices** – At the top center is a place for notes or messages that apply to the current shift or campaign. An example might be a reminder that the local fire department will arrive for a readiness drill today at 1450 hours.

- **Log type** – Multiple log types are supported through a drop-down menu; types might include operations, fatigue management, or abnormal operating conditions. Along the right-side area is a pick list for all the areas that comprise the current responsibility of the controller.

- **Area of operation** – The log can be arranged by hierarchical operating area providing for the appropriate level of specificity (unit, skid, instrument).

- **Flag** – There is a check box to flag the entire log entry as a required item for including in the upcoming shift report. When the [operator] gets ready to prepare the report, all log entries can be filtered to show the ones that are relevant. In this case, API RP-1168 is used as a guide.[31]

- **Drop-down pick lists** – The inset is an example of the allowable choices for the particular selected box for which data are currently being entered. Each data area will have its own pick list. This reduces repetitious keying and enforces more uniformity.

- **Log details** – In the central gray area is a form for all log entries. This frame ensures that no relevant topic or point of information is missed. There is also room for freeform notes.

- **Alarm and event data** – The bottom two horizontal sections contain alarm and event data that can be referred to as the log entry is being made. They can also be clipped to the log.

Handover

The shift handover involves the successful completion of seven activities. These activities need to be designed so that they are both effective as an individual responsibility and, when done well, reinforce each other.

1. **Documentation preparation** – The leaving operator will assemble all procedurally required information regarding the current shift that must be contextually conveyed to the arriving (next) operator. This material may be in hard copy, soft copy, or a combination. The enterprise will lay out which and how it

---

[31] API RP-1168, *Pipeline Control Room Management*.

should be prepared. The preparation process must include the ability to annotate all information to create a correct record as to what was actually discussed during the later handover together with any spoken or expressed concerns, questions, comments, corrections, and omissions.

2. **Plant or enterprise preparation** – This includes the completion of all tasks including appropriate documentation and other reporting that are not part of the handover but are necessary for it. The operator will ensure that the operational state of the plant or enterprise is stable enough to devote appropriate attention to the actual conduct of the handover. If not, this situation must be escalated.

3. **Handover communication** – This includes carrying out the full handover requirements for information exchange and understanding between arriving and leaving operators. It includes all site requirements for formal and informal discussion and annotations and documentation. The *engaged handover interaction* (see the following discussion) promotes effective communication. It is highly encouraged.

4. **Readiness assessments** – Appropriate evaluations of the state of readiness of both operators should be conducted. The arriving operator is tasked with evaluating the leaving operator (fitness performance during actual handover and fitness to self-travel to the intended destination for recuperative activities). The leaving operator is tasked with evaluating the arriving operator (fitness to assume full operational responsibilities for the expected duration of the shift).

5. **Transfer of operational responsibility** – Both operators acknowledge that operational control has transferred between them. From this point forward, the operator in charge is the arriving one.

6. **Joint operation** – The leaving operator will remain actively at hand for the prescribed period to assist the arriving operator as needed during the actual physical assumption of his hands-on duties. This period of joint operation is usually brief (5 to 15 minutes) unless the plant or enterprise is upset or abnormal. If the plant is upset or abnormal, the leaving operator would be the perfect resource to help the arriving operator garner other resources (therefore, he may remain for a substantial period of time). This is also where the *engaged handover interaction* is extremely valuable.

7. **Relinquishment of all duties** – The leaving operator goes on relief (e.g., goes home).

Engaged Handover Interaction

An *active handover* means that all participants (usually two, but keep in mind that more may be involved, even teams) are engaged. It just works better that way. Assume that there are just two: the arriving operator and the leaving operator. Being present is not enough! They need to be engaged. *Being engaged* means that the participants are jointly focusing their attention.[32] As you might expect from the words, this means that any time one operator provides information (or asks a question), both operators are focused on that specific item. It means that they are both looking at the same screen or lab report or artifact from the field. They make eye contact. One points or reads, and the other acknowledges or affirms or questions to ensure a shared understanding. Neither moves on nor goes back without taking the other with him. At points where conclusions are appropriate, they jointly discuss them. Agreement is not required. Any agreement, though, must be affirmed and owned. This means that the arriving operator may use the leaving operator's information, observations, and conclusions as data to aid in his shift, but he alone must ensure that all of it is correct and appropriate for his operational duties on his shift.

This ability to observe and respond to nonverbal clues is a vital component of face-to-face handovers. A joint focus of attention is not one participant doing a "data dump" to the other (no matter how complete or how accurate, entertaining, or engaging). It is not both participants sharing the same information but not checking or acknowledging what is being discussed and what it means. It is not one participant being the superior conveyor and the other being the inferior recipient, nor the other way around. It is a shared, peer-to-peer activity. At the end of the handover, both participants should feel heard, respected, and comfortable.

Face-to-Face
The best form of handover is face-to-face.[33] There is so much going on during a face-to-face event that is nonverbal and completely missing in all other forms. From the basics of pacing (easy to see whether the other person is keeping up or not, and to seamlessly make adjustments, ask questions, or offer clarifications) to the ability to point to and associate what is needed to join the discussion with any documentary or display information, to the ability to readily sense how the discussed information is affecting the parties (concern, relief, confusion, detachment), the role of nonverbal cues is key to a successful handover.[34] Even if the handover documentation and structure are well done, and sufficient time is available and used for the process, the nonverbal aspects provide a critical success bias to the activity. Different individuals

---

32 P. Mundy, L. Sullivan, and A. M. Mastergorge, "A Parallel and Distributed-Processing Model of Joint Attention, Social Cognition and Autism," *Autism Research* 2, no. 1 (2009): 2–21.
33 Richard M. Frankel et al., "Context, Culture and (Non-Verbal) Communication Affect Handover Quality," *BMJ Quality and Safety* 21, no. 1 (2012): i121–i127, doi:10.1136/bmjqs-2012-001482.
34 Parke and Mishkin, "Best Practices in Shift Handover Communication."

have different skills in recognizing and understanding nonverbal communications. Management should not presume any level of skill here. Consequently, appropriate training and skill-building activities should be conducted as part of basic operator training.

This also means that all handovers that are not done face-to-face between operators will require specific procedures, protocols, and processes in place to ensure the intended exchange is effective. At a basic level, each individual communication packet of information might require checking and verifying: the arriving operator specifically acknowledges his understanding of what the leaving operator is conveying and its significance for the next shift—each time, item by item. Of course, this must be structured to be efficient and not monotonous or overly demanding. All of this is done to ensure the handover is an active process, not a "data dump."

Figure 3-8 illustrates a purpose-built tool to track the handover process and serve as a checklist (it will have been customized by each site to meet their requirements), an annotation mechanism, and an audit trail to document the handover itself.[35]

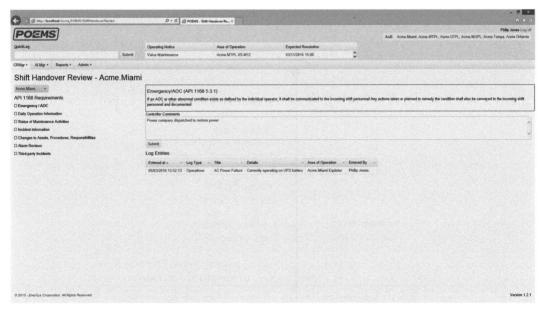

Figure 3-8. Example of a shift handover tool designed to follow API Recommended Practice RP-1168, *Pipeline Control Room Management*, including a checklist of steps on the right and an area for a controller to make specific shift handover notes.

Source: Reproduced with permission from EnerSys.

---

35 Treat, "Discussion Document."

Field Operators' Handover

Plants and enterprises may also have handovers done by outside or field operators. The purpose and general activities mirror what is done by control room operators. We will not go into details here. This type of handover is shown in Figure 3-9.

This handover timeline identifies the leaving field operator preparing the handover report, sharing the report with the arriving field operator, and transferring operational control to the arriving field operator to assume full ownership of the shift. After the handover, the leaving operator leaves unless the plant is upset and extra hands are needed.

We can relate the field operator's shift handover to the control room operator's shift handover both in time and coordination. The field handover can occur at one of four time categories: (1) simultaneously with the control room, (2) field first, (3) control room first, or (4) randomly without any regard to the timing of the other. Let us rule out both simultaneously and randomly. If they were conducted simultaneously, then any problem in one would likely create more load for the other. Yes, there would be more hands available, but the advantage of more hands is outweighed by the added

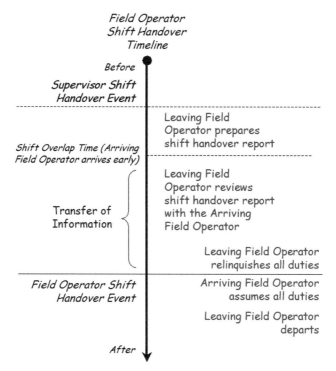

Figure 3-9. Detailed shift field operator handover timeline identifying the leaving field operator preparing the handover report, sharing the report with the arriving field operator, and transferring operational control resulting in the arriving field operator assuming full ownership of the shift.

risk, complexity, and distraction of the handovers. The other choice of random creates its own set of awkwardness in addition to being difficult to create useful procedures in case of problems.

We are left with deciding either field first or control room first. This choice is not as clear as in the earlier two cases. Considering that the control room handover is likely to be more complex, it is suggested that one be done first. Let the new control room operator get fully settled. If any issues arise, the field can assist without the distraction of being in handover itself. Figure 3-10 depicts the timing of this mode of handover.

Combined timelines for both operator and field operator handovers represent the same individual shift handovers but coordinated at the green (barbell) line. It is after this time that the field operator begins preparing his handover using the information as needed from the operator shift report and handover results. This enables the arriving field operator to better understand what has happened in the control room (the primary activity) and put it in perspective with what is going on in the field.

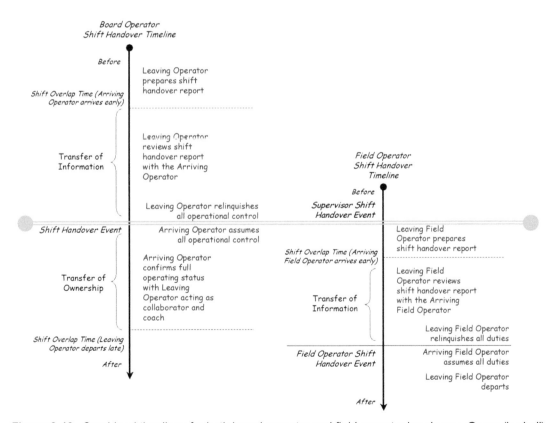

Figure 3-10. Combined timelines for both board operator and field operator handovers. Green (barbell) line depicts how the alignment of the handovers relates to the handovers. This arrangement allows the control room to become fully settled before the field does its handover.

Shift Handover for Supervisors

Plants and enterprises may also perform formalized handovers by operations supervisors. The purpose and general activities of this type of handover mirror what is done by control room operators but at a different enterprise level. We will not go into those specific details here. We will just focus on the timing and any coordination aspects. The handover is shown in Figure 3-11.

This handover timeline shows that the leaving supervisor's first responsibility is monitoring (and assisting where needed) the control room operator shift handover.

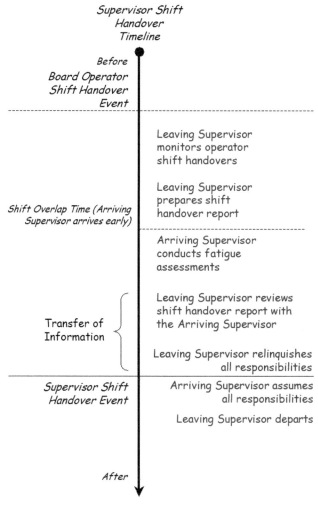

Figure 3-11. Detailed supervisor handover timeline noting that the leaving supervisor first monitors the operator shift handovers and then prepares the handover report. The arriving supervisor first monitors the fatigue levels of both arriving and leaving operators. Then both supervisors review the supervisor handover reports. Once that is done, the leaving supervisor departs, leaving the arriving supervisor in charge.

After that is complete and with the benefit of both the control room shift handover report and the experience of the control room operator handover, he prepares his own handover report. This activity may continue a bit after the arriving supervisor comes on shift, because the arriving supervisor's first duty will be to assess and/or confirm the fatigue levels of both the leaving and arriving control room operators. Once this is done, the leaving supervisor reviews the handover report with the arriving supervisor. After the handover, the leaving supervisor departs unless the plant is upset and extra hands are needed. Figure 3-12 depicts the suggested handover coordination timelines

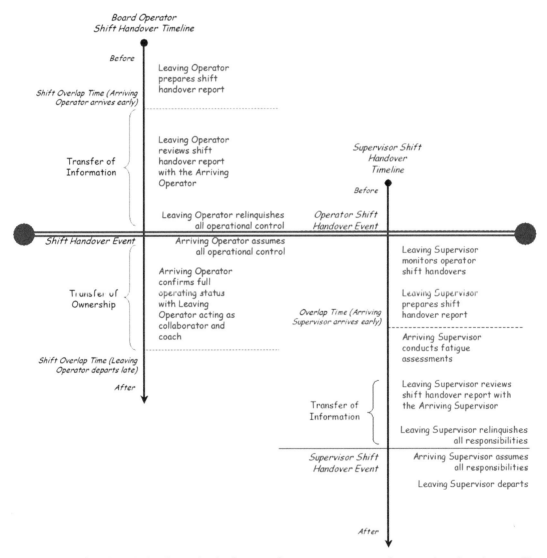

Figure 3-12. Combined timelines for both control room operator and supervisor handovers. The magenta (barbell) line depicts how the alignments of the individual handovers relate to each other. This arrangement allows the supervisors to monitor the joint operations phase of the control room handover, assess the fatigue levels of both operators, and assist in the handover in case of problems.

of control room operators and supervisors. Note the coordination timing alignment of the two activities at the cyan (barbell) line.

Figure 3-13 depicts the suggested coordination timelines of all three handovers. Here the control room operators and the supervisors coordinate the same way as before (Figure 3-12) with the follow-on of the field operators. The control room operators' handover is complete before the field operators' begins. The supervisor's and field operator's handovers align at the orange (barbell) line. This permits the supervisor to monitor or assist the leaving field operator with his handover report and then monitor and assist the handover as needed.

Special Case for Maintenance

Coordination of maintenance activities goes considerably beyond a simple hand-off. It involves shifting physical and operational control of equipment from the control room operator to maintenance personnel and then back after work is complete. These changes require following a strict handover process to ensure that both are done properly. This handover, while appearing to be quite different from the ones from control room operator to control room operator, is really not. While it is more limited, it is not less important. Incident investigations have shown that time after time, this handover or handback was flawed, and when it was, serious consequences resulted. An example that is close at hand is the spare pump on Piper Alpha that was taken for maintenance. Had it been properly taken, or had it been properly taken back, no disaster would have resulted. None whatsoever. All maintenance handovers require the close coordination between control room operator and maintenance. Recommendations follow.

Handover

There is equipment that the control room operator must remove from service and place in the hands of maintenance. The control room operator is responsible for directing everything necessary to effectively remove the equipment from functional service. *Functional service* means that no longer will the equipment be necessary or used for operation. In such a state, all proper work by maintenance will not affect current operation. All procedures for doing this will be referenced and followed. Procedures are set by the enterprise. This is a checklist of functionality, not a procedure.

1. Control room operator identifies all equipment to be taken by maintenance.
2. Control room operator reconfigures and adjusts current operations to remove the need for the equipment to be taken (and all ancillary parts like piping or instruments) from production.

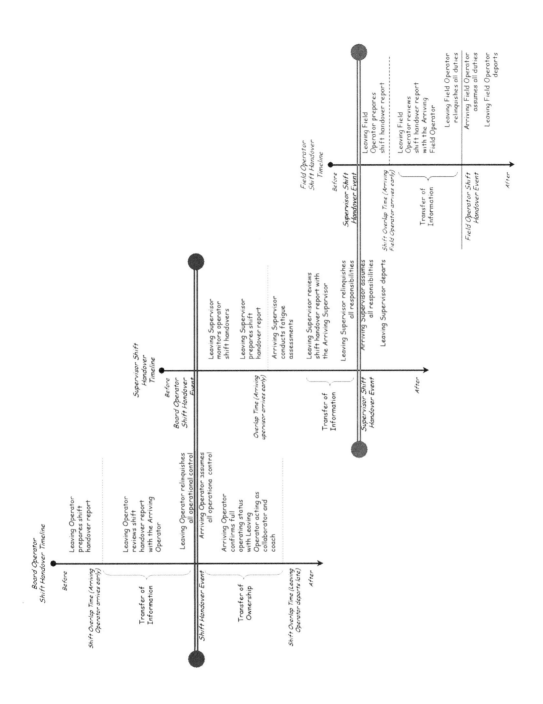

Figure 3-13. Combined timelines for control room operator, supervisor, and field operator handovers. The magenta (barbell) line depicts how the control room and supervisor handovers relate to each other. This arrangement allows the supervisors to monitor the joint operations phase of the control room handover, assess the fatigue levels of both operators, and assist in the handover in case of problems. The orange (barbell) line depicts how the supervisor and field operator handovers relate to each other. This permits the supervisor to monitor and assist with the field handover preparation and activity.

3. Maintenance verifies using independent means that the equipment to be taken has been properly removed from production.

4. Maintenance takes (temporary) ownership of equipment.

5. Maintenance takes the proper extra steps (not related to operations) to render the equipment safe and ready for maintenance activities, for example, fitting of blinds, electrical isolation, or erection of physical protection barriers.

Handback

Here equipment that the control room operator needs for operations is to be taken from the hands of maintenance to be placed back in service. The control room operator is responsible for directing everything necessary to effectively return the equipment to functional service. Functional service means that the equipment can be used for operation. Any limitations of full use are identified and accepted by the control room operator. This is a checklist not a procedure.

1. Maintenance identifies exactly which equipment is ready to be returned to service.

2. Maintenance affirms that the equipment is properly repaired and/or evaluated and ready to be placed back in service, for example, removal of blinds, electrical restoration, or clearing of physical protection barriers.

3. Control room operator verifies, step by step, that the equipment is ready for service.

   a. This is an active process in which the control room operator and maintenance check, test, and otherwise ensure that the equipment is ready to be back in service.

   b. Note that this involves limited or no verification that proper repairs have been made, but just that from what the control room operator can see and test, the repairs seem correct and the equipment should operate properly.

4. Control room operator takes back ownership of equipment.

3.10 Information Content of Shift Handover

The final part of shift handover is exactly what must be communicated or verified as part of the actual handover. It goes without saying that all of this must be compiled by the leaving operator and discussed with and then owned by the arriving operator. Much of this information would be contained in the shift log. Shift logs are the first resource, of course. However, and this is important to observe, a log format is generally lacking in "connecting the dots," and it is not sufficiently useful for separating the necessary information regarding the shift from the information just "for the record." A simple illustration of a "for the record" annotation would be an entry in the log that the mid-shift samples were taken on time and sent to the lab. It is expected that routine activities that are "logged" are only notable if there was a problem or deviation or other irregularity that the next operator needed to know. The following list includes both the obvious and the not so obvious. These are suggestions; the list is not necessarily complete. Each plant will specify what is required.

1. Demographics: operators involved, location, date and time of handover.

2. All standing orders or requirements that amend or alter usual and regular shift operation.

3. General operational status during shift, including products being produced, rates, and any weather conditions that might have affected operations.

4. All specific incidents, or unusual operation, regardless of their apparent effect on operations.

    a. All incidents (including spills, releases, and other abnormal or unusual operation).

    b. All log entries.

    c. All active alarms and any cleared alarms that might have a bearing on the next shift.

    d. All maintenance operations ongoing or completed during the shift.

    e. All permits (e.g., work permits, entry permits, or lock-out/tag-outs). These must be cross-referenced to ensure that related operations are properly understood and tracked.

    f. All personnel injuries.

    g. All near misses.

5. All operational concerns for which no cause or impact has yet been ascertained, but nonetheless the concern has not been resolved (see Chapter 9, "Weak Signals").

6. Next shift production requirements or changes that are needed, required by the operational plan, or are possibly anticipated by current operations.

7. Shift log(s) to be used as a checklist type of review to pick up anything that might have been missed or might benefit from additional discussion between the operators.

Again, this list may be incomplete or use terminology different from yours. Please understand that it is provided for guidance. Your site requirements will govern.

3.11 The Mobile Operator

Mobility for operators is only important if being able to move about is beneficial. The mobility we are talking about can be several feet or hundreds of yards. We even consider operator mobility over tens and even hundreds of miles. To be mobile, the individual operator moves about at distances that are meaningful for the job at hand. But movement is not what makes an operator mobile. To be mobile, the operator must have seamless access to the information, communication, and controls as he moves. Care must be given to understand what this mobility is all about. If it is just to avoid extra personnel, that objective must not overshadow the corresponding complexity of the operator's tasks and the inevitable competition for operator resources (e.g., time or problem focus ability).

There are three general categories of mobility. All are based on distance. Of the three, control room mobility, done well, should enhance the operator's performance. All of this presumes that the primary location of the operator is in the control room. Operators without a control room are a special case. Material here might help, but that case is not covered in this book.

Control Room Mobility

Most control room operator stations are designed for a tethered operator. The displays are generally in a fixed location (although the horizontal platform they are mounted on may raise or lower), the interaction devices (keyboards, mice, or other pointing devices including touchscreens) are fixed or tethered by cords or not, and the communication devices (telephone or station intercom to plant radio system) are also tethered by cords or not. Yet the ability of the operator to move about may be advantageous. In a single-operator control room, the simple ability to be yards away from the chair location may offer important benefits. The operator can attend to personal needs, cook or eat food, access exercise equipment, or engage in light maintenance or other chores. For multioperator control rooms, the benefits might increase substantially. The operators would be able to support each other without the need to entirely absent themselves from their primary responsibility. We do not for a moment endorse multitasking—that does not work. However, if operators can move a primary task to the background for a short amount of time, they might be able to competently pick up a new primary task during that period.

This level of mobility would be relatively easy to achieve, once we understand the requirements and options. First is the ability to see and manage display content.

A simple switch of the current operating screens at the control desk up to the large overhead or wall displays could do that job handily. Operators could do that at the touch of a control, or, better yet, the operator's location could be sensed and the screens automatically switch. Such sensing technology is already widely embedded in smartphones. From a human factors perspective, the specifics of how the screen switches progress (duplicating screens versus switching them) might well be different when the operator is moving away from the control desk as opposed to moving to the desk. The second need is to have all communication devices mobile. Telephone and plant radio systems might be on a headset with appropriate controls. Pointing and clicking actions might be switched to voice activation rather than tactile via a cord or short-distance wireless link.

You can see where this might go. It is important that the ideas and concepts are not dismissed out of hand. For only after we remove physical constraints once thought to be fixed can we examine types of functional capability we did not even imagine before.

Plant Area Mobility

Once we consider allowing the operator to move beyond the vicinity of the control room, technology suddenly becomes important. The voice part is easy. Hands-off equipment with adequate range and appropriate noise controls and connection security are readily available. Obtaining appropriate and useful visual information becomes a real challenge. Not only must the equipment be mobile, but it also must be in a portable format that is easy to transport and use without tying up hands too much. And it must be safe to use in hazardous locations or when the operator cannot devote a hand to hold or interact with it. Chapter 5 on the human machine interface (HMI) discusses the needs and challenges of small and very small displays. Their design will require a reinvention of how real-time related information is accessed and displayed for limited format devices. So far, that is mainly a research task. No demonstrated guidelines or proven experience are in place. Just think how an operator might use a smartphone as a remote HMI. Yes, it is handy and familiar. No, it is not well suited for viewing coordinated information, nor for making control adjustments. Equipment vendors are just beginning to approach how to work this out.

Large Geographical Mobility

In this arrangement, operators use their mobility to arrive and manage at locations quite distant from each other and any central operating facility. No longer will operators be able to wait until they return to the control room to complete tasks. This level of mobility will require new concepts and procedures for collaboration and specialized interaction, visualization, and controls tools. Interestingly, once these new capabilities are developed and demonstrated to be useful, they should also fold back into enhancements for existing control rooms.

Requirements for Mobility Support

Now that your operator is going to be mobile, you need safeguards, and the operator needs proper tools to do it right.

Safeguards

The moment your operator becomes physically mobile, everyone on the team must know that. Operators at the board are the baseline. Operators near the board, so long as everyone in view of the board knows where the operators are (and why), are fine. If aid and assistance are needed, it can be summoned by a glance, hand gesture, or quiet voice. Nothing extra is needed. An operator away from view is a different matter. Everyone else on the team must know where he is, why he is away, and how to communicate with him. The operator's location is important for physical protection in the event that his location puts people at risk due to operations of others. It also helps when the primary means of communication with the operator does not work. Knowing why he is away aids others in understanding what might be going on as well as predict likely next steps or locations he might visit.

Tools

The primary design objective of control room operations is the suitable accommodation for the operator's physical presence. To that end, physical security, environmental comfort, and information access are all included in the design of the room. If operators are control-room based but can be mobile, no control room design considerations are traded off to facilitate that potential mobility. This means our focus is on what must be added. And what is added must not interfere with what exists. There are two areas for addition: (1) information and communication and (2) protocols and processes.

Information and Communication

Mobile operators need access to all required operational information necessary to support their being away from the control room, as well as the specific information to support any remote tasks. This information includes but is not limited to alarms (if they retain alarm response responsibility), process conditions and states, communications delivered to the control room (that need to follow the operator), all controls necessary to change operational conditions (e.g., set-point changes or start/stop), and information and communications to perform/participate in shift handover if they will be away during that event. Consideration must be given to ensure effective communication channels are maintained to support all HMI devices and voice communications, including those in hard-to-receive locations likely to be encountered. Special attention may be needed to accommodate low-light or no-light situations, high-noise situations, high-glare situations, and situations requiring no hands or limited hand use (both for equipment holding and interactions). Where interactions with others are needed, the equipment must properly support that activity. This means not only shared mobile devices and communications, but also the ability to annotate and "point" to displayed information in order to focus the interactions. Refer also to the discussion in Chapter 5, "The Human Machine Interface," about off-workstation small displays.

Protocols and Processes
The very nature of being remote from others requires careful attention to the procedures and personnel needed for coordination, collaboration, and escalation. The mobile operator requires knowledge of the locations and availability for appropriate individuals. Explicit procedures need to be in place to support this activity. When severe operational stress, including weather, is taking place, everything must be able to properly function in those conditions. Refer also to the discussion in Chapter 8, "Awareness and Assessment Tools," and Chapter 10, "Situation Management."

3.12 "Long Arm" of the Operator

More often than not, operators are physically at a distance from the actual equipment they are managing via their controls. Being away from the operating site eliminates lots of sensory information. Operators at a distance cannot directly hear the sounds of the equipment working (or not working), sense the vibrations of things, smell the odors there, or see the actual equipment operating. Missing is the emotional feeling of being there to "take in the operation." Certainly, advances in video surveillance (with audio) can provide a measure of seeing and hearing, but not that much. In all cases, if the equipment under question has been handed over to the operator to operate (e.g., by a shift change, start-up, or maintenance return), it is the operator's sole responsibility. When other personnel are at the remote site, they should clear their presence with the operator. The process of "clearing their presence" is determined by appropriate enterprise procedures, and is beyond the scope of this discussion. The case where personnel are present for maintenance purposes was discussed previously.

A short-distance separation of control room and equipment was first used to provide a more comfortable and safe location for the operator in the same protected space that housed the mechanical, electrical, and electronic instruments and devices used for control. Later, the distances were significantly increased to provide a blast-safe location for the operators to protect against operational mishaps. In addition, some operations require the operator to be located at significant distances from the equipment and plants. Examples include pipelines and other transportation networks. In these situations equipment is commonly unmanned except for occasional checks and maintenance. No matter the distance, personnel at the site for operational purposes are acting under the direction of the operator. Often, their role is to check or verify some operational aspect, such as the position of an unmonitored valve, or whether there are visible signs of operation for unmonitored operational aspects, such as the position of a lid of a floating lid storage tank. Just as often their role is to make some sort of operational change that the operator needs but does not have remote operational control for. The command and control of the coordination between the operator and at-site personnel parallels somewhat that of a maintenance operation. The operator retains ownership of equipment at all stages of this process. The following is a framework checklist.

1. Operator identifies all equipment to be observed or adjusted by remote site personnel.
2. Operator reconfigures and adjusts current operations as needed to ensure safe and appropriate adjustments by the remote site personnel.
3. Remote site personnel verify using independent means that the equipment to be adjusted has been properly identified by them.
4. Remote site personnel take the proper steps to adjust the equipment.
5. Operator verifies step-by-step that the equipment is actually ready for remote operation. This is an active process in which the operator and remote site personnel check, test, coordinate, and otherwise ensure that the appropriate adjustments are done

When complete, the remote site personnel check out with the operator. Once checked out, all adjustments and presence within the physical operational equipment perimeter should end.

3.13 Goals, Roles, and Culture

It is midway through your operator's shift: do you know what he is really trying to achieve? How in the world are operators able to work effectively through the long hours of each shift? What keeps the operator going day after day? How is it that he keeps doing it as a career? What are his internal goals? And to be sure, how do they incorporate what the plant needs him to do? How can these goals be personal and important in a way that either makes each shift a good thing or relegates it to just another interference between days off? If your operators are to be good at situation management, you will need to understand and respect those internal objectives. Unfortunately, most enterprises pay little attention to personal objectives, much less participate in shaping them. This will need to change.

This section is about activity goals and what operators might really be up to while sitting in your control room. It is about understanding what makes a good operator, how to help him develop acceptable and effective internal goals for job performance and satisfaction, and how to enable him to successfully sit through long shifts day in and day out. This section introduces the importance of goals and offers a bit of insight into how to achieve results.

Basics of Motivation

Begin by looking at personal needs. Figure 3-14 is a colorful diagram of the levels of personal needs. The levels and concepts should work for all of us regardless of the color of our "collar" and the extent of our education and experience. No one is less or more entitled.

Figure 3-14 is the work of Maslow in the 1940s.[36] It starts with the first level of physiological (physical) need. We need these things to be alive. The next level is safety. Interesting how this is going. You might be thinking that so far these needs are obvious, so why cover them? Such needs must be discussed because when they are satisfied, we tend to take them for granted. When one is not satisfied, it places an enormous burden on life and living itself. Think about yourself. If you are living in a safe society

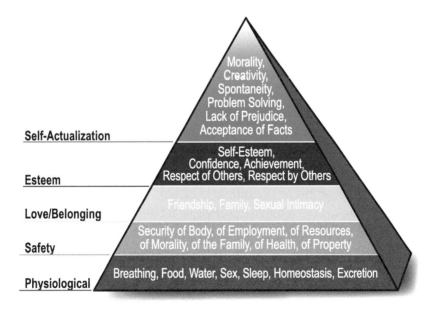

Figure 3-14. Personal needs hierarchy diagram that shows the importance of providing for the full self.

---

36 "Maslow's Hierarchy of Needs," Wikipedia, last modified June 4, 2018, http://en.wikipedia.org/wiki/Maslow%27s_hierarchy_of_needs.

and in a quiet and safe neighborhood, think about what it must have been like to have lived a thousand years ago in parts of Europe rampant with disease, thieves, and marauding warriors. The local landlord could forcibly take your wife or daughter for his own wishes. To keep it even simpler, consider that the basic task of getting fuel to make a fire to cook whatever meager food was obtainable was often impossible. The landed nobility owned all the forests. The word *windfall* literally meant what the wind blew off trees and fell to the ground. That alone was for the taking, as if just getting food was not hard enough.

As important as these lower levels of needs are, our focus will be the upper three. Remember, this part is about trying to understand what operators need and do to keep them on the job for a shift, shift after shift, day in and day out. Please do not for even the briefest of moments think that they will be able to be on that job and do their work because of a fear of making mistakes or for the pay. Sure, they should expect to be fairly paid and it is important to operate correctly. Both are a given. Alone neither is enough.

Belonging
Focus on friendship for the control room. Each operator will need to feel that he is part of the team. Each operator will need to believe that his part in the team is individually important and valuable, but not necessarily irreplaceable. This implies a lack of bullying, practical jokes, and "proofs" of belonging like initiation rites and membership "tests." Instead, a real sense of belonging results when there is open communication between peers and between peers and supervisors. The control room is not a social entity, yet sociability is essential.

On the practical side, this includes providing proper secure storage for personal belongings, including private possessions and items issued to each individual for the job. It includes ready access to ancillary facilities such as restrooms; exercise areas; training areas; kitchens or break facilities; meals, if provided, that respect dietary desires and preferences; and extensive descriptive labeling of storage areas and equipment that are shared with others. Everything must be kept where it belongs and in excellent repair with ready access to user instructions.

Esteem
This is about self-esteem, confidence, achievement, and respect of and by others. For these to be met in the control room, there is much to consider. This level of needs includes comprehensive and competent training programs so that everyone has the tools to do a proper job. It means training operators and training managers. It means testing and coaching so that everyone knows their level of competence and any areas

in need of more work. It means honest rewards for success and respectful criticism of failure. It also requires the setting of career advancement paths and open opportunities to participate. This gets to the heart of a corporate culture.

Self-Actualization
The top point of this discussion (and the triangle) rests almost entirely on the individual. Self-actualization can be achieved only if everything below is done well enough to be depended on. Operators must not be arbitrators of a struggle that puts personal and society mores in competition with unyielding enterprise requirements. It is up to the senior leadership to ensure that successful operation is good citizenship. Nothing here is in opposition to profit and good business.

Successful operation includes encouraging operators to engage in understanding and problem solving with the full expectation that the enterprise will be honest, fair, and responsible. This means providing operators with clear lines of responsibility, ready access to mentoring and advice, and adequate safeguards against things that might go wrong. They go to the heart of successful policies for escalation and operational authority (permission to operate in Chapter 10, "Situation Management"). As you will see throughout this book, the ability to operate successfully over the long haul will be shaped by the sum total of everything your enterprise does. Long-term success is skillfully crafted and carefully maintained.

Tenets of Operation
One of the best examples of how the ideas introduced in the previous section might work is found in the concept of Tenets of Operation first laid down by Chevron.[37] The organization started with two key principles:

1. Do it safely or not at all.
2. There is always time to do it right.

The reason for these tenets is basic. With all the many rules and procedures and other requirements an enterprise requires to ensure safe and reliable operation, it is quite difficult to keep everything in perspective and usefully at hand. How do operating personnel know which rules should take precedence over others? In many situations more than one applies and doing all would be either impossible or contradictory. It is not at all straightforward which one may conflict with another, or which ones

---

37 "Tenets of Operation," Chevron Corporation, accessed March 22, 2013, http://www.chevron.com/about/operationalexcellence/tenetsofoperation/.

must be followed in place of which others in specific situations. To avoid this ambiguity and ensure that operators and other key personnel always maintain a safe and responsible operating posture, the following 10 top-level rules recommended, *always* and no matter what. Always follow these first.

1. *Always* operate within design and environmental limits.
2. *Always* operate in a safe and controlled condition.
3. *Always* ensure safety devices are in place and functioning.
4. *Always* follow safe work practices and procedures.
5. *Always* meet or exceed customers' requirements.
6. *Always* maintain integrity of dedicated systems.
7. *Always* comply with all applicable rules and regulations.
8. *Always* address abnormal conditions.
9. *Always* follow written procedures for [all] situations.
10. *Always* involve the right people in decisions that affect procedures and equipment.

It is highly recommended that you include these principles, or ones suitably worded for your enterprise, in the foundation of operations guidance. They will reinforce your personnel (from maintenance technicians to operators to plant managers) to do their jobs without placing undue risk on responsible operation. See also the discussion in Chapter 10, "Situation Management," on permission to operate. Before moving on, let us clarify that it is management that is responsible if the operator follows these faithfully and decides that shutting down or significantly curtailing operations is required. Management expects the operator to do that. This means that if the operator should have shut down or curtailed operations but did not, management holds him accountable for the error in judgment.

Operator Objectives

We turn our attention to what operators need to be effective and make their work satisfying. You might be thinking that operator objectives should be a perfect match with the enterprise objectives outliner earlier. Not so fast. For sure, operators will appropriately complete all assigned tasks. They will do them following the established procedures using the approved resources. The real question is how best to ensure that this happens. Yes, you will lay down rules and check carefully every step of the way. You will recognize good behavior, discipline poor behavior, and release bad. That is part of your management duty. And you know you are doing a good job when accidents and incidents are low, profits are in line with expectations, and morale is good. But you do not check the numbers of accidents and incidents, the profit levels, and morale every few hours during your work day, do you? You measure your daily success in other ways. Those other ways are better left up to you. Operators are going to measure their success in other ways too.

Currently, personal operator goals are largely unexplored. Unexplored does not mean unimportant. Operators will inevitably develop them. Doing so is a part of being involved and in the moment. With a bit of help from the enterprise, some or most of those personal goals might be useful and productive. In this section, we will offer ideas about what might be useful to consider. Bringing this topic out into the open should foster consideration, discussion, exploration, and respect.

Preferences: Declared versus Revealed
We all have preferences for much that we do. Often what we say we like and what we actually like can be quite different. Part of the reason is personal. Part of it is a natural desire to please or conform. Part of it is that most of us have preferences but have not given much thought to identifying them or categorizing them. Why do it? We just follow them.

This changes in situations where one is either asked for preferences or feels that they need to be told to others. If you are the one asking, understand that most individuals feel obligated to respond even when they might not have very many preferences, or especially when they have a great deal of them. Asking is generally thought to be a good idea, but care is in order. Once you ask for preferences, you are usually more obligated to do something about what you hear. That is okay except that some of them do not make for good operation or policy. And it is problematic because of another reality of life: most of us declare preferences (the expressed ones) but have others that can be quite different (the unrevealed ones). Sorting all of this out, and it must be sorted out, requires patience and skill. Sort carefully.

Consider Changing the Frame
Change the frame from "the boss says to maximize profit, have no accidents, and minimize costs" to "operate as best as you can so long as you work on getting better, stay safe, and feel challenged but clearly in control." Now your operator is working for success instead of against failure. These are goals that he can directly work toward, easily measure progress toward, and decide how to better achieve.

Don't List

Sure, everyone needs a good list of what not to do. It is usually short and at the top of the operator's personal responsibility list.

List of Don'ts That Operators <u>Should Not Use</u>

- Don't ask for help unless absolutely necessary (personal bravado!).

- Don't cause an accident or serious situation. (Reputation preservation leads to too strong a bias for failure to act appropriately, at the expense of action.)

Don't lists can be problematic. They all too often tend to lead to a preoccupation with damage control as opposed to proactive involvement. They also foster a culture of risk aversion (don't think outside of the box). At the end of the shift, meeting them produces a feeling of relief, not accomplishment. Striving for relief is often stressful.

List of Don'ts to Actually Use

- Don't make a bad situation worse.

- Don't have a near miss that can be avoided.

- Don't operate unengaged or on autopilot.

Seeking accomplishment is self-affirming and satisfying. You want your operators to grow and feel good about coming onto shift, and not always wishing their shift would end without trouble. The rewarding list is the *do list* discussed next.

Do List

Do lists are quite different from don't ones. The difference is that most can be quite rewarding. Getting the list items right feels good; getting them close to right also feels good. Two brief stories can help us get into the spirit of how this works.

Pay It Forward

The movie and book *Pay It Forward* take a well-known principle and turns it on its head.[38] The well-known principle is that when someone does something nice for you, you should look for a way to pay that nice thing back. Return the favor. Pay it back. The "turning it on its head part" is that when something nice is done for you, you should look for ways to help someone else and make things easier for that person. Instead of paying back, pay it forward. Good where it was needed begets good where it is needed next.

---

38 Catherine Ryan Hyde, *Pay It Forward* (New York: Simon & Schuster, 1999).

The Stumbling Stone
> And when one of you falls down he falls for those behind him, a caution against the stumbling stone. Ay, and he falls for those ahead of him, who though faster and surer of foot, yet removed not the stumbling stone.[39]

Kahlil Gibran articulates this "pay it forward" illustration. Upon seeing an impediment or a hazard, each of us chooses whether to move or remove it for those who follow. Each individual, each operator, is a part of a larger society. This idea forms the heart of continuous improvement. Yes, we benefit from those before who made the pathway better. And just as important, we gain a measure of satisfaction for doing our individual part to add to these collective benefits for those who follow. Find every stone—for all of us.

"Do List" Examples

- Document and report all problems. If you have ideas for a good solution, provide them (but doing so is neither required nor the real objective here). Finding problems and other things that need attention is not easy. Problems often appear during times of urgency and stress—making noting and reporting all the more difficult. The silver rule here is that if one does not do his part to improve things, he has no right to expect improvement.

- Learn something new about what you do every shift. It does not have to be monumental or extensive—just new and useful. It can be something you may have started to learn earlier, or thought you knew but think now you are not so sure. If someone is trying to teach you something your already know, reinforce that interest and share stories.

- Teach someone something new every shift. It does not have to be dramatic or earthshaking—just new and potentially useful. It can be something you know and find useful. It can be something you just learned and want to share. If it is something that the individual already knows, share why you thought it was important and how you use it. And be ready to have the tables turned, so you learn from the intended student.

- Ask for others' understanding or advice about one thing every shift. Even if you think you fully understand what you are asking, it is likely that others can add to that knowledge.

All of this is aimed at changing the frame from failure avoidance toward achieving success.

---

39 Kahlil Gibran, "On Crime and Punishment," in *The Prophet* (New York: A.A. Knopf, 1924).

Message

By now it should be clear that operator success is not a check box on a list. It is not a task to delegate to someone and make sure it is done. Rather, it is the sum total effects of the strength and completeness of the enterprise design and operation. You will be able to have a competent operation if the organization provides what is needed to do so.

3.14 Close

Operators are the eyes, ears, and hands of situation management. They are key to day in and day out successful enterprise operation. This book is devoted to understanding their needs and requirements and encouraging the enterprise to organize itself to deliver proper operator management and sufficient operator support. This chapter has introduced important topics around readiness, training, tools, and shift handover. It also covered operator responsibility boundaries including the mobile operator. It closed with goals, roles, and culture.

We now move to the larger setting of control rooms and the culture and infrastructure for operating the enterprise for effective monitoring.

3.15 Further Reading

Gawande, Atul. *The Checklist Manifesto: How to Get Things Right*. New York: Picador, 2011.

4
High-Performance Control Rooms and Operation Centers

We shape our buildings; thereafter they shape us.

Winston Churchill (British Statesman)

We require from buildings two kinds of goodness: first, the doing of their practical duty well: then that they be graceful and pleasing in doing it.

John Ruskin (Art and Architecture Critic)

Architecture begins where engineering ends.

Walter Gropius (Architect)

When your operator enters his control room, you expect him to be in charge and successful. The more the control room form and design matches his needs and expectations, the closer he will be to getting into the roles and working out his tasks. That is no different from what your and my reactions would be to entering an amazing building and feeling its awe and inspiration. Sure, control rooms are not building lobbies, or museum exhibits. But that is not the point. The message here is that the care of design, the thoughtfulness of purpose, and the completeness of execution will make a positive difference. Let us do it right. Remember that this book is not the be-all and end-all of control room design. It is intended cover enough so you can get a good start on your own work.

So, what are control rooms and operation centers? What characterizes them as different, important, and necessary? What are their differences? Why are they needed?

And how will knowing the answers be helpful? It is all about what the room is for. It is guided by who is inside. The room is where the operator, your operator or perhaps you, work. It is where operators do the job of keeping things right. Industrial control rooms are used for operating petrochemical plants, electrical power generation stations, food processing plants, pharmaceutical manufacturing sites, and the like. They also include operation centers, for example, electrical power dispatch centers, mass transit monitoring and control centers, telephone monitoring and routing stations, security monitoring and dispatch centers, military command and control centers, and many more.

A control room is the setting for successful cooperation between the human operator(s) and the interface equipment provided to help do that job. This includes the spatial arrangements of equipment; the environmental management of the work space, who goes in and why; and the designs of the equipment and other related tools. Proper design will enhance operators' working experience and provide the necessary confidence that their tools are going to be helpful and trustworthy. When we understand and follow best design practices, operator work spaces are built that greatly enhance operators' success.

Control room discussion is offered because an effective control room or operation center is a basic requirement for enabling situation management to work. Without a comfortable and efficient control room, it would be difficult to adequately support operators for situation management. Careful control room design is essential to promote effective work practice. The design of the control room must not get in the way. We do not fully cover the topic to the extent that you can extract material from this chapter, give it directly to an architect, and ask for a specific control room in return. But, and it is an important *but*, almost everything in this chapter should provide background and alert you to issues to bring up with your control room designer so together you can figure out how to include the capabilities you require in your design. It is here to inform and empower your efforts to improve your operators' work spaces. Here is a brief look.

4.1 Key Concepts

Chapter Purpose	This chapter outlines what to think about and care about to provide the physical tools and space that we call a *control room* or an *operation center*.
	The chapter is designed to set the stage for operator success

Chapter 4 – High-Performance Control Rooms and Operation Centers

Control Room; Operation Center	A dedicated separate space for the sole purpose of providing the operator of a plant or operation with the sufficient operational state information, adequate operational objective information, and appropriate operational tools to effect necessary operational adjustments in order to meet all operational objectives and requirements.
	It is distinguished from other control spaces by the fact that the operational state information is conveyed by electronic or other remote surrogates about the actual operation, and the operational adjustment ability is effected by electronic or other remote operational surrogates to implement actions ordered.
	The space is for operators and only operators for the purpose of operating and closely related activities. This precludes the space from being used as a convenient rest stop, toilet access, or temporary gathering center for others.
Design	The "gut" feeling felt when the control room space is entered can make or break the operator's connection to and respect for the control room. In many ways it will affect his dominant behavior while inside.
	When many individuals share a control room, how each control room design contributes to the essential needs of collaboration, individuality, and inclusion is vital. Illustrative attributes include line-of-sight visibility versus back-to-back sitting, visual access by standing over partitions versus access by walking around partitions, sound isolation between operators versus sound cues that are not distracting, and control space replications for shared work areas as needed (e.g., situational advice, crisis management, and special task coordination not able to be done individually).
	The equipment location and how it is used must be carefully designed and managed. The seemingly simple height of the operator desk can assist work or produce unnecessary fatigue. Nowhere is the concept "a place for everything (needed) and everything in its place" more important or impactful.
Control Room Environmental Parameters	The following environmental parameters are vital.
	Lighting: All illumination in control rooms should support the operations tasks and the duration an individual works in the control room.
	Sound: The room design should support the auditory requirements of the user of the control room, allowing clear and unobstructed communications, absence of distraction, and evocation of serenity and stability.
	Vibration: The sensing and/or feeling of external vibration is a clear stressor and must be eliminated. Subliminal effects can be alarmingly uncomfortable for individuals.
	Comfort controls: Provide a uniform, stable, and comfortable temperature and humidity control without drafts.
Physical or Virtual Control Room	As technology evolves and our understanding of how operators are best utilized to manage operations, it might be that the best control room may just be no control room at all. Or even a virtual reality control room. But that lies a bit ahead of this story and will have to wait.

4.2 Introduction

This chapter discusses the considerations for providing the physical tools and space that we call a control room or an operation center to set the stage for operator success.

No matter what the size of your plant might be, what the finished products being manufactured are, or what the provided services might be, all control rooms have similar issues and related requirements. They must adequately support human users of automation systems. Early control rooms were designed to display and store instrumentation equipment with an afterthought to providing work space for people to interact with that equipment. That has all changed.

Figure 4-1 illustrates the visual and mechanical complexity of an early central control room. The individual dials, alarm panel, switches, lights, electrical indicators, and other elements are all mounted on metal wall panels. The arrangement is designed to facilitate one's ability to grasp an overview of the entire production operation. Each indicator represents an important aspect of knowledge or control for situation awareness.

Figure 4-1. Early control room, circa 1950 design. Note the presence of important indicators yet few controller stations. The operator was expected to be able to gain a good overview of the entire process by "eyeing the board."

Today's effective control room design is all about making it a purpose-built tool to house the individuals and equipment in a way that supports effective operations managed by people. Control rooms need to be much more than spaces we fill with people and their stuff. They must be fit-for-purpose designed and built. Successfully addressing these issues requires insight into human factors engineering and the best practices around ergonomics. A great deal of this forms the backbone of affecting the operator's ability to do his intended job. Part of the backbone includes management systems, such as procedures and checklists as well as the design of the control room itself.

Figure 4-2 illustrates a popular design of individual work areas for operators arranged around a shared segmented partial video wall. Later, we will see why this design is not recommended. Rather than being a bigger personal view, video walls are used more for overview and collaboration functions. The figure shows what it might look like spatially.

Figure 4-3 shows an extensive meteorological control room with many displays for relatively few operators. It suggests how a control room design fits the operational purpose.

Figure 4-2. Contemporary control room circa 2014 illustrating the human-machine interface (HMI) configurations for each individual operator as well as the large shared displays for coordination and overview.

Source: Reproduced with permission from California Independent System Operator corporation (CAISO).

Figure 4-3. Meteorological control room for the European Organisation for the Exploitation of Meteorological Satellites with many displays arranged in a wide sweeping arc. Note that there are few operators.
Source: Reproduced with permission from EUMETSAT.

4.3 A Note about Scope

By plan, this book does not include a discussion about the physical safety aspects of control room design or location. There is no extensive discussion of angles of view and other useful and important ergonomics that all control room design requires. Those considerations are extremely important and impactful. To design a control room properly requires expertise well beyond the intent of this book. You are advised to seek expert guidance in this area. Useful resources include the Engineering Equipment and Materials User Association (EEMUA),[1] International Standards Organization (ISO) 11064,[2] Edmonds,[3] Hollifield et al.,[4] and the Abnormal Situation Management (ASM) Consortium.[5]

---

[1] Engineering Equipment and Materials User Association (EEMUA), *Process Plant Control Desks Utilising Human Computer Interfaces – A Guide to Computer Interface Issues*, Publication Number 201, 2nd ed. (London: EEMUA, 2010).
[2] ISO 11064 Parts 1–7, *Ergonomic Design of Control Centers* (Geneva, Switzerland: ISO).
[3] E.J. Skilling, C. Munro, and K. Smith, "Building and Control Room Design." in *Human Factors in the Chemical and Process Industries*, ed. Janette Edmonds (Oxford, UK: Elsevier, 2016), 187–202.
[4] Bill Hollifield, Dana Oliver, Ian Nimmo, and Eddie Habibi, *The High Performance HMI Handbook – A Comprehensive Guide to Designing, Implementing and Maintaining Effective HMIs for Industrial Plant Operations* (Houston: Plant Automation Services, 2008).
[5] Abnormal Situation Management (ASM) Consortium, "ASM Consortium Guidelines – Effective Operator Display Design" (ASM Joint R&D Consortium, Phoenix, 2008).

4.4 Control Room and Operation Center Requirements

By now you can appreciate that a good control room or operation center is an important enabler of effective situation management. But that benefit only comes from understanding what these rooms or centers must provide to the operator. We begin this important discussion with the essential (or proscriptive) requirements. These requirements lay down the needs. At this point, we do not discuss how to deliver those needs—that is part of the rest of this book. First, let us understand and agree on requirements. Later, you will be able to pick ways to meet them that respect your operating culture and personnel. Here are the requirements.

Physical Protection and Security

The enterprise must provide the operator with all the design, construction, and operational infrastructure to ensure the following:

- Appropriate and effective physical protection from danger and harm resulting from the production or operations gone wrong

- Appropriate and effective physical protection from the errant effects of nature, both for direct protection and well-being of the personnel as well as the ability to maintain critical operations

- Appropriate and effective physical protection from violence, intimidation, and related nefarious activity

- Appropriate and effective cybersecurity to ensure that unauthorized personnel are prevented from operating, modifying, or otherwise affecting the operation of the enterprise or the underlying support activities

Environmental Controls

The enterprise must provide the necessary infrastructure to ensure appropriate health, comfort, and absence of distraction for the operator. This includes the following:

- Appropriate and properly managed facility temperature, humidity, and clean, fresh air

- Appropriate and proper management of noise and vibration

- Appropriate and effective design and operation to ensure effective visual, voice, and other means of communication for person-to-person, person-to-equipment, and equipment-to-person interactions

- Appropriate and effective design and management of illumination to support tasks and well-being during the entire operation event (coming on duty, during duty, for collaboration, and going off duty)

Information Support Tools and Technology

The enterprise must provide sufficient, fit-for-purpose support information to the operator that ensures the following:

- Ready, accessible, and current information about the approved production management targets and any operational constraints or imperatives to be met during production and operation

- Ready, accessible, and current information about the true nature of the production or operation being managed (primary information such as sensor values and control parameters and settings; secondary information including information integrity; and support information including weather effects and maintenance)

- Ready, accessible, and current information about the approved procedures and protocols to be followed or consulted during the entire scope of operation, which includes start-up, normal and abnormal operation, and shutdown

Sufficient Process and Operational Controls

The enterprise must provide the necessary design, construction, and operational infrastructure for the operator have ready access to sufficient controls (the "handles") for proper and effective production. Requirements include the following:

- Access to reliable and effective direct tools and controls required in order to properly and successfully perform the full range of operational duties expected (e.g., start/stop; open/close; production or operational parametric adjustments, such as flow rates, level settings, or temperatures; and reordering or resequencing production operations, production steps, or material pathways)

- Access to reliable and effective indirect tools and capabilities needed or chosen to support production or operation management (e.g., surrogate personnel operating under the authority or direction of the operator designed to extend the physical reach of the operator to influence remote or separate physical or functional areas not directly accessible by the operator located in the control room or operation center)

Operational Support

The enterprise must provide to the operator ready access to appropriate support technical, managerial, or operational support personnel to assist in the understanding and/or control and effective operation, including the following:

- Ability to share information via both informal communications and reliable primary mechanisms (e.g., control room display replication and procedure databases)

- Ability and mechanisms to reliably and effectively support the ability to collaborate in real time (including communications technology and physical work areas such as crisis management facilities)

Control Room Access Management

The control room or operation center is not a social center or any form of clubhouse. It should not be a convenient location for outside restrooms, coffee breaks, guest meals, cold-weather warm-ups, or anything else peripheral to operations. Think of it as a "members only" facility to the extent that the only personnel who should be present are those on the direct operating team (operators, supervisors, and the occasional manager). Others on the maintenance teams, quality control teams, safety teams, and other support services are invited guests, present and welcome for specific contact and purpose during the course of that purpose. Past invitations should not be considered a free pass for future admission.

This might sound a bit harsh and unsociable. Neither is the intention. These recommendations are not advanced to suggest that operators are a special class and are due this as an obvious sign of respect, though a bit of both is true. Proper access controls reduce distraction to enable the needed level of focus for those inside. Operating can be a very demanding task. Most times, for a long time, nothing happens, and then in a "coming ready or not" atmosphere, things go south and present an enormous challenge to even the best abilities and coolest hands. Few will have difficulty understanding the need for focus when challenging situations present. Yet most of us underestimate the need for focus when nothing appears to be going amiss. It is precisely at these times when a good watchful eye is needed all the more. As you will read later in this book, our ability to multitask is a misconception. The word exists, but people are really poor at doing it. A visitor in the control room who is not directly supporting a current task is a distraction at best. The operator is pulled both to keep current tasks on course and to interact with the visitor on a purely social or other off-task basis. It does not work well for either.

Special Operating Situations

While the control center is specifically for operations, there are special uses that must be factored into its design. Specifically, significant plant or equipment start-up, cut-over, or shutdown require that added personnel be present and have access to certain controls equipment. One-of-a-kind commissioning activities need to be accommodated. Times of operational stress such as significant abnormal operation or emergency operation also require added personnel to be present. Other uses involve the operators to be engaged in additional activities that support their operational role, such as training and certification. For facilities without separate training areas, this activity must be done in the control room or control center. All of these added uses are important and need to be factored into the control center and control room design, including operations protocols.

While not specifically a special operating situation, all designs are improved by including those small aspects that allow for current operating flexibility and for future growth. Planning might include multipurpose areas that can easily accommodate visitors, injured personnel, or temporary emergency storage of vital equipment or supplies. Planning would suggest that utilities infrastructure be a bit oversized. Planning might also include specific architectural design to facilitate structural additions for growth. While unusual, it might be important to include design aspects that are compatible with operational reduction. This has been unusual in the past but is becoming more common. Effective planning will allow for the reductions without extensive remodeling or having the control room appear almost abandoned.

Permits, Personnel, and Visitors

All activities within and affecting the operating area are the first responsibility of the shift operator for immediacies and shift supervision for continuity between shifts and over time. This means that there must be communication (see Section 10.16 for a formal definition) of all activities with the operator. This includes all deliveries, all withdrawals (e.g., removal of material or nonincidental equipment but excluding normal production output), all maintenance, and all other activity that places personnel within the battery limits of the operating area or places their effects on the equipment or operation. This discussion only considers how this must fit into the design and function of the control room or operation center, not the detailed requirements.

Permits

The control room should provide for effective communication and exchanges in a way that does not disturb the others present nor unduly disturb the recipient operator as well. This is normally handled by setting specific times for these activities to occur and ensuring that any face-to-face interaction be done without actually entering the main areas of the control room. A permit window or an electronic conferencing facility might work well for this.

Visitors and Other Personnel
Visitors are a distraction no matter whether their presence is for public relations or enterprise improvement. Where public viewing of the control room is an important part of community relations, the control room is designed for such viewing without involving physical entry. Partial glass walls, galleries, and the dual use of an emergency operations room adjacent to the control room with a communicating window are the predominant choices. Where enterprise improvement is the purpose, appropriate door and traffic pattern design should be used to facilitate physical entry, and visitor chairs or at-hand but out-or-the-way observing areas should be set aside. All of this is not to suggest isolation. Rather, it is to ensure that operator distraction is low and that the operator at his choosing can easily manage interactions.

Scope Note
It is easy to see that the previous lists are only part of the infrastructure necessary to provide appropriate situation management. Hand-in-hand would be the operator (including qualifications, training, and readiness) and the other support mechanisms (including proper maintenance, effective plant design, and much more). Most are discussed elsewhere in this book.

4.5 The Control Room

The control room is the physical place that houses the individuals charged with the responsibility of managing a plant or other operation and the equipment they employ to do it. These rooms can encompass the complete operational control of a plant or just manage a small portion. They come in a wide variety of different shapes and sizes. They can be staffed by one operator or up to nearly a hundred. They can be multifunctional to include both a board operations team and a field operations team. They can be multipurpose, having a combined responsibility for production, quality, and management interactions. Sometimes control rooms are local to the production process area and other times they are miles away, or anywhere in between. Some are elaborate, complete stand-alone structures. Others are little more than rented space in a university administration building. Today's local control rooms for hazardous materials production are strengthened and have a blast rating and isolation rating. The higher the blast rating, the more robust the building will need to be. Such protection can require walls with 1 foot or more of concrete designed to absorb and deflect a blast wave. Some are buried below ground and fitted with hermetically sealed access doors to ensure safety during the worst that Nature can deliver.

A Control Room Is Remote (but Not Necessarily Distant)

What differentiates control rooms from the ordinary is the task or job being performed inside. A control room is important because of what must be going on inside. Going on inside is the conduct of work intended for safe and effective production operations. What makes control room design challenging is the responsibility to provide that space in an appropriate and responsible way. What makes the design vital is that all the incoming information and all the outgoing management activity is done functionally remotely, separate from any physical plant or entity. The activity is not necessarily geographically distant, but away in the sense that the operators inside are only able to perceive the process, plant, or activity indirectly. There is almost nothing real inside—everything is intangible. In the real world, one can reach out and touch a pipe to determine if it is warm or check if it is vibrating, and sense the nature of it all. There is no ambiguity about what and where that pipe is. In the surrogate control room world, one must navigate an electronic device, examine sheets of paper, or do something else far removed from anything as concrete as a pipe. Once navigated there, it is only possible to sense what is provided. What is provided may be stale (due to equipment operational frequency), may be inexact (due to measuring standards), and may not even be what was sought (due to system construction or navigation errors). Success requires extraordinary care. Good control room design requires such care.

Design Evolution

The traditional control room for production processes started local to the manufacturing process with inside/outside operators monitoring and manipulating instrumentation both inside the control room and outside. These at-the-point-of-manufacturing operators manipulated the production equipment, often loaded raw materials in and off-loaded finished product out, usually did much of the maintenance of the equipment, and generally ran most everything there. As the production operations became more complex and the inside equipment became more sophisticated, the competencies required to manage things changed. Dedicated inside (panel or board) operators became necessary. This role eventually evolved to today's process control operator. Some companies rotate all operators (board or inside, and field or outside) through all the operator positions, while others have dedicated competent inside operators managing just the controls and dedicated outside operators managing everything in the field.

Safety concerns have identified people located too close to the process to be at risk during process upsets gone badly. In response, regulators have encouraged moving the process control operators away from the local control rooms to remote control rooms. These remote centers are designed to be safe havens and may be located near administration buildings for the convenience of personnel. This leaves the field or outside operators within the old existing control rooms to manage maintenance activities, monitor the physical equipment for problems, and manipulate manual controls. These are obviously no longer control rooms and have been renamed *field shelters*. Modular blast-resistant buildings provide a safe haven inside in case the abnormal situation turns dangerous have replaced many of these older buildings. Many old control rooms sited local to the actual manufacturing operations are being systematically demolished or mothballed.

Architectural Aspects

The shape of the control room has been changing with the development of technology and, to no small extent, fashion. Manufacturing control rooms used to be small, narrow buildings with a pneumatic instrumentation panel and room for a chair or two. These have evolved into specialized computer control rooms as a result of the transition of instrumentation from pneumatic to electronic and finally to microprocessors. Some buildings built in this transition period were actually round! These were not very successful. This aesthetically pleasing shape might have been enticing, but neither the interior space allocation nor the work-space effectiveness was aided in any way. Most current designs are rectangular or square. Consoles are placed in a horseshoe or circle, and operators sit with their backs to each other, or tiled in such a way as to be viewable by each other. This evolution continues.

As the technology has continued to evolve and prices for new technologies have become more affordable, large theater-type video walls have become popular and the shape has changed yet again from square to longer rectangular rooms with operators facing forward to these large screen displays (LSDs). There is an in-depth discussion of display types, arrangement, and use for operation and coordination in Chapter 5, "The Human-Machine Interface." At the time of writing this book, theater style is the most dominant style in the industry. Notwithstanding trends, the foremost requirement that must shape your design is what you will use the control room for and the needs of your operators. No one size or one design fits all.

Figure 4-4 depicts a small four-chair control center where extensive use of hardcopy resources is required.

Figure 4-4. Example simple control center design illustrating the presence of supporting documentation, individualized operator HMI displays, and overhead shared large displays for overview and coordination.

Source: Reproduced with permission from Winsted Control Room Console Company.[6]

4.6 Operation Centers

As mentioned earlier, the operation center is focused on more than just process control and is usually a facility used by the person or persons in charge to command, control, and coordinate all activities. Operation centers are often referred to as *command and control*. Today the differences between operation centers and control rooms are minimal. For purposes of situation management, they will be treated in much the same way. Both have a need for good situation awareness, and both have elements of control without direct field operations activities but often coordinate with other multiple field operations centers. They have more management involvement and can take on many responsibilities including security, supervision, and planning.

Their design is predominantly theater style; classic examples are the NASA control room or Air and Space Operations Center, the Theater Air Control System (TACS), military command and control centers, and electrical transmission centers. They often take advantage of a central large screen, configured with multiple windows.

The console furniture for both control rooms and operation centers is for the most part identical and the environmental considerations are similar, so only size and functions differ. However, certain rooms and centers may have specialized furniture and interface equipment designed for their unique roles. Most no longer have physical

---

[6] "Control Room with Binder/Literature Storage, Slat-Wall Control Console," Jay Stanley & Associates, https://www.jaystanley.com/products/furniture/command-consoles-desks/control-room-with-binderliterature-storage-slat-wall-control-console/.

control equipment, computers, and instrumentation interface equipment as they are located remotely. For the most part, hardwired switches have been replaced by soft keys (clicked on a mouse or pushed by a screen touch).

4.7 Collaboration Centers

Collaboration center is a relatively new term for a purpose-designed room (or center within a larger space) that is specifically designed to help individuals jointly and concurrently work a common problem. The individuals share a common space and need to extensively share points of view, link with others outside of the center for their points of view and "soft" resources, and share their own "soft" resources. The design of the center includes audiovisuals and communications. The first thing to clear up is that they are not for managing minute-by-minute real-time activities. That is what control rooms or operation centers do. Those are set up to provide the massive amount of real-time measurements and controls handles to do that job. Collaboration centers are set up for people to share their expertise and perhaps disparate information in order to either reach a consensus about the meaning of a given situation or reach a consensus about what might be important for others to consider as they reach that consensus. Their work can be ongoing, but usually it is for a specific campaign or other limited circumstance.

Requirements

Physical Requirements
The proper architectural and other appropriate and necessary design aspects for the center are listed below.

- Adequate physical security to manage admission and ensure the safety of attendees
- Proper room conditions including lighting, noise isolation and control, capable air conditioning and ventilation, hazardous substances intrusion prevention, and access to toilets
- Comfortable and sturdy furnishings

Functional Requirements
The functional requirements to facilitate proper collaboration activities are outlined below.

- Seating space and "connecting ports" for all participants so that they are within view of each other and considered part of the group
- Seating space for observers who are not collaborators unless invited to the table

- Shared audio system so all can hear; all audio ported to the observers

- Shared video space so that all can view all visuals ported to the observers

- Ability of each individual to port audio and/or video to the observers and to an external group in a seamless and attributed way

- Ability of each individual (via provided or bring-your-own equipment) to access all necessary not-yet-shared personal information for personal reference or possible sharing

- Ability of predetermined individuals to access outbound or two-way audio and video outside communications in a manner that does not disturb other individuals

Configurations

Preconfigured with Workstations

The function of these centers requires the use of preconfigured workstations. This might be due to the special nature of the collaboration needs or the fact that most participants will not have or are prohibited from bringing their own. Backup control centers are one type. Secure rooms are another. These centers will have specialized equipment appropriate to the needs of the participants and must be preconfigured. Where both preconfigured and bring-your-own workstations (e.g., laptops and tablets) are permitted, refer to the next section.

Bring-Your-Own Workstations

The purpose of these centers includes the necessity for a variety of workstations where the choice is mainly that of the participants. The potential for variety places additional requirements on proper cybersecurity equipment and procedures being in place. It also requires sufficient flexibility for space on the work tops for placement and use of the equipment as well as note taking and use of other reference materials. Appropriate interconnect equipment connections (e.g., Ethernet, infrared—IR, and Bluetooth; audiovisual; printers, faxes, and scanners; and power) must also be present and robust.

Figure 4-5 illustrates what a small bring-your-own collaboration center might look like. Note the large flat screen visible to all and the individual connection stations.

Importance

Collaboration centers can be useful facilities for teams in their own right. They are included here because their design and technology make them a good model as an adjunct to the control room for crisis management, potential backup to the actual

Figure 4-5. Example of a small bring-your-own workstation collaboration center showing the dedicated nature of it, electronic power and connectivity for individual work tools, and large shared display for collaboration.
Source: Reproduced with permission from Winsted Control Room Console Company.[7]

control room, and for training. Collaboration is a key part of all shared control rooms. Many of the attributes and capabilities of collaboration centers should be seriously considered for direct inclusion into control rooms and operation centers to support operators as well. This discussion offers a nice introduction to our next topic.

4.8 Advanced Technology Control Centers

There is a new and exciting technology to support the control room operator that is advancing so rapidly and extensively that it would be difficult and awkward to include it within the current control room functionality. This capability of video tools for surveillance and plant condition monitoring support is now considered nearly essential to provide the operator with important and timely information. The current and near-term requirements for these tools to be fully manned demand more skill and attention than could be expected from existing operators in their present control room or operation center. There is a growing realization that the best way to resource these capabilities is by using special-purpose advanced technology control centers.[8]

A good example is for operating video surveillance of airborne drones. This center would require specialized pilot interface controls for managing the flight of the

---

7 "Winsted Catalog," Winsted, https://www.winsted.com/2015catalog/files/assets/basic-html/page-76.html.
8 Jack Pankoff, private communications, November 2015 and continuing.

equipment coupled with extensive video imaging and display capabilities. More than one individual may be needed to manage things. A pilot would be managing the flight aspects of the machine, including launch and recovery, flight path, altitude, and attitude. One or more surveillance operators would be controlling the various instruments aboard the craft. Instruments might include ordinary video feed or additional specialized analytical devices such as acoustic, infrared, and nuclear. The operators' duties would include targeting, resolution, image size, and image coordination. In this situation it would be easy to understand the need for all team members to fully coordinate and support each aspect of monitoring.

So far, so good. Now, consider that the entire video surveillance mission is to support an operator needing to know whether an unmanned offshore platform might be leaking. Such a leak might be fugitive air emissions, oil or other liquids on the platform, or oil or liquids in the water. This would require extensive understanding of how the platform is constructed and operated. It would involve close coordination of that knowledge and the flight and monitoring operations. Both the control room and the advanced technology control center would need equipment and protocols to support that close coordination.

Another example of close coordination is the control room aboard a nuclear submarine (see Figure 4-6).

Figure 4-6. Control room of the attack submarine USS *Seawolf* (1997). Note the proximity of coordinating operators and supervision located directly behind.

Source: Reproduced with permission from Dive and Discover.[9]

---

9 Photo by Renso Amariz, March 31, 1999, "Dive and Discover: History of Oceanograph," Dive and Discover, https://divediscover.whoi.edu/archives/history-ocean/battle-slideshow6.html.

There are other better-known centers that support industrial production and operations. These include severe weather centers, seismic monitoring centers, national (homeland) security centers, and an entire web of product distribution centers. Yet the nature of the relationship is more background support than close tactical operation support. Thus the coordination loading on control room operators is less and little specialized equipment is needed. This is not to say that they are any less important, just that their interactive role is more informative and less interactive, thus requiring less specialty or specialized control room equipment.

4.9 Design of Effective Work Spaces

Let us get right into the "work" part. Sure, operators will be working there. You want that work to be made easier and more effective by how the space and all elements within are designed and built. To do it well you want to make the work part focus on the jobs that needs doing, so not to be hampered or frustrated by the physical space and its components. Remember how you felt the time you opened the door to a fantastic workshop, hardware store, or pantry? Remember that it seemed to magically transform you into someone who wished to belong there. You wanted to pull stuff off the shelf and out of the drawers and get started. You were part of it. This is why we are going to see what it takes to put the operator in a similar frame of mind.

Until now, most of what you have read about control rooms and control centers talked about lighting, noise control, visual distraction, and the ergonomics of everything from the operator chair to the desk to the layout and availability of essential amenities. All of this is extremely important. We will get into it later in the chapter, and we'll do it in a way that will make it useful and comfortable. That list is essential but not nearly enough. How often have you been in a place that was thoughtfully designed and tastefully and carefully built, but felt sterile? Everything was there and purposeful, but after a while you felt that you would like to leave soon. Certainly you would not want to spend 12 hours of hard work there. You could feel things were not right, but it was hard to put a finger on what.

The Concept of Space

Functional is not enough. It falls far short of encouraging operators to stay there and work. In addition to functionality and the inclusion of a bit of aesthetic embellishment, we will add more. We add the concept of space. By respecting and exploiting the full design of space, not only is being inside useful and effective, it feels good to be there. Being there opens the occupant to greater empowerment and energy. We manage space by understanding and managing spatial introduction, geometry, expanse, collaboration, and retention. What follows is an introduction to these concepts and the reasons they are important. These concepts might be new to you but are important. Some of them are new to control room design experts as well. All of us are just beginning to understand the power of architecture and how it can and should be applied to control rooms. However, please do not expect this chapter to be a design guide for architects—far from it. Its purpose is to encourage you and control room designers to go beyond bare functionality and leverage the power of design to produce control rooms and control centers that are successful work spaces.

Spatial Introduction

Spatial introduction is the specific way (pathway) that the space design invites one to enter. Introduction is all about creating a promise. The promise is what the space will "deliver" to the entrant. Promise initiates or facilitates the individual to prepare to accept the space as "inviting" or "dangerous." Depending on purpose, it is the initiation of the reason for entering. Retail shops, online home pages, an artist's studio, and store windows, all strive to promise something good, often very good, if only you were to enter. On the other hand, a city's protected water works, a military base, or a recluse's lair all strive to announce that trying to get inside would not be a safe or comfortable idea.

Control room introduction must follow a carefully defined fine line. On the one hand, it wants to effectively invite the operator to come on shift. It should prepare him to make the transition from someone else to the role of responsible operator. Entering the control room should perfect that transition. On the other hand, we want that invitation to extend only to operators—not in an overt or confrontational way, but in a functional and respectful way. One of the more successful methods for doing that is to have a way for the visitor to "visit" without entering. Having an observation wall or gallery does that. The "visitor," be he a coworker in a support role or a visitor to the company, is invited to visit in a way that is open and inviting for that role but in no way disturbs control room operators. This really is useful and does work.

Geometry

Geometry is the tool the designer uses to shape the space to deliver on its intended feel and purpose. A successful geometry will produce the desired *expanse*, facilitate the transition into *introduction*, foster *collaboration*, and encourage *retention*. Geometry is the tool kit, the art, and the sizable craft of the architect. It asks what room shapes or designs or measurement ratios contribute to an inviting and productive space. It suggests discovering and avoiding those aspects that detract.

Expanse

Expanse is the emotional feel that being inside the space gives us. Expanse is all about feel, but feel is intimately connected to the purpose for the space. And purpose is the answer to why we put the operator inside control rooms.

Keep it simple; imagine that there are four levels of expanse:

1. **Inside** – Neutral; a space, but just a space. Being there is just a location; you do not gain anything by entering and you do not leave anything behind after you leave. Maybe you did leave footprints but certainly nothing you would ever miss.

2. **Inviting** – Our space; a place that you and I go to work together. We can work on the same task. We can be working on different tasks but want to share the energy of working. Remember the time when you went into a study area, found your nook, and got right to work? Work wanted to be done; the energy and focus and need to work came at you from everywhere.

3. **Intimate** – My space, for me. I can invite others to be in here, but I do it for me or for them in a way that I want them to experience.

4. **Reverential** – What a space! It was mine, yours, ours, and everyone's at the same time. Entering was a transition. Being there was an experience. Maybe the place told you something new. Maybe the place let you teach yourself something you were missing. Being there caused you to be in a different place both inside and with your feet. Leaving was hard, rewarding, and necessary. Returning after leaving was high up on the wish list. All in all, the experience was an elevation of the self with the experience of being a part of something much larger.

Lest we forget, I remind you how often you might have seen a space from outside; formed a wonderful (or repulsive) opinion of what it might be to be inside; and on entering, discovered how misleading those earlier feelings were. Depending on what preceded, you may have been pleasantly surprised and happy to be there, or could not wait a minute more to leave. Expanse is all about delivering on the promise. The promise is what the space requires to be a good control room or operation center.

Collaboration

Collaboration is the way a space affects its occupants' interactions. Does being in this space together encourage individual operators to think of themselves as a team? Does the space facilitate their working together in a way that is progressive, proportional, and respectful?

- **Progressive** – Allows for easy start without either a strong commitment to continue or a penalty for discontinuing; yet encourages what is needed

- **Proportional** – Allows for the right amount of collaboration that appears appropriate for all

- **Respectful** – Allows for participants to decline to collaborate without interfering with others, without suggesting to others that collaboration might be wrong; yet permits the noncollaborator to join in later without bias

Retention

Retention is the ability of a space to retain its occupants for the purposes intended. We want the design to reassure occupants that they are welcome for as long as they remain—not only welcome but wanted. Control rooms will encourage operators to be there, on task, and effective. If you need motivation for this feeling, please recall places where you happened to be and *could not wait to leave*. Getting "out of there" was the only thing you wanted to do as soon as you could! Now that you remember what this felt like, try to imagine how important it is to have the reverse. Of course, you really do not have to imagine; you have such places. They are likely private ones. Now associate that ability for encouraging occupants to stay, to help them to want to stay—not forever, just for their turn at duty. Ask the control room to provide this feel.

A Few Thoughts

Please approach this discussion with an open mind. When you do, you will see a lot of value here. By intent, there is little in the way of concrete methodology on how to deliver on this. The author apologizes. All of this is important, but it takes serious professional experience to do it right. That experience, unfortunately, exceeds mine. The topics are included for a purpose. They are important. Control rooms that otherwise appear to be perfect in delivering all the check boxes might not be so nice to work in. Rather than leave it at that, I made a choice to expose you to the "rest of the story" with the hope that it would be helpful. Now you have learned a bit more about what to ask for. Please work with qualified professionals who understand and can assist you.

4.10 User-Centered Design

User-centered design must seem euphemistic. It replaces an earlier somewhat fancy term, *human factors*. *User-centered design* means that whatever we provide ("things") for people (operators in our plants) to use should be designed with all their capabilities and limitations in mind. One should be able to walk up to the thing, pick it up, and start using it effectively right from the start. This means that the thing must, as part of its fabric of being, instruct us as to how to use it. Easy things must be obvious. More difficult things must be readily discernable. Impossible things should be evident. That is simple enough. We call it *affordance*.

Affordance is the property a physical thing has as a result of purposeful design. It suggests, or teaches, the user how the thing is to be used. It is built in in such a way that when anyone picks up or starts to use this thing, the thing suggests, in an almost persuasive way, how it wants to be used. As soon as you begin to wonder how you might find a way to have the thing operate slower, or quieter, or with more power, the thing has already presented you with handles to start to do just that (and more, of course, because the thing has a lot of other stuff you can ask it to do). A pail with a handle sticking up would "suggest" to anyone walking up to it that it can be picked up by the handle. A button means "push me to make me do something." A knob says, "twist me." A doorway that has deliberately been made to look like an invisible part of the wall would not suggest how it is to be used. A thing that requires reading a manual before one can figure out how to turn it on would not. Your list is longer than mine. You are getting the idea.

In addition to affordance, we ask plant control systems to possess additional properties that make them usable. Notice that I did not say *useful*. That is another issue. Let us save it for later. We generally think of six important items for usability: environment, scaling, compensation, understandability, implementability, and unified feel. Of course, now we are talking about operators, control rooms, and such. We leave the abstract behind.

Human Factors Details
Review the key components in Figure 4-7. Start at the 12 o'clock position and progress clockwise around the diagram. Starting at the top is a natural way to approach something like this.

Environment
The control room must provide appropriate temperature and other environmental controls such that no operator feels a need to even think about them. Included here are noise controls, visual distraction elimination, management of seating, viewing distances and angles, and all the rest that make up the operator's work area. We include the proper design and sufficient supply of storage for equipment and resources that might be needed and used. Nothing interrupts and interferes with a task more than

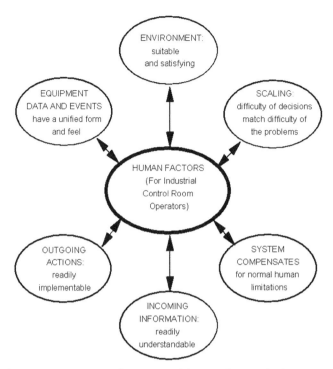

Figure 4-7. Human factors components of a successful control room design.

having to stop and find a scale, stylus, replacement battery, and all the other stuff operators use and use up on the job.

When applied to the operator's ability to gain situation awareness, this implies that the display, keyboard, and other operator view and entry equipment have been chosen and arranged for comfortable, easy, and convenient use. Operators coming back to shift after long weekends off, vacations, and even extensive jury duty or sick leave must feel welcomed by the equipment. Everything should be where it is expected to be and work the way it is expected to work.

Scaling

Proper *scaling* means that the general difficulty for the operator to perform certain operations should be affected by the basic difficulty of the task itself. Easy things should not be hard to do; hard things should not be impossibly difficult to do. To avoid scaling problems, a "one size fits all" approach to design and implementation rarely works. That approach usually is either biased toward the easy-to-do things, making it much more involved and difficult to do the harder things, or it is designed around the more difficult-to-do tasks, therefore placing an undue burden on the easy tasks. We see this commonly in decision-making aids. Often, they are designed to work their way down a set of sequential queries, with the appropriate answers chosen at each step.

Because the problem may be a difficult one, each step along the process must be finely honed to elicit the proper amount of thinking to arrive at the correct choice. Then we move on to the next query. In this manner, the operator can work his way toward the needed support. Such a laborious process might be useful for a very involved and difficult situation, but it will be extremely vexing for simpler ones.

Compensation

There is an old story that describes the attendants at an "ideal" control room:

> The only two occupants are a man and a dog. The man is there only to feed the dog. The dog is there only to make sure that the man does not touch anything.

The message is clear enough. Giving in to the temptation to "tweak" things a bit, the operator often contributes to operational abnormalities—not the production improvement intended. Almost as often, the problem was due less to "tweaking" or "tinkering" than to the foibles of human error. We are not perfectly created engines of production operation support. None of us can shoulder the mantle of invincibility just by entering a control room to manage an important and perhaps dangerous production operation. We come as we are, to borrow from an interesting social party event. It is the responsibility of the equipment and technology designers and providers to ensure that normal human errors are accommodated. *Accommodation* means that terrible things should not happen just because someone makes a small mistake. Two illustrative examples come to mind. The first example that has occurred more often than we would like to admit centers on operator consoles with several stacked video displays. Each pair of displays is managed by a shared data entry device (e.g., keyboard and mouse). Often enough, the operator's eyes were on the upper display, but the keyboard was on the lower display. The second example relates to the practice of directly entering controller set point or output values via a keyboard. Our operator enters "1" <+> "." <+> "5". But the control system missed the decimal (maybe the operator hit that key a bit lightly) and reads the entry as "15" instead. The controller cannot recognize intent. The new set point immediately becomes 15.

Understandability

One of the best characteristics of modern process control systems (PCSs) is their ability to provide a great deal of data to the operator. In addition to the control variables and measurements surrounding control loops, they can provide elaborate temperature profiles, imbedded analyzer values, valve positions, and a host of other seemingly important and relevant data. And, as you suspect, the operator has too much data and perhaps too little information. User-centered design principles suggest that much fewer data are needed; much more context and interpretation are required. Moreover, the method of presentation itself can either obscure understanding or provide just the necessary context. A simple way to do this would be to provide graphs with goal marks located at appropriate levels on the same scale instead of tabular data with numerical targets.

Implementability

You might suppose that this part of the user-centered design would be the least offended requirement. Operator workstations have been designed and built around providing all the necessary "handles" for operators to manipulate the things operators do during their time in the chair. Let me suggest that such is not always the case. All too frequently an operator has to "hunt around" for the right valve to check, or the needed process value to observe, while he manipulates another process value—but neither can be found on the same display. And there is the situation where the operator is required to shut down part of a process by gradually reducing the set points for several controllers in strict unison to avoid thermal or other operational stresses. The movements of set points are synchronized by the changes in appropriate process measurements. Not only are the controllers on different displays, but also many of the variables needed to check the progress of the work are not colocated either. The operator has to do a lot of shifting around and remembering.

The message for achieving good design is to ensure that routine operations and strategically important operations can be carried out in ways that ensure success and provide for a minimum of operational risk.

Unified Feel

When we ask for equipment to possess a unified feel, we are suggesting that once operators learn how to interact with some of the equipment, they will know how to interact with all similar equipment. An excellent example of this requirement can be found by approaching any conventional personal computer. Whether it be run under a Windows, UNIX, Apple Mac, or other system, it would be obviously clear to the user how the keyboard is used and what should happen when a mouse or other pointing device is moved. We know what to expect when one "right clicks" the pointing device. This is also to say that the activities that are used by employing the keyboard and pointing devices are also going to be interacted with according to an expected feel.

Example of Mixed Technology

The Russian nuclear power plant control room shown in Figure 4-8 provides an excellent example of using different forms of technology to assist the operator to manage

Figure 4-8. Control room of a Russian nuclear power plant illustrating the use of a wide variety of interface technologies deemed suitable to ensure situation awareness, assessment, and operator intervention for managing a critical plant.

the plant. In this control room there is a dedicated control wall (curved) with embedded single case controllers and recorders, physical discrete instrument analog display mimic of the reactor core, critical alarms, and electronic graphic displays. Below the wall is a dedicated control desk with physical controls and additional analog indicators. In the center is a modern computer HMI station.

4.11 Control Room Design

This section discusses the specific elements important in the design of a control room or operation center. This discussion introduces the important elements that impact *situation awareness* and good ergonomic design principles. Note that the topics here need to be interpreted together with the earlier topics on effective work spaces (Section 4.9) and user-centered design (Section 4.10). In other words, take things as a whole. This discussion is one part. It should be useful for both improving an existing control room and designing a new one.

An international standard, ISO 11064—*Ergonomic Design of Control Centres*—not reviewed here in any detail, covers the basis of design.[10] ISO 11064 follows a life-cycle model that begins by identifying the requirements and location of control rooms by

---

10 ISO 11064 Parts 1–7, *Ergonomic Design of Control Centers.*

referring to relevant standards. Next is the conceptual design phase that identifies several alternative designs with their strengths and weaknesses, allowing the user to select the design that is the most functional, affordable, and acceptable. Next, the detailed design phase adds practical design and ensures that the control room meets all applicable required codes. The final phase is a verification and validation phase that is often completed at the end of the construction.

Important elements of control room design include:

- Location
- Security
- Building style
- Layout
- Design considerations
- Principles and ergonomics
- Console design
- Large-screen display design
- Life cycle

Location

The selection of the location of a control room includes many considerations. Physical locations include the following:

- **Proximate to the plant** – Just outside or even within an operating area (subject to the requirements for personnel and equipment safe operation, of course).

- **Remote from the plant** – Well outside the physical operating area (for safety reasons or to provide for convenient personnel access ["follow the sun" options]).

- **Satellite or field** – Proximate to an operating area but having an operating responsibility for a small plant or supplementary operation to the main control room.

- **Redundant** – Usually quite remote from the operating area, designed for emergency operations (due to weather, threat of physical danger, or actual physical damage at the primary control room), or designed as a remote command or observation center for coordinating strategic operation in response to threats at the main control room.

- **Convenience** – This explores the distance of the control room to process units, to offices, or to other facilities that have a functional relationship, such as an adjoining local laboratory.

- **Organization of the layout for lines of sight** – Does the operator need a physical view, for example, like an air traffic control tower? Lines of sight associated with other views such as large screen displays and any direct speech communications need consideration.

Security

Security provides protection from injury from both the inherent dangers of the enterprise as well as danger from outside—"bad actors."

- **Physical safety** – This means personnel safety from the production operations themselves, such as concerns about potential explosive areas, toxic gas, or fires.

- **Physical security** – This is about selecting a building location considering threats from terrorists, thieves, or industrial espionage, and how visitors', contractors', and emergency responders' access will be managed and what gates or barriers need be provided for access.

Building Style

We generally think that building style would be based on functionality. Is the building going to provide a common viewing gallery, large-screen displays, or functional layout of consoles based on site geography or organizational departments? Is the control room to consolidate other control rooms? What other administrative functions or buildings, such as laboratories, are associated with the design? Today many companies choose to have their engineering and managerial offices close to the control rooms. When doing so, they need to ensure that this geographic convenience does not interfere with the needs and function of the operators. The building design can also reflect the enterprise's desire to project a public image. Though this consideration would be added on, it is not a determining factor that affects other essential requirements.

Layout

Layout is considered for the individual control room as well as what other facilities will be within the same building, what adjacencies are required, and any constraints to minimize distractions. These additional rooms can include meeting rooms, exercise rooms, libraries, offices, mechanical rooms, electrical rooms, or server rooms. The layout should consider facilitating teamwork such as communication and collaboration requirements. Also considered are the circulation of personnel and maintenance access. The selection of space must consider ceiling heights, wall thickness, floor types (computer floor), and floor space per working position. For additional information on this topic, consult the appropriate resources.

There is a clear distinction between control room layout (just discussed) and design (briefly discussed next).

Design Considerations

The first requirement is to work the design from the operator out. This means that all aspects and needs of the operator take precedence. These considerations must account for ergonomics; anthropomorphics; building standards and codes (including fire and security); industrial guidelines; company guidelines; lighting levels; vibration requirements; acoustics; environmental controls for temperature, humidity, and ventilation; personnel and equipment traffic flow; and compliance with the US Americans with Disability Act requirements, among the long list. How is the building protected during a toxic gas release? For example, by closing and sealing vents after auto detection of hydrocarbons in the air? Civil construction requirements include soil samples from the identified site to assure building integrity as well as to avoid polluted landfill. How are cables and piping to be brought into the building? How best should we accommodate future planned or known expansions? What are the compatibility and consistency needs between existing buildings within a company or a site?

Principles and Ergonomics

To help operators to perform at high standards, it is important that the monitoring and control equipment be designed to the latest ergonomic standards. To achieve this, ISO 11064 calls for a task analysis to identify operational requirements for start-up, shutdown, process transitions, abnormal operating conditions, and, finally, emergency operations requirements. Other requirements that must be taken into consideration are the HMI requirements, alarm management, standard operating procedures, training, and the tie-in to other systems requirements. These requirements also spell out the ergonomics around the console design, use of large-screen displays, and seating or standing requirements. Basically, we are talking about the effective application of the user-centered design principles introduced in Section 4.10. Staffing requirements must be based on workload studies identifying the number of people required within a shift team.

Console Design

The console design should have been specified based on the HMI design. This dictates navigation techniques, the number of screens on the console, the number of keyboards and mice, and communications equipment (phones, radios, and video conferencing). Further consideration should be given to body posture, fatigue countermeasures, layout, maintenance availability, adjacency requirements, redundancy, and hardwired switches and lights. Finally, we need to consider the requirements for storage space and any secondary displays.

Life Cycle

The life-cycle model must be established to identify continuous and long-term maintenance requirements, replacement strategies, and any change control mechanisms. This is very site specific.

4.12 The Mobile Control Room

A mobile operator will likely not need a mobile control room in the traditional sense. A physical control room (or operation center) provides a complete framework for including the operator in a suitable environment with sufficient information and appropriate physical operations management capability (e.g., actually change a valve position, start a pump, or reroute telephone calls to a different trunk location). But what if your operator could do everything needed without being tied to a control room? Or what if your operator needed to be out and about because he is responsible for most of the physical hands-on things, detailed look-see's that are not conducive to video monitoring, as well as all the rest of the control and operations monitoring that requires use of electronic or other forms of remote controls? What then? Are we to disregard using all those functional and useful design tools and requirements just because we might not have a "room" to use them in? Well, of course not. We will just need to understand the essential aspects for the new mobile setting. This topic was introduced in Section 3.11, "The Mobile Operator." Those foundations and concepts clearly apply here.

In addition, each of the requirements noted in earlier in this chapter will need to be reviewed to make sure that your mobile control room is capable for the job. The fact that it is mobile should not in any way suggest that an expensive, awkward-to-design, or difficult-to-implement capability can be skipped. You will need to carefully review the items listed in Section 4.4 and Section 4.10 of this chapter. Where the mobile control room supplements an existing physical control room, there is a tendency to forgive limitations in the mobile setting. Such forgiveness must not get in the way of proper design. If your operator is expected to monitor and manage from either, then the mobile setting must cover the equivalent scope and provide equal functionality. There is no room for "well, if the operator is having a problem doing it remotely, he can always return to the control room to do the job right" unless that is always the case. If another operator supplementing the one in the control room covers the mobile part, there is some wiggle room. In that case, clear lines of responsibility need to be in place to prevent situations from being missed and avoid conflicting management of operations.

One thing is clear: industries have been fairly early adopters of technology and have been willing to take advantage of new and innovative ways of implementing control and supervision. The bottom line is to make sure to respect the difference between having some neat technology to do something with and having the proper technology to do the required job.

4.13 The Role of the Control Room

Early control rooms were a focal point for mounting equipment on large overview panels or mimic panels. They evolved to a place to store distributed control system (DCS) equipment. The operator part was not explicitly included in design. Certainly, it was a place for operators, but beyond this there was little else included that was not dictated by the needs of the equipment it housed. Control rooms were used as a convenient meeting room for all manner of personnel who happened to be in the area, for whatever reason. A ready toilet was available and handy for all in need.

Today, proper control room design is all about the operator. Its design is one of the main influencers on people performance, managing fatigue, and reducing human error. It is not only about getting the lighting levels to a value that stimulates people and acts as a fatigue countermeasure, it is also about how people work and how the design features that we create influence behavior and performance. We have a habit of designing control rooms from the outside in: we pick a location, then a building style; pick a room layout; add consoles and screens; and then do the HMI design. If we really want to do the job correctly, we should do the reverse. The operational requirements will suggest the appropriate HMI. The HMI should dictate the console layout, the console layout leads to the room layout, the room layout proceeds to the building, the building shape picks the location, and so on until the job is done.

The control room design should have just one major consideration: situation awareness. This is the one element that has led to so many major accidents, incidents, and major losses. With the knowledge that has been gained from reading the rest of this book, the reader should be able to understand and provide a better design for the control room. We have listed the major considerations for the design of a control room, but this text is not intended to be a control room cookbook. Your chosen design should be based on these considerations but customized by your requirements and operations experience.

4.14 Looking to the Future

So far we have covered what is common practice for control rooms. This material is based on years of experience in the processing industry as implemented in hundreds of control rooms around the world. The design here does work and will work for you. But what of the future? Will handheld devices dominate and make control rooms superfluous? Will other technology critically shape the design and use of control rooms? Will the desire to eliminate shift work propose an entirely new way to site and use control rooms? Expect changes, perhaps ones that cannot be envisioned. For now, the high-performance control room is a best practice and should provide real and substantial benefit for a sufficiently long time in the future to warrant investment and use.

Research done on handheld and heads-up display type devices has confirmed their usefulness for working on individual equipment and identified that they are not so effective at monitoring the big picture. So it is unlikely that outside operators will take control back to the field. If they did, it would probably be quite a while before the majority of industry would make that transition. We should be fairly confident in predicting that control rooms will be around for a while.

We also see the changing face of large video screens with touch control similar to the iPad, and then there are the awesome displays used on many TV shows that show the flexibility of throwing a picture onto a video wall display. This is our now. We are seeing a rapid drop in pricing, and this technology is becoming widely affordable to industry. The future is about improved HMIs, more awareness and better implementation based on best practices for situation awareness, and sound ergonomics. Interestingly enough, there are interactive large-screen displays that show old, traditional instruments on the screen and have the ability to manipulate them as if they were discrete devices.

The operator console or desktop is changing and is now ergonomic; it has improved display capability and the power to easily and rapidly be configured to any server, providing the power of many different systems to the operator desktop displays. We now have solutions that allow a single mouse and keyboard to interact with multiple screens and multiple servers providing a powerful HMI. The control room is also a place to fight fatigue while being on long shift work. We have seen the introduction of exercise equipment into this environment to allow operators to fight fatigue by raising their adrenaline levels, and in the future many more fatigue countermeasures will be used in the control room. Today, furniture with sit/stand features is prevalent. Furniture manufacturers will likely get more aggressive in addressing fatigue, making the furniture ergonomic: no more screens that cause repetitive stress injuries because they are too high or have too wide a field of view thus requiring a lot of head movement to use them. There is a larger trend that sees control rooms moving closer to operation centers and thus becoming more like command and control centers.

4.15 An Architect Weighs In

We approach the end of this chapter by pausing to take a brief look into what formal control room architect Karen Smith looks for and advises her clients. She provides a different and useful perspective of how engineering and design basics fit together into the control room design activity.

> The drivers for a control room are many: Safety is undeniably the most compelling business case. In addition to an opportunity to integrate proper ergonomics, there is tremendous potential for a well-designed control room to improve operations, communication, productivity, vigilance, and worker retention. Intangibly there is a sense of pride. A building can act as a catalyst to improve morale, and subsequently performance, job retention, and overall wellness of the operations group.
>
> **Process**: Control building design should follow a pragmatic approach. ISO 11064 outlines a credible process. Following it will increase the opportunity for a successful ergonomically compliant new control building. Depending upon in which continent the control building is situated, it may or may not require ISO compliance. Alternatively, if 11064 is not a project directive or requirement, there is still a compelling case to consider it and develop the control building with an iterative approach.
>
> **Participants**: Once a business case is made and there is a new building or re-build on the horizon, it's important to bring a cross section of key stakeholders to the table. Control building design takes years to develop. Subject to the usual market fluctuations and other kinetic factors, this means the opinions of the individuals driving decisions are subject to change. This is manageable and can be a great team building and learning experience for the plant. Having the right individuals dedicated to the process throughout the duration is critical to the success of the overall project. Management and Operations should have [an equal] place at the table for the duration of the design process. This can vary depending upon the individual plant and one size does not fit all. Having an open mind to change, and assuring buy-in from key stakeholders is critical to the success of how the new building will be perceived as it comes to life. Recognizing that trade-offs must be made and that not everyone will get precisely what they want is key, albeit a tall order. Ultimately allowing the building to act as a positive element for change developed by the voices of those that will live within it is the best formula for success.
>
> **Life Cycle**: Control buildings are not easy to change. Once built, the decisions and design for better or worse will need to endure for approximately 30 years, or even longer. Buildings exist today all over the world that house process control operators in deplorable conditions. Lighting is poor, acoustics inhibit proper communication, layouts are not conducive to support appropriate interactions; the list goes on. And yet this is not unusual. While ISO is a game changer, it's still a hard sell and not mandated—how do you adequately measure a return on investment for a properly designed control room? Perhaps

you have experienced this first hand? It's in the best interest of the plant to consider the future when making changes to the physical space that surround[s] the operators. Think about how fast the technology changes, and align that with the design of the control room. Note also that ISO calls for planning for a 25% growth to be incorporated in the initial design of a control room. This will impact the entire building or suite, as it should.

A word of caution. The idea of designing a new control building is an opportunity for change. It may be that a floor plan or concept emerges early on. This can be generated by any number of individuals with good intentions, however it can become a liability. The floor plan should emerge ideally AFTER the proper questions have been asked and answered. Accept that there is a tendency for many to "draw" up their version of a layout, no doubt in a nice, rectilinear box. Engaging qualified expertise and an objective facilitator for the control building is key. Try to limit the armchair architect from the plant and engage HFE industry expertise as well as a proven, qualified control room architect. The results will be quite different, and far better.[11]

I recommended that you seek out and work with experienced and qualified architect/designers to guide and assist you.

4.16 Close

A weak design for a control room poses a risk of low operator morale, decreased performance, increased operational errors and errors in judgment, and more. Please realize that high performance it is not just about lighting, acoustics, and walls; it is the whole house for HMI including good alarm management practices, and it is about workload and user-centered design based on task analysis. There is a lot more to it than just designing and building a control room. Getting it right the first time is important. Operators are more likely to endorse the changes when they can see positive improvements. These boost morale and engender operator confidence.

There are many new options to exploit but none as important as getting the basics correct. When we consider control rooms, we should seek to design them as "high-performance" control rooms. The correct designs will lead to performance improvements in the operators' ability to monitor and control. We have hundreds and maybe thousands of examples of bad design, poor ergonomics, and downright bad practices in the design of control rooms. Many companies just delegate the design to furniture manufacturers or architects who have little knowledge of this topic. You want to avoid ending up with bad designs and poor ergonomics. They foster human error, poor performance, and extensive losses. History has shown poor performance (due to the loss or compromising of situation awareness) leads to the loss of life and billions of dollars in losses. Your organization will not be competitive and will be always dealing with complaints from operators.

---

11 Karen (Lebovitz) Smith, personal communication, April 2017.

5

The Human-Machine Interface

The designer of a CRT display must study and work ... .
Know the user of the display; design the display for that user.

Richard S. Shirley (Foxboro Co.)

Operators cannot manage what they cannot see. The HMI delivers most of what they need to see. A control room must be fitted with enough ways to see where things are, what the current situations might be, what might be going wrong, how wrong it may be, and where and how to intervene to make things better. Operators can then manage what they can see and understand using expertise and appropriate tools. The modern video display provides the platform for this to happen. This chapter provides a solid foundation for the design and implementation of the best practices for video displays, the HMI. These *displays* are the primary way to provide operators with enterprise information and situation awareness inferences. They are also the vehicle for providing operational documentation and task guidance. Screen design has evolved from the early versions that stressed the colorful, dense, and flashy, into a technology that is capable of providing appropriate information in a manner that facilitates user appreciation, understanding, and task-supporting interaction.

At first, video displays seemed to represent a significant step forward when they replaced panel boards. It was a step forward in technology but a significant step backward in operator support. This was an unintended result of the evolution of screen design, not the result of intrinsic faults or limitations of displays, that led us down the wrong early paths. Once the faceplate barrier was broken, so to speak, the world of graphic design seemed to open up. When color made the scene, all the process control system (PCS) manufacturers started a race to see who could use the most appealing and flashy colors to preen in front of prospective buyers. Three-dimensional looks and animation made the situation seem even more appealing but cut into usefulness in a big way. What was missing from all this new technology was meeting the objective of what the video display should do and how best to do it.

Not a One-Stop Shop

You will not find a one-stop shop for HMI design here! Sure, there is a lot here. This wealth of information gives you a good framework to understand what the HMI is all about. This material is intended to supplement the designer's extensive knowledge of enterprise, culture, and locale. What you see here is intended to add to what you already know. Or, if you do not know something, to properly introduce it. There are topics and aspects that are not covered here. Everything has a finite and reasonable limit. This chapter has limits. Readers who are deeply familiar with HMI design will find many items that they may not have considered before, may disagree with now, would explain differently, want to use, or would never use. More breadth and a greater depth of the detail, hardware, and technological implementation are well covered by others.[1] They provide extremely useful interpretations, guidelines, and tools. This chapter is designed to establish a richer understanding of what the whole HMI is about. It positions you to read and understand most other material. Here you will find broad, clear basic design principles functionally linked to *situation management*. You will also find discussions of useful and interesting concepts to assist you in appreciating the various tools, what their capabilities might be, and what to look for as you build or design your own equipment. Knowing what to do with the objectives and processes for good situation management will enable you to identify your HMI needs and work out practices that effectively support them. Each section in the chapter has been written with that objective in mind. Those of you who are familiar with HMI design will easily conclude that rather than this being material in conflict with traditional design, it firmly assists in shaping that design to better fit for purpose.

Chapter Coverage

Covered topics include display screen configurations for operator stations; display screen layout and construction; navigation and operator interactions; styles and style guides; trends; icons, dials, gauges, and dashboards; very large and very small displays; use of sound and video; and methodology for evaluating effectiveness. Again, as a reminder, the specific technology for detailed designs is well outside the scope of

---

1 See the Further Reading section for more information.

the coverage here. This chapter is rich with a discussion of specialized display screens and their associated *pages*, *windows*, *formats*, and *elements* (special HMI nomenclature is explained in Section 5.3). The goal is to provide overviews and insights, not background tutorials. They are here rather than in the next chapter on situation awareness and assessment, though they are directly intended for awareness, assessment, and management. This was done to provide a natural flow through the fullness of HMI design. As you discover information and situations that operators need to understand and manage, you will have a good idea of effective ways for delivering it to your control room or operation center, or to your clients engaged in those activities.

5.1 Key Concepts

Fundamental Guide	Put no information in front of the operator unless the designer knows (a) what it is for, (b) how it must to be understood, (c) how it is to be used, and (d) is complete enough for that purpose.
Concept Design	Understanding and employing the critical design concepts for the HMI provides the power to see and understand developing abnormal situations as well as manage each situation effectively.
Evaluation	Users are effective at evaluating their HMI design for local style adaptations and usability after design and/or implementation. However, they are unequipped with the tools and design concepts to perform that design themselves.
Data vs. Information	Presenting only data means that the importance and significance of those data values are blatant, obvious, or trivial. Otherwise, it is the designer's responsibility to recast the presentation to show information relevant to the task and situation.
Concurrent Visibility and Accessibility	Operators should not need to write down or remember information and controls needed to coordinate operations and manage situations.
Color	Color is used *exclusively* to convey information. There is no other acceptable use; when used it must be globally consistent (same conventions everywhere).
Truth	All information provided to the operator must be accurate, appropriate, and framed. This means that measurement noise is sufficiently rejected, significant digits of display are the least necessary, the right signal is used to convey needed information, and the appropriate maximum/minimum or historical variations are provided to understand the current value(s).
Workload	Builders of displays need to assess the time to understand and interact with each display to factor in the overall operating load on operators.
Intuition	Any design for the HMI based largely on intuition and appearances will not work. The irony is that it may seem to work for a while and be reasonably comfortable, but it will fail due to excessive user effort needed for routine tasks and will be ineffective for abnormal situation management.
Video and Animation	All movement of information or other display screen content distracts. Animation and video should not be used *unless* it is under the direct control of the operator and is used to sequentially reveal information that is necessary to understand or manage.

5.2 Introduction

The first thing you will notice as you dive into this chapter is that you will not be asked to use much in the way of intuition. Everything, well almost everything, you might have been using before may need to be put aside, at least for a while. This does not mean that your earlier work was done wrong. Rather, we are going to take a second look. And by looking again, we will develop a better understanding of inherent purpose and usability. In all likelihood, previous designs were done with best intentions and honest effort. Much did appear to work, to give credit where credit is due. The operators either liked it or did not complain much. Hopefully, they did provide timely feedback to smooth over any rough spots and make things more convenient. Yet there were real problems:

- Operators had to work at looking around to ensure that they really understood how the actual operation was getting along.

- Operators found that doing routine tasks appeared to take more effort than they thought necessary.

- Operators found that during attempts to manage operational upsets, they were too much "in the dark" with respect to needed tools and information.

- Managers noticed that new operator training seemed to take much more "remembering" and "false starts" than "observing" and then "figuring things out."

- Incident investigations often pointed out that the operator response was delayed, had ineffective results, or even had downright errors.

All of this points to problems. This book is about helping to turn these problems into opportunities that can then assist delivering HMIs that work better. It is about the real needs of the operator and how to meet them in ways that work. Let us define the most-used terms for HMIs as we start things off. Take care here as few are used consistently in the field.

Physical Differences and Preferences

No physical design guidelines or recommendations for the size and location of control room equipment are included in this chapter. However, the importance of ensuring equipment is compatible with a wide variation of operator stature, reach, and other aspects of hands-on interaction must not be underestimated nor inadequately accounted for. This variation includes a person's physical size and weight, handedness, preferred positions (standing, sitting, varying, etc.), and other limitations and preferences due to physical condition, illness (temporary or long-lasting), gender, age, and more.

5.3 Nomenclature for Display Screens and Components

We use British Standards Institution (BSI) Standard ISO 11064 for nomenclature in this book.[2] Figure 1 from the Standard is reproduced as Figure 5-1 on the next page. For example, the physical hardware device that would be used to show all the content is termed *display*. All of that content would be conveyed to the viewer on *display screens*. The content builder would be free to design and provide at will so long as it was within the capabilities and limitations of the display hardware and the display screen software. Any individual display screen (or as many as you need) would show content utilizing these aspects:

- **Display** – The physical hardware (on which content display screens are shown).

- **Display screen (or screens)** – The actual content (information, real-time data, text, etc.) that is available for operator selection and showing on the physical display.

- **Page** – The totality of viewable content, of the moment, that the display screen contains.

- **Window** – Any of several (usually) individually managed visual subsets of content on a page. A page may contain a single window (occupying either the entire viewable area or a part of the viewable area with the rest without any content) or many separate windows (with due regard to visibility).

- **Format** – Any of several content clumps (we will leave that undefined for a while, but think charts, pumps, diagrams, status text, etc.) that occupy any given window.

- **Element** – Any needed basic content items used to build a format.

Later in this chapter we will need to have a group label for all the places where content (stuff) is placed for the operator to see on the HMI. Let us call them *display screen components*. Everything must be on a display screen. If there is no structure to a given screen, then the entire display screen is built as one piece. Information and

---

2 BSI, *Ergonomic Design of Control Centers.*

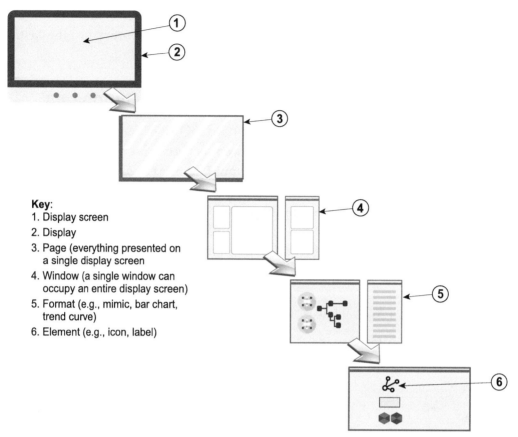

Figure 5-1. ISO 11064-5 nomenclature diagram for graphical displays. This level of detail permits specific reference and discussion. Starting from the upper left, each successive component will be found as continuing a part of the component above.

graphics and all the rest of the stuff we want to show the operator will be placed on display screen components: pages, windows, formats, or elements. A page may have windows by themselves, formats by themselves, elements by themselves, or a mix of any and all. In similar fashion, each window may have none or as much as needed of the HMI parts that are below it in the construction hierarchy. And the same goes for each component going down the hierarchy.

Let us be sure we understand how this works. Using this terminology, no content whatsoever (visible or not) can be on a display screen unless it is on a page. Nothing can be on a page that is not contained in (one or more) windows that occupy any of the various pages available for view. Nothing can be contained within any window that is not part of any format. All formats (there are many) are built from as many elements as needed. Any or all of the various pieces can be reused as needed; that is, no element is restricted to a given format (unless the designer has special purposes in mind); no format is restricted to a given window, and so on. Moreover, any given piece (element, format, window, etc.) can be derived from any other similar one and modified as desired.

So why devote this attention to something that may seem to be overly complicated and not easy to use? The reason is to allow you to establish a vocabulary and construction methodology that will work. You have here the proverbial "double-edged sword." Swinging one way, the display screen can be configured to provide an almost limitless design and array of content. It would be enhanced by the designer's ability to visualize graphical elements and formats (from Figure 5-1). It would be as powerful as the designer's ability and knowledge of the information and data content required. Swinging the other way, depending on what was selected to show and how it was designed to be shown, the viewer could be well enabled to find, visualize, and understand the content, or be completely frustrated, or worse, dangerously misled. You are well aware of this dichotomy. Our discussion is to inform you about what is recommended and the power of why it is beneficial.

The next several sections will provide useful "cornerstone" guidance about what to consider as you evaluate and design useful and functional operator screens. Section 5.4 will lay out the overall requirements. These are basic and should be in the front of your mind as you do the work. While you may consider various styles, your choice will need to deliver on all these requirements. You do not need to remind me that you know better than I that these requirements, as important and useful as they are, do not give much in the way of specifics. Section 5.5 provides an introduction into the specifics by providing two important lists of "do's" and "don'ts" to follow as you begin to work out your display designs. Section 5.6 moves the general recommendations up a level. It respects the reality that operator screens do not exist in thin air. They are integral to the operator workspace. The section provides useful ways for you to integrate your display design into your workspace design.

5.4 Four Underlying Requirements for Operator Screens

There is an extraordinary power given to the designer of operator screens. Let us be blunt here: everything the enterprise will need the operator to know or find out about the equipment under management must be either available for view on the HMI or obtainable some other way (external documentation, checking with someone else, etc.). Here we deal with what is put on the HMI.

Requirement 1: Purpose

All information displayed for the operator must be displayed for explicit purpose. That is evident, of course. Although it is relatively easy to just say, "The operator needs to know all of this," doing so is not a good idea without more (much more) purpose. The designer needs to know specifically what use will be made of the information intended for that *screen*. Will the information be needed for start-up? Will the information be important and needed to check for proper equipment operation? Is the information going to be used for shift handover? Is the information on various screens important for troubleshooting? This list goes on for as many needs as your operators have. As the designer, you must know them all to make sure that your screens are consistent with all the needs. This can cut two ways. The first way is to shoot for the ability to meet all the needs for all screens. This is not always possible or desirable. So, when necessary, you will be designing alternate screens to cover the other purposes and making sure the operator knows which to use when.

Requirement 2: Understanding

Now that the purpose part is on the table from requirement 1, the designer must to be sure that the information is displayed in a way (or with enough context right there) that enables the operator to understand the context and frame for it. Things like *normal*, *abnormal*, *unusual*, and *dangerous* come to mind. Sure, we expect the operator to be able to generally sense these aspects, but—and this is the important *but*—it is not enough to rely on the process of "sensing" to do the job. This means that context and frame must to be a part of the information. This does not suggest that most screens are cluttered with lots of "context" information. But, in some way or the other, the operator's ability to know context must be there.

Requirement 3: Use

Satisfying the first two requirements gets us ready to make sure to design *screens* that actually enable the operator to do what needs doing. Here we would ensure proper navigation or coordination between screens, have the needed control "handles," have procedure documents and drawings accessible, and the like. At the end of the day, everything the HMI depicts to the operator must support why it is there and help make sure the operator can do what is needed to properly manage things.

Requirement 4: Complete

The final requirement is to make sure there are no gaps between design and execution. Test concepts. Test scenarios. Fix or amend where needed.

5.5 Principles of Display Screen Design

Here are several lists of useful guidance. There is some overlap. Please use the repeated parts for reinforcement. Where there appear to be differences, please use them for clarity; it just may mean a different way of describing something. Think about it as having two jars. One jar is for all the items you want to use. The other is for all the items you will work to avoid. Use the stuff from both jars.

Wickens's 13 Principles

Christopher Wickens and his colleagues defined 13 principles of display screen design in their book, *An Introduction to Human Factors Engineering*.

> These principles of human perception and information processing can be utilized to create an effective display [screen] design. A reduction in errors, a reduction in required training time, an increase in efficiency, and an increase in user satisfaction are a few of the many potential benefits that can be achieved through utilization of these principles.
>
> Certain principles may not be applicable to different display [screens] or situations. Some principles may seem to be conflicting, and there is no simple solution to say that one principle is more important than another. The principles may be tailored to a specific design or situation. Striking a functional balance among the principles is critical for an effective design.
>
> **Perception Principles**
> 1. **Make display [screens] legible (or audible).** A display [screen's] legibility is critical and necessary for designing a usable display [screen]. If the characters or objects being displayed cannot be discernible, then the operator cannot effectively make use of them.
> 2. **Avoid absolute judgment limits.** Do not ask the user to determine the level of a variable on the basis of a single sensory variable (e.g., color, size, loudness). These sensory variables can contain many possible levels.
> 3. **Top-down processing**. Signals are likely perceived and interpreted in accordance with what is expected based on a user's past experience. If a signal is presented contrary to the user's expectation, more physical evidence of that signal may need to be presented to assure that it is understood correctly.
> 4. **Redundancy gain**. If a signal is presented more than once, it is more likely that it will be understood correctly. This can be done by presenting the signal [simultaneously] in alternative physical forms (e.g., color and shape, voice and print, etc.), as redundancy does not imply repetition. A traffic light is a good example of redundancy, as color and position are redundant.
> 5. **Similarity causes confusion:** Use discriminable elements. Signals that appear to be similar will likely be confused. The ratio of similar features to different features

causes signals to be similar. For example, A423B9 is more similar to A423B8 than 92 is to 93. Unnecessary similar features should be removed and dissimilar features should be highlighted.

Mental Model Principles

6. **Principle of pictorial realism.** A display [screen] should look like the variable that it represents (e.g., high temperature on a thermometer shown as a higher vertical level). If there are multiple elements, they can be configured [*spatially*] in a manner that looks like it would in the represented environment.

7. **Principle of the moving part.** Moving elements should move in a pattern and direction compatible with the user's mental model of how it actually moves in the system. For example, the moving element on an altimeter should move upward with increasing altitude. [*This principle is not to be confused with animation, which is not advised.*]

Principles Based Upon Attention

8. **Minimizing information access cost.** When the user's attention is diverted from one location to another to access necessary information, there is an associated cost in time or effort. A display [screen] design should minimize this cost by allowing for frequently accessed sources to be located at the nearest possible position. However, adequate legibility should not be sacrificed to reduce this cost.

9. **Proximity compatibility principle.** Divided attention between two information sources may be necessary for the completion of one task. These sources must be mentally integrated and are defined to have close mental proximity. Information access costs should be low, which can be achieved in many ways (e.g., proximity, linkage by common colors, patterns, shapes, etc.). However, close display [screen] proximity can be harmful by causing too much clutter.

10. **Principle of multiple resources**. A user can more easily process information across different resources. For example, visual and auditory information can be presented simultaneously rather than presenting all visual or all auditory information.

Memory Principles

11. **Replace memory with visual information**: A user should not need to retain important information solely in working memory or retrieve it from long-term memory. A menu, checklist, or another display [screen] can aid the user by easing the use of their memory. However, the use of memory may sometimes benefit the user by eliminating the need to reference some type of knowledge in the world (e.g., an expert computer operator would rather use direct commands from memory than refer to a manual). The use of knowledge in a user's head and knowledge in the world must be balanced for an effective design.

12. **Principle of predictive aiding.** Proactive actions are usually more effective than reactive actions. A display [screen] should attempt to eliminate resource-demanding

cognitive tasks and replace them with simpler perceptual tasks to reduce the use of the user's mental resources. This will allow the user to not only focus on current conditions, but also think about possible future conditions [*in the right context*]. An example of a predictive aid is a road sign displaying the distance to a certain destination.

13. **Principle of consistency**. Old habits from other displays [screens] will easily transfer to support processing of new display [screens] if they are designed in a consistent manner. A user's long-term memory will trigger actions that are expected to be appropriate. A design must accept this fact and utilize consistency among different displays [screens].[3]

Engineering Equipment and Materials User Association's 10 Principles

The Engineering Equipment and Materials User Association (EEMUA) provides a useful discussion on the basic principles for display screen design. At first glance, these seem obvious—how could anyone not do it this way? They appear to suggest that you are free to pick and choose the ones you like. Well, you can, but if you do, you forfeit important benefits and play into the hands of many of the human foibles that good operational practice struggles so hard to avoid. Work out a way to use them all.

- **Clarity** – It should be self-evident what the purpose of the display format [ISO: display screen] is. Individual items on the display format should be obvious. Graphic, text, and numeric displays should be clear, easy to read, and it should be obvious to what they refer. Cool and subdued colors should be used in normal operation, to allow bright colors—which will stand out—to be used in abnormal situations [mostly for alarms]

- **Consistency** – All graphic indicators, text and color coding should be consistent, not only within a display format, but between display formats in the hierarchy. The operation of a control (e.g., hotspot/hot-link) should have the same effect on all display formats.

- **Variety** – A limited variety of display techniques should be used, so that operators can become easily familiar with them.

- **Feedback** – Any control or other action on a display format should give feedback to the operator to give awareness that the system is performing the requested task or indeed is giving an error. The feedback needs to be near instantaneous. Feedback on completion of the action may also be required (or on failure of the action).

---

3 "Human computer interaction," Wikipedia, last modified June 1, 2018, http://en.wikipedia.org/wiki/Human-machine_interaction. Christopher D. Wickens, John D. Lee, Yili Liu, and Sallie E., Gordon Becker, *An Introduction to Human Factors Engineering*, 2nd ed. (Upper Saddle River, NJ: Pearson Prentice Hall, 2004), 185–193.

- **Robustness** – Where interface action is taken by the operator, the system should be designed so that it can cope with incorrect key strokes/mouse-clicks and so that the operator can return to the original position if need be.

- **Failure** – Failure of a display [screen] or of items on the display [screen] should be immediately apparent to the operator.

- **Redundancy** – Multiple information displays should be avoided, unless this is required to achieve specific reliability requirements or as determined by task analysis.

- **Demand versus status** – Indications should make clear what values are indicating actual plant status and which are indicating set points or demand values.

- **Spatial variation** – If situations exist where several displays depict similar process units, then it can be beneficial to place similar objects in different positions, to help differentiate between the units. Note the [intended] clash with the concept of consistency.

- **Detailed Design Guidelines**

    o It is essential to provide overview display formats.

    o Access to schematics and detailed information must be quick.

    o Ease of use must be considered at all times.

    o Methods of schematic display must be flexible.

    o Integration of all [necessary] plant data is required.

    o Enhanced information display is desirable.[4]

Five Design Principles

Even the best conceptual design must pass the usability test of actual site acceptance. Different manufacturing environments and different geographic locales may require a slightly different style of design. Such style variations are certainly possible and even

---

4 EEMUA, *Process Plant Control Desks*.

advisable to promote a more natural feel for the users. But let us not forget the aircraft cockpit lesson presented in Chapter 1, "Getting Started." Consistency, logical design, and clearly implemented functionality are vital. Let us take a look at the important design principles.[5]

1. Focus early on the actual users, asking questions such as: Is this a task to be done by cooperation or individuals? What is the minimum level of platform familiarity required? (Do not forget emergency modes of operation and assistance.) What is the frequency of repeated operations? Are separate operations really linked (either sequentially or back-and-forth)?

2. Take advantage of the available technology and versatility of user-centered design (human factors).

3. Ensure that the new design contributes sufficiently to overcome the broad familiarity and acceptance of the technology it desires to replace (e.g., Dvorak versus QWERTY keyboard).

4. Ensure that the design is respectful of the cultural values and the foibles of human nature by considering the following types of questions: Are instructions framed as recommendations (even though all understand that they are required)? Are the tone of the text and texture of the visuals respectful to the locale? Are confirmatory relationships readily apparent for cultures where confrontation is usually avoided?

5. Test the design (both at build and ongoing) to ensure it remains fit-for-purpose—take care to understand the differences between levels of skill and experience and operational responsibility of the various users.

ISO 9241 Seven Design Principles and Five User Guidance Principles

The International Organization for Standardization (ISO)[6] has put forward several useful items. These lists are taken from Jakob Nielsen:

---

5 "Human Computer Interaction," Wikipedia.
6 ISO 9241-210:2010, *Ergonomics of Human-System Interaction – Part 210: Human-Centred Design for Interactive Systems* (Geneva, Switzerland: ISO, 2010).

Seven Display Design Principles

Clarity – The information content is conveyed [depicted on the screen for the operator] quickly and accurately.

Discriminability – The displayed information can be distinguished accurately.

Conciseness – The users are not overloaded with extraneous information.

Consistency – The design is unique and conforms with the user's expectation.

Detectability – The user's attention is directed towards information required.

Legibility – The information is easy to read.

Comprehensibility – The meaning is clearly understandable, unambiguous, interpretable, and recognizable.

Five Display-User Guidance Principles

Prompts indicating explicitly (specific prompts) or implicitly (generic prompts) that the system is available for input.

Feedback informing about the user's input timely, perceptible, and non-intrusive.

Status information indicating the continuing state of the application, the system's hardware and software components, and the user's activities.

Error management including error prevention, error correction, user support for error management, and error messages.

On-line **help** for system-initiated and user-initiated requests with specific information for the current context of use.[7]

5.6 Seven Principles of Workspace Design

Now that we have the basic guidelines for the HMI display screen construction process, we turn your attention to what to build with display screens. Not the specifics of any display screen, but what you would need to keep in mind as you build the ones you need. Remember, every display screen must have a purpose that is always clear to the viewer and is designed to fulfill a task or assist in a responsibility. Just because you can build display screens does not mean you need to create more than the operator requires. Each one will require engineering, building, testing, and maintenance. Each will also require instruction and training. This is a lot of overhead for something that is not necessary. But if it is, then by all means get it done.

---

7 Jakob Neilsen, "Usability 101: Introduction to Usability," Nielsen Norman Group, January 4, 2012, https://www.nngroup.com/articles/usability-101-introduction-to-usability/.

In a paper on visualization, Stephanie Guerlain, Greg Jamieson, and Peter Bullemer developed this set of useful design principles that tie together the functional aspects of why an operator uses the HMI.[8]

1. **Create a workspace that supports monitoring, diagnostics, and control.** A *display screen* may be split into functional areas of monitoring (both overview and some detail).

2. **Support periodic monitoring of the [process] through design of an overview display [screen].** Status includes current values, proper targets, production status, alarms and warnings, and the information update status (when were the last set of readings done).

3. **Support direct navigation from the overview display [screen] to more details**. No amount of information exists in isolation. Make sure that the rest of the story is easily accessible. Ensure that proper navigation is a two-way tool—provide easy, clear notice of where the navigation left from, where it got to, and how to get back.

4. **Use representational aiding to map domain properties onto corresponding graphical elements**. This is a fancy way to say that you'll need to use a consistent relative scale to depict the magnitude of all variables on the same *display screen* or closely related ones. Also use the same symbology (icons or other *elements*) for the same purpose. Boxes around variables always mean the same thing. Solid shape versus outline shape always means the same effect or state.

5. **Use consistent color coding throughout the display [screen] to represent the same thing**. Remember the cardinal rule of color: color must be used to convey (important) information. Red color (text, box, highlighting, etc.) always means the same one thing that red was chosen for. Use the same consistency for all of the other colors everywhere all of the time.

6. **Show variable information relative to limits**. Showing information relative to limits provides a vital scaling mechanism to directly convey the operating situation in context to the viewer. If limits are variable, this reference is even more important to keep everything in useful perspective without requiring historical recall or constantly noting specific operational values.

---

8 List items taken from Stephanie Guerlain, Greg Jamison, and Peter Bullemer, "Visualizing Model-Based Predictive Controllers," Proceedings of the IEA 2000/HFES 2000 Congress, 3-511 to 3-514, 2000, http://www.sys.virginia.edu/hci/papers/VizMPCHFES2000.pdf.

7. **Show important context information when the user changes a limit, including past operator changes**. Where manual intervention is taken, it is important to provide all related aspects to the operator including direct and likely potential effects of the action. This may be accomplished by utilizing "tool tips" to warn or advise, and yoked subordinate displays to verify or modify the intended or taken action.

5.7 The Human-Machine Interface

The human-machine interface (HMI) includes devices, distinct from actual physical conditions or direct indications, that electronically display to a person and allow interaction by the person with values, states, and other information that comes from real (or hypothetically real) equipment.[9] These displays are remote from the physical or virtual world things that we can see and touch. Examples are mechanical dials and gauges, switches, video monitors, keyboards, touchscreens, voice-sensitive input devices, and a broad number of other devices. For this reference book, we will restrict the uses to those for operators of industrial plants and operations centers. It is important to understand HMIs and use them appropriately and effectively.

As mentioned in the opening of this chapter, this is an overview. It will lack completeness. There is a wealth of good information available from others, and the basic coverage is excellent. Those other resources go in depth into the craft and technology of building the actual screens. The discussion here assumes the exercise of those construction and deployment skills. Here we augment the current accepted practices with new information and concepts that go to the heart of enhanced situation management. A brief background into HMIs answers questions such as: What it is for? What it is not for? and What hardware types are available? Navigation types and when they are useful/acceptable are covered. Operator needs when using a fixed control room may differ from those for a portable device the operator can take outside. These differences are briefly discussed.

Viewing access is discussed not as anthropomorphic measurements (although they are important and well covered in other resources) but rather as the aspects of alternate display screen modes to support multiple viewing distances for different purposes: single operator coordinating with multiple desks, multiple operators collaborating, single operator managing an upset, operator monitoring when the process

---

[9] "Human Machine Interface," Wikipedia, last accessed May 25, 2014, http://en.wikipedia.org/w/index.php?title=Human-machine_interface&oldid=606339419.

appears normal, operator monitoring when the process appears abnormal, and operator monitoring when the process appears to be seriously in trouble.

A Wartime Story Sets a Stage

This story is a powerful illustration of the importance of the human-machine interface in real life. It is an unfortunate but true aspect of war that fosters insight and innovation. For the Second World War, it was the coming of age of aircraft and underwater submersibles. We follow a story of aircraft. At that time all aircraft required pilots. Their training was vital and urgent. A great deal of investment went into that activity. But soon, a dramatic and paradoxical situation arose: the US Army Air Corps, Navy, and Marine Corps air arms were losing more pilots to training accidents than were being lost due to enemy action. While training might have its risks, they must certainly pale to those of war. Something was dramatically wrong. This is that story.

It is the nature of most training to start simple and build from there. There is another way, called *trial by fire*, but we put that aside for now. Starting simple meant that the first activity for a pilot trainee was as an observer in an unpowered glider. Then the trainee moved on to piloting the glider. Next was a small single-engine training aircraft. As proficiency was attained, the trainee moved up to the next complex aircraft. Eventually, the men were trained on the aircraft they would engage with, such as a fighter, bomber, or transport.

In general, training involved starting with the basics: ground school on what piloting an aircraft was all about and how to do it, how to enter and egress (both normally and in emergencies) the aircraft, how to use the instruments and navigate, how to take off and land, how to engage armaments, how to engage or evade the enemy, how to do it all at night and in bad weather, and how to handle emergencies. As the trainee moved from one aircraft type to another, the new aircraft cockpit and handling were mastered. This was an effective program except for one major flaw. It was observed that all the training fatalities occurred during exposure to emergency activities (both planned and unplanned). The root cause was as dramatic as it was stealthy.

Each type of aircraft was outfitted with similar function instrumentation, but the instrumentation's designs and cockpit locations varied. As the pilot trainee moved from aircraft type to aircraft type, he had to learn each cockpit anew. When conditions were routine, those differences were annoying but easily manageable. But when conditions were dire, there was no time for contemplation—everything was reflexive. It is the nature of reflexive reaction that when experiences are obtained through a variety of differing events, the individual is far less able to segment the different requirements and fumbles with the equipment. This robs time and energy with deadly results.

Fixing the situation was as simple as it was entirely effective. All aircraft, regardless of its design, purpose, or sophistication, was outfitted with the same central cockpit instrumentation, in the same spatial arrangements, and with the same visual design and performance. The choice of those key instruments was made based on the essential knowledge the pilot required to understand what the aircraft was doing and what needed change or close monitoring during emergencies. This central instrument package is all that is needed during both routine and emergency handling of the aircraft. Learning its purpose and function in one aircraft meant learning it for all. While the aircraft might differ in design and performance, the pilot interaction information remained the same. The number of training accidents was remarkably reduced.

Notice that all cockpits contain six basic instruments (shown in Figure 5-2 at the upper left). While more may be added, the additions augment rather than replace the

Figure 5-2. Examples of the six basic instruments for all aircraft. From the top left (clockwise) are the basic "T," Schempp-Hirth Janus-C glider equipped for cloud flying, and Slingsby T-67 two-seater. By standardizing critical instruments this way, every pilot can fully orient to any aircraft in the world.

basic ones. In an emergency or in confusing situations, paying attention to the six basic ones and adjusting the aircraft to ensure that they are reading appropriately is all that need be done. The takeaway from the example is twofold:

1. Design the core part of an operator interface to convey all essential information in a concise and clear way that also demonstrates what needs to change to keep things right.

2. Design the operator interface to follow the same principles of design and interaction regardless of the scale of the plant or operation. As operator focus moves up or down the hierarchy, the interface depictions must follow the same visual and functional design.

Components of an HMI

An HMI consists of one or more of the attributes and capabilities that enable a human operator to see elements and aspects of the real or virtual plant or enterprise that is under observation and/or management. The technology may encompass the entire spectrum of devices or technology employed to provide any capability. However, the focus is a control room or operations center. Although equivalent or even identical technology might be used for field use as is used in the control room, the concern here will be the control room. However, most of the functional design (with the exceptions of environmental requirements for weather protection and illumination) would be shared.

Technology for Devices

HMI devices may be built or constructed as follows:

- **Analog** – Consisting of real discrete components that articulate, indicate, and respond to direct physical stimuli. The stimuli may be purely mechanical, purely electrical, or a convenient combination.

- **Digital or electronic** – Consisting of specialized video display units (cathode ray tubes—CRTs, flat screens [e.g., plasma or LED], holographic, or acoustic).

The operator interaction is:

- **Visual** – Eyes on the device to see as well as modify or control.
- **Tactile** – Physical touch (to command) or physical feel (to observe)

- **Acoustic** – To receive, command, or to convey information

- **Advanced technology** – Proximity sensing with or without additional interactive information transfer

Screen Concepts
- **Physical mimicry** – Constructed of real or virtual things that appear to the viewer similar to what they might look like if the viewer was observing the real thing. An example is a storage tank with the product level depicted as a moving level as if the viewer was looking at a transparent tank. Such use would not have to preserve relative scale of effects. For example, if the pH is of important interest, its effects may be accentuated to allow small differences to be clearly observed.

- **Abstract mimicry** – Constructed of virtual things (CRT or flat screen) that depict situations differently from how they appear if one was observing the actual plant or operations center object. This allows the depiction of real entities in a different information framework that exposes certain parameters or relationships free of their physical relationship. An example is depicting a portion of a plant showing the production costs, the revenue portion gained, and the environmental loading for each step of the operation. Such a depiction must be accompanied by instructions on how to ensure that any operating decisions made from it are made using principles and requirements set out by the enterprise rather than left up to the contemporaneous interpretation of the operator.

Combining these concepts is the norm. Care needs to be exercised to keep everything in balance. For example, operator screens work best when the overall framework readily suggests the real plant, but only to the degree necessary to minimally do so. Do not use three-dimensional depiction, avoid overly graphic designs, do not use animation, and skip any use of color not supportive of clear operational risk.

HMI Design Philosophy

Here we are using the word *philosophy* in its rudimentary sense. It means a sufficiently complete descriptive plan for what the HMI should look like, how it needs to function, and what else is necessary to fit it into the enterprise. This plan respects the industry practices while it allows specific site needs to be realized.

> The purpose of an HMI Philosophy is to define the principles that guide design of all control system HMIs implemented within an enterprise. Use of the philosophy maximizes consistency of the HMIs at all levels by standardization of hardware and

consistent look and feel of all operator screens. The philosophy guides the efforts to achieve the following goals:

- Improve operator situational awareness – The ability to detect, analyze and correctly respond to changes in process conditions.

- Improve focus on Key Performance Indicators (KPIs) – Those critical operating parameters that indicate quality of unit operations.

- Re-establish the "Big Picture" view for the operator – It is important to be able to view and manage the whole process effectively.

- Actively establish "Specific Condition" views for the operator – Simplifying the operation for critical situations and in particular those situations that happen rarely.

- Incorporate lessons learned through recent improvements in industry standards concerning screen technology, ergonomics, and graphic design, such as ASM Consortium[10] recommendations and the ISO 11064 standard.[11]

- Lay out the change management and approvals process.[12]

Style Guide

A style guide is a preferred instantiation of the requirements of a philosophy done for better usability, performance, and uniformity. It covers all aspects of *screen design* including the following:

- Hierarchy of navigation organization

- Segmentation of information into pages, windows, formats, and elements

- Layout, backgrounds, line styles, text styles

- Design of trend screens, glyphs, icons, gauges, and dashboards

- Behavior of graphic objects that will change or be modified by operational or equipment state (e.g., shutdown or in alarm)

- Uniform design for embedded linked information including video, documentation and procedures, reports, and other

- Uniform design for screen labeling text and numerals, messages including alarms and events, online documentation, approved word glossary or dictionary, abbreviations, and such

---

10 Bullemer et al., *ASM Consortium Guidelines.*
11 BSI, *Ergonomic Design of Control Centers.*
12 Russel Treat, "Discussion Document—Abnormal Situation Management—HMI" (Version 1; April 22, 2014), personal communication.

- Any platform specifics necessary for proper implementation including approved work-arounds to accommodate existing limitations, and any vendor-provided specials

This guide will be customized to meet site- and vendor-specific needs.

Graphics Library

A *graphics library*, also called a *graphics symbol library*, is a compendium of predesigned small components (elements, formats, and windows) used to build display screens. The construction tools will rely heavily on the *style guide*. Each item is carefully conceived and constructed for a specific purpose. Proper design will enable each component to compactly, efficiently, and effectively convey the needed information in appropriate situational context. Uniformity ensures ready understanding for the operator. Examples include depiction of process variables (flows, temperatures, pressures, analytic values, etc.) in graphics, depiction of individual process elements (pumps, reactors, switching stations, on-ramps, etc.), and quick-entry operator forms or comments boxes. Figures 5-3, 5-4, and 5-5 depict a few examples.[13]

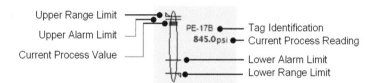

Figure 5-3. Pressure symbol icon depicting in one compact unit all important aspects for the variable being monitored. This ensures a proper frame for the information.
Source: Reproduced with permission from EnerSys.

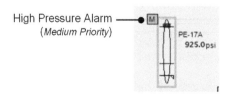

Figure 5-4. Pressure symbol (from Figure 5-3) showing an active medium-priority alarm (reserved color for medium priority alarms is yellow). Note the redundancy of the alarm rectangle around the icon in yellow and the solid yellow square with an "M" inside.
Source: Reproduced with permission from EnerSys.

---

13 Treat, "Discussion Document."

Figure 5-5. Solenoid valve icon. This valve is energized (note the filled "operator" portion) and closed (note the filled "butterfly" portion).
Source: Reproduced with permission from EnerSys.

Everywhere these variables or symbols are needed, they are used exactly as depicted in the library. In this way, standard symbols are developed for all screens as well as manuals and other uses that rely on uniformity of presentation.

5.8 Display Screen Design

Everything anyone can and would want to see in the control room that is not visible directly from the physical pieces of equipment themselves must appear somewhere on some display screen. What is displayed, how it is depicted, and where it is located will determine everything. Effective practices will present information that has a clear relevance, is placed in perspective, associates with operations in an expected way, and strongly facilitates understanding and any needed intervention.

Overview of Display Screen Design

How the display-based digital interface is conceived, designed, executed, and maintained will go a long way toward providing the operator with good tools to do the job at hand. There is a comprehensive methodology for doing HMI design and implementation effectively. The approach is to start with the basics and the reasons something might be recommended to be done in a certain way. Then alternatives are provided that you may wish to consider as you make decisions for specific applications and uses. Display screen design incorporates the following three aspects:

1. **Display screen structure** – How the entirety of information destined to be content on displays is divided up between the various display screens. The amount of and way information may be duplicated or found on multiple display screens is included.

2. **Display screen construction** – All the design aspects of color; sound; tool tips, zooms, and contemporaneous annotation; layout, text design and use; and graphics design and use.

3. **Navigation** – How the user finds the needed content or is directed to content to decide whether it is needed.

Display Screen Structure

All enterprises, be they manufacturing plants, pipelines, or transportation monitoring and control entities, have a comprehensive amount of data and other information that needs to be readily available to operators. Knowledgeable individuals intimately familiar with the enterprise choose the specific data and information needed. Additional important design aspects are covered later in this book in Chapter 6, "Situation Awareness and Assessment," and Chapter 10, "Situation Management." The discussion here is about how the data and information are organized (displaying it comes later) on the set of all display screens available to the operator on the HMI. Said another way, given all the data and information that an operator will need, how will it be available for view? What will go on which display screens? What should show via a "right click"? Summarizing, this discussion is about which HMI display screen shows what. Specifically, what are the alternative ways of doing it and how would the best (or better) way(s) to organize the layout and content of a suite of graphic displays and display screens be chosen? You will want answers to these questions:

- What information (content) is needed for display to the operators?

- What information comprises each display screen (given that a display screen requires limitation of content amount to be useful)?

- How should the information (content) be arranged on the individual display screens?

There are two parts we should understand for successful design: *hierarchy* and *segmentation*. They are introduced next.

Hierarchy

Hierarchy is the part of the selection of display screen information that groups information by level of view. Typical levels are: overview, secondary, and tertiary (see Figure 5-6). At any level of view, all the display screens at that level contain data and information consistent for that level. A three-level display screen hierarchy delivers a robust structure that encourages ready access to information while at the same time keeping important situation context and promoting efficient navigation to go deeper. The first two levels follow an expected progression from the general to the more detailed. But the third is a departure. The third level does not provide a more detailed view. Rather, it is presumed that if the first and second levels are not capable of providing the needed detail, another paradigm is needed. That paradigm is support, not more detail.

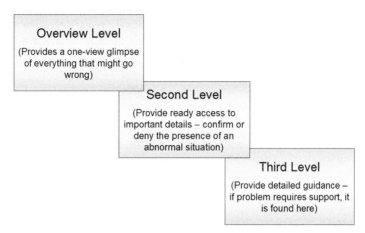

Figure 5-6. Three hierarchy levels for display screens. This structure permits all information to be located directly and then accessed without excessive "clicking." This structure also encourages uniformity and appropriate co-location of related information.

This structure is a very clear departure from the widely accepted norm for navigation and structure: "clicking to oblivion" that characterizes most personal computer operator interface exchanges.

Clicking one's way down a "tree" of choices to locate information is a tedious, time-consuming activity. If the tree arrangement logic differs from the mental expectations of the user, it can be distracting. Even if there is a good match, it takes time and requires a constant stream of locating the click target area, using pull-down menus to provide the list of acceptable choices, deciding which choice matches the current need, and selecting it to go to the next choice. Often when there appears to be no good match, one must click down and back to attempt to find something close, if one knows what appears to be a potential match. Or one might use the reverse process of looking for the least bad match, working from what feels as if it would not be a match, but no other alternatives are at hand. As you might imagine, this approach is frustrating and often becomes completely unproductive during times of operator stress. At the very time it is most needed, it fails. This is a big reason why efficient operator machine interface interaction requires as little clicking as possible.

Overview Level
An overview must provide a way to take in the operator's entire area of responsibility. It must show summarized alarm statistics, grouped status information, and important conditions of both upstream and downstream operation, and provide ready navigation to everywhere. It must communicate whether or not the enterprise is operating well. If well, how well? If not well, then how not well, why, and where? It must do this without requiring the operator to search it out. In the special case where an operator is responsible

for several unrelated or unconnected entities, his overview may consist of two layers. The first layer is the overview of everything. On this view, if anything is wrong or seriously suspect anywhere in his entire span of responsibility, it will show up. The next layer of overview will depict overviews of each composite piece of responsibility.

Secondary Level
The secondary level is for details. It shows alarms and process details, provides for monitoring and adjusting where needed, and again, includes navigation to other needed places. Control handles and specialized information components may appear as "pop-ups" here, not as another level below. Pop-ups have the advantage of maintaining situation perspective, which we require. But they can be distracting if not done carefully. A change in level does not mean that the design paradigm cannot be modified, so long as the use and style is consistent.

Tertiary Level
By the time the operator needs to go to a tertiary level, more detail will not likely help. Level three is not designed to provide even more detail than the secondary level. We must presume at this point that the operator needs to better understand things. Usually, better understanding is best supported by advice and amplification. This situation is not unlike initially reading a newspaper. From far away, all we can see is that there appears to be a paper. Moving closer, we can easily read the headlines and perhaps key lines. Moving closer still, we can read the text. Moving closer still just provides larger text; there is not much more to gain from it. If our operator needs more than what is provided by the secondary level, we know that something different is required. Here is the place for highly specific assistance. Examples of more and different assistance include access to procedures and design documents or historical incident documents, detailed laboratory reports, and more. Again, you will need easy, direct, and purposeful navigation to all other levels.

Important Key Point
Operators have many displays in their operator station. It is important to note that different displays can have display screens depicting different hierarchal levels. In fact, this should be considered expected practice. This way, the operator is able to maintain an overview situation awareness while exploring more detailed situation awareness suggested by the overview, such as preparing for a shift handover, following up on a potential problem, or working a weak signal (see Chapter 9, "Weak Signals").

Segmentation
At any given level, there may be quite a bit of data and information that you will want to make available to the operator to see via the HMI. *Segmentation* is the word we use to

describe the process dividing or segmenting the total plant data and information into the content of each display screen. This is important because there is only so much that can be properly placed on each. Yet most tasks the operator needs to have information about are more than can fit onto a single display screen. Even though there might be enough displays for the operator to use at the same time, each display screen needs to be sufficiently complete to provide a coherent view. A *coherent view* means that enough of the essential information for understanding what is on view is present on the current display screen. All important information needed to properly grasp a situation should be either on view or suggested for verification by what is on view.

Here are a few of the ways to consider dividing up stuff for each display screen:

- Put information pertaining to separate individual physical portions of the enterprise on separate display screens. Examples: reactor, tank system, major on/off ramp.

- Put information pertaining to related physical portions of the enterprise together. Examples: conveyor system, loading docks, hydrogen plant.

- Put information pertaining to the interconnected physical portions of the enterprise together. Examples: pipeline or electrical power distribution system, raw material to finished product quality tracking.

- Put related information needed to be understood concurrently, together. Example: current weather storms with operational vulnerabilities and required delivery objectives.

"Dividing up" or segmenting is not as easy or obvious as it might first appear. Here are some of the problems:

- The necessary clumps of data and information for different display screens can be quite different in amount. Display screens do not work well if they are too crowded; they also do not work well if they are too sparse.

- Grouping in physical clumps might not be very useful for understanding and managing operational problems. Most problems affect a wider portion of the enterprise than a single clump.

- Related information needs to be available and in relational context so that the operator can see it without having to remember it, write it down, or mentally closely link between looking at several display screens (either all in view, or not yet selected for view).

Examples

Please keep in mind that these examples are *not mutually exclusive*. Therefore, it is expected that there will be a primary overall architecture for the basic or default operational display organization. This display organization defines a design that provides the everyday tool for operators. The tool comprises most of what the operator will use to access information, decide problems, take remediation steps, and verify their effectiveness. The additional view structures shown in the following examples are just that. They are pre-engineered alternate views of the operational information and control handles. They might be used to prepare the shift handovers (see the responsibility-based display screen organization in Figure 5-7). They might be used to closely monitor maintenance in a given area (see the geographically based organization in Figure 5-11). They would certainly be used when any part of the operation comes under threat, for example, when particularly disruptive weather threatens the operation (see the risk-based organization in Figure 5-8).

Responsibility-Based Display Screen Organization: Hierarchical Overview Display Screens

Responsibility-based HMI display screen organization is primarily used to track status and manage operational risk. Users are able to identify when and where to intervene. It is best used where problems occurring in different portions of the operator's span of responsibility normally do not spill over from one area to another. Figure 5-7 illustrates a traditional hierarchical structure.

The overview level would typically be a page providing complete coverage of all essential aspects of the entity being overviewed. This generally means the entire

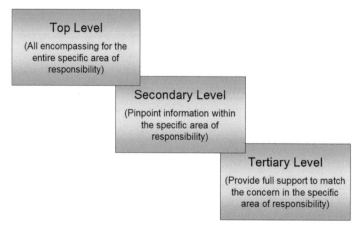

Figure 5-7. Responsibility-based display screen organization. This provides ready access to all information and controls within the given operator's specific area of responsibility. The arrangement of content of each level is governed by the relationship of the parts to the overall responsibility of the operator without undue regard to exactly where the equipment is physically located.

operator span of responsibility is depicted. Ordinarily, it would contain sufficiently abstracted information to convey operational conditions of everything important enough to merit attention. It would be general practice to ensure that this display screen remains always in view. The levels below the overview would be pages for viewing content on any display needed. The detailed construction of any given page (windows, formats, elements, navigation, etc.) would be up to the needed requirements, the extent of coverage required, and the display screen design and deployment requirements being followed.

Risk-Based Display Screen Organization: Process Causality Display Screens
Operational, risk-based HMI display screen organization would be primarily used to show the flow of influences required to manage operational risk. Users would be able to know which operational area might be affected and who should be tasked to intervene. Figure 5-8 illustrates a causality flow structure.

The complete causality flow may fit on a single page thereby providing coverage of all essential aspects of the entity being examined. The causality flow also may start on a home page that has a sufficient number of windows to help the operator follow the complete flow by opening the needed ones. Or the causality flow may be represented by a series of pages with clean and clear navigation to link them in the direction of flow (up and down). The detailed construction of any given page (windows, formats, elements, navigation, etc.) would be up to the needed requirements, the extent of coverage needed, and the display screen design and deployment requirements being followed.

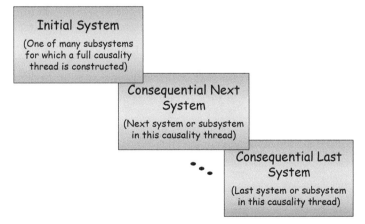

Figure 5-8. Risk-based display screen organization. This provides ready access to all information and controls within the given operator's specific area of responsibility focused only on operational risk without undue regard to exactly where the equipment is physically located. The arrangement of content for each level starts with the highest-level risks. Each subsequent level supports access to understanding and working needed problems or issues. This organization would be important during significant operational problems.

An example of risk-based screens would be a set designed to ensure that an extremely valuable product is kept extremely close to specifications. The information would clearly track the critical specifications; any threats or pressures against meeting the specifications; and all "handles" the operator might employ to tweak things or make more significant moves to reduce the likelihood of threats impacting production.

Task-Based Organization

A task-based HMI display screen organization is used to manage and facilitate specific tasks. This organization provides a coordinated set of display screens with instructions, information, and goals. Users would be able to know where the work would be done, how any work might interfere with other operational entities or aspects, and detailed instructions or directions for doing the work and returning things to proper service status afterwards. Figure 5-9 illustrates an all-in-one (focused, self-contained) type of structure.

This represents a single page or an opening page with links to successive pages or windows until completion. There are two column "threads." The left one is physical: what equipment, how is it arranged, and how is it interconnected. The right one is operational: what to do and how to do it. For each thread, there would be several additional windows that flow from each other in a natural progression or arrangement. At all points on each thread, direct links may be used to align them so that the physical part on the left is at the right location on the operational part. The details of how that is structured would be guided by the complexity of the task, the need for guidance, the availability and usefulness of real-time information, and any other resource materials incorporated.

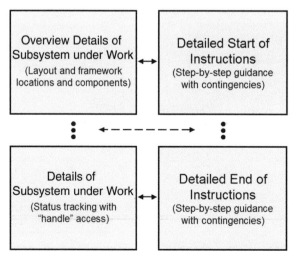

Figure 5-9. Task-based display screen organization showing two parallel paths (vertical portions) and how access to the next level down (horizontal links) provides detailed instructions and/or information to support understanding and management of the situation at hand. This arrangement allows direct focus on the task at hand and avoids undue navigation to locate needed supporting information.

Similarity-Based Display Screen Organization: Collection of Related Similar Entities for Comparison

Similarity-based HMI display screen organization is used to expose differences related to a predefined set of operational status conditions. Users would be able to identify which, if any, of several production entities may be operating out of the ordinary. Figure 5-10 illustrates a window designed to show all pumps of similar type and service to identify those whose mechanical operation may be at risk. Note the use of icon elements to show both pumps and the two important operational variables (horizontal stack of two bar charts, with value and alarm conditions shown for each). Moving clockwise from the upper left, the first pump has both variables in high priority alarm (red). The second pump has none in alarm (gray bars). The third pump has one in medium-priority alarm (orange) and one in high-priority alarm red). A careful observer will notice that the upper critical variable for the third pump has a lower value (and is in alarm) than the corresponding variable for the second pump. While the pumps may be quite similar, even alike, there is something about the third pump's operation that causes the alarm that is different for it than for the second pump.

Because comparisons are the primary task, care must be done to ensure that all are contained within a single page. If there are subsets with little overlap, each set of subsets may be shown on separate page or different windows within a page or group of pages.

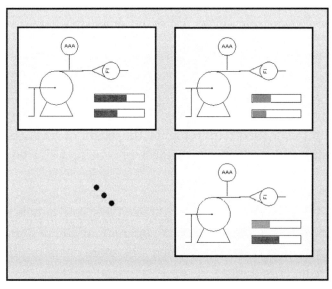

Figure 5-10. Similarity-based display screen organization used to highlight operational differences, including subtle ones, between very similar process and operational units. This type of display makes it easy to understand where problems may be developing by employing the same "mental model" of the process for all of them.

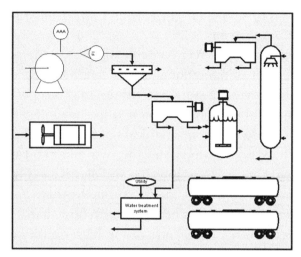

Figure 5-11. Geographically based display screen organization making it easy for the operator to see all the "moving parts" that should be working together harmoniously because they are colocated. It is presumed that any operational problems would quickly cascade to other closely located parts on the display screen.

Geographically Based Organization: Collection of All Colocated Entities
Geographically based HMI display screen organization is used to expose risks related to physical proximity. Users would be able to identify which, if any, of several entities may be at risk due to a geographically located abnormal situation. Such a situation may be the abnormal function of a nearby entity. Or it may be due to a shared risk such as local flooding or terrorist activity. Figure 5-11 illustrates a window designed to show which things share the same local geography.

Because locations and juxtapositions are the needed information, this would usually be a single page providing complete geographical depiction, a home page with sufficient windows to cover the complete area, or a series of pages with clean and clear navigation to link them in appropriate spatial directions (north, northeast, east, etc.). The detailed construction of any given page (windows, formats, elements, navigation, etc.) would be up to the needed requirements, the extent of coverage needed, and the display screen design and deployment requirements being followed.

Visibility and Viewability
Windowing and the powerful ability to do overlays can enhance the delivery of context-related information to the operator. An operator might view an overview screen, see something of interest or concern, and then seek more detail or information without ever leaving the screen. This ability adds useful navigational tools that maintain the current focus point. Yet this requirement is not without issues. We must understand that there may be hidden windows or formats. The most carefully designed displays cannot ensure the continuous visibility of information, at times vital information. Therefore,

the current acceptable practices for display screen construction must carefully ensure the proper visibility of operational information through use of viewability controls. The simplest way is to reserve a certain portion of the display screen for transient information and prevent overlaying elsewhere except for transient information like tool tips or pop-ups that only stay visible briefly. This may be more difficult than first thought if the HMI tools do not support proper controls. Additionally, enterprises that are required to preserve screen view records for later recall must ensure that all stored views maintain their relative visibility so that it is clear what the operator was able to see as any event unfolded.

Flash

Flashing a component on a display requires special attention. A component is flashed to significantly raise the success for rapid situation awareness of something abnormal. The flash part of a flashing component is visible (flash on) for some brief time and not visible (flash off) for another brief time. A momentary glimpse by an operator might catch the flash on or flash off. There is no way to be sure. Therefore, it is necessary to design flashing components to always tell the truth. That means that everything that flashes must be identifiable as flashing whether the flash cycle is in flash-off or flash-on. Moreover, the flashing component must not hide underlying information during the entire flash cycle. There are two aspects to manage.

1. **Flash cycle** – The entire flash cycle time should be fast enough to ensure that even brief views will expose the flash as active. This usually means that the sum of the flash-on time and the flash-off time is around 2 to 3 seconds. Moreover, the flash-on time should be much greater than the flash-off time. This adds to the operator's ability to readily see a flashing component.

2. **Flash visibility** – To ensure that the flashing component is always visible, no flashed content can be hidden. For text messages, either flashing the background or flashing a frame around the text can easily do this. Make sure that the flash-on view is not too busy or distracting. Where the frame is flashed, the actual frame would cycle between very visible and not so very visible. This way, it is clear that it is flashing even during the flash-off time of the cycle.

Here are two examples. Figure 5-12 shows how a text message might be flashed. Flashing the actual text will not work because the entire message will be hidden during the flash-off portion of the cycle. So we place a red frame around the message. Red is used because the situation we are conveying has red as its reserved color attribute when it is in this condition. The red frame line is narrow during the flash-off portion of the cycle and heavy during the flash-on portion.

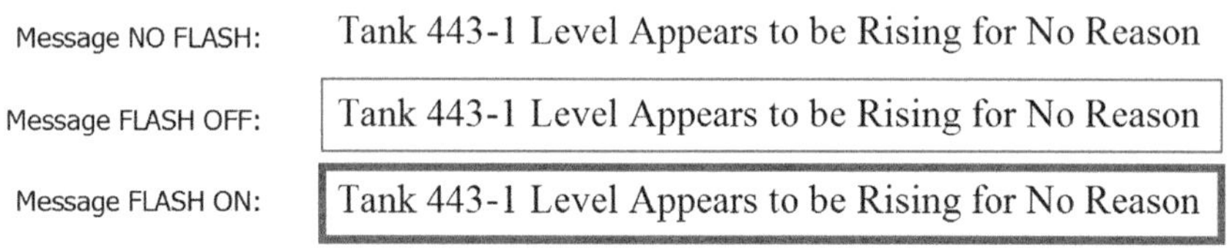

Message NO FLASH: Tank 443-1 Level Appears to be Rising for No Reason

Message FLASH OFF: Tank 443-1 Level Appears to be Rising for No Reason

Message FLASH ON: Tank 443-1 Level Appears to be Rising for No Reason

Figure 5-12. Illustration of the flash cycle for a text field. The thin-lined red box identifies the flash-off cycle. The broad-lined red box identifies the flash-on cycle. A red line is used to indicate an alarm condition that uses the color red. During the entire period where the message is flashed, a lined box is always visible. If the message is not needed to be visible when it is not being flashed, then the line identified by "Message NO FLASH" will not be visible.

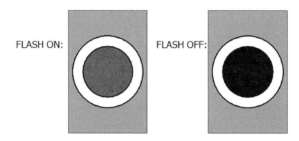

Figure 5-13. A flashing red light on a rural highway. The illuminated white outside ring is visible at all times. The inner red light cycles between being on (red) and being off (unlit), depicted as black. With this signal depiction, it would be very hard to miss the signal being present and working properly (flashing or not).

Figure 5-13 depicts a flashing highway stoplight in rural Mississippi. Most of us can readily admit preparing to stop upon seeing a yellow traffic light, only to recognize that the light was actually flashing and not just yellow getting ready to change to red. It is easy to imagine a situation where a flashing stoplight might be needed at the intersection of two rural roads. The problem with any traffic control device on a rural road is that it is mostly unexpected. It is easy to come up to a nondescript crossroad and not see the flashing stop light because it was glimpsed during the flash-off cycle. To remedy this situation, an illuminated white ring surrounds the flashing red light. It would be hard to miss a lit white ring even if the red light were in the flash-off part of the cycle.

Display Screen Design

This part is not pretty. That one-liner is often used to announce a dramatic moment in a novel or movie, usually one that shows how awful or unappealing a situation is. It is used here to emphasize literally that display screens should not be pretty. Something is made pretty to convey a pleasing whole. Each part of something pretty is meant to blend together. Operator screens are the extension of the visual senses onto a virtual reality that brings necessary data and situations into view. The better we do it, the more the operator is able to find, see, and understand what is really going on. We need the operator to see the details, spatial relationships, and other information without distraction. Pretty distracts. This is not to say that we are shooting for ugly or unattractive.

Color

Color has a single vital purpose: to accentuate the unusual or abnormal. All color distracts.[14] It is all too easy to create mental tension between what is visible to the eye and what is intended for the mind to perceive. You will want to design as far away from this problem as you can. Each screen, page, window, and format should have a neutral gray background. We look for just enough gray to allow the clean black text and graphic lines to show up softer, without a harsh black/white contrast. The display should be easy on the eyes hour after hour. Lines are made only as intense as necessary to convey the message. Where some information is more important from the operator's understanding and need-to-know perspective, increase the line width or intensity. Element shape and construction should be as simple as possible without sacrificing clarity and differentiation.

Abnormal or alarms will be depicted in color only of sufficient intensity to gain the operator's attention. Avoid undue drama—attention will stand out without being overdone. Differing colors (the palate should be quite small and always consistent) are used to indicate different levels of abnormal—this allows the operator to clearly see what poses the higher risks. To drive this principle home, let us look at the distractive power of the poor use of color.

Early on, the builders of HMI graphic display screens sought to exploit the use of color. See Figure 5-14. You can easily see hints of color being used for identifying purpose navigation targets (cyan buttons at figure bottom and green labeled buttons at figure top), process lines (green), certain process elements (yellow), ordinary measurement values (white on gray), abnormal measurements (white on red), and most subsystem labels (white). There appear to be color use convention deviations: measurement set boxes in cyan, some labels in yellow, equipment status in red (not indicating a problem), and some process lines in blue. The most important message of the figure is simultaneous lack of visual focus against the backdrop of complete distraction. The viewer does not know what current message the graphic is meant to convey.

As the evolution of technology for HMI design advanced, the problem of overuse of color was compounded by 3-D shading and using relative sizes and physicality of construction as opposed to functionality. Figure 5-15 shows a problematic example.

In Section 5.13 you will see how color should work.

---

14 C. James Goodwin, *Research in Psychology: Methods and Design* (Hoboken, NJ: John Wiley & Sons, 2010).

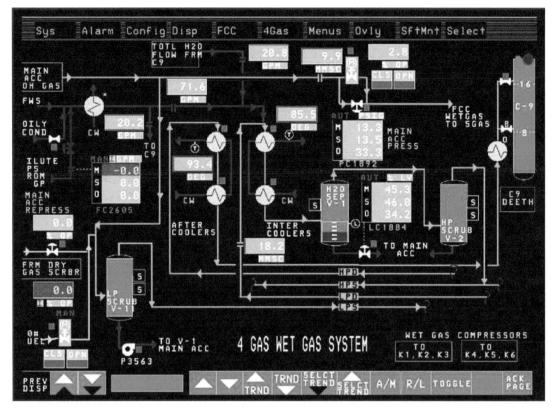

Figure 5-14. Example of the inappropriate use of color in an outdated HMI design. While color clearly separates the various parts of the screen and is interesting to view, it actually fails to highlight any but the most primitive abnormalities. Notice how the bright yellow of the heat exchangers appears to jump out of the view, but their actual operation is of very little interest. Even when the eye is looking for things, the colors get in the way by their distractions.

Source: Reproduced with permission from User Centered Design.

Display Screen Arrangement/Layout

This part will go over the recommended arrangement for display screen components. Each guideline applies to all components.

- **Origin** – The top left corner is considered to be the "beginning" of the eye's flow through the component.

- **Flow** – Components should flow left to right, top to bottom; avoid right-left and bottom-top reversals.

- **Connections** – These should be direct and without visual artifacts that distract (see Figure 5-16); the depiction seeks to convey relationship, *not* geometry or physicality.

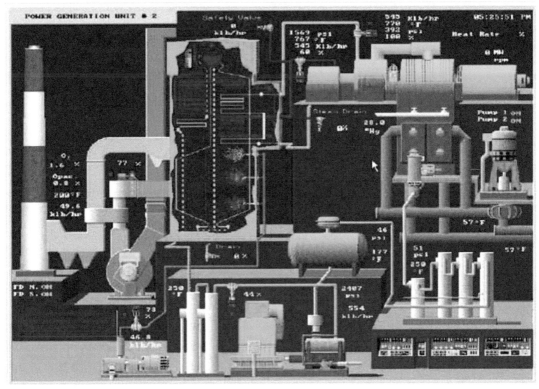

Figure 5-15. Example of the overuse of color, 3-D shading, and physicality. Notice how distracting all of this is. The attempts to add physical reality, in addition to color, do even more damage to the viewer's ability to see and understand the information. You see the picture. You do not see the information.
Source: Reproduced with permission from User Centered Design.

Figure 5-16 depicts the same mechanical configuration of a pump and its connected vessel. Operationally, the two depictions are identical. Graphically, they give very different information. The system on the right shows a pump directly connected to the vessel. The system on the left shows a pump with an offset connection to the vessel. The physical piping is of no importance. Yet, the depiction of the offset visually suggests that the offset would be important (e.g., to ensure that the operator knew that the pump had a potential to gravity drain into the vessel). Graphical depiction can inadvertently distract from the rightful message. Misleading aspects include relative size, relative position, and degree of artistic care in rendering each object. Careful attention should avoid these misleading aspects.

Dynamic Page Assembly

Dynamic page assembly means that the operator would have the ability to pick whatever he thinks might be useful to create a custom page view. He would be permitted to

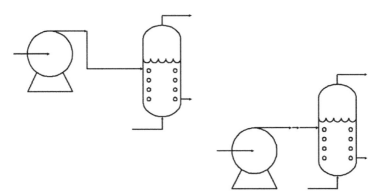

Figure 5-16. Illustration of unintended consequences of the specific way graphical artifacts are drawn. The two views represent the *same* process units. Notice how the connection from the pump to the vessel at the left has a drop in the leg. The view at the right does not. The very presence of the drop distracts by suggesting that it represents something important. We know it does not.

choose windows, formats, and elements to create a custom page view. On the surface, this would seem like a really good idea. After all, we realize that having the needed information together on a page or viewable at the same time via several pages on several displays is essential. So why not let the operator create his own? Because we should not. The reason exposes the heart and foundational message of both this chapter and the book.

Successful situation management depends critically on having everything thought out. We require procedures that are general enough to apply across the need yet specific enough to be spot-on useful. We require display screens to contain all the appropriate information to be understood in context, and enough display screens to provide the needed depth. We require training to be thorough enough to equip operators to do their jobs as effectively as an enterprise knows how. If we give an operator the ability "on the fly" to assemble pages as he wishes, it is unlikely that he will assemble them so that the chosen information is the correct needed information. Important items may be missing, included items may provide misleading information, the arrangement may be confusing, or many other things may not be right. If the operator really thinks he needs his custom display, either he is mistaken or the original design is incomplete. If he is mistaken, then either he is thinking about the problem incorrectly or he is not aware that what he needs is already available. It is better to have the operator believe that the HMI design is good so he might reassess the problem or escalate it to someone who can assist. If the original design in incomplete, then the operator should work as best as he can with what he has and later make the case for improvement. As you are well aware, any improvement will require good engineering design and proper management of change. The operator would use neither if he built something on the fly!

Display Complexity and Minimum View Time

Each display is built to meet specific information and control needs. The organization and content should be selected to best help the operator understand what is depicted to support that need. The content should be necessary to perform that specific activity. For example, if the activity was to find operating or performance irregularities for the pumps in a pumping station, we might use similarity-based content. If it were to make sure a tank was functioning properly, we might use geography-based content. If the operator was performing a complex task, we might use task-based content. Let us leave the construction themes to focus on the results.

Each screen with all its pages, windows, formats, and elements, composes a view that the operator will need to use. The screen will be built with extensive care following competent guidelines. That being said, it will take time for the operator to gain the needed information and interact with each screen. The more complex the screen, the more time the operator will need. The more important the task, the more information and time are needed. Builders of screens, need to assess the time to understand and interact with each screen as part of the design and to factor in the overall operating load on operators. This becomes even more important when the number of displays increases. Each display holds a screen. The more screens that are visible at the same time, the higher the base load on the operator to keep up. This is a trade-off. It means that the addition of every screen has an overhead on the operator (and yes, a benefit too) that must be clearly factored in.

Color Blindness

> Color blindness, or color vision deficiency, is the inability or decreased ability to see color, or perceive color differences, under normal lighting conditions. Color blindness affects a significant percentage of the population. There is no actual blindness but there is a deficiency of color vision.[15]

Color blindness is generally thought to affect less than 10% of the general population to some degree, with men affected substantially more often than women.[16] Most color-blind people are quite mildly affected.

> Being color blind is officially considered a disability, however there have been studies documenting certain advantages including penetrating some camouflages. For instance—it was found during World War II that analysis of aerial photos yielded better results if at least one member of the surveillance team was color blind.[17]

---

15 "Color Blindness," Wikipedia, last modified June 13, 2018, http://en.wikipedia.org/wiki/Color_blindness.
16 "Color Blindness," Wikipedia.
17 "Color Blindness," Wikipedia.

Legal Standing

Color blindness may make it difficult or impossible for a person to engage in certain occupations. Persons with color blindness may be legally or practically barred from occupations in which color perception is an essential part of the job (e.g., mixing paint colors), or in which color perception is important for safety (e.g., operating vehicles in response to color-coded signals). This occupational safety principle originates from the Lagerlunda train crash of 1875 in Sweden. Following the crash, Professor Alarik Frithiof Holmgren, a physiologist, investigated and concluded that the color blindness of the engineer (who had died) had caused the crash.[18]

Assessment of Operators for Color Blindness

The current US interpretation for operators (and pipeline controllers) is that all affected personnel undergo testing to determine whether or not existing HMI colors and usage as well as other conventional use of color in the control room are sufficiently distinguishable for each individual. Where appropriate adjustment or added redundancy can be done safely and without confusion, it should be considered prior to barring individuals from operating positions. Any "adjustments" will be established in the HMI design basis documents and implemented uniformly.

5.9 Navigation

Navigation is the "science of locating the position and plotting the course of ships and airplanes."[19] Navigation is as important to an operator during a situation as it must have been to an explorer or ancient mariner on the high seas. Besides the obvious lack of water or desert sands that go on forever, your operator has one important advantage over those explorers: he should know where he is at any given moment. What is left is the need to "plot a course" and actually go where he is needed. That skill is only part of the real navigation story. Navigation stands squarely on two basic pillars: (1) powerful tools and capabilities built into the HMI for easy yet accurate use; and (2) the careful, expert use by the operator. And the operator must be able to do it all in a way that is intuitive and natural—whether he is young or old, experienced or not, highly alert or normally fatigued, calm and collected, or anxious or tense. Let us lay out what is necessary and why HMI navigation is so important.

Purpose of Navigating

The purpose of operator navigation on the HMI is to obtain needed information and execute operational commands. The ease with which the operator is able to use the

---

18 "Lagerlunda Rail Accident," Wikipedia, accessed August 4, 2018, https://en.wikipedia.org/wiki/Lagerlunda_rail_accident.
19 David Bernard Guralnik, ed., "Navigation," in *New World Dictionary of the American Language*, 2nd ed. (New York: Simon & Schuster, 1980).

tools provided by his HMI will significantly impact his speed and effectiveness. The best navigation is much like airplane flights: direct and nonstop. This means that very few operator decisions are needed to go from where he is now to where he wants to be. Of course, if the desired location needs to be discovered, then each step along the way should directly follow the threads of finding what is at each step along the way. That is, the operator should be directing the jumps rather than the HMI design requiring small jumps at a time. This requirement is poignantly illustrated by the Microsoft Windows navigation strategy that requires mouse click after mouse click (or screen touch after screen touch) to work one's way down a pathway. We can unaffectionately call this "click forever" navigation.

The Navigating Cycle

Operators never stop navigating. Even a perfect job of designing HMI screens and artfully laying them out on the available displays only goes so far. There is always more to see or follow up on based on what is happening or not happening at the moment. Sure, good design will dramatically reduce the need to find other stuff when operations are going to plan. Still, good navigation is a necessity. To help us understand it, we can view navigation as if it were a step-by-step procedure. It is not, but let us look as if it were. Approach this as if you are looking at stop-motion video. Each frame represents a navigation step. The frames go in sequence. The purpose is to ensure that checks and confirmations are done at every step of the way. We do not want to get lost. And we want to get to where we want to be in good order. These considerations should be part of basic operator training.

1. Confirm the current location. Is the navigation starting from the place intended?

2. Decide what is needed next (the purpose of the navigation). Before clicking, decide what is being looked for.

3. Decide where it is. Knowing what is being looked for, decide where to find it. (At this point, either the specific "location" is known or there is a good idea of what might be near what is needed.)

4. Decide how to get there. Using the existing tools and infrastructure, determine the specific way to navigate (e.g., use a map, do a guided search, or follow a planned idea).

5. Mark where you are and get to where you want to go.

6. Confirm where you went is where you needed to have gone. Did the navigation step(s) take you to the intended destination?

7. Ascertain that what you needed is where you went to find it.

8. Get what you need.

9. Return to step 2 and repeat the steps as long as needed.

Right about now, you might be thinking that this process is way too analytical. Navigation should be a simple thing; operators (and the rest of us) do it almost instinctively. This is all true. However, in the control room you will want to ensure that this really works. Navigation failures can make a difference between an awkward situation and a disaster—not always, not even often, but almost always when accidents and disasters are examined for cause. This discussion will be about formulating a competency. Using this tool is a part of being a professional. Here is how the capabilities are laid out.

Current Location
Operators usually know where they are (in the HMI navigational sense). Although from what we will see later in this book, checking on that and confirming it should be a routine practice. As you will also see as we go down the list, being able to mark the current location can be quite important. You might have wished you had done so on your smartphone when you parked your car in a large parking lot, returned via a different entrance, and could not find your car.

What Is Needed
Not always does the operator know what is needed next in the way of information. If this was always true or even mostly true, situation awareness would be a trivial problem. It is not, of course. Therefore, operators continually depend on other aspects to identify this. Refer to the next chapter on situation awareness and assessment. Regardless of how, the operator will need to know what is being sought.

Where the Needed Information Is Located
Knowing where to go to find the needed information is arguably the core of navigation. Everything must be organized around an understandable and natural framework. When the operator learns where to find one thing, related things should be readily and naturally found. Once the operator is on the trail, the trail should be easy and natural to follow. This requires a careful plan, responsibly carried out with lots of clues and confirmations in every graphic and every other component on display screens.

Decide How to Get to the Needed Information
There is no single right way to navigate. Just like when bicycling from Point A to Point B, there are a lot of different pathways to travel. If you are new to the neighborhood,

you will stick to known roads, even if it means going out of your way just to make sure. Later on, you will be taking shortcuts left and right. In between there will be a lot of wrong turns and longer than necessary trips, and a few "got lost" moments. This might be okay for a bicycle ride, but not so for trips that must be efficient, quick, and accurate—such as operating an industrial plant.

Operators will need accurate clues and carefully designed information "filing" systems. All related navigation destinations should be present on the current *pages* or in flat (as opposed to hierarchical), ready lookup tools.

Mark the Current Position and Navigate to the Needed Information
A useful tool for navigation would be the ability to place a virtual bookmark on the current *page* and view so that the operator can return to it directly from wherever he navigates to. Unfortunately, this capability is yet to exist in commercial HMIs. The best work-around is to place the screen to be "bookmarked for a possible return point" on a dedicated (for the moment) display and use other displays for other pages. This is another important benefit to having enough displays for ease of work. Just to make sure this point is clear, neither a "mark-page" button nor a "back" button is supported by control HMI platforms. Until they are, your operators will have to develop a procedural work-around.

Confirm the Target Location
Just because you navigated to a location does not mean it is the one you intended or needed to arrive at. All navigation jumps need to be confirmed. This places an important requirement on display screen component designers. There must be an identifier placed in a predictable location on each page that sufficiently describes exactly what the display screen components mean in relation to the physical entity they are intended to show.

Confirm the Needed Information Is Present
You navigated to a certain location to get specific information. Look for what you need, not what is there. If, for example, you seek the current measurement value for a pressure control loop, make sure the value seen is the actual measurement and not something else, such as a set point or an alarm activation point.

Obtain the Needed Information
Observe what information you seek. See what it is and place it into the context you need to have to understand what you see. If the information was sought to prepare a report (e.g., the shift handover), then immediately place the proper values or your interpretation into the report. Reliance on memory is usually not a preferred tactic.

Jotted notes take time and at times get misplaced. Sometimes yesterday's note values get mistaken for today's. Check back to ensure that what you have taken matches what is there.

Navigation Tools

Navigation is getting from Point A to Point B. These locations can be anywhere. However, the ways of getting between them are limited. The best ones are planned for.

Dead Reckoning

Dead reckoning is the simplest and usually least productive way to navigate. To use it, start off and try to think (guess?) where Point B might be. Start clicking. After each click, think (guessing again?) where the next place Point B may be. Keep this going until you get to Point B; find out that Point B is not the place you need after all; start looking for a different Point B; or get lost, give up, or get saved by a shift change. To be fair, there are some operators who are really good at dead reckoning. Most are not. It is best to depend on other ways.

Direct Enter Search

The operator clicks open a search box and enters "Point B." Only the lucky few get this to work. First of all, the only Point Bs that will be found are control system configuration tag names or, less likely, names of pages, windows, or formats. You have to know *exactly* what the name is and key it in without error. No modern controls platform search has ever heard of the phrase "close is close enough." If this were not enough to hope for a day off tomorrow, most of the references to the searched-for-location are inconsistent. The database ID is different from the HMI label, different from the labels on the drawings, and different from any physical label on the actual field device. Who has not had the experience of being in the field at a valve, reading the valve label, and talking with the operator on the radio asking him to stroke the valve so you can watch, only to hear unrepeatable words and have long waits? Often you will be asked to suggest where the line is coming from that connects this valve with something the operator knows.

Table Lookup

Table lookup solves the spelling problem and gets around the "close is close enough" limitations. However, the table has an organization that was logical for the designer, but likely not necessarily so for the operator. Finding the place in the table to start may be harder than dead reckoning—or may be just the same. You might be thinking that this one is not one of the sharp tools in the drawer. You would not be far wrong.

Targets

Navigation targets embedded in pages, windows, formats, and elements are the workhorse. The HMI designer was aware of the intended use for each. He knew most of the places the operator would need to visit. He would have configured them on the current display screens that contain the views of Point A. It is normal practice for pages to have a set of more universal targets and a set of targets that relate to the present location page. Moreover, the HMI design spells out where each class of target should navigate to. Now the operator knows what target to look for to get to the Point B he needs. Things are much better for the operator. Before we leave this method, please take note that it will only work if your site's HMI design specifications are rich and carefully done and the HMI is built with diligence and accuracy.

The next topic discusses two navigation methods that are facilitated automatically by knowing the current location and operating situation(s). They are extremely supportive of the operator and at the same time able to reduce the number of displays taken up by navigation.

Yoking

Yoking is a display coordination concept. It is used to navigate to a coordinated set of related lower-level pages from the higher level. It is this coordination that gives it the *yoking* name. The term is used analogously from the device used to couple farm oxen together so that they may be used as one unit, for example, to pull a plow. Display components are yoked when they appear together as a set, either built on the fly to comprise a single page or displayed on multiple displays at the same time. Figure 5-54 later in this chapter shows an example.

Focus

Focus is the design of a human interface device to reflect the current state of things. A simplistic example is the practice of graying out all portions of a manual data entry page that are not active for the current task. Thus, if all needed information in a form is not filled in, the "enter" button will be grayed out. Focus is used to set up the right page components to reflect the current state. The Control Area shown in Figure 5-54 was used because the operator navigated to the particular secondary display by clicking on the alarm for this loop there.

The "Product" of Navigation

Operators use navigation to position themselves in front of sufficient information to understand the present situation (the need for the operator to see something) and to take necessary action in response to that situation. Yet operators will rarely find all needed information and all reasonable "handles" for action on a single display screen. To provide the operator with such a display screen every time would require an unreasonable number of them. Everything needed that is not on the single display screen (or a set of simultaneously visible ones) disregards a fundamental HMI requirement: operators should not need to write down or remember necessary information to access controls needed to manage a situation. Ensuring that we properly provide for the operator is done using one or both of the following ways: provide enough displays, and/or provide complete situation-based display screens.

Enough Displays

The best and most efficient method to determine how many displays are needed for an operator area is to find out how much display screen information is needed to be simultaneously viewed by the operator for the range of situations he is responsible for. This number includes both managing the current situation *and* managing the overall plant. Usually this means that there will be both an overview graphic display and an alarm management display. The remaining number is determined by the needed information and control requirements. The reader's attention is called to the obvious issue: depending on the care and skill used in designing display screen content, the number of displays needed could vary from just a few to a very large number. As a designer, this provides very useful feedback on whether the design approach you are using is working. If most tasks require two, three, or four displays, you are probably on the right track. If only one display appears to work, either the task is fully localized so that everything can be put together without crowding or confusing, or the design is missing quite a bit of needed information because the designer does not properly understand the situation. If most tasks require more than five or six displays, it is likely that the display architecture has problems with hierarchy or segmentation, or both.

We assume that your display screens have been properly specified and built. The navigation problem now becomes finding the right display screen for each display. Finding them needs to be supported by navigation tools directly resident on the primary display screen used to identify the problem or the next one in the logical progression.

Situation-Based Display Screens

Situation-based display screens are ones that are designed to dynamically configure the content of the display to provide the needed information, control access points, and provide navigation links for the operator to assess and manage the current situation(s) under observation. They are particularly useful for situations that are quite similar in understanding and management yet appear at several different locations in the process or plant, for example, tanks in a tank farm, reactor efficiencies for several unrelated reactors, late commuter trains, or a congested link in a communication system.

5.10 Glyphs, Icons, Dials, Gauges, and Dashboards

The video display unit (VDU), HMI, or *display*, as we will term it here, is the hardware equipment used to show display-based information to our operators. That information is depicted on any number of display screens (the individual views for the displays). Each display can show only one display screen at a time. The display screen content varies from the predesigned and carefully managed to the contemporaneously constructed. Earlier discussion has addressed how much of the display screen content is designed and managed. This section provides added dimensional content to those display screens. This content can be dramatically informative. Using carefully crafted modules of well understood visual-based information can do this. You will see how each packet of information is captured and shown to the operator. This adds the ability to provide clarity and context to the repertoire of the graphic interface designer.

Glyphs, icons, dials, gauges, and dashboards are added to existing display screen content. For example, the operator may be viewing a process overview of a portion of his area of responsibility. Typical overviews contain physical or virtual equipment layouts and key interactions between the components. Also present are the key or major influential process measurements (e.g., flows, temperatures, and qualities). The system designer provides context to these measurements by depicting them using icons. Where grouping of information is important, these individual icons can be morphed into a composite dashboard. Dashboards may be self-contained on an entire display screen view, or just a portion of one.

Relationship to Style Guides

The design, symbology, navigation, and deployment of the following elements should be carefully set out in the HMI design and covered in the appropriate style guide for it.

Glyphs

We begin our coverage with glyphs. Glyphs are predesigned, fixed graphical "stamps" (see Figures 5-17 and 5-18). While a color may change and some variation in size is seen, for the most part they are static and show the same item and convey the same message every time. They are used extensively by modern computer software as navigation targets. You are probably familiar with their use in most modern product operation

Figure 5-17. Glyph that means "Do not dispose of in ordinary trash." It is clear and easy to understand exactly what is intended.

Figure 5-18. Glyph traffic sign indicating to drivers to slow down vehicle speed on the road ahead.

manuals, quick start guides, and everyday traffic signs. They quickly identify content as *important*, meaning "please pay close attention to this"; with *caution*, meaning this part describes operation that may harm individuals or the equipment if done without proper care; and so on.

Glyphs are useful in a very limited way for operator graphics. They offer a quick and ready way to identify the same thing everywhere it is needed. The primary use will be in the mode of "tool tips." These are the symbols that are displayed when the mouse or other type of cursor is hovering over a target. The tool tip directly conveys what action or alternative will result from "clicking" on the target. Examples are illustrated here.

Figure 5-19 shows the tool tip explaining the acronym HTML in a text document. Tool tips are often combined with a characterizing glyph to quickly advise the viewer of the nature of the tip.

Figure 5-20 shows a few examples. Note how easy it is to categorize the type and importance of the tip from the glyph design. Its intent is clear.

Figure 5-21 shows glyphs used as navigation buttons. Note the redundancy with both text and visual image. Strictly speaking, the visual images are gratuitous decorations, yet they provide a "softer" look that suggests thoughtful consideration of the user. Using them in process control graphics would be distracting.

Demonstrations of tooltip usage are prevalent on Web pages. Many graphical Web browsers display the title attribute of an HTML element as a tooltip when a user hovers the mouse cursor over that element. In such a browser you should be able to hover over images and hyperlinks and see a tooltip appear.
A web browser tooltip displayed for a hyperlink   Hypertext Markup Language

Figure 5-19. Example tool tip showing "Hypertext Markup Language" displayed when the cursor is hovered over the HTML link. Where abbreviations or other short codes are used in notations or messages, tool tips can be used to provide more information for understanding should it be needed.

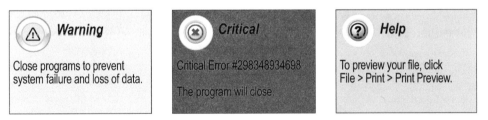

Figure 5-20. Example tool tips combined with glyphs. Only the appropriate one is displayed as they are used one at a time.
Source: Adapted from www.webfx.com.

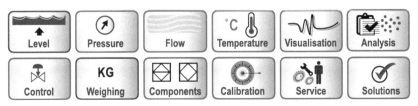

Figure 5-21. Glyphs as navigation buttons illustrating 12 different buttons. Only the appropriate one is displayed as they are used one at a time.

The most popular glyphs are the ones used to identify social media types and links. While they are almost universally referred to as *icons*, strictly speaking they are just simple glyphs. Figure 5-22 shows the more common ones.

Icons

Icons are powerful codes. Icons can be selected to evoke a ready understanding. These codes are not hidden (as in spy and intrigue stories). They are meant to be in-your-face obvious. That is what affordance is all about. Icons speed up understanding and minimize confusion. They both evoke a meaning and confirm that what is evoked is correct. While it might seem to be redundant, such redundancy is important. Our minds are always testing truth. When two points come together and the message is the same, we tend to believe.

Figure 5-22. Social media glyphs illustrating 16 different types or links. Only the appropriate one is displayed as they are used one at a time.

Figure 5-23. Example icon illustrating a normal temperature. In addition to the label *Normal*, this icon depicts a normal temperature reading because it does not look abnormal. That is, nothing about the icon is visible except the blue bar showing the relative value of the current reading. Contrast this with the icon in Figure 5-24.

Icons are more than placeholders and announcers of warnings. They communicate status. That is, they contain context. Take the thermometer-like icon shown in Figure 5-23 above.

We call these context icons. First, the obvious: these icons are used to convey temperature. They are used on graphics to flag important temperatures that the operator will want to observe. They serve as indicator measurements of normalcy or pending abnormal operation. Looking at the icon, note that all that is visible is the outline shape (a temperature-like thermometer), a range (the white vertical box), and the current measurement (the blue bar about halfway up the range). What you "conclude" from this is that the displayed temperature is within its normal range of expected values. We know it is normal because it does not look abnormal. No additional information is displayed. This is done to ensure that everywhere this temperature icon is displayed, any viewer will immediately conclude that the temperature is normal. No additional concern need be invested in it.

When that same temperature is abnormal, the icon form remains the same, but context is added. We easily see in Figure 5-24 that our temperature is abnormal. It is clearly too high. First, the normal range box now has all its abnormal areas identified, from slightly abnormal to grossly so. We see that the current value, again a simple blue bar, is in a red area. Red means trouble. So this temperature must be badly amiss. And to reinforce this, we have the current actual value displayed against a red background. At times added context would be there to show relief valve settings or emergency shutdown (ESD) limits. Taken together, it looks clearly abnormal. The context information was always available in the system database, but to show it to the operator when the temperature was normal would have been a distraction that would serve no purpose.

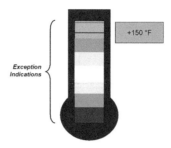

Figure 5-24. Example abnormal temperature icon, often referred to as a *contextualized icon*. Note the change from the previous icon. Now showing is the relative value (Exception) regions in the temperature range showing the "sweet spot" for normal (the white region) and the increasingly abnormal regions (the gray areas of increasing darkness). The current value (wide blue bar) is in alarm (red region) with added alarm indication (red boxed value). The damaging region value (thin blue bar) is slightly above.

Figure 5-25. Sample progress bar used to depict both the current progress of a task (blue squares), the total progress the task will take (black rectangle), and a rotating arrow to show that the task is progressing despite any slow accumulation of blue squares.

Figure 5-26. Sample progress bar with more progress shown (more blue squares than the progress bar in Figure 5-25).

Consider another very common icon: the progress bar used extensively by software programs to reassure the user that a task he is monitoring is progressing (or not). See Figure 5-25. Before their general use, and interestingly enough when software programs were much less reliable, the user just had to wait until the task either completed or he ran out of patience. More reliable software might have reduced the need to use these; however, their presence is so expected that their absence is annoying. The bar contains three parts. The frame shows the total task required progress. The actual task progress is shown as either a solid or a broken bar. See Figure 5-26.

The current progress amount is assumed to be at the last visible part of the bar. The third part is a motion indicator and is a recent addition. It is required when some tasks progress so slowly that the completion bar moves too slowly to reassure the viewer that things are still working properly. The motion indicator is constantly moving. This motion reassures the viewer that the progress, while slow, is moving along.

It is all about information, rapid assessment, and clarity.

Dials and Gauges

A dial or gauge is a useful and compact vehicle for displaying contextual information in a graphical or picture format. This section is designed to fully complement the earlier HMI section, building on its principles, yet providing important extensions to the ability of the operator to see, understand, and manage. In a dial you usually depict a single entity. In a gauge, you are depicting more than a single entity. At the same time, each of the entities in the gauge is depicted in a similar way.

Example Gauge: Nuisance Alarms

As an example, see the nuisance alarms impact gauge of Figure 5-27. Note that each alarm is shown as an arrow line. There is importance in the length, width, "arrowhead" type, and the order. What may appear at first glance to be a complicated starburst of lines is actually a compact snapshot of how the operator is being distracted by nuisance alarms. Nuisance alarms are ones that the operator does not follow, yet they get in the

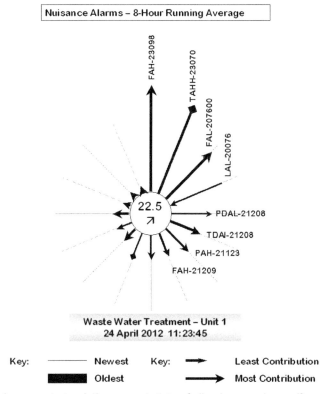

Figure 5-27. Example gauge to track the current state of all nuisance alarms (for a wastewater plant). At a glance you can see that the (hourly average) number of nuisance alarms is 22.5 (hourly average) and increasing (within the center circle); which are the oldest and newest ones; the relative number of times in alarm (the length of the arrow) with the longest at the top (decreasing in numerical number clockwise around the wheel); and whether the individual ones are increasing or decreasing in number (arrow out or arrow in).

way of seeing useful ones. At a glance, as a hypothetical situation, the *best view* of this gauge, meaning that the nuisance alarms are of little impact, would be for the following:

- All arrows are about the same length (meaning that none is behaving much worse than the others).
- All arrowheads are either pointing inward (meaning that the number of the individual alarm activations is decreasing) or displaying a diamond (meaning that the number is holding steady).
- The number in the center is below 2%.
- The arrow in the center is pointing down (meaning that the total contribution of nuisance alarms to the total number of alarms is decreasing).

Think how useful this might be for a large plant or operation where each component part has such a gauge. Any nuisance alarm problem would be clear.

Let us take a closer look at how this gauge is constructed. The example is Unit 1 of a wastewater treatment facility. Depicted are the top (by occurrence) 16 nuisance alarm tags that have been tracked over the past 30 days (as a running window of data). They are arranged clockwise from the 12 o'clock position, with the highest contributor at the top. Additionally, they are coded in order of entry into the top 16 "club" by line width. The broadest line represents the longest time in the list. Note that once a variable makes it into the top 16, it carries the entry date until it leaves. In the center of the diagram is a cumulative figure that represents the percentage contribution of the top 16 into the total contribution of all alarms for the month. The arrow in the center indicates whether the number of alarms in the top 16 is steady, increasing, or decreasing. The arrowhead at the end of each contributor represents whether the given nuisance alarm contribution is increasing, decreasing, or somewhat steady.

There is quite a bit of content in this gauge, although it is very simple for the operator to ascertain whether the alarm situation is getting better or worse and where in his areas of responsibility any (alarm indicated) problems are. Purely from an alarm management point of view, if the number in the center is getting smaller, it means that fewer individual alarms are contributing to the top 16. This in turn means that the alarm activations are becoming more widely spread within the wastewater facility. This connotes that the actual alarm activations should more closely represent actual abnormal situations rather than an alarm system in need of attention. But all of this is an aside. The true benefit is in the content of the gauge. Let us take an even closer look at our example.

Alarm FAH-23098 (at the 12 o'clock position) has the highest number of alarm activations (it is the longest line, which is always at this position). It is also one of the longest of the top 16 that have been in alarm (it is one of the widest line widths). And the number of its activations is increasing (the arrowhead at the end points out from the center). When he has some extra time, the operator may look at this alarm and attempt to determine what might be the problem. He might find:

- A bad instrument transmitter

- Operation too close to the alarm limit (which suggests an unusual operating value, the need to adjust the alarm deadband, or add on-delay or off-delay for the alarm configuration)

- A faulty control valve

- Noise on the instrument signal due to faulty wiring or improper wiring location practices

Alarm LAL-20076 (located at about 2:30 on the dial) is the fourth top nuisance alarm (clock location and being the fourth longest line), is one of the newest additions to the nuisance alarm list (it has a narrow line width), and its contribution to the top 16 is diminishing (arrowhead points inward toward the center). Because it appears to be improving and the total contribution of the top 16 is at 22.5% and getting larger, it would seem that attention to this one might wait until other more important things are attended to.

Example Gauge: Pipeline Nominations Tracking

Natural gas or petroleum transmission pipelines move product from one geographical location to another through various segments of enclosed pipe. The product is "transferred" or "delivered" to a separate business entity, or from one internal entity to another, via the action of custody transfer. The process of *nomination* is used where the amount of product transferred must meet a predetermined contractual amount. In general, these delivery schedules are set 30 or more days in advance. The normal time frame for monitoring delivery is daily. Failure to satisfy the nominated value is serious, so every effort is made to avoid this. Traditionally, pipeline companies track the nomination value carefully during each daily delivery. They typically track how much has already been delivered against the target delivery requirement. We look at a liquid pipeline next.

We have a single controller (that is what pipeline operators are called) in charge of a 230-mile section of interstate transmission pipeline with several branches. The specific configuration is of little importance to our example. The custody transfer and therefore the satisfaction of the required delivery amount (nomination) is done at the entry to this pipeline. All receipt of product is under the control of others; however, delivery must not exceed the nominated amount. The receiving party can adjust how much delivery is taken for limited periods if requested by the controller. This is a bit unusual, but it keeps our example straightforward. There are only remote shutoff valves (ROVs) along this line. The nomination period is 24 hours measured from midnight to the following midnight.

To manage the amount of product he permits to enter the pipeline, the controller provides a desired flow value from a delivery system consisting of storage tanks, pumps, and control valves located at the entry of his pipeline but may be under operational management of others. To reduce pumping costs and increase the flow capacity of the pipeline, drag-reducing additives (DRA) are metered into the line.

Let us take a look at a candidate nomination "gauge" created as an example of the potential and utility of such a tool. First, we define the symbols in Figure 5-28.

Figure 5-29 shows how this fits together for a nominations gauge.

The view of the nominations gauge shown in Figure 5-30 depicts our controller fully managing the nomination. The cycle has 15 hours and 43 minutes left to run. Three hundred and forty-five million barrels of product remain to be delivered during that time. If all goes to plan, the target will be met. This is indicated by the position of the target-met symbol at the target product end value. Because the target will be met without any change, there is no need for the "reachable goal after operational

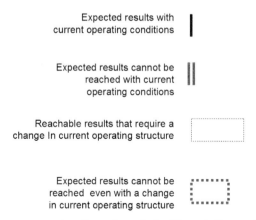

Figure 5-28. Nominations gauge explanation of symbols. Though they might appear to be complex, they are not complicated. Once mastered, they will permit the viewer to see the overall progress and completion potential.

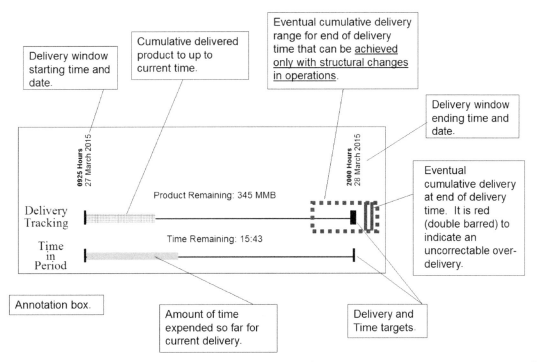

Figure 5-29. Design structure for nominations gauge identifying all component parts. Although none of these labels will be shown on the gauge they might be coded as tool tips if desired.

change" symbol and to show it would provide an undue and unnecessary distraction. Notice that the Delivery Tracking bar (the upper hashed one) is about the same length as the Time in Period bar (the lower solid one). It is possible that the earlier rates were adjusted a bit, but the similar lengths of the bars suggest normal and proportional progress. Remember the guideline: if it is normal, it should not look abnormal.

Suppose the situation were different for this same time. Figure 5-31 depicts the new situation. As it stands, with no changes, the nomination will be exceeded.

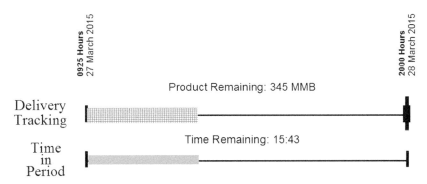

Figure 5-30. Nominations gauge illustrating that without mishap, the current nominations (345 MMB) will be met at the required time (15 hours and 43 minutes from now).

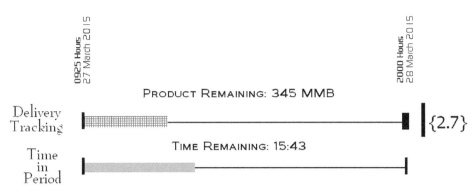

Figure 5-31. Nominations illustration requiring a flow decrease in order to satisfy the requirements. Unless a decrease is made, the actual delivery will be over by 2.7 MMB.

However, it should be possible to meet the nomination with proper adjustment and monitoring. If the controller would reduce the flow rate by about 2% and hold it for the remaining 15:43 hours, the job will be done. Notice that the distance scale used to display the *goal symbols* along the delivery line is amplified by a factor of 10 over the normal delivery scale in all display screens where there is a target miss. The reason should be readily apparent: it is not expected that the controller be far off of the target; therefore, we amplify small differences to clearly indicate any irregularities. For example, in Figure 5-31, if nothing changes in the delivery schedule, there will be an excess of 2.7 MMB of product put into the line. The black bar is shown at a scale as if it were 27 MMB over delivery (for visual effect), thus having the visual effect of a ×10 zoom for only that symbol relative to all other information depicted.

The gauge in Figure 5-31 looks similar to the earlier case. However, note the added item, {2.7}. It is present for this situation because now it becomes important. It is an indication to the controller that unless a change is made, delivery at the end of the delivery period will be 2.7 MMB over nomination. On the other hand, the delivery progress bar is noticeably shorter than the time in period one. This suggests that left unchanged, the pipeline would under deliver product. So what must have been going on is that the controller was underdelivering at some earlier time but noticed it and made a recent increase in flow. This is a good time to explain that our controller cannot simply substantially reduce delivery near the end of the delivery period or stop delivery at the moment he reaches the nominated value. To do so would disrupt operations at both ends of his pipeline. His pump station operator would not like this. The customer would very much not like this. Moreover, stopping the inlet flow without simultaneously stopping all outlet flow will almost always result in the pipeline not being full, a condition termed *slack line*. To avoid this, the controller would gradually modify things to ensure proper delivery.

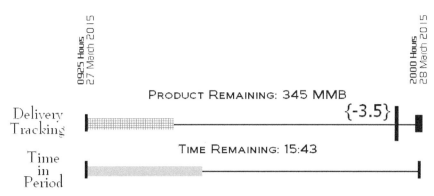

Figure 5-32. Nominations require increase of flow. Unless an increase is made, the actual delivery will be under by 3.5 MMB.

For our next case, instead of requiring a decrease in flow to meet the nomination, an increase is required. Figure 5-32 depicts a situation in which the nomination will fail by being short 3.5 MMB if the current delivery rate is maintained. A proper increase now should do the job.

Suppose that after increasing the flow for a short while, the pipeline reached a condition where no further flow increases can be made that would meet the nomination. Figure 5-33 shows that situation. It looks almost the same as Figure 5-31 with some important differences. The end of nomination flow value looks similar but is shown in red. Because no additional flow increase is possible, the best target for the final delivery will be short by 2.9 MMB.

The next example case for our pipeline is the situation in which the nomination cannot be met during the operating period at the current rate, nor can it be met at any rate. Figure 5-34 illustrates the nominations gauge for this. It shows the final delivery will be 4.3 MMB over nomination. The red dotted box indicates that if a pipeline

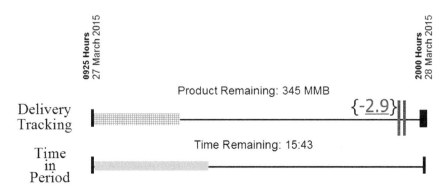

Figure 5-33. Nominations require an increase in flow, but no increase is possible. Without any resources or room to make the necessary increase, the actual delivery will be under by 2.9 MMB.

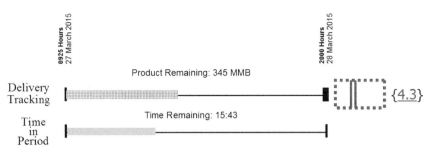

Figure 5-34. Nominations requiring a decrease in flow, but a sufficient decrease cannot be made. While there is room to make a flow reduction (red dotted box), no reduction level is available that can eliminate the overage. Without any change, the overage will be 4.3 MMB.

configuration change were made (e.g., by changing a pump or piping configuration), it would be possible to reduce the overage to less than 1 MMB (the value 1 is visually estimated [as the left-most edge of the red dashed box] from the red vertical delivery bar being at 4.3).

One might be tempted to suggest that this case is an unrealistic situation. After all, the controller is very experienced. Surely there are operating tables that show what delivery flow rate would be needed to meet a target nomination value. They do. So why would the controller not use them. He did. Why did they not work? Any number of things may have gone wrong. But the basic situation is that the flow could not be reduced enough. A most likely cause is the need to avoid slack line. In our example, something did happen but somehow remained unnoticed by the controller. In our example, as soon as the problem became apparent to the "gauge," the controller could see it.

Dashboards

A dashboard is more than just a pretty picture or an interesting graphic. It is a coordinated view of data and situational frame. According to Stephen Few, "A dashboard is a visual indication of the most important information needed to achieve one or more objectives; consolidated and arranged on a single display screen so the information can be monitored at a glance."[20]

Few's definition needs a bit of clarification so that we all understand the differences between a dashboard and any other well-designed display screen. A dashboard uses a single information display screen format structure. It is composed of building blocks, but it is a compact single entity. Dials and gauges form the component building modules for it. Each part contributes to the understanding and meaning of

---

20 Few, *Information Dashboard Design*.

the others. Remove one and something important will be missing from the message. Expect most dashboards to be highly related dials and/or gauges located on the same display screen.

Example Dashboard of Social Media Users

Figure 5-35 depicts (of unknown purpose or origin, therefore as example only) the geographical locations of social media users. Notice the overlay of data to a large-scale street map. It shows a total of 1,666 collective users: Linked-In 793, Facebook 201, Google Plus 29, and Twitter 643. The color-coding is unknown. If we speculate that the colors yellow, orange, red, and green stand for each of the four media types. It is clear that the dense user locations are all clustered in a pattern on the locale map. Speculating more, suppose this dashboard reflects social media positive respondents to a particular issue (e.g., support of bicycle helmet use). It would be clear where door-to-door canvassing could be concentrated to solicit signatures for a local ballot initiative in support of the issue.

Using the information in this dashboard, a limited campaign could be better managed to maximize results for a limited canvass.

Example Dashboard for Pipeline System

Consider that the situation represented by the nomination gauges (see Figures 5-29 through 5-34) is for one pipeline controller's area that is part of a central operations center for an entire pipeline system. He is one of six controllers. Showing all six areas would be a dashboard that the overall operations manager should find useful. Refer to Figure 5-36. At a glance, the manager would be able to spot delivery concerns anywhere in the system.

Figure 5-35. Social media respondents showing their physical location on a street map backdrop. The respondents are using various media engines (Linked In, Facebook, Google Plus, and Twitter) to support a local issue (unknown). The deeper gray region shows the center of mass.

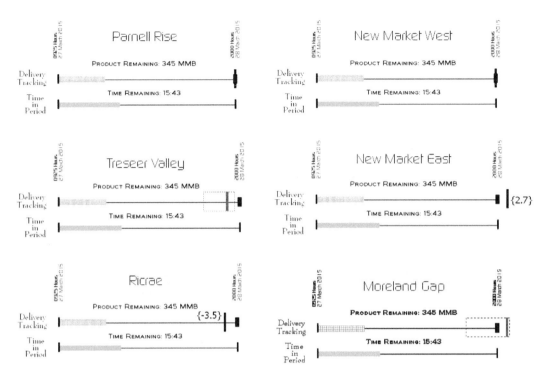

Figure 5-36. Entire pipeline system nomination performance dashboard. The system contains six segments. One segment (Treseer Valley) is low and cannot make up the needed difference. Another segment (Moreland Gap) is a bit over and has plenty of room to reduce flow to meet the requirements. All others are on target.

Dials and gauges are not simply copied into a dashboard from their individual occurrences in other places (other display screen locations). They are ordered or displayed in a manner to suggest appropriate operational or geographic relationships. There are information changes. In Figure 5-36, the individual recommendations for remediation or advice are not shown. Any changes to implement the needed adjustment would be made from different operational display screens, not this dashboard. The individual pipelines are identified on the dashboard, whereas they were obvious in the individual gauge view.

Example Dashboard: The Deviation Diagram

Another type of dashboard is called the *deviation diagram*. One for a chemical plant is shown in Figure 5-37. It is simple but informative. The purpose is to provide immediate visibility of what is abnormal and the location of these abnormalities in the plant processing stream.

The Foxboro Company first developed this type of display screen. It accompanied the launch of the company's entry into the distributed control system (DCS) race with a product called VideoSpec. Please keep in mind that this figure is not about time. The

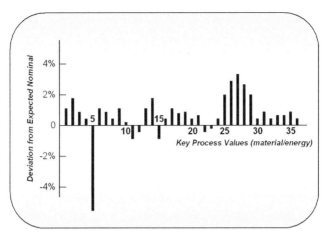

Figure 5-37. Deviation diagram showing an overview of an entire operator area. Shown are 36 critical operating values and their departure from the expected value. The arrangement is the order of the significant movement of product from area entry (at the left) to area exit (at the right).

order of the bars reflects the major production order or processing order of a particular plant. The order of the bars in the diagram represents the physical "order" of the key variables (of energy or materials) from the beginning of the plant (at step 1 on the left) until the end of the plant (at step 36 on the right). The bars are not process steps but key variable values within all steps. The diagram is a representational construct of the whole of a plant and its relationship within itself and to its operational goals. It is constructed by first listing the major plant-processed entities from entry at battery limits (from outside or another portion of the plant) to exit from battery limits (to another portion of the plant or shipping). For most continuous plant manufacturing facilities, we track mass (raw materials, intermediates, and finished products). Each significant mass is listed as a single vertical bar for a single aspect (pressure, temperature, pH, etc.). The example contains 36 bars (from 1 to 36), each one shows the current averaged value of the entity difference between what it is and what we expect it to be. An example of an individual bar would be a propane flow; next the polymeric reaction temperature of the propane with other constituents; and still later, the conversion rate of that reaction. Multiple precursor chemicals will each have their own bar. Significant processing variables would each have a bar, for example, a critical separator temperature and next its pressure. Adjacent bars mean that the entities they represent are functionally (if not physically) closely related in their processing steps. A bar to the left of another bar means that the one on the left comes before the one to the right in the normal "flow" of production. For each bar, the height shows how far away it is from a proper target. A "target" is not normally the related controller set point. Rather, the target represents the expected value of the particular attribute needed for effective production. Right on target would be a bar of zero height. Values higher than target would be a bar height above zero; values below target would be a bar below zero.

Now we get to the part about what the diagram is intended to illustrate. First and foremost, everywhere there is abnormality in the plant is clearly visible. Noticeably high or low bar heights clearly point to a notable difference from normal. At those places, we see bars deviating very noticeably from the zero value. The higher (positive) or lower (negative) a bar is, the more the deviation—and therefore a greater cause for concern. A single or isolated highly deviating bar (e.g., at position 5 in Figure 5-37) means that something is clearly wrong but its effect is strictly localized, for example, a failing transmitter that is not a part of a control loop. Groups of noticeably deviating bars (at positions 25 through 29 in our example) suggest that a broader area of the process is abnormal. Deviating bars at the left side of the diagram mean that our problems are in the earlier part of the plant. Deviating bars to the right mean the problems are near the final part of the plant's production. With a glance, the operator has a broad overview of the entire process. If a good job was done selecting the variables to display, a good understanding of the overall process will ensue. It is a simple yet powerful visual agent! It is only one of many.

Dashboards Require Unity of Presentation

Dashboards are used to provide clear and compact knowledge to the operator about important performance for an area of his operation. The information contained on the board needs to be assimilated easily and properly. To ensure that this will happen, our dashboards will require care in design. Here are some useful principles:

- Dashboards are not meant to be studied—they are meant to be observed with immediate salience.

- Only show information that is required for the operator to see and process at this specific in-view level of display screen hierarchy.

- Shape and relationship between individual items (dials, gauges, composites) needs to be consistent throughout the dashboard and consistent with other HMI components.

- Positions of the individual items on the dashboard should convey functional relationships and/or physical relationships. Connections should be clear and causal.

- Information displayed on the dashboard is intended to be observed and processed by the operator. It must be clearly discernable, clearly understandable, and clearly comparable.

- Multiples of complex gauges or dials are not conducive to being included on the same dashboard.

- Avoid "loading up" the dashboard with content because there is a lot of related items at this level or because there seems to be room. If there are many related items to display, then the dashboard is being formulated at too low a conceptual design level. Either move up a level or use more dashboards. If more are used, carefully distinguish each one so the operator maintains clear context.

Examples of Poor Dashboards

Overloaded Dashboard. Figure 5-38 was intended to provide a Level 2 overview.[21] However, with the heavy use of dials and gauges, it does not work. Compare it to a proper dashboard. We do find an effectively planned layout with like systems grouped. The sliding indicator scales are appropriately used for like components within each system group. Normal expected values are clearly indicated, and there is an overview sliding indicator scale for the subsystem that is easy to understand. The trend plots for the key performance indicators are useful, but the expected targets are not indicated. There is good use of spider diagrams. Their shape irregularities clearly indicate where

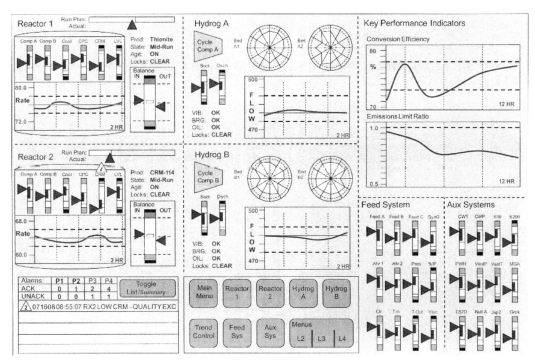

Figure 5-38. Overloaded dashboard display screen. Clearly, there is a lot of information and it might all be related. There are many reference marks for normal and acceptable limits. However, there is so much that it is of too little use unless closely studied. Dashboards are not intended to be studied. All important inferences should be visible at a glance and readily understandable.

Source: Courtesy of PAS Global LLC, www.pas.com.

---

21 Hollifield, Nimmo, and Habibi, *High Performance HMI Handbook*.

the imbalances are. Unfortunately, there is no clear indication of how abnormal the irregularities depicted might be. An alarm summary area is also shown, but alarm conditions should be observed differently on a dashboard or on an overview display component. It is just entirely too much for any screen. Dashboards are not intended to be closely studied.

Taking a step back from this rich detail, observe that nothing comes right out and provides what we expect from a dashboard. Successful dashboards employ similar display screen paradigms for each area. And those paradigms clearly and easily explain the overall health of the monitored system. What we do see more resembles a full control board. But for that use, this must be studied rather than just observed. In the final analysis, the designers have simply used different symbology to create an organized but highly cluttered display screen.

Excessively Complicated Dashboard. We now turn our attention to another form of difficult dashboard. Figure 5-39 is an example of an excessively complicated dashboard.

The dashboard depicts a highly structured view of nuisance alarms for an operator of three water treatment plants. Each water treatment plant's alarm view is clear and well structured. However, when the three plants are placed together, we

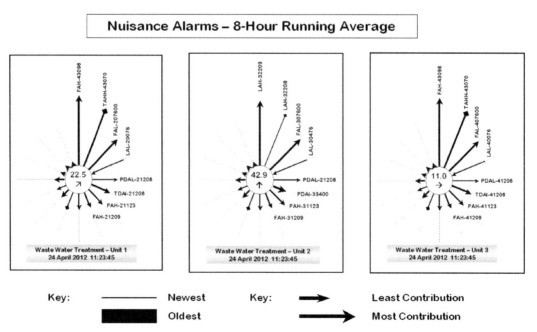

Figure 5-39. Example of excessively complicated dashboard screen. It is composed of three nuisance alarm gauges, but it is just too dense to be conveniently useful.

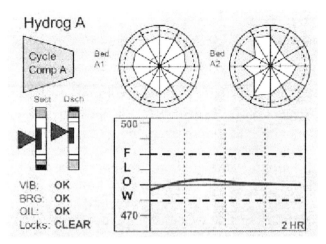

Figure 5-40. Clear and effective dashboard. Everything is related. Each dial, graph, glyph, and gauge is simple and understandable. Quickly, a viewer will gain all intended information.

Source: Courtesy of PAS Global LLC, www.pas.com.

see no synergy, just complication. There is no synergy due to an interesting quirk of the individual components. Recall that the nuisance alarms are ordered clockwise from the 12 o'clock position from most to least contributing. Therefore, shape alone is telling only in that the frequency of the offenders may be dropping off either faster or slower between the three units. But the visual effects appear to overwhelm the information. Consequently, we are left with three very similar views that each have a great deal of detail. Reading this dashboard requires too much work for too little benefit.

Better Dashboard. A better dashboard is illustrated by extracting a portion of the overly complicated dashboard example in Figure 5-38 (see Figure 5-40).

The dashboard likely represents a hydrogenation plant (one of two). It shows the essential components. Note the two beds shown as spider diagrams with normal region (within dashed circle) and the 12 components scaled so that the normal pattern would be closest to circular. Also, a trend is shown with both upper and lower normal regions (within the dashed lines). There are icons for suction and discharge for the compressor together with the four critical operational checks.

5.11 Design Fundamentals for Icons and Dashboards

You have seen many examples of icons and dashboards. Now it is time to consider their design on a more technologic basis. This way you can depend on your design working in the way you needed and intended.

Foundations

What follows are the important design requirements and guidelines.

Design Fundamentals

The following are requirements for displaying abnormal situations. The requirements for normal ones are generally very sparse, depicting only the type of signal or attribute and a relative location within the entire operating space of that attribute.

- **Message** – Must communicate something of importance and relevance in relation to its time and the display screen and/or display location.

  1. Clear indication of the single (could be collective variable) condition or attribute being conveyed to the operator. Sample conditions include *normal, abnormal, critical*, and such.

  2. Clear indication (usually by graphical location or proximity) to the physical/virtual depiction of any related process component(s) and condition(s).

- **Value** – Must communicate the current qualitative and quantitative aspects of the message.

  3. All information must be current and accurate. Where there is latency in obtaining proper values, that situation must be clearly indicated to ensure that the operator knows that it may be stale and by how much (if possible).

- **Function** – Must warn, inform, or confirm.

- **Frame** – Must set a reference with values for normal or expected, out of normal, and extreme.

  4. Clear indication of the current threat level *distance* from the current (not normal!) value to the threat (e.g., critical). This distance should be depicted with a scale and aspect to clearly indicate a distance that the operator can directly relate to. Nonlinear scaling is perfectly acceptable if it promotes that objective.

  5. Clear indication of the velocity and/or acceleration (if more appropriate) of the movement to or away from the current threat level. Here we generally use a small number of quantized values. The information is conveyed visually with symbols rather than via text. Examples might include:

     a. Slowly increasing, increasing, and quickly increasing

     b. Slowly decreasing, decreasing, and quickly decreasing

     c. Hovering, departing, failing, or restoring

- **Action** – Must suggest the behaviors that are likely to achieve the desired function; which is most likely, versus least likely, versus in the middle.

  6. Clear indication of whether operator action is suggested or the operator is expected to remain a spectator to this aspect. We do not want to confuse important notifications with required mediation responsibilities.

  7. Clear indication of the degree of attainable restoration or degree of improvement if full restoration to *normal* is not likely. Examples include:

     a. Attainable

     b. At risk

     c. Not attainable

Design Types of Dashboards (and Dials and Gauges)

All dials, gauges, and dashboards are intended to clearly convey one of two important messages: either risk or resolution. Simple concept. Powerful tool.

Risk-Based

Risk-based display screens clearly and proportionally indicate to the operator the current risk to operational integrity that his process is facing. These risks come in all the usual flavors: safety to personnel, damage to the manufacturing equipment, financial loss to sales or product, reputation damage, environmental damage, or other challenges. The different aspects or categories of loss are important, of course. They can be identified on the display screen when that knowledge is useful to the operator. Displaying might be of little value where the risk is a composite of several categories. This would be a case-by-case determination, of course.

Each display screen should contain the following aspects:

- *Frame* (see earlier discussion on frame) for the risk (the palate for all the other parts), which clearly shows what the risk categories are and how extensive unmanaged risks can become

- Current position (value of present risk exposure)

- Risk boundary or boundaries (clearly identified) for how far is too far

- Scaled distances between components (that bear some relationship to relative risk, the likelihood of the risk escalating, and the expected speed of progression from the current position to the likely risk boundary)

- Direction of likely risk progression

The purpose of this type of display screen content is to put the operator on notice that a part of the enterprise he is responsible for is likely to get into trouble if things progress, as they seem to be doing. At this juncture, specific alarms that may point to this problem may be activated. If so, this display screen would contribute to the alarm processing activity. If not, then the display screen item would be important as an early notification.

Note that any risk-based display screen component need not necessarily spell impending doom. That status would ordinarily be reached much later in time when more risk is exposed. Although for rapidly expanding situations, this might be the first and nearly only notice before the onset of serious problems. Remember, this is a dashboard or other related component, not an alarm system.

Resolution-Based

Resolution-based display screens are purposed to clearly and directionally indicate to the operator the current problem threatening operational integrity and the suggested pathway(s) toward remediation. The page would be constructed using a similar list of items for risk-based screens.

Each display screen should contain the following aspects:

- *Frame* of the magnitude of the resolution problem (the palate for all the other parts)
- Current position
- Operational boundary or boundaries (clearly identified)
- Scaled distances between components (that bear some relationship to the current position and the expected speed of recovery from the current position to the likely safe or normal boundary)
- Direction of the best pathway to restore normal operation

Figure 5-32 illustrated the current situation for a pipeline controller. His required delivery objective will be below that required if the current operating condition remained unchanged. If delivery flow were increased, the required product delivery can be met.

Salience Requirements

Salience means the ability to see and readily understand a situation or message. The real benefits from icons and dashboards are their compactness and ready understanding. Salience is like a language in and of itself. To achieve salience, the following are needed:

- At-a-glance connotation of what is being depicted with the current condition level as a *frame* (normal, abnormal, critical, etc.). The viewer must be able to actually see what is intended whether he is in a usual viewing position, looking over-the-shoulder, or a bit more distant. Care must be taken to ensure that any view will communicate the required message. This means that it should not be possible to miss something when viewed at a reasonable distance. Visibility needs to be an all-or-nothing thing.

- Items depicted in such a way that it is always clear that what is being shown is an icon or dashboard and not a part of the operational framework. This is not to suggest that both cannot be co-located. It is that they should not be confused with each other in any way.

- Single-dimensional or multidimensional depiction, so long as each dimension meets the first three fundamentals of resolution-based (earlier page) and the collection meets the remaining two.

5.12 Trend Plots

A *trend* (or *trend plot*) is a visual representation of values for a variable of interest over time. It belongs to the family of analog indications. That is, trends look like something taken from the real world and shown in a way that suggests something real. They can be graphically represented as dots along a timeline, bar graphs along a timeline, or a continuous curve evolving over the same time line. Figure 5-41 illustrates the simplest of forms. Note that both the vertical axis (showing the scale for the variable of interest) and the horizontal axis (depicting time) have a "broken line" symbol near the point of the two axes coming together. This convention is used to

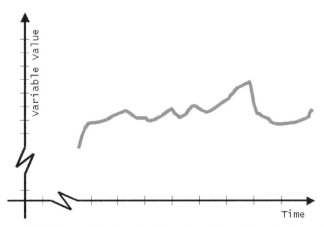

Figure 5-41. Single trend. This is useful when the time-changing values of the variable (the history) are important to see in addition to the current value.

indicate that the scales do not start at zero. They allow the graph to be conveniently shown close to the axes.

Trends are extremely compact mechanisms for conveying quantitative information: what the actual values are for the variable of interest over the time frame of interest. They are best at suggesting qualitative information: how the variable of interest is changing over the period of interest. It is easy to see how steady a trend is, whether it is increasing or decreasing, whether it includes bumps and dips, and a great deal more. It should be easy for the operator to see if the trend looks typical or different. However, all this information and the ease of grasping it must not get overvalued. Please refrain from overtrending and overcomplicating the trends that are used.

Build-on-Demand Trends

Most controls platforms include built-in trending capabilities. They allow the operator to select a point (or multiple points) to trend together with some scaling formation (e.g., scale all to fit or scale in engineering units) and time window selection (e.g., from now on, for a 10-minute period, or where the last current real-time value becomes the current trend value or starting point for the trend). These usually begin from the current time and build forward. This is a bit unfortunate because some history cannot be easily displayed. It is always possible to trend from historical records, though it is rare to be able to populate a contemporaneous trend with a starting history unless this type is prebuilt and available for selection from a list.

There are various useful tools for trends. One is the ability to select a time on the trend and see actual variable values. Some packages allow multiple time selections with values that refer to the same reference measurement scales so comparison is easier. Most packages allow zooming in and out as desired.

Continuous Trends

Continuous trends are usually ones built into an element, format, window, or page and are always visible. These are used where the trend is an important and consistent part of the screen information. They have a fixed time window consistent with the rest of the screen. This type of trend rarely has the full set of examination tools, such as zoom and others.

Pop-up Trends

Some continuous trends are reserved for pop-up use for the operator to request as needed. In this form, the trend is always refreshed and prebuilt to match the needed information. A sparkline trend (discussed later on) is one example.

Complex Trends

It is possible to build wonderfully complex trends with lots of potentially useful content and elegant formatting and display characteristics. Unfortunately, they are quite unsuitable for the control room. Control room trends are useful only when they provide in-context information in a form that is readily understood and useful. They are not meant to be studied and fretted over. As a good example of this bad example, refer to Figure 5-42, courtesy of StockCharts.com.

It is apparent that there must be a great deal of interesting and possibly important information here. It will not divulge itself without significant concerted effort. Something important will almost always be missed.

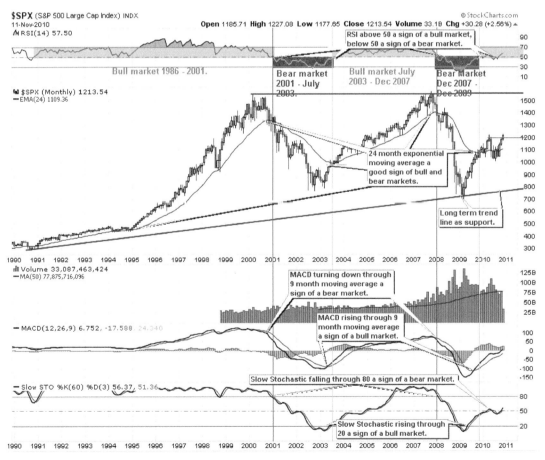

Figure 5-42. Overly complex trend combination. Not only is each one dense, the combination appears overwhelming. Any screen element that requires study and close attention is problematic: doing so takes a lot of time, and something important will almost always be missed or misinterpreted.

Source: Chart courtesy of StockCharts.com

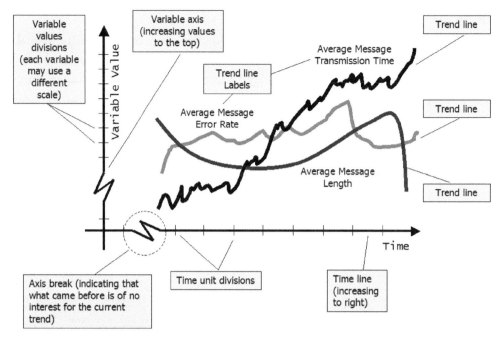

Figure 5-43. Trend plot components identified. All are important to ensure the viewer correctly understands the intended information.

Trend Components

Let us break down the essential components of a trend format (recall that *format* is an ISO 11064 term).

Refer to Figure 5-43. All trends should be labeled whenever there is a possibility for misinterpretation or misreading. Where possible they should be graphically positioned with other relevant and related content on display screens to facilitate ready association and context. Keep them as simple as they can be to do the job. With all the labels shown, this figure is very busy and distracting. Figure 5-44 shows the normal view without distraction.

Special Types of Trend Charts

Here are a few special types of trends or charts that you might find useful.

Fan Charts

A fan chart is used to suggest likely future results projected from past results. It starts with a trend graph up until the present. Then, using various statistical calculations, it projects ahead for the requisite amount of time. As you might imagine, the farther into the future the projection, the more likely it is to be different from a

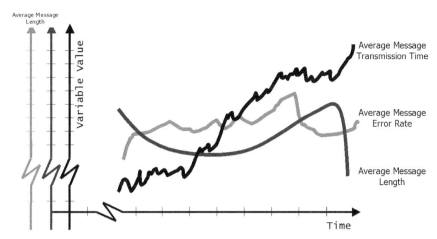

Figure 5-44. Example of superimposed time-related trends. This type of graph is used to depict how several variables relate to each other in real time.

straightforward projection. Figure 5-45 is a fan chart.[22] This form of chart would be useful to display to the operator a range of things that might happen to an existing trend in the short-term future. It might be built using statistical estimating of values. It might expose forward-looking operational risk so that the operator has a visual idea of what might happen.

This chart type was named by the Bank of England (circa 1997), which first used it as an "inflation report." In this illustration, from 2003 to the end of 2005, the graph uses actual values. Then for each of the second quarters in 2006 through 2008, it forecasts what the likely inflation values might be. Each of the four bands of prediction is based on increasing standard errors. The fanning out is due to the nature of prediction: the farther into the future you ask it to go, the less certain the values become.

Sparkline Charts

A sparkline is a compact snapshot format of a graph of a single variable over a predetermined time frame associated directly with its contextual use.[23] It is shown without scale values. The purpose is to provide an in-context view of data to support understanding of an associated visual display. Edward Tufte quite appropriately coined the phrase "data-intense, design-simple, word-size graphics."[24] The sparkline chart is a

---

22 "Fan Chart (Time Series)," Wikipedia, last modified June 12, 2014, http://en.wikipedia.org/wiki/Fan_chart_(time_series)/.
23 Edward Tufte, "Sparkline Theory and Practice," Edward Tufte forum, last modified May 27, 2004, http://www.edwardtufte.com/bboard/q-and-a-fetch-msg?msg_id=0001OR&topic_id=1.
24 Edward Tufte, *Beautiful Evidence* (Cheshire, CT: Graphics Press, 2006).

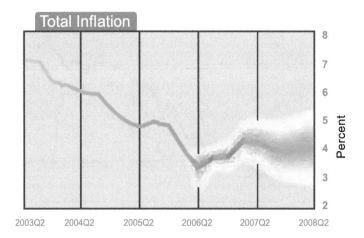

Figure 5-45. Illustrative fan chart (based on the Bank of England "Inflation Report") extremely useful to compactly depict where future values are likely to lie. The "fan" part begins at the beginning of the uncertainty period (which may actually be the present, before, or a bit later).

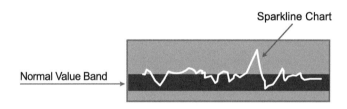

Figure 5-46. Sparkline chart showing key. This is an effective way to show a trend in a way that mimics an icon or gauge.

standard tool in current editions of Microsoft Excel. Figure 5-46 illustrates what one looks like.

Observe how easy sparklines would be to use in graphics. For the most part, they would be on-demand pop-ups. Though their use in overview graphics to depict key variables would likely have them resident on the display screen.

5.13 Example Display Screens

In this chapter, you have seen a great deal of things to keep up with. There are design principles, display screen designs, icons and dashboards, navigation, and the shapes and sizes for physical displays. Now we take a good look at how all of this lays out by way of an example set of display screens. You will see how each of the page components

(Section 5.3) fits into the overall construction process.[25] Observe how seamlessly the tools and other information and navigation aspects seem to flow together.

There is a "sweet spot" for the density and content amount of "stuff" (to use a highly technical term) that will be contained on a single display screen. This is relatively easy to do at the very detailed level. We can adjust items for that specific part onto the page. However, as we move up to less detailed views, we quickly run out of room to include all the lower-level content (nor should we do so if we could). Appropriately dividing the information into displayable pieces requires careful understanding of how everything works and finding the aggregate indicators of abnormality (and normality) to use instead of having to use everything. It is the reason for segmentation and hierarchy. It is the power of overviews and the challenge of design.

Overview Page

At the overview level, we provide the operator with sufficient information about the operator's complete span of responsibility. Here the designer must understand what the operator needs to worry about and investigate and what not. Figure 5-47 shows one concept of an overview page. There is only one overview for each operator. This is not to say that there might be other pages where information is shown that cover the operator area for specific purposes. That is mostly geography. The overview is about having one location for all key information and awareness. Notice that this example contains most of the piece parts used for proper page design (recall Figure 5-1 from ISO 11064). Examine the component building blocks of the overview, moving from left to right. This overview incorporates many of the HMI building components introduced elsewhere in this chapter.

- **Key upstream variables** – Shown as the set of seven trend gauges arranged vertically along the left border of the overview. Each variable comes into this operator area from some other part of the plant or from the outside. Each was selected because knowing its value (and recent history) provides an important frame of reference to the operator to identify abnormalities coming into the plant from outside. Sure, there are many other variables coming in. But indicating when these are okay or not okay provides significantly more knowledge than the others.

- **Aggregated alarm indications** – Shown as stacks of four boxes within each of the five subsystems (here shown empty because there are no active alarms at the moment). Any active critical-priority alarm(s) would show up with their actual number in the top box colored red. If any were unacknowledged, the box

---

25 Douglas H. Rothenberg, *Alarm Management for Process Control—A Best Practice Guide for Design, Implementation, and Use of Industrial Alarm Systems*, 2nd ed. (New York: Momentum Press, 2018).

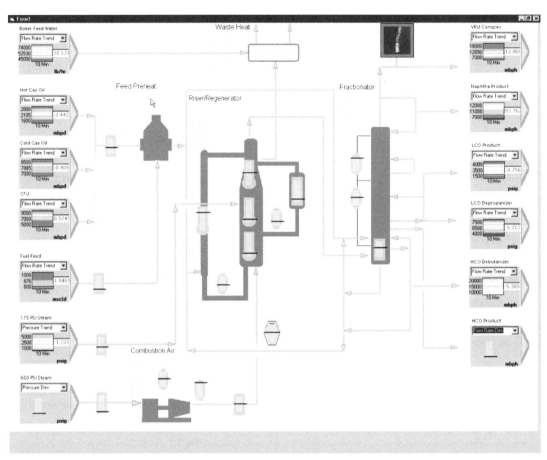

Figure 5-47. Overview display screen showing all the important aspects of an operator's entire area of responsibility. Key variables are shown as icons. Alarms are summarized so their locations and criticality are apparent. Trends are present for key inflowing variables coming from other plants. Similar trends are for the important product variables of this plant that flow to other plants or for shipment.

Source: Reproduced with permission from Honeywell ASM.

outline would flash. The rest of the alarm indications for *high*, *medium*, and *low* priority are shown in a way similar to the critical ones but using their proper color backgrounds and locations.

- **Subsystems** – Shown in the general flow of the diagram; each with a label. The five are (1) Feed Preheat, (2) Combustion Air, (3) Riser/Regenerator, (4) Waste Heat, and (5) Fractionator.

- **Key variables** – Shown within each subsystem as icons. Flows are shown as vertical rectangles on a horizontal line. Pressures are shown as vertical ovals. Temperatures are shown as stylized thermometer bulbs. Levels are shown as vertical rectangles on a vertical line. Analytical variables are shown as elongated vertical hexagons.

- **Video feed** – Shown as actual live video (or in some cases, special sequences of video from a live loop) of significant views the operator is required to monitor. This is not usual; its use would be very limited.

- **Key product variables** – Shown as the set of six trend gauges arranged vertically along the right border of the overview. Each variable represents a product produced in this operator area that will go to some other part of the plant or to the outside. Each was selected because knowing its value (and recent history) provides an important frame of reference to the operator to identify abnormalities produced within his plant. Sure, there are many other variables inside. But indicating when these are okay or not okay provides significantly more knowledge than the others.

Figure 5-48 shows the breakdown labels of those individual parts and components with keys to the graphic location. This operator is responsible for all five major subsystems {a}: Feed Preheat, Combustion Air, Riser/Regenerator, Waste Heat, and Fractionator. Within each subsystem, each key variable is identified and depicted by an appropriate status icon, {b}, with one icon for each key variable. Also shown is the aggregate alarm status, {c}, for each subsystem. Because alarms are important for identifying abnormal operation, each alarm priority is separately shown with the number of active alarms indicated for each. What goes on inside this operator's area can be

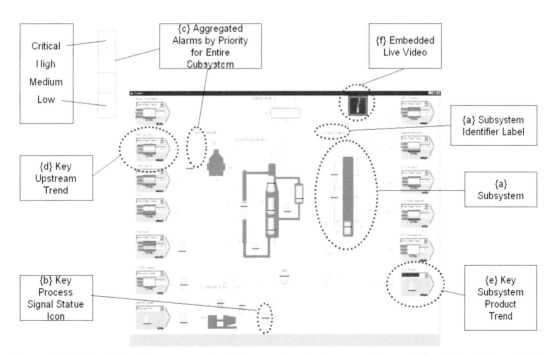

Figure 5-48. Identifying the components of the overview display page (shown before this as Figure 5-47).
Source: Reproduced with permission from Honeywell ASM.

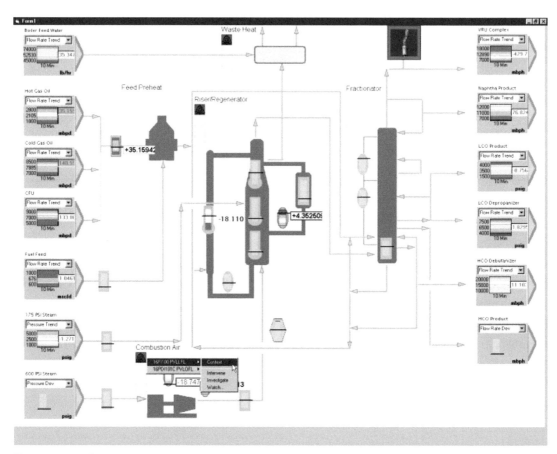

Figure 5-49. Overview display screen now showing alarms that are summarized (see four-box stacks depicting one or more of the highest priority alarms) so their area locations and criticality are apparent. Note the pop-up dialog box with suggested actions in the lower center, and the "tool tips" values for certain variables.

Source: Reproduced with permission from Honeywell ASM.

very much impacted by what happens upstream. Even though his plant may be operating effectively, if an important upstream input changes to a value that has a strong effect, his plant is likely to experience abnormal operation. Therefore, all key upstream effects are short-term trended, {d}. In like manner, we show all key product trends, {e}, for what goes on in the operator's own plant. These trends track the essential targets for the plant. Any that are abnormal, even if the other key variables are not, indicate trouble. Finally, any live video, {f}, is included.

Figure 5-49 is the same overview display illustrating alarms, their process area location, and a recommended first set of operator actions in the lower center pop-up.

Almost every object on the page has navigation behind it. Clicking on one of the five subsystem identifier labels will navigate to the appropriate subsystem (as a secondary level page). Clicking on any alarm box will navigate to the appropriate secondary level page with the alarms shown or to the appropriate lines on the alarm summary screen (depending on the specific HMI philosophy design requirement).

Secondary Page

Operators view secondary pages to follow up on specific tasks or in response to understanding and/or investigating abnormal situations. These pages would be visited during periodic assessments of operation during a shift. They would be visited during the preparation of a shift handover report. They would be visited at times on a whim, just to ensure that the operator is correctly interpreting what he sees on the overview pages. There is very little aggregation of information at the secondary page level. Summarized alarm states (in alarm, acknowledged in alarm, unacknowledged but not normal) are visible. Most process variables or icons for them are visible. Operating states are visible.

Several pages are needed to display all this information, so good navigation is important. There are two navigational concerns. First, the operator needs to be sure that the current location is the right one. This is usually done via tab buttons that also serve as navigation targets and by verifying the labels and content seen. Second, the individual sensor readings, valve positions, equipment status, and other data are already visible. Clicking on them can link to any number of places specified in the HMI design plan. Figure 5-50 shows an example secondary page for one display. There are four other secondary pages for this operator area and several subordinate secondary pages for each. See Figure 5-51 for the key to the individual components.

Figure 5-51 shows the identifying components for this page. Along the top are the navigation tabs, {a}, one for each of the major subsystems for this operator. A low-temperature alarm, {b}, is indicated. All important sensor measurements are indicated with icons, {c}. There is a conventional flow control loop, {d}, and a low-select pressure control loop, {e}. Important manual block valves are shown, {f}. And the major upstream inputs, {g}, and downstream products, {h}, are identified. Finally, links to other related display screens are shown, {#}.

Subordinate Secondary Pages
There are usually more layers in secondary pages to help the operator keep the perspective and to be able to "drill down" to the exact locations for the instrumentation and controls. In this example, the Riser/Regenerator is composed of three portions: riser, regenerator, and disengager. See Figure 5-52 for their locations.

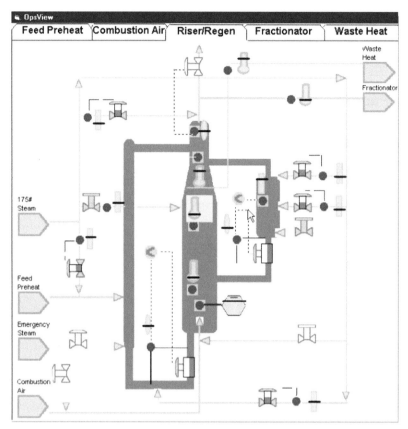

Figure 5-50. Secondary page for Riser/Regenerator. This is one of the five major components of the entire plant. This level provides all details including alarms and access to controls. The icons and other components use the same convention as the overview.

Source: Reproduced with permission from Honeywell ASM.

Figure 5-52 shows the portion for the regenerator. The graphical components are the same as shown before with the exception that actual measurement values are present.

Situationally Based Secondary Pages

Summarizing what we have covered so far, operators use the overview page to stay aware of any problems in their area of responsibility. The operator expects to be able to see established problems for which alarms will be active, or budding or potential problems that experience suggests are from irregularities or raised suspicions. He will follow up on all that he sees. Following up means finding the needed information and taking remediation where necessary. This "following up" using the HMI will take place on secondary pages and tertiary pages (see the next section). Figure 5-53 shows a focused and yoked secondary page composite (useful to conserve the number of displays used for navigation). The operator got to this figure by clicking on an alarmed icon in the Riser/Regenerator unit of Figure 5-49 (alarm shown in the figure depicted).

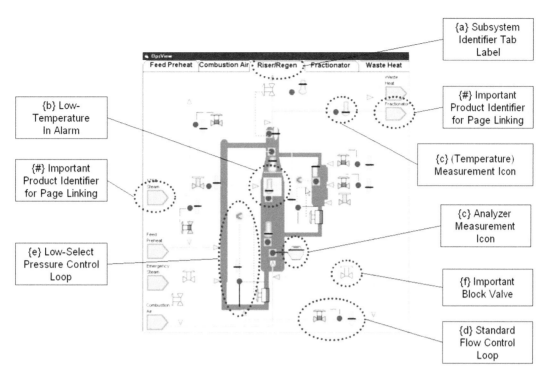

Figure 5-51. Identifying the components of the secondary display page (shown earlier as Figure 5-50).
Source: Reproduced with permission from Honeywell ASM.

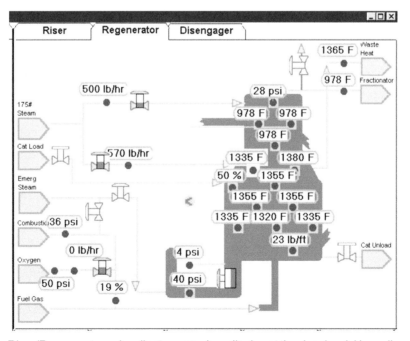

Figure 5-52. Riser/Regenerator subordinate secondary display at the riser level. Here all useful process variables and alarm conditions are shown.
Source: Reproduced with permission from Honeywell ASM.

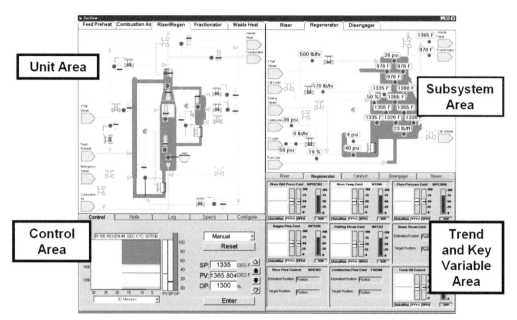

Figure 5-53. A situationally based secondary page that is pre-engineered to provide information to help the operator understand what the situation is and how to manage it. This display combines the window (shown earlier as Figure 5-52) together with three additional windows to form a coherent view. It is constructed by the process of focus and yoking.

Source: Reproduced with permission from Honeywell ASM.

This same information can also be shown on multiple displays via an appropriate number of screens.

The HMI navigation process automatically set up this single page (Figure 5-54) because the low temperature alarm clicked at the overview level was in the Riser/Regenerator. There are four windows on this particular page. This combined view is designed to aid the operator to respond to that alarm. Because the alarm is in the Riser/Regenerator unit, it is the selected unit. Because that alarm location is actually in the Regenerator portion, it is shown as the Subsystem Area (see Figure 5-52). This demonstrates the concept of yoking (discussed in Section 5.9). The two remaining windows will demonstrate the concept of focus.

Because the alarm is located in the Regenerator subsystem area, only the important key variables and states for that area are shown in the Trend and Key Variable Area (lower right of Figure 5-54). Finally, because the alarm is specifically a riser temperature alarm, the specific controller block is shown in the Control Area (lower left). Figure 5-54 shows how the links are used.

As you can see, this is a useful and efficient navigation process. Note that if the operator were to navigate to another point in the Riser/Regenerator (Figure 5-51), the

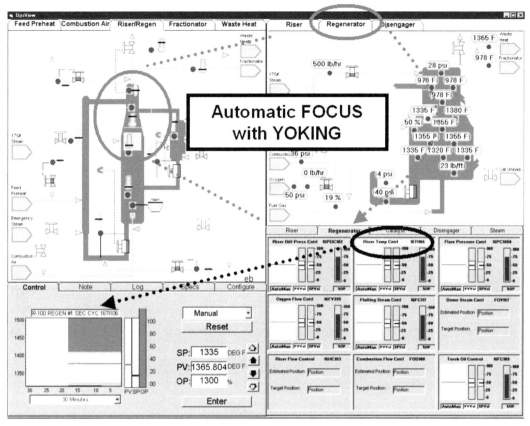

Figure 5-54. This figure explains the (on-the-fly) construction details that use focus and yoking to build a situationally based secondary page. This specific view was set up because of the (in red) low-temperature alarm in the regenerator. It adds reader-information labels to the information shown in an earlier screen (Figure 5-53).

Source: Reproduced with permission from Honeywell ASM.

new figure would be like Figure 5-54 with the exception that the Control Area would change to the appropriate one. Of course, if the selected item were only a measurement, the Control Area would depict a measurement, not a control loop.

Tertiary Page

Tertiary pages are support to the operator's objective when working on secondary pages. If the operator is working an alarm, then the alarm response sheet (refer to Section 8.3 for an example) would be extremely useful to the situation. On the other hand, if the operator involved is assisting other operators in carrying out a complex procedure, a procedure status would be most helpful. Figure 5-55 shows an example.

Like the secondary page example, this one is composed of several coordinated (yoked) windows. Figure 5-56 identifies them and explains some of the icons used.

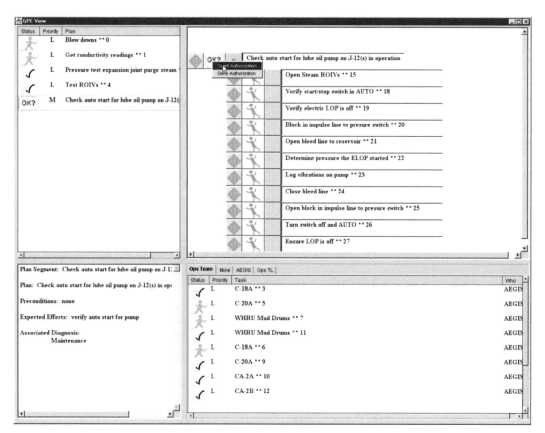

Figure 5-55. Procedure tracking display screen. Each window depicts a different level of activity for the work being done. Figure 5-56 identifies each.

Source: Reproduced with permission from Honeywell ASM.

There are four coordinated windows that comprise the page by focus and yoking:

- **Window 1** – Tasks at hand to do (upper left). This is the list in order of first task to last task that this operator is responsible for. Note the current focus line (with OK?) in light yellow.

- **Window 2** – Detailed steps for the selected task (upper right). These are all the specific steps that need to be done by the operator to fulfill the requirements of the current task (from tasks at hand to do).

- **Window 3** – High-level action for the selected task (lower left). This is an abbreviated summary of the objectives and intent for the current task.

- **Window 4** – Tasks that the rest of the team is doing (lower right). Our operator is one of a crew working on the execution of a larger activity. This window lists where the rest of the team are in the plan.

The page uses glyphs to convey the status of each individual activity. We begin at the left of the figure and work our way clockwise. The current task needs to be verified

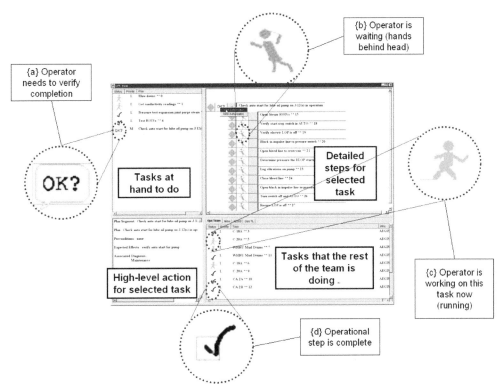

Figure 5-56. Identifying windows and glyphs for tertiary page.
Source: Reproduced with permission from Honeywell ASM.

for completion; therefore, the yellow OK? glyph, {a}, is present. Where an action has been initiated but the operator is waiting for it to complete, we see the "operator sitting with hands behind head" glyph, {b}. For an action step that is being worked on at the present time, we see the "green running figure" glyph, {c}. For all completed tasks, the check-mark glyph is used, {d}.

Pop-ups

All pop-ups, including cursor enlargements and tool tips, are considered subordinate pages at whatever level they are found.

5.14 Mass Data Displays

Mass data displays are a relatively recent form of graphic information display. Early work was led by Carsten Beuthel et al. around 1995.[26] The word *mass* refers to the word *massive* to connote the ability to graphically depict a lot of data attributes in a way that

---

26 Carsten Beuthel, Badi Boussoffara, Peter F. Elzer, K. Zinser, and A. TiBen, "Advantages of Mass-Data-Displays in Process s&c," *Proceedings of the 6th IFAC/IFIP/IFORS/IEA Symposium on Analysis, Design, and Evaluation of Man-Machine Systems* (Cambridge, MA: IFAC), 439–444.

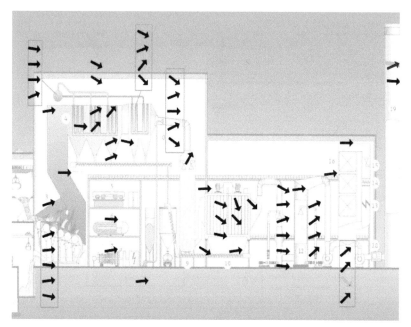

Figure 5-57. Heat exchanger example of a mass data display. Depicted is a single measurement class of "heat efficiency" that is computed at every location shown by the arrows. The inclination of the arrow indicates whether the efficiency is increasing (the more up, the more increasing), steady (horizontal), or decreasing (the more down, the more decreasing).

Source: Reproduced with permission from ABB.

preserves their relative geographical context and shows current values. Its usefulness is predominantly for a single type of variable that is measured over a large number of spatial locations where it is useful to know what the readings are and where they are located. Viewing an example might be the best way to begin. Figure 5-57 illustrates what a mass data display looks like. "A key concept of Mass Data Displays is to show normalized values. Normalization is an important element of recent initiatives such as High Performance HMI. The idea is to focus (not only) on the absolute value of a given process signal but also to show its value relative to the desired operating ranges of the process."[27]

Figure 5-57 depicts the heat efficiencies of the heat exchanger at many different spatial locations. It is intended to provide early notice of slowly developing effects. Each arrow line represents a (observed if available, otherwise calculated) value of the efficiency at that physical location. Each arrow has several possible values over a specific time window, such as the last 4 hours: steady (horizontal arrow), slight change (arrow sloping slightly up for increasing, slightly down for decreasing), modest change (arrow sloping at + or − 45°), or predominant change (arrow sloping steeply up or down).

---

27 Martin Hollender and Moncef Chioua, "Assess Complex Process Situations in the Blink of an Eye," ABB Company, Document 3BUS095744 (Ladenbeug, Germany: ABB, January 2013).

Arrow slope is a clearer and more readily visible indicator of relative value than arrow thickness. Note that where it is important to denote significant values, the arrow may be colored (see the green steeply downward sloping arrow at the lower right. It is quite easy to feel comfortable with how this device is performing and where to look for more information. There are two general types of mass data displays: (1) steady-state changes, and (2) normalcy of current values. Only one half of a circle of change is used. The half used is usually aligned with the major direction of material or energy flow.

Departure from Steady-State Value Mode

In the first type, steady-state change mode, the values for the arrows are shown as the current departure from an average for a given time period window. A horizontal arrow represents a current value that is the average value for that variable over the time window. This type is depicted in Figure 5-57. The period window and the angle departure from horizontal of the arrows are carefully picked to show small but potentially important deviations in the making. Good starting values for a full-up (conversely, full-down) position would be 3% to 5%. You would be looking for a value of, say, 30° to be unusual enough to warrant attention.

Departure from Normal/Expected Value Mode

In the second type, normalcy mode, the position (angle) of the arrows represents the difference between the current window average and the (theoretical) proper value of the parameter at the arrow location. The proper values do not change over time unless the equipment is modified or some other processing effects are modified so that the monitored device operates differently from before. This change would involve an engineering, construction, or significant operational change. Those sorts of changes would be most infrequent and unusual. The time window is selected to be longer than the dynamic settling time for the variables being tracked. This ensures that most process noise is rejected. The tilt of the arrows is selected to represent increasing (or decreasing) deviation. *Horizontal* means pretty much right on (within 1/10 standard deviation) the design value. Fully vertical would be 1 standard deviation from the design value. The angles in between are proportional. Again, you would be looking for arrow movement to effectively focus the operator's attention on what is unusual.

Things to Keep in Mind

There are a few things to keep in mind as you consider how useful mass data displays might be. The first is that all the arrows (or *needles* as they are sometimes called) must represent the same thing: efficiencies (highly complex calculations are just fine, of course) or pressure or temperature deviations from normally expected (that is where the normalization part of this is important). The second is that the display need not be

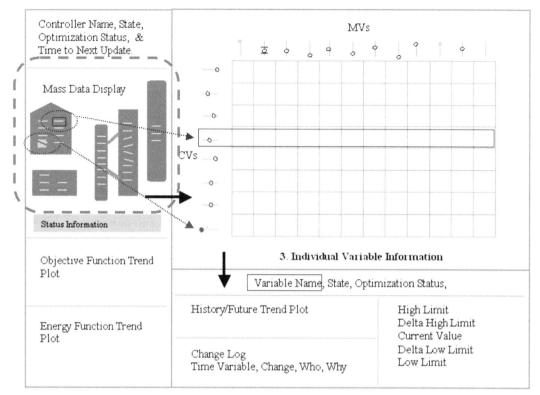

Figure 5-58. Mass data display *format* included within an advanced control monitoring display. Certain locations in the furnace are control points, so the added elements provide deeper detail into the understanding of the particular performance of the controls.

Source: Reprinted with permission from SAGE Publications, Inc.[28]

large in the sense that it takes over a lot of the page. This means that mass data displays could be useful as a window or format on a page. Figure 5-58 shows an illustration used for tracking an advanced control loop.

The mass data display is included as a format at the top left of the left column showing overview information. The right column depicts various diagnostic information.

> [The entire view shows the] Layout of the redesigned MPC [Multivariate Process Control] workspace. This is a schematic representation of the MPC workspace, not the actual display. There is a relationship between the three functional areas of the screen: The Overview on the left hand side of the screen, the more detailed diagnostic information in the top right hand side of the screen, and the detailed, individual variable information shown in the bottom right. A variable selected in one view will be highlighted in the other views.[29]

---

28 Guerlain, Jamison, and Bullemer, "Visualizing Model-Based Predictive Controllers," Proceedings of the Human Factors and Ergonomics Society Annual Meeting, 44, 22, p.4 copyright © 2000.
29 Guerlain, Jamison, and Bullemer, "Visualizing Model-Based Predictive Controllers."

Extending Mass Data for an Overview

The structural display symbology of mass data can be extended beyond the depiction of how a single variable varies spatially. Now each "arrow" would depict a different, yet significant process variable or composite indicator. Use the same capability to incorporate each condition arrow into a spatially relevant frame. This way, the condition of the "arrows" taken together depicts the overall condition of the portion of the process.

5.15 Multivariate Process Analysis

Process plants and operations are comprised of many components. They are interconnected. They are dynamic and variable. So it is natural to think that to properly understand their functioning would require keeping track of a great number of values (measurements, control points, and such). It turns out that this is far from what actually happens. We know that all the many pieces and parts are not connected randomly. Each component is linked to others by a small number of physical things (pipe, wire, roadway, etc.). Each component will do only what it can do physically, such as add energy or react components. And it will do so in accord with the "laws" of chemical reaction, mechanics, and the rest. So what might seem as a huge number of different possibilities will narrow down to only a few.

> The multivariate control charts represent one of these emerging statistical techniques successfully used to monitor simultaneously several correlated characteristics that indicate the quality of a single production process. The use of graphics in the industrial environment has increased in recent years due to many resources of information technology now available to reduce the complexity of modern industrial processes.
>
> In order to reduce the number of variables Principal Components Analysis (PCA) was adopted making it possible to consider all of the original variables in only two or three dimensions. Thus, most of the variance of the process is represented by the dispersion of the points on the main components.[30]

What Is Multivariate Process Analysis?

Multivariate process analysis applied to industry was pioneered by John MacGregor and Theodora Kourti through the work of the McMaster Advanced Control Consortium at McMaster University Chemical Engineering Department in Hamilton, Ontario,

---

30 Elisa Henning, Custodio Cunha Alveres, and Robert Wayne Samohyl, "Multivariate Process Monitoring and Control with R," accessed December 15, 2014, http://www.r-project.org/conferences/useR-2009/abstracts/pdf/Henning+Alves+Samohyl.pdf.

Canada.[31] The methodology involves measuring the entire large set of variables over time. Using established mathematical processes found in principal component analysis (PCA) and projection on least squares (PLS), the data is reduced to a much fewer number of important pseudo variables. The pseudo variables have values at each point in time that the original ones had. Each pseudo variable, in order, contains the most important variations of the entire process. Often, the first two contain a significant amount (e.g., 60% or more). This is extremely useful because we can visualize them in two dimensions. But the usefulness is even greater. It turns out that if we examine the region that the pseudo variables hang around in when the real process is normal, if the actual process goes astray, the pseudo variables leave their normal region. If the process is operating around normal but with normal variations, the pseudo variables will be observed to be in clusters. Each cluster represents a variation of normal. But there is an additional important benefit. For each data point of the pseudo variables, it is possible to reverse calculate what the actual variables were doing in the order of their largest contribution at that same instant. Not only is there a way to identify abnormal, it is also possible to identify what is most likely causing things to become abnormal. Lest you decide that this is *the* answer to your operator's concerns, please understand that this is only a statistical tool. It is better used to suggest concerns than to decide exactly what might be at fault.

This clustering and movement can be powerful indicators to where the process might be operating at any moment in time. Moreover, as the real process moves out of normal, it is easy to observe the pseudo ones moving as well. The technology is often used graphically. The overall monitoring is shown as Figure 5-59.

The hundreds of actual process variables (upper left) are reduced to a small set of pseudo (called *latent* in the figure) variables (upper left center) where each one, in descending numerical order, contains the most information possible for each. The most important two variables are graphed for each time instance present in the original independent variables (center). Note the apparent clustering. When needed, any cluster value can be reversed mapped (lower right) back into the entire set of actual variables to see which one(s) are most likely causing the value to be where it is.

The following example should provide additional insight.

---

31 "Professor John F. MacGregor: A Pioneer of Multivariate Statistical Process Control and Recipient of the Fourth Herman Wold Medal," *Journal of Chemometrics* 17 (2003): 1–4, https://doi.org/10.1002/cem.781; T. Kourti, "Multivariate Dynamic Data Modeling for Analysis and Statistical Process Control of Batch Processes, Start-Ups and Grade Transitions," *Journal of Chemometrics* 17 (2003), 93–109; McMaster Advanced Control Consortium (ACC), http://www.macc.mcmaster.ca/.

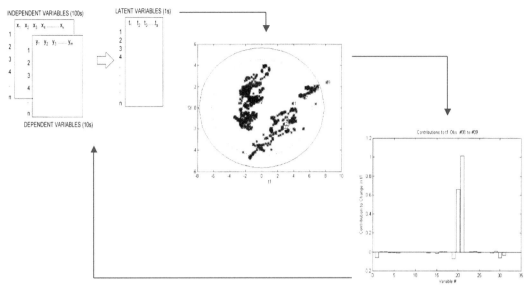

Figure 5-59. Illustration of multivariate process monitoring construction. The center scatter diagram-like element shows the locations for the key information variables over time. Analysis permits the construction of a normal region of them. Movement of the "dots" in the diagram out of the normal region signifies that the process itself is moving away from normal.

Bender Treater Example

A Bender Treater is a process unit used in crude oil downstream processing. An example is shown in Figure 5-60.[32]

This unit has hundreds of variables. Keeping track of each one is relatively easy, but knowing all the regions of good operation is not. Certainly, there are variables that must be closely controlled. When those go astray, and if they are important enough, alarms will activate. For all the rest of the operations, keeping a sharp eye may not be easily doable. Figure 5-61 shows the resulting chart over several weeks of operation.

The figure shows the process operation graphed against the two most important pseudo variables, t_1 and t_2. The Bender Treater shows three clear clusters of operation: C_1 from March 18 until March 23, C_2 from March 23 until March 28, and C_3 from March 28 until March 31. Notice how compact each is. C_1 is between the 99% and 95% confidence interval (a measure of statistical certainty as to the reliability of the data to mean what we see). As time progresses we see it move to C_2, which is well within the

---

32 Douglas H. Rothenberg and Robert J. Sadowski, "Lima Refinery Bender Treater Analysis Project" (BP Internal Report, BP Refining Engineering, 1996).

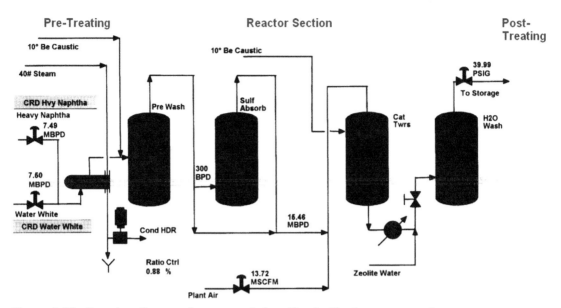

Figure 5-60. Overview diagram of components for a Bender Treater process unit.

99% confidence interval. Watchers would be encouraged to think that operation was getting better. A careful operator would ask what changed and determine what it suggested. Six hours later the process moved abruptly to region C_3 outside the 95% region and remained there for the next 3 days. As soon as the movement left C_2 and moved to the new region (to later become C_3, but because it was the only one, that region would not have been seen for some time) it would be clear to an observer that something has changed. We have a tool for situational awareness.

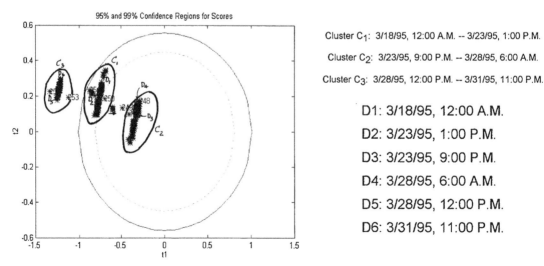

Figure 5-61. Bender Treater showing clusters of operation. Movements between clusters are important to observe and understand. The diagram makes it easy to see this movement.

Important Note

As interesting as multivariate process analysis is and no matter how well one might develop an application, experience has shown that this form of information is only one aspect of tracking normal and abnormal. You will always want to combine this information with other corroborating information and activities.

5.16 Displays Large and Small

Displays are the real estate for the information content shown on the display screens of all computer-based control and information systems. Where human intervention needs to be done remotely, it forms the vehicle for that as well. Size and location do matter. Purpose and capability matter more. This section is about purpose. But first, let us lay out some physical size parameters. Think about three actual size/purpose ranges: large, working, and small. Because we are talking mostly about electronics and flat screens, you know that size seems to be more about cost than dimensions. It certainly seems to be that way at home. Even though our room sizes do not seem to change that much and the furniture arrangements are not that fluid, as the price falls our flat-screen TVs (now also content portals) get larger. Yes, the image clarity gets better too. But that is probably not the point. When the displays get bigger, we get to have a different emotional relationship with them. We are moved from being an outside observer to an inside one. It feels different. We feel that presence. It is probably hardwired into our visual brain processing circuits.

The size-versus-cost thing is to some extent less linked to display selection in the control room. Compared to the other equipment costs, the cost per square unit of display really does not enter into the scheme of things. We select size for other, more functional and purpose reasons. Selection by function is the reason for this discussion. The workstation (WS) is where the operator sits in a control room. Off workstation (OWS) is everywhere else.

Type	Label (OWS = off workstation) (WS = workstation)	From Size (diagonal in inches)	To Size (depending on view angle)
Large	OWS Large	42	Full Wall
Medium	Workstation (WS)	24	32
Small	OWS Small	5	12

Table 5-1. Work station general sizes for control room usage.

Notice that there are gaps in the sizes. We are not classifying the marketplace; we are selecting for use. Also note the limit for OWS Large is based only on physical room size. Remember those mimic walls from yesteryear's control rooms? We will have more to say about this aspect later. The most important part of this to understand is that size and purpose are not interchangeable. Each size category has a specific use. Migrating the same use to a different display size is not going to work very well. Yes, putting a lot of careful design and engineering into a specific migration might seem to work, but that is not going to be useful across the board and it will cost too much and be harder to use.

The size scales are different and so their intrinsic basic use will be different. That does not mean that their uses cannot be coordinated. They really should be. Each size can be used to provide a reference (looking up in size) or detail (looking down in size) for the others. This means that for any given task or need, the operator might have some or all size scales focused on the same activity supporting him.

- **Large displays** – These displays are really large. They provide the ability to show overview content with a high emotional emphasis and in spatial context that might be actual or virtual. They excel at creating a "feel." They are well-suited for collaboration.

- **Workstation displays** – These are the workhorses of getting a job done. They gather all the information and provide all the handles to do that work efficiently. There are enough of them to keep a visual focus at close hand and facilitate smooth situation management.

- **Small displays** – These are the ready tools for supplementing many of the other activities that are going on. They can be used to explore threads of specific inquiry to check on something related or to communicate with someone who is needed to assist or must be kept informed. And small displays do it all without changing the focus of or relationship with the other displays in use. And, of course, they are completely portable so the operator can take things with him (information and/or control handles) as he moves about in the control room or out of it.

Workstation Displays

Workstation (WS) displays are most often used in sets of four to six often double-high rows of various configurations. Workstations comprise the operator's primary workspace. Information is viewed here; control actions are taken here. Normally, the operator remains within ready view and convenient access during the entire shift.

Workstation size reflects the need for sufficient individual display room for each piece of physical equipment balanced by the desire to keep everything within a proper viewing envelope and hand reach. Almost all the display screen technology and standards have been designed with this balance in mind.

Each display size is chosen to ensure that the workstation displays are a set. They are arranged in the control station (where the operator is located and generally "camps out") to fit within acceptable viewing angles and hand-reach distances. The fewer the number of displays, the larger each may be. Each individual display size is selected using the appropriate view range depending on where the operator sits and manages. Focus and distraction are important human factor issues to keep in mind. Where a display is intended to be used as a primary device, it is generally made large enough that (1) all the content is easily viewed and (2) it covers enough visual field to reduce distractions from the surroundings. We seek a proper balance between reducing distractions and preserving the ability to view the rest of the displays. Consequently, where two or more displays will be used with each other, their combined size needs to be managed to preserve a comfortable viewing field. Figure 5-62 illustrates a few workstation arrangement possibilities.

Note how the larger main displays are located centrally so that each may contain information in a view that occupies enough of the visual field to reduce visual distraction. Try to envision where you would locate amplifying materials (e.g., a procedure or an alarm response sheet) to aid the operator with a main task or to coordinate multiple tasks.

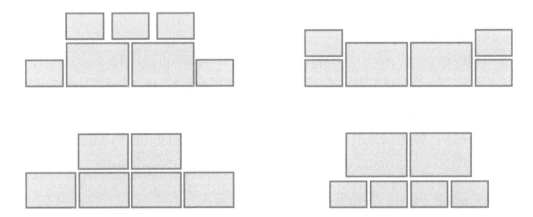

Figure 5-62. Workstation configuration options depicting a mix of sizes of displays and their physical arrangements. The main operator working displays are at the lower center. The upper ones and the side ones are usually preassigned so the information is always where expected. Because all are on-workstation, the relative sizes are for convenience and preference.

Figure 5-63. Closely spaced main workstation displays combined with poor use of large off-workstation display. Note the excessive use of color, content density, and small visible sizes that make them difficult to read and identify what is important.

Figure 5-63 illustrates a failure to understand the differences between the type of information, display design, and display structure best suited for the workstation displays and that best suited for a video wall or off-workstation large display. The more important difficulty for this control room is that the large off-workstation display is employed just as an extension of the display space of the workstation ones. They are too densely packed with disjointed views as the eye travels the width.

All this picking and arranging mess should disappear as soon as technology offers a single frame palate. The single frame will then be allocated and arranged to fit tasks and needs. We now have a single large-sized display. It is not an OWS large display. Within the overall size, one may create windows of any size to fit a purpose. From that point on, all the other design and functionality for this situation become the same as before. We just have more size and locational flexibility to adjust window sizes "on the fly."

Off-Workstation Large Displays

Off-workstation (OWS) large displays are not bigger workstation displays that encourage and allow the screen designer to put more content on. Rather their full utility comes from their ability to show about the same amount of information and relative size of graphics as a workstation display but significantly larger visually. Think of it as a movie screen—the same movie but a larger image. Being larger, and depending on viewing distance, it will occupy a significant part of the visual field. The very size has an emotional impact on us. Our natural tendency is to focus more directly. It seems to automatically raise the content's importance. And it invites more than one viewer. This makes it an ideal venue for collaboration. By the same logic, regular workstation displays are quite unsuitable for collaboration. Yes, two individuals could crowd around one and share the same display. Yet doing so either puts both viewers at an angle to the display or places one viewer centered and the other off to one side. It is a different energy. The one thing it does well is to enable finger-pointing (unless it is a touchscreen). Being able to point to items and to trace a finger path, such as we can do on paper, can be extremely valuable. Therefore, it would be nice to also have a "pointing" facility for the OWS large display.

Figure 5-64 illustrates one arrangement that utilizes six WS displays and an OWS large display (back on a nearby wall or a mounting pole located behind the workstation). In some plants, others responding to the situation of the moment replicate the central OWS large display for view in a collaboration center (see Chapter 4, "High-Performance Control Rooms and Operation Centers").

There is one other useful mode for the large display. It is a digital version of the old mimic display. In mimic displays, there is a certain minimum size for the graphical depiction, and a lot of depiction is needed. An example would be a pipeline system where it is necessary to show all segments, all major connections, and all storage or

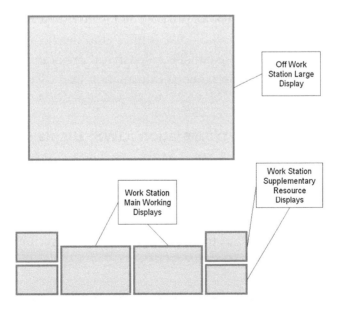

Figure 5-64. Workstation (WS) displays with a large off-workstation (OWS) display above.

Figure 5-65. Entire plant floor status mimic display arrangement.
Source: Reproduced with permission from ZPAS.

pumping stations. The dynamic requirements, say for fault identification or for alternate real-time configuration changes, require a panorama of visual background and status information.

Figure 5-65 illustrates a mimic display using several physical display units. The arrangement is presumed to mimic a physical or causal flow of a plant, yet there are no apparent connections between the screens linking them. This application might be categorized as an illustration of a personal video wall display. See also the discussion of video walls in the next section.

Figure 5-66 illustrates a physical mimic panel (a bit outmoded, but still found and useful in operating control rooms). Its purpose is to provide a complete overview of an entire plant. A current design would use digital graphics on an OWS large display. Note that while the physical mimic panel is extensive, there is a unity of design, a flow of purpose, and the absence of extraneous information and distraction. All of this is very important.

Requirements for Large Off-Workstation (OWS) Displays

Now that you have seen a few examples, it is time to lay down the important requirements for them. For without a clear understanding of how they need to be designed, you will not be able to use them effectively enough to improve operations. By following the requirements, large OWS displays should provide a valuable addition to both achieving situation awareness and fostering effective collaboration.

Physical Design

Large OWS displays must be sufficiently sized for the information on the display screens to be depicted so all detail is visible to operators without the need to reposition the information to see. The mounting requires rigidity to ensure a complete lack

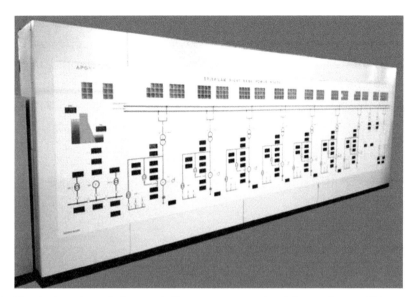

Figure 5.66. Physical analog mimic panel. Key measurement points are there within a superimposed line drawing of the entire power plant. The dedicated alarm panel "windows" are arranged at the top. While appearing sparse, this design afforded an excellent overview of the entire plant.
Source: Reproduced with permission from ZPAS.

of vibration from both the processing equipment in the plant or operations center as well as any structural movement of the control building itself. When several individual displays are tiled together, the bezel design should be as narrow as possible and the adjacent units close mounted to avoid any open space between. If open space is unavoidable, it should be covered to prevent distraction. The video specifications for the actual display images should meet or exceed those for the workstation displays. In no event should the display images exhibit even a hint of pixelation, color distortion, or image smearing.

Display Screen Visibility
Even though the display may be quite large and prominent, it still needs to be visible to the user (view angles, glare control, line-of-sight management, image intensity, and contrast, etc.). Any user interaction via pointing or other devices must be clearly identified on the screen, and the operator's use must follow all existing conventions consistent with workstation displays.

Unified Page Presentation
All display screen content must follow all existing HMI design for workstations.

Ability to Focus
Special care must be incorporated into the design to accommodate various, almost simultaneous users. The user of the moment must be clearly identified. The transfer

of control between users must be clear. During use, tools for proper focus by boxing, zoom, scribble, or other enhancements must be possible. Any such focus must not hide changes or other notifications. Due to the wide expanse of the display, ability must be provided to focus on multiple areas at the same level or on multiple lower levels. Temporary annotation must be provided to inform all viewers of the current focus if not a pre-existing screen.

Auditing
Full audit trails are required in order to replay the event for training, incident investigation, or other purposes.

Illustrations of OWS Large Displays

Figure 5-67 shows a set of flat panels arranged as OWS large displays. Note the orphan display between operator areas. Note that these flat panel displays appear to be simply larger-sized extensions of the WS displays. They are positioned for view, but for the sitting operator. The enlargement serves only to preserve the same level of image and text size as the WS displays due to the longer distance from the operator's eyes. This example is not an efficient nor effective design for collaboration purposes.

Figure 5-68 shows another instance of a set of flat panels used as OWS large displays. Note the extensive amount of small detail present. The operator has an extensive distraction load. Again, this illustrates a poor use of OWS large displays.

Figure 5-67. Two sets of flat panels arranged as OWS large displays with a third shared one in the middle. (The information content could be better selected.)

Source: Reproduced with permission from ABB. Extended Operator Workplace. https://new.abb.com/control-rooms. © 2018 ABB.

Figure 5-68. Operator using flat panels arranged as OWS large displays. This use is not considered a video wall because the displays do not depict a continuous view; rather, each display is a "tile" in a multi-tile mosaic view display. This has an enormous potential for significant distraction away from the operator tasks required in the control room.

Note that what distinguishes sets of flat panels used as OWS large displays is the bias toward use of them as individual displays rather than a coordinated single display. This distinction is important, although we are still working on how to fully exploit the single very large display.

Off-Workstation Small Displays

Off-workstation (OWS) small electronic displays are the most prevalent format in the world. Individuals who never touch a computer (personal or otherwise) look at a smartphone as a right to own. We are swipers, shakers, and fingers-on-the-screen keying-in experts. We expect the format to switch from portrait mode to landscape mode (although few use the terms) at the rotation of a wrist. Let us take a look at them strictly from a control room and operator perspective.

Formats

As you are well aware, there are basically two formats: Smartphone size with a screen about 5 inches diagonal and tablet size with a screen about 10 inches diagonal. The phone is easily pocketable. The tablet is a full handful and must be interacted with using two hands unless it is lying somewhere. Figure 5-69 shows smartphones.

Figure 5-70 shows a smartphone and tablet. Although these are familiar, this methodology is suitable only for small format devices. Without careful attention, it is easy for an entire screen to be filled without proper grouping and arrangement. As you see,

Figure 5-69. Typical smartphone screens. The very small display area requires special attention to content and links in order to be useful. Special content and formatting is required.

many HMI attributes are functionally similar but work differently. It is easy to switch back and forth without much attention because we understand and accept their specific differences.

Benefits

These are amazing devices. Their power comes from the ability to (almost) always be connected to a communications infrastructure. Thus voice, video, and document/data

Figure 5-70. Smartphone and tablet navigation icons. These are familiar to most of us, of course; however, this methodology is suitable only for small-format devices. Without careful attention, it is easy for an entire screen to be filled without proper grouping and arrangement.

are always at the ready. Their utility comes from their portability. The tablets especially are windows into the world. Purpose-built applications can be powerful and convenient.

Problems
For all their ubiquity and familiarity, however, small displays are not so easy to get stuff from or put stuff into. The general navigation protocol is "tap forever." Because of the very small screen size, there is little room for enough direct navigation buttons. Moreover, their basic design does little to support proper bookmarks and easy jumping to related information unless these functions have been previously designed together. Audit trails are nonexistent. It is next to impossible to recreate what the operator was looking at and what he did using the device. Most of the devices are fragile, and almost none are hazard rated for use in controlled environments. This limits their industrial utility. The bottom line is that their small screen size requires the user and designer to develop a completely different user expectation. Screen content is really small chunks. This places a heavy demand on designing the chunks to be what they are needed to be. And it places an even heavier burden on the operator to be able to use the chunks to extend his capabilities in a way that is worth the effort.

Head-Up Displays
Head-up displays are displays that place screen information within the line of sight of a person in such a way that both the head-up screen information and the primary view being looked at are simultaneously visible. While the concept is pretty simple and intuitive, doing this properly can be hard. The hardware is supplementary in that it provides the ability to view the head-up part in front of the primary view. Presenting the screen content on a transparent background does this. Head-up hardware can be fixed-mounted or head-mounted to move as the head moves to examine the primary view.

The reader is advised to appreciate that this discussion is not intended to either recommend or dissuade the use of head-up displays for the control room. As a need develops or the technology evolves, these may become useful additions to the control room. However, this utility must be firmly balanced against the added distraction and equipment needed.

Fixed-mounted displays are of use in a control room where they are able to cover a view much larger than any single existing display comprising the primary view. For if the viewing angle is similar to that of a single display, all benefits of a head-up display can be almost trivially obtained by combining the view into the (primary) display screen view. Where the fixed head-up display viewing angle is extensive, it would permit the screen content(s) to follow the head view without the need for operators to wear anything.

There are important design concepts to keep in mind. The head-up screen content must meet the following requirements:

- It must be directly relevant to the primary view at all times the head-up display is in view.

- It must be completely contained so that all the content is clearly part of the head-up display part and not the primary display part.

- It must be capable of existing (in the foreground) without any visual distraction of the primary view.

- It must be simple to the most basic level so it can be fully understood just by being in view; no reading, comparisons, or other interpretation should be needed.

- The eye focus point must be such that the head-up screen content and the primary view content are compatible. The viewer must be able to see the head-up content and the primary view content simultaneously in proper optical focus.

The examples that follow illustrate some of these requirements.

Examples

An automotive example of a head-up display is shown in Figure 5-71.

Figure 5-71. A head-up display example for an automobile. Only the white content is the display portion. It appears directly in the driver's line of sight, making it less necessary for the driver to take his eyes off the road. This one includes the vehicle speed (90 kmh) and notification of an approaching turn (300 m) ahead.

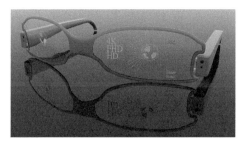

Figure 5-72. Eyeglasses version of a head-up display. This has a very small display overlaid within each lens.

The visual quality of this image is not perfect (a photographic focus limitation). But it is clearly intended to show the primary view the driver sees out of the windshield. The head-up part appears to float magically over the top. It contains two simple messages (left and right fields) and a graphic. Without actually knowing its purpose, we invent something: there is a turn 300 meters ahead. Your vehicle is approaching at 90 kilometers per hour. Had nothing been required by the driver, no head-up content would be displayed.

For completeness, Figure 5-72 depicts a conventional eyeglasses version of a head-up display. Note that it appears as a small overview inside the center of each lens of the glasses for the image display.

5.17 Video Walls

Conventional Video Walls

A video wall is pretty much just that: a large wall-type display that is capable of displaying digital video (static images and/or motion video) images. This form of display is commonplace as background for televised news shows. The cameras will show the background as newscasters present the news. The broadcast image might shift from the view of the newscaster to the video wall for specific content or just to vary the feed to the viewer.

Figure 5-73 shows the wall constructed of a dual set of probably 110 flat panels arranged 11 across by 5 high. Note the pronounced bezel frame around each. Preferred composite displays are almost seamless.

HMIs in the control room are not for impact or variety, but for specific purposes. We take our lead from the well-established practice of mimic boards once common in upscale control rooms in our analog past (see Figure 5-66). They were always within view. Because the displayed material was designed to show only the key information,

Figure 5-73. Video wall as background.

the distraction factor was minimal. As something notable appears, it is readily visible. Once something is recognized, it was clear how impactful it was and where in the process it was happening. On the other hand, without extreme care, the presence of video wall displays can be distracting.

Figure 5-74 shows the operation room at the Meteo-France Toulouse site, named as Meteopole, outside the city of Toulouse, France. The operators are at their separate workstations and not at all positioned for easy line-of-sight to the large display. There is a variety of content on the display but the sharing of the display screen real estate suggests no apparent plan or sense of relationship between elements.

This is probably enough motivation to understand that using a video wall display in a control room will need careful, purposeful design. Properly mastered, this form of display should have important operational benefits. Let us see how this might work.

Video Walls for Control Rooms

There are two general modes for using video walls in a control room setting: (1) a video wall as the main HMI and (2) a video wall as an augmentation to the workstation HMIs. Each use will be quite different from the other.

Main HMI

A video wall display as the main HMI represents either a step backwards into the world of walking the analog board or a step into the future that puts the operator in motion as opposed to largely using fingers. Figure 5-75 shows how an extensive display, creatively designed provides a full-person interaction modality. The seeing

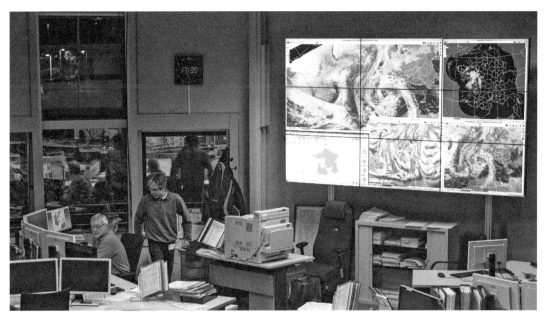

Figure 5-74. A view shows the operation room at the Meteo-France Toulouse site, named as Meteopole, outside the city of Toulouse, France, November 3, 2015. With two of the most powerful supercomputers in the world, French national meteorological service Meteo France participates in the international scientific work on climate change, on which is based the negotiations of the Climate Conference (COP21). Paris [hosted] the World Climate Change Conference 2015 (COP21) from November 30 to December 11. Picture taken November 3, 2015.

Source: REUTERS/Fred Lancelot.

Figure 5-75. Video wall as the significant view into the process.

Figure 5-76. Futuristic video wall display as a main display.

of what needs to be seen, the requests for alternative information, and the entering of commands can are located spatially in the process screen area. This facilitates operators to connect their virtual location to actual process physical locations to gain and retain situation awareness. Figure 5-76 illustrates a futuristic version where the selection points for interaction and the responsive display views are dynamic and dramatic.

All of this exciting display ability must of course aid the operator; not entertain or distract. It's too early in the technology deployment to fully understand how to deploy this form of display. However, the same design principles and operator-use tasks for conventional HMIs covered earlier in the chapter would apply here.

Video Wall as Augmentation to Workstation HMI

The video wall display as an augmentation to the traditional HMI workstation is the expected manner of use in the near future. Examples of this use are depicted in the following images. Figure 5-77 shows a small-scale, somewhat personal arrangement in what appears to be a Honeywell concept view. This design suggests a tablet being used as the primary hands-on part of the HMI. Notice that this illustrates a change of paradigm. Now the WS display acts only as a navigation and data entry platform. The OWS large displays revert to depicting information at the current operational level of focus rather than at the overview level. This is not to say that this design may not have merit; I am certain that it does. However, the operational need for an overview remains. And there are fewer real estate and interaction handles available and no extra viewing angles open to also have OWS large displays farther in the background.

Figure 5-77. Not a particularly good use of a video wall–type display. Depicted here is a personal workstation arrangement; however, it really is used as an extension to augment a very limited tablet display. The overview function is entirely missing. Note the sit-stand capability of height adjustment. Collaboration could be facilitated, but its provision must be supported with appropriate screen content in this configuration.

Source: Reproduced with permission from Honeywell.

The MBTA Boston Transit System main control room represents an interesting use of the video wall display concept. Here the entire transit system is represented so that problems and concerns at a high level can be shown and shared with the other operators and monitors. Figure 5-78 from Winsted shows a view of the room.

We see a mixture of uses for this video wall. Its current use may not be the best (the overly complex wall screens), now that we understand how to better exploit its potential. Observe the segregated two-level observation or command center region to the right and raised from the main floor. The main theme is a linearized depiction of the transit routes. Boston is broadly two-dimensional. However, to display the system, the routes are broken down into their component areas of responsibility with the controllers near their routes. Video feed is interspersed in the wall. Each controller appears to have a full component of WS displays. Attention is also called to the upper-level area to the extreme top right of the view. This may be either a command center or a visitor center.

Figure 5-78. MBTA Boston Transit System main control room. There are several individual control pods as well as a separated interior area with a viewer's balcony above.
Source: Reproduced with permission from Winsted Control Room Console Company.

Georgia System Operations is an electrical power distribution cooperative. Its main operations center makes use of a fully shared video wall. Figure 5-79 depicts this prominent use of a shared video wall. As can be seen, it easily integrates with conventional control desks. And there is also a raised observation, command, or visitor center to the right.

Using a video wall need not require a highly engineered or complex architectural control room design. Figure 5-80 shows a more conventional use. This video wall is massive and appears to show weather information overlaid on geographical map backgrounds. Note the alternating sections that display performance and other data. Doing it this way requires the content to be located in the area of local need and therefore not in view of all operators.

5.18 Paper versus Electronic Screens

The glass control room (a play on words from airplane *glass cockpit* terminology) is the current norm. Everything is becoming electronically based—controls, reports, checklists, procedures, shift handovers, schedules, and other odds and ends like work orders. Most of this movement away from paper has been driven by real and

Figure 5-79. Another example of the use of a video wall. Here the shared element is the entire power distribution grid. Each operating pod manages a portion of this grid. A raised view and command center overlook the entire operation.

Source: Reproduced with permission from Winsted Control Room Console Company.

Figure 5-80. Video wall in an uncomplicated control room design. Shown here are main operators as well as 80 or support personnel at the rear. Notice this use of the video wall has content segmented. Doing it this way requires the content to be located in the area of local need and therefore not in view of all operators.

Source: Reproduced with permission from Winsted Control Room Console Company.

important reasons. It is a current view that electronic is better. It is assumed that paper is fraught with problems and issues and lost trees. This discussion matters, yet reality is often different from trends toward change. There remain some drawbacks to electronic forms as well. Studies have shown that reading from electronic display devices is "slower, less accurate, more fatiguing, [and] decreases comprehension."[33] Unless we carefully understand what this means to the individual user or "reader," we might be assuming our way into unrecognized problems—problems that can materially affect quality and outcome. Let us look.

Pros and Cons

Pros: The benefits of electronic forms are that they

- are readily available (not tucked away in some folder or binder that who knows where it is),
- are always up to date and approved (by definition),
- can be linked to specific tasks or display screens so context is preserved,
- can be specialized to apply only to specific situations at hand (where the documents have been designed for this purpose), and
- can be immediately (electronically) shared with others.

Cons: Electronic formats also have limitations and are not suited to all users or uses.

- An operator cannot find an electronic form if it is not where it should be. (Most operator search engines are nonexistent or primitive at best; a Google-like search capability is not the norm.)
- It is difficult to see an overview (both document text view and quick whole-page-at-a-single-view that is readable).
- It is difficult to annotate (either directly or via Post-it notes) and archive the working copy "as worked" and flagged with a "push" to those who need to do more work on it or approve it.
- Electronic formats are tied to electronic viewers (portability, pocketability, and viewability).

---

[33] Andrew Dillon, Cliff McKnight, and John Richardson, "Reading from Paper Versus Reading from Screens," *The Computer Journal* 31 (5), 457–464.

- They are less tolerant of weather conditions and to viewing conditions like sun or darkness.

- It is not easy to pass them on directly to another individual or to his device close at hand.

- They are difficult to "cut and paste" on the fly.

- They can inadvertently be sent to unauthorized others (unless specific controls are in place).

Naturalness

Two general age groups staff control rooms: the twenty- to thirty-something-year-olds and everyone else. The younger group has been raised with electronic teething rings. They take mobile communications, personal music, and small screens for granted, at times even preferring them to conventional displays. Email and text messaging are important and integral to their primary interacting skills. Give them a query task, and a Google search is an almost reflexive response. They are comfortable with interlinked views of almost anything—click here for more information is almost a "litmus test" requirement for acceptable content flow management. If it does not have a mouse and a touchscreen and, more recently, voice recognition, there is no use for it. All in all, this group is ready, trained, and primed for electronic information.

On the other hand, the everybody-else group is different. Sure, they may be familiar with electronic tools—and exploit their use for advantage. They generally know how to interact with the devices and are comfortable doing so. But they are also comfortable with the predecessor mediums. They are more likely to search a long electronic document for a specific item and then pick up the printed version to look into it. They hope for a physical touch interaction when no other means is available. They are less likely to expect a touchscreen type of interaction when they see a mouse. They expect much more format flexibility (paper and other fixed form documents) than their younger counterparts.

The bottom line is that it is important for designers to understand these differences and not force one mode over another as the required information access form. They should also provide appropriate alternates for screen and paper interactions.

Readability

It is settled that the screen electronic and presentation design must be high resolution and without functional artifacts and visibility distractions (glare, contrast, etc.). Let us list these and the rest of the static items requirements for electronic displays:

- Sufficient pixel resolution to render all displayed items perfectly clear

- Sufficient display size to show needed information in an appropriate size for viewing and enough space for the depiction of related content without need to switch pages

- Comparable aspect ratio to content and everything properly rotated for viewing

- Comfortable fonts and graphic conventions

- Character and symbol spacing, line spacing, and format enhancements (underlining, italics, etc.) that are perfectly clear, conventional, and comfortable

- Color rendering, background controls, and contrasts that minimize reader fatigue

Following the Thread

Electronic forms by their nature are flexible. Portable document format (PDF) documents are fixed in format but easily permit linking into and out of them. Other documents and screen content usually contain lots of navigation targets and real-time refreshed content. This flexibility can be important and useful for some tasks, but not all. Electronic documents provide a significant likelihood for out of order content, out of context content, stale content, and missing content. Clear construction and use controls need to be in place to prevent this from happening.

Procedures

Procedures require performance in strict order with nothing left out and nothing added. During the execution, it may be necessary to access other information and perform specific controls manipulations. Access to these other items must be provided that does not interrupt the procedure information presentation or flow. Both order and context must be preserved. They must support delay, leaving, and return and do it seamlessly for concurrent or successive users.

Documents

Documents are meant to convey information. This information is conveyed by the content (the words used, word order, and word juxtaposition with other words). It is also conveyed by the order of presentation: paragraph order and document section order. Therefore, the reader must be aware of the expected reading flow path through the document and diligently follow it. Any "click-outs" must be directly relevant and necessary for clarification or remedial understanding. They should be there for necessity, not for convenience or subject completeness. All click-outs must support direct return of the reader to the exact place he left. Long documents (more than three or four pages) should contain clear content lists and labels so that the reader is constantly aware of where in the document flow he is and where he might want to view next, all the while preserving the intended thread flow.

Content presentation should be placed into context with the larger whole. Anything snipped out should be referenced-lined back to where it came from (much like how a small area blowup of a graphic has a boundary line around it and a line connecting it to a much smaller boundary line showing its actual size and location in the source). Content should be displayed with sufficient surrounding content to clearly denote how it fits into the document. This suggests that paragraphs show at least the one before and the one after (or as much of each as reasonable given the size of the focus paragraph).

Content Change Management
This is not intended to be a discussion of change management in the traditional sense of management of change (MOC). Rather, it is literally the control of content changes. Documents in this category are commonly referred to as *live documents*. The most familiar would be Wikipedia. The most common control room example is operating displays with up-to-the-moment equipment status indications, operating values, and alarm indications. We expect this sort of freshness and clarity. What are less expected are procedures that may be updated during use. However, appropriate posting controls can control any conflicts.

The issue is the document controls that the operator (or viewer) of content has available during explicit document use. There is a need to be able to fix the content of a display or a related group of them. "Fixed" content might be needed for reporting (status documentation, problem identification, etc.). It might be needed for understanding (freeze a display so that all the parts and information on it can be viewed at the same time and with the same values). When "fixed," that status must be evident and audit trailed.

Related Content
Very little information is used in isolation. Where content refers to, is related to, or is needed to manage other things, those other things need to be visible to the operator at the same time. Where existing display screen real estate and resolution permit it, the other items may be shared in other pages or windows to be visible simultaneously. If it cannot be visible on a shared display screen, other displays are needed. They must be configured and available. If the other displays cannot be dedicated for the term of the need, then they must be capable of departing and returning to the same configuration

as before departure. The concept of yoking is useful here. The end result is for the operator to be able to see *at the same time* enough of the content that is needed to understand or manage without navigating around to find it.

Personalization and Annotation

Personalization means the reader's ability to take his personal copy of the approved content and move it onto his virtual desk for use. On his desk, he has a full set of markup tools. A representative list follows.

- Underlining
- Highlighting
- Striking out (words and phrases, and whole paragraphs or pages)
- Adding marginal and in-text notes and diagrams and other annotation
- Replicating and editing parts of existing annotations for reuse instead of needing to completely reenter items
- Linking of text (e.g., circling one area of text or a diagram and drawing a line [visible electronic link] to another place or circled area in the document to where it is related or linked)
- Adding Post-it type notes
- Adding other differently formatted documents (pasting, or inserting whole pages) so that they can be utilized as an integral part
- Adding labeled bookmarks so that the user can return to parts as needed
- Being able to contemporaneously share the personal document with others as modified, but clearly identifying it as a working markup from a specific individual (useful for collaboration or approval)

Of course, these personal documents are for the operator who uses them. However, there is a need to preserve them both temporarily and often permanently as part of the operating record for later recall and understanding. This capability must be part of the embedded operating protocol for document control.

Comparisons to Think About

We have briefly looked at some important aspects of electronically displayed content. This discussion concludes by returning to the strengths of paper content to further appreciate the important things electronic information display should master.

Paper is clearly ordered in flow (page numbers, stapling or binding, etc.) from start to finish. It is much easier to position a paper document in the hands to find a comfortable viewing angle (devoid of reflections and glare), with appropriate lighting, at the right viewing distance, in a comfortable and convenient spatial location (desk with supplies, chair that is close to the kitchen), in a less distracting environment, and on a to-do pile. Reading and examining complex or lengthy content in paper form is less tiring, and there is a tactile connection that we humans seem to be sensitive to and like.

This discussion is not provided to convince anyone to adopt or abandon electronically displayed content. Rather, it is to advance understanding of some of the use realities and afford an opportunity to both appreciate them and plan to maximize their advantages and minimize negative effects.

5.19 Fire, Gas, Safety Instrumented Systems, and Security Systems

Fire, gas, safety instrumented systems (SISs), and security systems all require monitoring and, potentially, intervention. They are manned to the extent that personnel are tasked with observing them when they announce problems. Control room operators only consult them during events announced by their separate internal alarms or to check their status where other problems or issues come up that may relate to them. Where information from these systems appears on control room or operation center displays, or alarms appear in the control room alarm systems from this equipment, their presentation and design in control room equipment should follow existing standards and practices for the control room or operation center. All other design, implementation, and use are beyond the scope of this coverage. The reader should consult specific references and statutes for guidance.

5.20 Sound, Audio, and Video

There is a great deal of technology available to incorporate audible sounds, audio messaging, and audio announcing into control room designs. There are also many options for incorporating full-motion and animation video into the same settings. This section will cover how audio and video might be considered and for what specific purposes. The reader is reminded of the sage advice: just because we can do something does not mean that we should. In the control room or operation center, less is always more!

Sound

Sounds are very powerful stimuli. Humans appear to be hardwired to instantly recognize familiar sounds, even those from a distant past. Before the invention of symbols (pictures, writing, etc.), our ancestors accumulated a vast store of knowledge and experience by listening to stories, hearing advice from others, and retelling. Our minds and hearts seem to be custom made to soar to the delights of favorite music. Music and rhythm have their own special avenue into our inner minds. It is little wonder that sound can be an important medium for operators. Caution is the guide. Sound is so easy to overuse.

Cues

Mica Endsley's example of the incorporation of audible cues to reduce startle is a good illustration of considerate practice.[34] She presents the situation of a shopper in a grocery store near the fresh vegetable shelves. A water spray is used at regular intervals to keep the produce moist. To "prepare" the shopper for the onset of a spray event, she suggests using the sound of thunder from a thunderstorm. After the thunder, the water spray will commence. A cue is well considered, but this specific one is weakly chosen. Yes, thunder is an integral part of a thunderstorm. So is lightning. So is strong wind. Consider using the more direct cue of the sound of rain preceding the water spray. The rain sound implies one and only one thing: falling rain. Thus the shopper is alerted without ambiguity and quite a bit more gently.

This idea of cues is important. They are often used as an advance warning of an announcement over a public-address system. Because the announcements are usually loud and unexpected, they often startle the public. None of us like being startled. And when startled, we often miss both the beginning and the importance of the announcement. The larger picture here is that cues are an important tool. In this section we develop a broader concept for their use. The cue is used to both eliminate the startle factor *and* prepare the listener for appropriate understanding. By careful construction of cues, we focus the operator on the key aspects of the current situation to the exclusion or reduction of the extraneous.

Cue Purpose
The following is a list of the considerations for a cue. A cue should do the following:

- Gently alert the hearer that an automatic action or audible message is about to begin.

- Convey the importance of the situation and any urgency of action represented by the situation (e.g., "danger," "important," or "just to let you know").

---

[34] Mica R. Endsley, Betty Bolté, and Debra G. Jones, *Designing for Situation Awareness: An Approach to User-Centered Design* (Boca Raton, FL: Taylor & Francis, 2003).

- Announce what is about to happen.
    - Automatic operation will start or stop.
    - There is a physical action to consider.
    - There is a thought to consider.
    - A watchful eye is advised.
    - End of concern; go on to something else.
- Suggest a change or clarification of thought or process.
    - Situation is unusual or abnormal but remains as expected.
    - Situation is unusual or abnormal and has now become not as expected.
    - Situation has become entirely indeterminate.

To be clear, cues are not alarms and they are not messages; they are just a polite mechanism for preparing for something audible in a way that has as low a "startle" factor as reasonable.

Cue Construction
Cues should exhibit the following attributes:

- Be unobtrusive.
- Provide a clear audible notice of more to follow.
- Provide a reference frame for the type of content to follow.

Announcements

Announcements discussed here are those that are intended for the operator. The control system or plant is communicating something that the operator needs to hear as opposed to see (as on a display screen). Announcements should be preceded by appropriate cues.

In general, they would be used where the message is brief and the understanding is both clear and not easily put aside when heard while performing other tasks. Unfortunately, during the press of daily operation and especially during abnormal operations, an audible message is likely to be temporarily or permanently put aside. Consequently, all announcements should be posted to an announcement board readily visible on an appropriate display screen. This display screen should have a search capability. The content should be appended to the standard shift report.

Messages

Most messages should be used to confirm or deny the expected. If it is important to audibly inform the operator that something has started (normally), has ended (normally), or has been bypassed (where bypass or not are both normal), an audible message may be used. However, this use should be very infrequent. An audible message should be used only where this information is not likely to be readily at hand during other operations and the operator should know about it.

Messages should be preceded by an appropriate cue.

Video

This section discusses embedded video used within the HMI. Video on separate displays should be carefully located in the control room to eliminate distraction. All movement distracts, which is why animation on the HMI should be entirely avoided. The most ubiquitous use of embedded video is in enterprises that monitor effluents that discharge into the air or waterways. Unless the viewer is carefully attentive to it, those video-depicted changes mostly go unobserved until something catches the eye. Unfortunately, often something meaningless catches the eye, which leads to distraction.

Current good practice for use of video includes operator-controlled cameras that are located so that the operator can direct them to display conditions out in the plant or enterprise. Typically, this includes actions such as directing field operators to a needed location, inspecting for leaks or smoke or physical damage, and verifying personnel and/or large equipment locations. Where practical, the video should include audio. It should be noted that all video should be archive recorded and of the best available quality. Use of video should be limited to a "need to see" and "see it now" use rather than always available and active. In the world of situation management, motion will always overly engage and distract viewers. Distracted operators will find it difficult to carefully observe and engage in thoughtful activities.

5.21 Evaluating Effectiveness

Users are effective at evaluating their human-machine interface design for local style adaptations and usability after design and/or implementation. However, they are usually unequipped with the tools and design concepts to perform that design themselves. Use evaluation can be a tricky activity. First of all, most operators (and most of us reading this book) do not want to change too much. Operating is demanding, and mistakes can have extensive consequences. So once something is learned and potentially mastered, it takes effort and risk to learn a new way of doing it. If the new way is clearly better than the old one and the operator knows this and buys in, the change is easier. If the new way is not obviously better (to understand, to learn, and to use), there are problems. Many of us have witnessed this firsthand. In the early days of DCS installations, many operators quit or retired rather than have to learn and use the new paradigm. One refinery bought an early version of DCS that did not include graphics and then was stymied for years by operators who did not want to upgrade to graphics. But if the new is better and it must replace the old, regardless of operator preferences, it should be employed. Care must be used for deployment.

Evaluating HMI effectiveness is a complicated activity that must be carefully understood and done. All the steps for doing it should be carefully constructed to provide adequate operator evaluation without any hints as to what the correct or preferred answer might be. The timing should be appropriate to plant or enterprise needs and good operational practices. The information should ascertain the following:

- Perceived evaluation
    - Comfort level of the operator to use the tools and technology
    - Confidence level of the operator that the tools and technology are effective
    - Expressed concern about errors, inaccuracies, or nonworking items
    - Expressed need for more training
    - Expressed need for more, different, or modified tools
- Quantitative evaluation
    - Results of training exercises and simulations
    - Check tests on how operators perform specific tests or operations that have been especially designed to identify HMI problems, and ways of their proper use
    - Reviews of operating statistics for incidents, alarms, and other events and activities to identify whether they are reduced or otherwise affected by HMI design and/or use
- Performance reviews
    - Review of near misses
    - Review of incident reports
    - Review of pending modifications or repairs and the work process to ensure all are done properly and in a timely manner

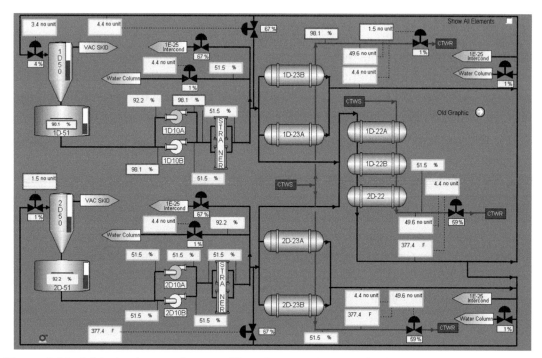

Figure 5-81. Original unimproved screen. There is excessive use of color, yet it fails to identify only abnormal aspects. There is the distracting use of 3-D. The mix of identifying labels, process values, and process icons is confusing and lacks focus.

Source: Images provided courtesy of Emerson Automation Solutions.

Let us take a quick concept review of what good screen design is meant to provide the operator. The two following images are from Emerson.[35] The first (see Figure 5-81) shows a conventional screen designed with good intentions but before the advances in understanding. Note how obvious everything is but how nothing useful stands out. Yes, there are lots of red-boxed values; some are important, some not. The black lines seem to compete with the blue ones. The blue labels compete with the vessels and other items. Color here is used for decoration—not the primary purpose of conveying information.

Now look at the next screen (see Figure 5-82). This screen contains the same plant area as the first, but the design uses most of the new aspects of proper style. Note that now the basic process components appear to be shadows on an unobtrusive and easy-to-view background. Alarms stand out, and they are fewer than the red-boxed ones in the original. Observe that important process variables are no longer shown as text

---

35 "DeltaV Operate Themes," Emerson Process Management (white paper, August 2016), http://www2.emersonprocess.com/siteadmincenter/PMDeltaVDocuments/Whitepapers/WP_DeltaV-Operate-Themes.pdf.

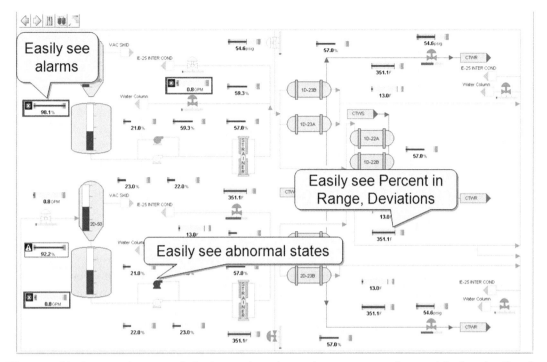

Figure 5-82. Improved screen. Color is consistently used to represent abnormal (red and yellow for alarms, blue for equipment. There is a consistent use of icons (although the duplication of actual readings is unnecessarily distracting). The equipment depiction is clean and simple, thus providing just the proper backdrop for the other content.

Source: Images provided courtesy of Emerson Automation Solutions.

values but represented by icons that clearly indicate the value and where it is in the proper range. To be sure, there is a bit more distraction caused by the use of color. And the screen designer should revisit how much needs to be on this screen and why. All in all, the view got much better for the operator.

The bottom line here is that the HMI is an evolving tool that needs to be evaluated and improved throughout its life cycle. Plants and enterprises change, expected performance targets for operation change, people change, equipment changes, and economics change. This central tool for successful operation needs to be fully up to the job all the time.

5.22 Loss of View and Key Variables

This is a true story. *Loss of view* is a technical term for a problem with systems that use electronic displays to show information and such to the operator. Loss of view describes the situation where all view screens or displays are suddenly blank or blue, appear "frozen," or are otherwise unable to provide any useful information to the operator. At this moment, and for as long as the situation lasts, the operator is functionally unable to obtain any information from the HMI. Nor are there any "controls" by which the operator can make any interventions (auto/manual changes, start or stop a pump, etc.). The control room operator is effectively cut off from his plant or operation.

Let us get personal. Imagine that you are in the control room when a loss of view occurs. When it first happened to a major refiner, it made company news and led to a fundamental new understanding of how to handle this situation in the future. But we are getting ahead of the story. The following depiction is a recreation based on the author's recollection. Imagine that we are looking over the operators' shoulders during the event.

> In a control room for a major oil refinery with a number of operators, suddenly everyone sees nothing. As far as they can tell, the screens are the only equipment affected, but who can tell what! After a short time, our control room fills up a bit. Visitors come, from the operations superintendent to plant manager and some in between. Eventually, the senior operations team decides that the situation is too out of control to do nothing and orders the plant shut down. So without the aid of any controls (read that PCS controllers accessible by screens), operators were sent out into the plant and tasked with shutting it down by hand.
>
> They manually opened or closed valves. They started and stopped pumps and fans. They tripped out package units. They brought everything they found in the plant to a shutdown status. They managed to do it without personal injury, but when things finally cooled, there was a lot of equipment damage. And there was lost production as well. The whole thing cost a very large amount of money—so much, in fact, that a major company investigation was launched to understand what really happened and decide how to prevent it in the future.

What the follow-up investigation determined:

- When view was lost, the electronic displays were totally unavailable, and no workaround existed that would have restored the view quickly.

- There was no secondary method of remotely monitoring any important aspect of the process; the operators were totally blind.

- All other PCS functions were found to be working. This means that all control loops were operating and all valves and other outputs were being appropriately managed by the control system. All alarms were functional, but of course the operator was not aware of their status. All PCS interlocks and other logic were fully functioning.

- All safety-related shutdown systems were fully operational; all physical protections (overpressure relief valves, dikes surrounding tanks, etc.) were in place and fully functional.

Based on these findings, here is what the investigators concluded:

- The plant had been carefully designed to operate, the PCS controls were fully operating, all safety interlocks were in place, and all physical containment and all code requirements for safe operation were in place and functional.

    o *The best thing to do was leave things alone and let the control system do its job, then just wait for view to be restored. The worst thing to do was to do what they did—try to shut down without the eyes and hands of the control system and operator.*

- Even with the existing PCS, the operators lacked an effective overview of the essential process operation. In the "old days," anyone could come into the control room, walk to the panel board, see at a glance the essentials of the process, and determine what was going on.

    o *Therefore, they required that each process unit be examined and the four to eight fundamental variables identified. Once identified, they placed electronic analog chart recorders with a pen for each variable next to their appropriate PCS electronic display screens. It was then possible for the operator to know at a glance that his process was basically sound or, if something significant was amiss, to know what it was and where to look for a resolution.*

The loss-of-view lesson for situation management is this: it is possible to identify key fundamental aspects of a process, which, if monitored, would provide a robust measure of the health of the process. This robust list of key variables is not usually extensive. Taken together, the variables will provide a sufficiently complete picture. Remember the depiction in Figure 5-47 in an earlier section? The real lesson from the loss-of-view story is how it reinforces the concept that fewer, but fundamental, pieces of information can do the job. The lesson for situation management is to design screens and content to place this information in front of the operator together with the proper frame.

5.23 Building Effective Screens

Effective screens contain enough operational information for the operator to be aware of something not going well enough. The navigation and display arrangement must enable the operator to carry out sufficient follow-up or other investigation to do a proper situation awareness and assessment. Building effective screens requires deep knowledge of the enterprise, how it works, and how it gets into trouble. Effective screens can be designed using this knowledge and the HMI tools reviewed earlier in this chapter. The following enterprise knowledge resources are also helpful:

- Enterprise design requirements
- Operating procedures
- HAZOP and PreOp safety reviews
- Failure mode analysis
- Layer of protection analysis
- Near-miss analysis

For your site, this will all be spelled out in a proper functional requirements document prepared for this purpose. Your site will prepare this with the aid of experienced experts. The next chapter introduces situation awareness and analysis as we build your capability to provide situation management. You now have a pretty good foundation to move ahead.

5.24 Further Reading

Bullemer, Peter, Dal Vernon Reising, Catherine Burns, John Hajdukiewicz, and Jakup Andrzejewski. *ASM Consortium Guidelines Effective Operator Display Design*. ASM Consortium, 2008.

Edmonds, Janette, ed. *Human Factors in the Chemical and Process Industries*. Oxford, UK: Elsevier, 2016.

EEMUA (Engineering Equipment Materials Users' Association). *Process Plant Control Desks Utilising Human-Computer Interfaces—A Guide to Design, Operational and Human Interface Issues*. EEMUA Publication No. 201. London: EEMUA, 2002.

Few, Stephen. *Information Dashboard Design: The Effective Visual Communication of Data*. Sebastopol, CA: O'Reilly Media, 2006.

Friedhoff, Richard Mark. *Visualization: The Second Computer Revolution*. New York: Harry N. Abrams, 1989.

Hollifield, Bill, Dana Oliver, Ian Nimmo, and Eddie Habibi. *The High Performance HMI Handbook – A Comprehensive Guide to Designing, Implementing and Maintaining*

Effective HMIs for Industrial Plant Operations. Houston, TX: Plant Automation Services, 2008.

ISO 9241-12:1998. *Ergonomic Requirements for Office Work with Visual Display Terminals (VDTs) – Part 12: Presentation of Information.* Geneva, Switzerland: International Standards Organization.

ISO 11064-5:2008. *Ergonomic Design of Control Centers, Part 5: Displays and Controls.* London: BSI (British Standards Institution).

NAMUR Standard AK 2.9. *Human Machine Interface (HMI). Potsdam,* Germany: NAMUR.

Shirley, Richard S. "A Fog Index for CRT Displays: How Good Is Your CRT Display?" *Advances in Instrumentation,* v 40 part 2, proceedings of the ISA/85 Conference and Exhibit, Philadelphia, October 21–24, 1985.

Tufte, Edward R. *Envisioning Information.* Cheshire, CT: Graphics Press, 1990.

Usability.gov. "Visual Design Glossary Terms." https://www.usability.gov/what-and-why/glossary/tag/visual-design/.

Part II

Situation Awareness and Assessment

6
Situation Awareness and Assessment

*Because you can predict does not mean you understand.
The ability to observe does not mean you will get it right.*

Unknown

*We learn so little from experience
because we often blame the wrong cause.*

Joseph T. Hallinan (Author)

*Prior to any major accident there are always warning signs which,
had they been responded to, would have averted the incident. But they weren't.
They were ignored. Very often there is a whole culture of
denial operating to suppress these warning signs.*

Andrew Hopkins (Australian National University)

This chapter introduces the reader to the dual importance of *situation awareness* and *situation assessment*. Together they play a vital role in the job of operating successfully. The overwhelming preponderance of experience and written work on these topics is from aircraft operations. A great deal of high-quality literature covers both topics. Therefore, this chapter is brief. This book is more about the processes, tools, and technology to successfully bring these concepts into the control room. The chapter lays out a foundation for situation awareness and assessment, discusses how individuals perceive the same events and aspects differently, and provides some acceptable ways

6.1 Key Concepts

Situation Awareness Question	How can the operator become aware of all the things needing attention that are going wrong or about to go wrong in a way that is understandable and manageable?
Situation Assessment Question	How does the operator determine all the reasonable and plausible explanations for the "data" observed from situation awareness, and from that list, select the one(s) he will "hang his hat on"?
Roles and Results	An effective functional balance between individuality and collectivity is vital for collaborative success. There is no success in the control room without collaboration.

6.2 Introductory Remarks

Start with everything normal. Good operation relies on situation awareness to expose any attention requirements by operators. Awareness is the state or ability to perceive, to feel, or to be conscious of events, objects, or sensory patterns. At this level of consciousness, an observer can see or sense without necessarily implying understanding. *Awareness* is defined as perception and cognitive reaction to a condition or event.[1] This awareness can be the process of observing or actively seeking, or just a feel, often referred to as our *sixth sense*.

Awareness provides the "data" for our operator to question. Situation assessment's responsibility is to produce judgments about those questions regarding operational conditions of production or operation. Situation assessment is what operators do with the "data" of situation awareness. To begin, you will want to get a sense of what goes on in the control room or operation center globally. Once you are aware of the effects of culture and expectation and how they pervasively influence behavior, you can be in a position to consider what you might do to make improvements. Without taking a hard look, it is too easy to gloss over a situation and decide there is not much to it. Another improvement opportunity that might be lost.

A principal purpose of this book is to provide a comprehensive perspective that helps you reach a good basic understanding. The book is designed to provide a useful

---

to influence both the individuals and their process of awareness in the directions of success. Later chapters will build your knowledge and confidence.

---

1 "Situation Awareness," Wikipedia, last modified May 20, 2018, http://en.wikipedia.org/wiki/Situation_awareness.

framework rather than list after list of things to keep in mind. This framework gives you the right amount of detail to wrap your arms around the concepts so you can feel comfortable with understanding, designing, and implementing your solutions. Then, as you need them, you can pick up the specific technology and tools for the task. These are here as well, of course.

6.3 The Situation Management "Situation"

Ask the questions that are on your mind. What does *situation management* bring to the table that is not already there, or if not there, not truly necessary? Why is it so important to read this book? You might think that your plant is doing well. You might wonder what you will gain by getting upset about gaps and about all the work this book is probably going to suggest that you must do and cannot operate without. These are valid questions. The answers are the reason for this book. If you take successful plant operations seriously, this material can better the odds for your operator.

I am not sure exactly what it takes for your operators to show up in the control room for every shift knowing full well that "today might be the day." What level of internal fortitude must they have that keeps them from stressing out or chilling out to the point that they more or less ignore the dangers and exposure? Yet they do not. For the most part, they do well. But the risks of not performing their jobs well persist. We are working to change the balance to the success side of things.

Let us clarify what operators do in the control room. They enter duty, work a shift on duty, and exit duty. Figure 6-1 depicts this. Every activity and responsibility must be supported by a full understanding of the activity and the degree of responsibility for success. Every duty must be backed up with appropriate operating procedures, effective training and evaluation, and proper tools to do the work right.

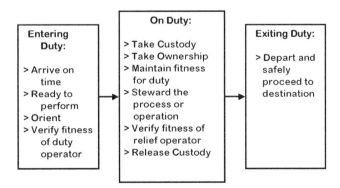

Figure 6-1. Operator roles and responsibilities and how they differ depending on where he is during the shift.

Operators are in charge. They are responsible. They are directly in the line of blame for most incidents, accidents, and disasters—and they know it. Let us change their odds to improve success.

The Operational Setting

We return to the operating setting by revisiting the operating regions introduced in Chapter 1, "Getting Started." Good operation is about "driving" production around the potholes, barriers, and missing guardrails so that it keeps working properly. Situation management is a big part of that. Figure 6-2 (from Chapter 1) again shows us the entire operator arena, but this time we see specifically where the control system is expected to do its job and where the operator must intervene to put things back to right. The illustration is a bit simplistic, but you can see how the operating arena works.

Let us start with things going almost perfectly: operation is inside the magenta dot at the center. No one asks for perfection. Staying "very good" is hard and expensive, so it does not stay there. All the normal variations in operation get to have a say in what is going on. Temperatures, pressures, and flows change a bit. Raw materials change a bit. Production rates or demands change a bit. So, things move out of the "sweet spot" to where they are still good. Nothing must be done here. The process control system is designed for just this goal. It will do that job until it cannot. Sometimes controls adjustments are needed. But if things go too far, direct manual intervention is a must. Alarms are the operator's normal notification of something going wrong. Good situation awareness must cover the rest.

If operation continues to degrade outside the safe operating region, we expect all the protection mechanisms (e.g., interlocks, relief valves, and emergency shutdown

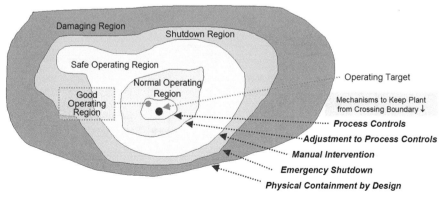

Figure 6-2. This illustration identifies the different operating regions and their "nested" nature as operation deteriorates from being on target through upset and potentially damage. Also shown at the right side are the parts of the infrastructure that are intended to manage upsets or, if they happen, to protect against the worst.

activations) will protect. We expect that level of good enterprise design and maintenance to be there. Our focus here is operations and operators.

6.4 Situations to Be Aware Of

"Situations to be aware of" may sound like a lot of carts coming before a single horse. Of course, you want your operator to know how the plant or enterprise entrusted to his care is really doing. Is it performing as expected? If it is not, how far astray is it, and where is this abnormal operation happening? Clearly, knowing where things are okay is important. However, one must also realize that situations can change in ways that might be subtle and hard to see. This chapter is designed to convince you of the importance of operators being fully aware of the operational status. Fortunately, there are powerful tools to handle these situations. The next chapter will lay them all out. Let us get our arms around the situations first.

Problematic Situations

Trouble does not usually come from nowhere. For all but the most unexpected events, there are always clues, innuendos, and other subtle suggestions and information that signal a problem. The operator's task is to find them and make sense of what is found. Some situations are more likely to result in problems than others. In such situations, the operator should be forewarned to actively look for trouble brewing. The following situations might appear on an operator's radar screen to point him to potential problems:

- Broken equipment or equipment whose operational range is restricted by not being in full working condition

- Maintenance operations ongoing, just completed, or soon to be started

- Procedures that are not working exactly as expected or not working the same way they did when they were last used

- Plant documentation with errors, omissions, or aspects that cannot be verified as either correct or incorrect but are of concern

- Data missing or having suspect values or behavior (laboratory data, HMI screen information, data forms or permits, etc.)

- Deliveries of anything, including routine and nonroutine, that have started or are soon to start in the plant or operation area

- Presence of unexpected or unauthorized personnel in the plant or operation area

- Substitute operational, maintenance, or support personnel presently working

- Any multi-step operation without a specific procedure or annotated checklist to follow

- Any individual whose workplace location is at the site but whose present whereabouts is unclear or unknown

- Any personnel injury

Our job to provide the information. This list is a partial one. Please use it as a starting point for your list to help understand the need to exercise more than ordinary care when these situations exist. You can add to this list from your own operational experiences.

Operational Situations

Trouble can also come when operation is somewhat near the outside boundaries of normal or usual. There is a built-in temptation to assume that things are okay and that nothing unusual is going on. Much of the time that is true, but not always. It is the "not always" part that operators need to be aware of and attentive to. Operational situations that might result in problems include the following:

- More than just minor changes in operation (10% or higher rate of production change, product changes, online equipment changes, etc.)

- Any alarm activation

- Any challenge to an interlock or permissive that did not result in a direct shutdown

- Any spill or loss of integrity even if it is water or some other seemingly harmless material (unless a part of and expected during routine operations)

This list is a partial one. Please use it as a starting point for your list to help understand the need for the operator to exercise more than ordinary care when such situations occur.

6.5 Strong Signals

Strong signals appear as either direct or indirect clear indications of operational problems to be resolved in a timely manner. The "strong" part is that there is nothing ambiguous or subtle about the fact that there is a signal. Operators must always verify that the problem they think is indicated by the strong signal is actually that problem. Properly resolving these problems requires deep understanding of the process and proper operating procedures. "Weak Signal Management" in Chapter 9, "Weak Signals," provides a more detailed discussion of this topic. For now, we will make the point that enterprises need to be carefully designed, maintained, and managed. Otherwise, the constant challenges to good operation will demand too much resource and dilute the ability to manage effectively. This need for whole enterprise responsibility is a critical recurring theme for sustainability and the essential enabler of the useful tools for ensuring it. Operators (and supervisors) need expertise and experience to keep things right.

Indirect Strong Signals

Unlike alarms that directly point to specific abnormalities with predesigned protocols for handling them, *indirect strong signals* announce that something is clearly wrong, but not what is wrong, why it is wrong, or how to fix it. Here is a brief listing of how indirect strong signals might appear:

- Missing procedure for a (nontrivial) task the operator must perform
- Malfunctioning equipment, sensor, or controls
- Important process conditions that are either unavailable or hard to know
- Procedure that is being followed but is not working as expected
- Required or recommended personnel not present when expected or needed
- Tasks that are falling behind or not completed because of operator workload or distraction
- Too many alarms activating
- Production quality or quantity that is not what it should be or is becoming problematic
- Outside conditions (weather, security, utilities, etc.) that are either not taken into consideration or beginning to require significant operator resources

So, what do operators do with indirect strong signals? For any situations that are covered by existing procedures and practices, operators should follow them. For the rest, they should begin by noting that each item is really a symptom of a likely much larger problem. Identifying them as symptoms is important. Proper problem remediation requires approaching the underlying root cause and resolving it. Sure, treating symptoms can be a stopgap. At times, both treating symptoms and identifying the root cause is advisable but not as policy. We still need to know what is wrong. An excellent process for root cause discovery would be *weak signals*; this topic is covered in Chapter 9, "Weak Signals."

Direct Strong Signals

A bit like alarms that announce an abnormality, *direct strong signals* announce that something is unusual or unexpected. The difference between them and alarms is that these abnormalities were not envisioned by the plant or enterprise designers because they were not reasonable possibilities or were expected but not important enough in their own right. Here is a brief listing of how direct strong signals might appear to give you an idea of what they might look like:

- A pump stopped but the spare started (either by auto-start or manually) without incident.
- The relief operator did not show up to take his shift.
- A spill is observed in the production area.
- Unexpected personnel are observed in the production area.
- Data on the HMI screen is present where expected, but the values seem to make little sense.
- All weak signals have been confirmed to represent actual problems unfolding.

Notice the last point? Even though we have not gotten to weak signals yet, once any hunch or clue about a potential problem is confirmed, it becomes a real problem and working it out is what operators do.

What do operators do with direct strong signals? Recall that indirect strong signals (discussed earlier) point to symptoms. Direct strong signals, on the other hand, point directly to a problem. Again, like the indirect case earlier, when situations are covered by existing procedures and practices, the operator should follow them. For the rest, the operator should exercise caution. The most important action to be taken is no action for immediate remediation. As the operator, please do not even think about picking up what you see as a problem and going right to solving it. The order of business is to ensure that the problem seen is in fact the problem that needs solving. To do that, the operator has two options

1. rule in the observed problem by confirmation without any reasonable reservation or contradiction, or
2. rule out any other possibilities for the observed problem by finding reasonable contradictions for the others.

Ruled-In Problem

If the clearly observed problem can be confirmed by a convergence of the evidence, and there exists no reasonable contradiction, the operator can proceed to remediate the problem. It is assumed that there will always be a protocol for remediation. Be that protocol specific, or general but useful, one must always exist. If it does not, operators are not inventors. They should escalate the problem to the proper resources, regardless of its apparent significance. Often what first appears to be insignificant ends up being a pivotal facilitator to incidents or worse.

Ruled-Out Problem

If the clearly observed problem cannot be directly confirmed, the operator should rule out all other possibilities. When that is done carefully, there is sufficient likelihood that the original problem is the one needing to be addressed. Remember the "logic" of Sherlock Holmes.[2] The operator should proceed to remediate that problem. It is assumed that there will always be a protocol for remediation. Be that protocol specific, or general but useful, one must always exist. If it does not, operators are not inventors. They should escalate the problem to the proper resources, regardless of its apparent significance. Again, often what first appears to be insignificant ends up being a pivotal facilitator to incidents or worse.

6.6 Situation Management Roadmap

> **Situation awareness**, or SA, is the perception of environmental elements within a volume of time and space, the comprehension of their meaning, and the projection of their status in the near future. It is also a field of study concerned with perception of the environment critical to decision-makers in complex, dynamic areas from aviation, air traffic control, power plant operations, military command and control—to more ordinary but nevertheless complex tasks such as driving an automobile or motorcycle.
>
> Situation awareness (SA) involves being aware of what is happening in the vicinity, in order to understand how information, events, and one's own actions will impact goals and objectives, both immediately and in the near future. One with an adept sense of situation awareness generally has a high degree of knowledge with respect to inputs and outputs of a system, i.e., an innate "feel" for situations, people, and events that play out due to variables the subject can control. Lacking or inadequate situation awareness has been identified as one of the primary factors in accidents attributed to human error. Thus, situation awareness is especially important in work domains where the information flow can be quite high and poor decisions may lead to serious

---

[2] Sir Arthur Ignatius Conan Doyle, *A Study in Scarlet* (London: Ward Lock & Company, 1886).

consequences (e.g., piloting an airplane, functioning as a soldier, or treating critically ill or injured patients).

Having complete, accurate and up-to-the-minute SA is essential where technological and situational complexity on the human decision-maker are a concern. Situation awareness has been recognized as a critical, yet often elusive, foundation for successful decision-making across a broad range of complex and dynamic systems, including aviation, air traffic control, ship navigation, health care, emergency response and military command and control operations, and offshore oil and nuclear power plant management.[3]

The Process of Situation Awareness

Situation awareness is knowing where the operation is right now and where it is headed. Both are important to frame whether or not what is happening right now is a problem. Situation awareness is therefore a process rather than an outcome. By that we mean that you engage in situation awareness activities, not that you are fully situationally aware. The message is that if you think you really know what is going on, it becomes much more difficult to keep an open mind to more information that might not support that knowledge. Let us start with the flow.

The first thing you will notice in Figure 6-3 is the natural flow from situation awareness (seeing what might be useful) to situation assessment (the meaning and importance of what you have seen) to situation management (hands-on intervening to ensure things go the way they need to). The next thing you will notice is that the flow goes both ways. The larger arrowheads show the natural progression. But there also is a regression—the arrows go back as well as forward. This means that every step of the way needs confirmation. Something noticed is not always noticeable enough to be understood. Something there does not always need attention. Something that appears to need attention may not be responding to that attention. This is not going to be a checklist activity. A great deal of focused infrastructure will be necessary. Situation management provides it.

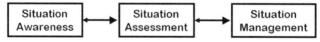

Figure 6-3. The interactive flow from situation awareness to situation management showing how the operator's knowledge must build by first observing, then understanding, and finally managing. The bidirectional arrows depict the fact that each step is a collaboration between the one preceding and the one next; any concerns in one must be examined in light of both.

---

3 "Situation Awareness," Wikipedia.

Active versus Passive Monitoring

At its core, operators achieve and maintain situation awareness in two related (of course) but quite distinct ways. The first is through their general responsibility of "keeping an eye out" for all the things that happen. And when we think about it, this is most of the job. Operators are expected to remain ever vigilant for things that happen but should not, and things that should happen but do not. Watching for things that have happened but should not have is easy when they are readily visible. The operator just has to be present and watching. For the most part this is reactive. Watching for things that are not readily visible is much more difficult. They must be searched out, located, and understood. This degree of active monitoring is tiring to maintain hour after hour, shift after shift. Soon operators settle into a personal style that appears to work—until something happens.

Looking for things that could happen but happen very infrequently is even more problematic. To do it well, the operator must thoroughly know the complete inner workings of the plant and be ever watchful to adequately compare the current status with the desired or needed operation. Save for the obvious, for example, a batch completion or a reaction phase change, this requires special talents, a great deal of focus, and lack of distraction. Given the numerous other tasks and interruptions that minute-by-minute operation involves, this sort of looking slips far down on the operator's list of tasks. Eventually, this becomes mostly a matter of luck. Chance is unreliable.

All of this responsibility, and training to carry it out, yields a heavy load of stress. Operators know that they are exposed and responsible. They sense the mismatch between responsibility and capability to meet it. That they are able to show up for work and do as well as they do is often a credit to their personal resiliency, good process design, responsible maintenance, and luck.

6.7 Situation Awareness Tools

Tools are the things that operators use. These tools need to be deliberately designed with their purpose in mind and possess clear strengths and known limitations. They are expected to be in constant use. They are of critical importance to the operator. Without them, operators use nothing more than hunches, intuition, and lucky guesses. How would a carpenter work without saws, hammers, rules, and levels? How would a chemical engineer work without unit operations, plant design practices, transport processes, and principles of chemistry and reactions? As you can see, some tools are physical and some are delivered in the form of accessible technology. Operator tools are similarly diverse. And as is true in other professions, the tools need use, practice, and sharpening.

This section briefly introduces the situation awareness tools and provides references to where they are covered in this book. The tools are outlined in the following list:

- **Operator interface (HMI)** – The playing field, to use a sports analogy, for almost all the activities: active looking and using tools, and passive awareness. Reference Chapter 5, "The Human-Machine Interface."

- **Alarm system** – The platform for notifying the operator of all abnormal situations (that have been predicted through effective process design, engineering, and operations understanding) that are in progress and must be attended to. Reference Chapter 8, "Awareness and Assessment Tools."

- **Strong signals** – Failures or lapses of existing operating procedures and protocols to readily resolve issues and problems. Reference this chapter.

- **Weak signals** – Identify the out-of-the-ordinary and evaluate their potential significance. Reference Chapter 9, "Weak Signals."

- **Collaboration** – The activity of seeking peer and supervisory interpretation and guidance for understanding current operations. This activity is a powerful way to ensure current thinking incorporates the larger enterprise collective knowledge and experience. Reference crew resource management in Chapter 10, "Situation Management."

- **Shift progress reports** – The structured and planned act of setting aside all distracting activity to perform a complete review of current operations. This would be done at several preset times during each shift. Reference Chapter 3, "Operators," and Chapter 9, "Weak Signals."

- **Shift handover** – The formal activity for assessing the current status of the plant and conveying it to the next operator. Reference Chapter 3, "Operators."

Each of these tools will be carefully crafted and fully ready for use by all operators. Moreover, they are expected to form the backbone of all activities performed by operators as they progress in their shift. Operators are not expected to do their work without using all these tools and using them appropriately. And like other professionals, they are tasked with keeping their tools sharp. This includes constant refresher training (both situational and at ever-increasing resolution) as well as participation in feedback on performance for continually improving both the tools and their application protocols.

6.8 The Psychology of Situation Awareness

We are not governed by what is true and what is not, but by what we believe to be true. What we believe to be true is often mistaken for the truth. *We* includes teenagers, bankers, car mechanics, clerks, engineers, and, yes, operators and supervisors. In our personal lives, being closer to the truth may make us happier, sadder, prouder, more ashamed, downright discouraged, or on top of the world. It depends on what that truth is really telling us. In the control room, being closer to the truth will mean the difference between a chance for success and heading full speed ahead into a problem. Let us think about ways that enterprise design and control infrastructure might be able to find a few "potholes" and either fill them in or somehow slip around them without falling in.

Ownership

An operator must only own the responsibility for action within his sphere of experience and delegation of responsibility and the preparation to do it competently. Operators do not own the outcome. Owning the responsibility happens at shift. Owning the competence happens by doing the training, arriving for the shift prepared, and being on the job during the whole shift. By not owning the outcome, operators will not be expected or allowed to even consider the wrong axiom, "The end justifies the means." On the other hand, the enterprise does own the outcome. This is why it must follow design standards and best practices. This is what safety equipment and protocols are for. This is what insurance covers.

"Relative" Prime Responsibility

Whenever an enterprise or an operator gets ownership confused, bad things happen. This discussion about relative prime responsibility is meant to reduce the chances of operators getting confused about ownership. *Relative prime responsibility* is the effect on the *operator* that will always be produced by *management* if management functionally directs behavior, not by dictum, but by fiat. That is, it is when management holds "feet to fire" or what management acts on that determines an operator's immediate goals and therefore remediation behavior. It is not any higher goal, regardless of how much or how often it is touted by training, slogan, or positive reward. The placement of prime responsibility must be at the sought-after risk level. Otherwise, the sought-after risk will not be the exposure that is managed. Examples include: (1) the operator in the Japanese railway disaster who was more concerned with being late than the safety of train, (2) the overemphasis on perfection of personal protective equipment without mention of operational safety in the BP Texas City disaster, and (3) Pacific Gas and Electric's diversion of funds for pipeline replacement to executive compensation.

This responsibility really does matter.

Accepting Reality

The saddest audit uncovers situations where there were failures due to selective acceptance of reality—for example, "I can do it" as a slogan, which is not always a confirmed reality. Some of those failures are illustrated here:

- Misunderstanding of facts (confirmation bias; see Section 7.9)

- Failure to always follow procedural principles (permission to operate, etc.; see Section 10.15)

- Failure to properly design (weak infrastructure including poor personnel selection, inadequate process design, poor maintenance, inadequate training, poor management of impairment of operators, and leadership failure resulting in demoralization and detachment; see Section 1.14).

Leadership and Cooperation

Our style of approaching things and ways of thinking and behaving are important parts of our individuality. We briefly visit some of these characteristics here. We do it for a better understanding, not to reshape personality, but to facilitate effective behaviors in the control room. The next chapter takes a deep dive into this. This is just a toe. The important message is that good-intended individuals acting in good faith will work better when they use good tools. Proper tools shape how we can better relate to each other. This is important because it will lead to the results good operation needs. Let us look at two aspects: leadership and cooperation.

Leadership

Leadership is the way groups are guided by one or more individuals. We think of it as a good thing. An effective leader is able to move the group in a collective, organized way. A good leader chooses that direction of movement for the betterment of the group. Oversimplifying just a bit, a leader might lead by using a thoughtful consensus of the group or by taking full and authoritative control of the reigns. If that authoritative control is taken in an overly bold manner, unless the leader is right the first time, disaster can result.[4] It is not the bold part that is the issue. It is whether the leader is right or wrong in his assessments. When the leader is wrong, everyone goes the wrong way. Bold leaders leave little room for question or dissent. That does not work for nations. It does not work in control rooms. So you will want to replace

---

4 Susan Milius, "Bold, Incorrect Spiders Mislead Groups—Wrongheaded Notions Prove Dangerous to a Colony's Health," *Science News*, July 11, 2015.

bold with *cautiously decisive*, which includes collaboration and cooperation with others who have experience and provide expertise. Collaboration and cooperation happen through the proper use of situation management. This is why you will want to build it that way. The following chapters will help develop your understanding of how to get that job done well.

Cooperation

One of the fundamental success factors for individuals is cooperation. Even the most rugged individualists recognize that no single person can do it all alone. Rhetoric aside, here is the fundamental concept: whenever an individual hits a hard spot, if the right tool, skill, or knowledge is absent, that individual is out of luck. On the other hand, if the individual is part of a cooperative group and someone in the group has the right tool, skill, or knowledge, the situation can be managed. This idea is the driving force behind crew resource management (Chapter 8, "Awareness and Assessment Tools"), escalation (Chapter 10, "Situation Management"), and shift handover (Chapter 3, "Operators"). Cooperation is an important part of the "why" of this book.

The Triple Package

You certainly are not going to reshape what makes an operator. But we are going to visit some human traits; they are learned traits that add to an operator's ability to successfully pursue outcomes. We open this discussion by referring to the *triple package* of learned culture that seems to work toward success.[5] When an individual possesses these three traits, he is (statistically) significantly more successful. These traits are:

1. A strong sense of (honest) pride and the feeling of belonging to something exceptional (but not exclusive, by any means)

2. The right amount of "insecurity" that ensures that success "must be sought"

3. Impulse control

Pride and Exceptionalism

When operators and supervisors identify with a responsible enterprise, maintained and operated responsibly, that is pride. When operators and supervisors are installed and interacted with as professionals, that is respect. When you give operators the tools and processes to be successful, that is responsibility.

---

5 Amy Chua and Jed Rubenfeld, *The Triple Package: How Three Unlikely Traits Explain the Rise and Fall of Cultural Groups in America* (New York: Penguin Press, 2014).

Insecurity

When you teach your operators and supervisors why they are there, they can better understand the risks and responsibilities of their positions. When you train and retrain your operators and supervisors, they know their strengths and weaknesses.

Impulse Control

When you stand behind the actions of your operators and supervisors, they can take the time to do things the way they need to be done.

Intuition and "Raw" Information

This topic was saved until last. Weak signals have always been present. But the *only* way operators were expected to find them was a combination of chance and intuition. The better trained and more experienced the operator, the better he should be at this. Or that is what we thought. Part of it is true, but the part that is not is that many of the experienced operators had years of experience but, to quote an aphorism, "a single year of experience, many times over." They were not as experienced as we had thought. Many of the others had enough experience but were too quick to resolve the problem. With this cautionary introduction, let us proceed.

6.9 Situation Assessment

Situation assessment means effectively being able to use the information about what might be going wrong (and going right) to understand what might be happening. Start with the situation assessment question.

Situation Assessment Question

The situation assessment question is: How does the operator determine explanations for the "data" observed from situation awareness that are reasonable in general and plausible for the current specific situation? Let us break this down. Operators (and supervisors) should ask of the data the following questions:

1. What are all the reasonable explanations for what has been observed (through situation awareness), without making any initial judgment in favor of or against any?

2. What evidence supports each of these explanations?

3. What evidence goes contrary to each of the explanations to suggest that it be ruled out as reasonable?

4. How does one rank the explanations in order of reasonableness?

5. What is the threshold of decision for selecting the most reasonable explanation? Is that threshold sufficient or even justified?

6. Has the threshold of decision been met by the current situation?

Please keep in mind that no action of any nature or any kind will have been selected. We are not there yet. We explicitly do not want to be considering actions here because doing so now would prematurely move the work process ahead. Doing things too early makes it all too easy to blur "deciding what might be wrong" into "here is what I will do about the problem" before it is fully understood. When we do that, we tend to not properly identify what is actually wrong. Working on the wrong problem (or not working on any problem because the right one was missed) has been implicated in all major disasters.

Here are a few of the more useful tools to help us lay a solid foundation for doing situation awareness. We want to keep it one step at a time.

6.10 Surrogate Models

Surrogate model is just a fancy term for a hypothetical situation to test your ideas and guesses. It can enable operators to get their arms around complicated and often-difficult problems. The best thing is to be able to grasp the causes of problems outright. Say that the operator sees something going wrong with the flow rates in an important and expensive part of a mechanical separation plant. Sure, if he understands gravity separators and centrifuges and both are part of that plant, he calls up his knowledge and gathers what he knows is the proper data to narrow down the possible problems. Once he sees what he needs to see, he will be able to figure out the one or two real problems. Situation assessment is not situation solution or situation management. It is just figuring out what is currently abnormal and likely in need of attention.

On the other hand, suppose that the operator is inexperienced. Suppose that he is still gaining understanding and experience. Neither is yet strong enough to enable him to narrow down what he must realize about the current situation. To him, the current abnormal situation is probably too complex or too hidden. In most of these situations, creating an analogy or "surrogate" can be a useful approach. A surrogate is a substitution—something that works much like the original but is, in this case, known and familiar. If what we find is good enough, that surrogate can be used in place of the real problem to determine what is going on. Let us be clear: a surrogate need not be like the original or even look like the original. It only must be similar enough in the operator's mind to help him discover what might be going wrong in the surrogate. Taking that information, the operator would imagine similar problems with the actual plant or operation. Examples will come; first we need to lay more foundation and explanation.

Sources for Surrogates

A surrogate is something that seems to be similar enough to what we might be seeing at the moment but appears too complex and/or too hard to visualize and understand. We tend to draw surrogates from interesting parts of our past. Please keep in mind that like any useful tool, there are benefits to reap and pitfalls to avoid (see Chapter 7, "Awareness and Assessment Pitfalls"). Keep a very watchful eye!

Surrogates from Our Immediate Surroundings

Surrogates from our immediate surroundings arise from two areas: immediate surroundings and proximate surroundings. The immediate surroundings are in terms of time. What has happened recently that might be useful for understanding the present situation? The proximate surroundings are in terms of physical closeness. What is located in the plant near to where the problem is occurring? Here are a few suggestions for how to look for both.

- **Recent happenings** – Watch any good crime movie or show, and you may hear the detective say, "I do not believe in coincidences." When events or clues appear close together, they are diligently associated. Once associated, they are looked into. If they do not pan out (an old gold rush term), they are put aside. So, an operator might first take a look at what happened on the earlier shift to see if it is continuing or could be affecting the current issue or problem. However, this "looking into" must be conducted to prove relevance, not disprove it. Start with the current thing being different. If you can show that it is similar enough, then you have a repeat; otherwise, it is different— and likely quite different.

- **Closely located happenings** – Here we are using things that might have gone wrong in the same physical area as a guide to what might be going wrong now. No matter whether the other thing went wrong yesterday, or months or years ago, if they are in the same general physical area, the operator may opt to use one as a starting point for understanding what may be amiss now. Again, this approach is a very simplistic starting point. Quickly use it and decide on the evidence whether or not it is useful to understand the current situation. Again, the requirement is to first prove relevance, not the other way around.

Surrogates from Personal History or Folklore

One of the wonderful things about working with an individual who has loads of experience and also the gift of storytelling is the way past dramas are told. You might hear about the time when everything in the plant as far as the eye could see was going wrong. Emotions were high. Panic was lurking at every new bit of information. Somewhere a light shined, and the key player or players managed to find the kernel and turn things around to narrowly avoid disaster. You will hear how finding out about one thing led the players to the next thing until, finally, something worked. And that story was one of many. Almost everyone remembers each one.

Those are stories in which you were center stage or to which you were drawn like a whirlpool draws in the surrounding waters. In some way or another, some of these stories can be ingrained inside you. But unlike folklore, some of these stories may have had not-so-happy ends. For the ones that worked out well, sometimes we think we see them again. When we do, there is the tendency to use them as a model for what to do now. With care, quickly evaluate them or quickly discard them and move on to the protocols. But for the ones that did not work out so well, also quickly evaluate them or quickly discard them and move on to the protocols. Act decisively, not reactively. Use the experience as a tool, not a plan.

6.11 The Four-Corners Tool

Deciding whether or not to act in a given situation might require action on the one hand. It might require not acting on the other hand. Deciding what to do poses a built-in dilemma. This situation demands careful examination and consideration. If the decision is about opportunity, our normal inclination is to care a lot about *the best outcome that I can realize if I do the thing*. Often, we do care about what might happen *if I do nothing*. However, we really should know the three other alternatives (1) *what is the worst outcome if I did something*; (2) *what would the best outcome be if I did nothing*; and (3) *what would the worst outcome be if I did nothing?* The *four-corners tool* is made to order.[6] It is designed to balance a plan that allows either action (decision to do something) or inaction (decision to do nothing) in a way that sharply exposes the relevant choices and consequences. Figure 6-4 illustrates the framework. A question is posed for each of the dots in each corner. Understanding the answers for all four questions leads to the best answer. Here is how it works.

---

6   Ben Carson, radio broadcast interview, National Public Radio, ca. 2009. *Author's note: I have been unable to locate even a semblance of a reference follow-up for this tool. I include it because it is quite valuable; without it something is lost. My apology to the original source.*

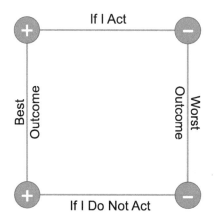

Figure 6-4. Four-corners decision framework that forces a proper balance between what might be good or bad for both the act of making a choice and the alternative of making no choice. With this information in hand, an informed decision is likely.

Acting, Outcome Questions

Along the top edge of the square – If I do what I am thinking of doing:

Q-1. What is the *best* that can be expected to happen?

Q-2. What is the *worst* that can be expected to happen?

Not Acting, Outcome Questions

Along the bottom edge of the square – If I do not do what I am thinking about doing (meaning probably doing nothing):

Q-3. What is the *best* that might be expected to happen?

Q-4. What is the *worst* that might be expected to happen?

Discussion

Choose the action (corner) that provides the right one in light of all possible four choices. This is a rather simple methodology. It can be very effective. Using it invites a look at all four logical outcomes. With the best of the best and worst of the worst known, one should have enough information to weigh the value of the best against the damage of the worst. Notice that the best is not necessarily at any specific corner. Nor is the worst. There is no need for any probabilities for the outcomes, so please do not estimate or assume any—it will certainly defeat the objectivity of the process. Be comfortable with reasonable knowledge. This should go a long way toward maintaining a balanced consideration. Going to all four "corners" tends to reduce bias or preconception. Sure, any biased corner is going to be visited. But the rest are examined before a conclusion is reached. Considering each part (corner) of an alternative one at a time provides an excellent procedure for others to participate with both the outcome predictions and the selected decision. In summary, the four-corners tool provides a compact and logical way to visit and document the process and decision.

6.12 Close

Some impending failures of situation management announce themselves. If we are fortunate, there are alarms on the screen. Those alarms might occur too often or too quickly when things go wrong. When many alarms activate, it is rarely clear what to do about them. Sometimes, the alarm system itself contributes to poor upset response. If we do not get alarms or are not able to see something amiss, bad things can, of course, still happen. Operators need effective ways to discover things going amiss in time to do something to remedy them. Successful situation management comes from enterprises having appropriate infrastructure and sticking to their operational protocols and practices—every day, all the time. This chapter has introduced why. The following chapters provide tools to do that job well.

6.13 Further Reading

Endsley, Mica R., Betty Bolté, and Debra G. Jones. *Designing for Situation Awareness: An Approach to User-Centered Design*. Boca Raton, FL: Taylor & Francis, 2003.

7

Awareness and Assessment Pitfalls

*"People look without seeing, hear without listening,
eat without awareness of taste, touch without feeling,
and talk without thinking."*

Leonardo Da Vinci (Renaissance Man; Italy 1452–1519)

This chapter provides insights rooted in anthropology, psychology, and sociology. Please do not think that because these insights are not rooted in engineering, they are not powerful and important. As you read about mental models, doubt, "how we think," and biases, you may wonder if such material belongs in a practical technology reference book about control room management. It does. These insights can be just as important as traditional design and engineering guidelines. In fact, social scientists have long been involved in investigating the many psychosocial contributors to disasters. They devote time, money, and reputation to this work. From their investigation, we gain important insights into understanding and managing such disasters. Drawing on their analyses and applying their lessons can help us avoid embarrassment, increase job satisfaction, and dodge or reduce the frequency of disasters. All are worthy goals. Each one gives us a heads up about what often gets in the way of successful *situation management*.

The material in this chapter speaks to the heart of situation management. And it respects the very nature of human nature. Operators are individuals with distinct personalities, histories, aspirations, preferences, dislikes, worldviews, and a host of other traits and opinions including having no opinion whatsoever. They perceive the world and who they are in their own personal ways. But their individual parts are not the only things they bring to the table. They also bring along deep-seated effects of our rich human nature past. Over the millennia humans have evolved important traits that have protected us and allowed humanity to evolve to where it is today. One of these traits is that when faced with a situation, we instinctually handle it based on our expectations and experiences to work out how the present situation matches what we expect and have experience with understanding and handling. This is why we immediately move out of the way of a falling object. No time is lost by considering exactly how far away it may be or how heavy or dangerous it might be. We just get out of there fast. But, just because it is in our survival nature to always compare what we see and feel to that vast yet mostly hidden instinctual "database" does not mean it works everywhere. One place where it does not work is in the control room.

Even though the control room has been designed to be an optimized workspace; the plant or operation has been designed, built, and maintained to high standards; the procedures and documentation are effective and accessible; the task and competency training of the operator is top-notch; and the personnel practices are effective to ensure readiness and fitness for duty, unless the individual operator (or supervisor) properly uses this infrastructure, good outcomes will not predictably result. As you will see in Chapter 8, "Awareness and Assessment Tools," control room management works very hard to provide appropriate warning and enough time to manage situations where operator management is reasonable, and it designs the rest to fail to safe. Proper use depends on two critical expectations:

1. Appropriate peopleware and technical infrastructure that are carefully put in place to accommodate the honest differences and likely foibles of our human nature.

2. The designer, operator, and supervisor who are carefully trained to be aware of how thought processes and natural biases will get in the way of careful judgment and deliberate actions, and then to work diligently to minimize these disruptive effects.

Our individual differences in thought and logic are important cultural distinctions that profoundly shape our well-being and life. These differences should not be perceived or judged as being better or worse than any other. Logic and truth are not universal, nor universally sought or praised. Behavior in the control room must accommodate the culture and norms of the community without sacrificing proper enterprise-responsible operation. Seeking the goal of safe and respectful operation is the best path toward universally understood good. In this chapter you will explore individuals' differences, foibles, and seductive natural biases. You will be able to better understand them and work to manage their untoward interference in the control room. It is that important. Let us take a look.

7.1 Key Concepts

Design	Procedures and policies must specifically take into account all the naturally occurring intuitive foibles and limitations of the human operator.
Doubt	No matter how thorough the training, no matter how important the job, and no matter what the reason, if an individual does not believe something, and proper action based on that belief is required, it will not happen.
Multitasking Is a Myth	There is no such thing as multitasking. Any time the thinking part of us must be engaged, it can only focus on one item at a time. Even if that other focus is devoid of emotion or lightly engaged, focus will not split. Multitasking should not be a permitted behavior in the control room.
It Is Judgment	Differences in thought, logic, and life are important cultural distinctions that profoundly shape well-being and lives. These differences should not reflexively be perceived as being better or superior to any other.
Which Time Zone	Logic and truth are not universal, nor universally sought or praised. All expected behavior in the control room must be gained in a way that accommodates the culture and norms of the individual and community *without sacrificing proper enterprise responsible operation*. For example, not all cultures accept a scientific basis for reason.
Danger of Fear of Failure	When fear of failure is perceived to be more important than the actual consequences of failure, operational failures will almost always result.
There Is Only One Logic in the Control Room	In the control room, it is essential that shared successful operating principles and beliefs be developed in a way that is fully competent, understood, accepted, and practiced by operators, supervisors, and managers.
We Look without Seeing	We filter everything our eyes take in through our mood, our culture, and other distractions in view. This has the powerful negative effect of blinding us to almost everything else.
The Reality of Complexity	As soon as we suspect something is complex, we most often balk at "going on to find out really what it is all about," and instead fall headlong into our biases and preconceived notions.
Managing Cross-Purposes	If two institutional goals compete with one another, such as productivity and safety, the higher goal must be clearly expressed. Otherwise, you risk control room chaos. Or, if an individual feels threatened by a loss of security, earning potential, or prestige by divulging rare or hard-won knowledge, key information might not be shared with newcomers.
Our Foibles Are Always Present	No matter how well the equipment is designed, how effective the procedures and operating instructions are, how well trained operations personnel are, or how fastidiously equipment is maintained, unless all of this is properly utilized by operators, successful outcomes are at risk.

7.2 Introduction

Your operator is on shift in the control room. Good operation relies on good situation awareness to expose operational irregularities. Awareness is the state or ability to perceive and feel so that we are conscious of events, objects, or sensory patterns. In this level of consciousness, an observer can confirm data without necessarily implying understanding. Understanding comes later. Patience. *Awareness* is defined as perception and cognitive reaction to a condition or event. This awareness can be the process of observing; actively seeking; or just a feel, often referred to as our *sixth sense*. Awareness provides the "data" for our operator to use. Situation assessment's responsibility is to produce judgments about the operational condition of the production in his care. Situation assessment is what operators do with the "data" of situation awareness.

We now look into what might get in the way of situation assessment going well. The issues and concerns are as real as incidents and accidents. Few major incidents occur without one or more of these impediments and limitations being the root cause of mismanagement into disaster.

7.3 Readers' Advisory

This chapter contains material that goes to the heart of how individuals are profoundly influenced and guided by their beliefs. These beliefs often come from cultural heritage, geographic locale, and other respectful influences. As you read, please understand there is no intent to judge any belief or personality. As members of the community of the world, we are all important and worthy.

A reader may be tempted to find an aspect of culture or personality that may get in the way of successful situation management. It would be most unfortunate if consideration of fitness for operators were based on any material here. This is not the intent, nor is it in any reasonable way necessary or responsible. The express purpose of this discussion is to advance understanding and practices of effective ways to manage important (operator) situations. In doing so, this book attempts to provide honest, respectful, and potentially valuable insight into ways of guiding the professional behavior of operators.

7.4 Why We Make Mistakes

A big mistake is to think that we do not make many mistakes. An even bigger mistake is to avoid understanding why we make mistakes. A way to introduce this concept and make it personal is through what I call "traveler's immunity." It is no accident that when people are out of their element, such as on a trip away from home, they often do more daring things. Why is that? Sure, there is a certain freedom that comes from getting out of regular habits and away from the daily grind. And that is a part. But dig a little deeper and something important comes up. Being away from the familiar seems to grant us special powers over physics. The ground there cannot be as hard as at home, the water cannot be as treacherous as the water at the local beach, the roads are not as dangerous as back home. It is as if being away can convey a special immunity from the ordinary dangers at home. Sure, the "back home" dangers are really back home. But physics is the same everywhere.

Let us examine the "physics" of why we make mistakes. The best list I can think of comes from Joseph Hallinan:

- We look but do not always see.
- We all search for meaning (before understanding).
- We connect the dots (even when they are not connected).
- We wear rose-colored glasses.
- We can walk and chew gum—but not much else.
- We are in the wrong frame of mind.
- We skim.
- We like things tidy.
- Men shoot first.
- We all think we are above average.
- We would rather wing it.
- We do not constrain ourselves.
- The grass *does* look greener.[1]

This chapter is all about understanding and building in ways to keep the items in this list from getting too much in the operator's way.

---

1 List adapted from the table of contents from Joseph T. Hallinan, *Why We Make Mistakes* (New York: Broadway Books, 2009).

Looking without Seeing

The bottom line is that operators use seeing as the overwhelming source of information for keeping track of things. Yes, there is an alarm system intended to interrupt business as usual and focus on an abnormal situation requiring immediate attention. But if the operator were able to see things going awry early enough, most alarmed abnormal situations could be worked on before the alarm activates. And for all the rest, we depend on the operator seeing it on the HMI screen or on laboratory reports and the like. The "uh oh" part of this is that it is really hard to look and see what needs seeing. Our eyes and brain are not set up well for this seeing part. So seeing can be really problematic. If nothing else in this book seems useful to you, this should be. Understanding and managing how the operator sees (because we cannot actually fix it) is why we spent time on HMI design in Chapter 5, "The Human-Machine Interface," and everything that goes into carefully building HMIs.

Let us review a few of the important design aspects:

- Information displayed for the operator must be for purpose—everything must be needed; observe it all.

- Displayed information must be placed in context—to facilitate needed associations.

- Information must be in a form that is usable—jobs that need doing are supported by where the job is located and how the information is shown.

Before leaving this section and the importance of "looking and seeing," let us prepare for the important ways for the "looking" part to work better. Operators are expected to be able to find situations before they become problems. Yet human nature gets in the way. So, we will need to find ways to look that can make the seeing work. What does not work well is the old-fashioned method of "looking around." "Keep an eye out" is not a good work plan. Using current best operator practices should help turn this around. Chapter 8, "Awareness and Assessment Tools," Chapter 9, "Weak Signals," and Chapter 10, "Situation Management," will show you the tools and how to use them.

7.5 Dangers from Automation

Automation is a powerful technology that enables operations to be done more safely, more reliably, and more efficiently. That's all well and good. On the other hand, where automation is hand-in-hand with human operations, certain dangers are exposed that must be recognized and managed. These activities mean that the design of the automation requires care to ensure that operators remain fully engaged, and specific procedural requirements must be established to improve and maintain operator engagement. Otherwise, the unintended consequence will be an operator who is more of an observer than a participant. This topic is not intended to be lip service or boilerplate. We are talking about real concerns with real needs.

Nicholas Carr has exposed these unintended dangers nicely.[2] The following topics raise real concerns that you should understand and provide meaningful ways to avoid. The discussion that follows uses Carr's terminology and includes a few of his examples.

The Substitution Myth

What if we wanted to change one small thing? If we could change it a little bit, making the change would not cause any undue effects anywhere else, would it? So, changing might be a way to go. In the front of our mind (as opposed to something in the "back of our mind") is the expectation that this change will not get overblown. Why should it? After all, it is just a small part. Substitution is the act of replacing one thing with another with the expectation that both are equivalent. The myth is: Whenever you automate any part of an activity, you are not just making a change; the act of automation fundamentally changes the broader activity. Small changes can have profound effects. These changes must be understood and accepted before being implemented.

Following that line of thinking, we might think about "bending a rule" here or there; for example, forgetting to report a bit of income on a tax form or overlooking someone shoplifting. We can and sometimes do. Let us leave the moral and legal issues behind; they are not where this topic needs to go. Where this wants to go is three other places:

1. How large is the sensitivity to the change or degree of importance of the effect of the change? Often, perhaps more often than not, small things have small importance, but not always—not always enough to ever be counted on. According to Carr in *The Glass Cage: Automation and Us*, more often than we can foresee, small changes introduce important, sometimes fundamental, effects.[3] Although each change in the following list may seem small, its effect can be significant, but not all the time:

   a. Not fully reading a procedure

   b. Skipping a step in the procedure

---

2 Nicholas Carr, *The Glass Cage: Automation and Us* (New York: W. W. Norton & Company, 2014); Nicholas Carr, "The Glass Cage: Automation and Us," YouTube video, 55:54, from a lecture to Google employees on October 8, 2014, posted by "Talks at Google," October 14, 2014, https://www.youtube.com/watch?v=Mt8ooCms4sE&feature=youtube_gdata_player.

3 Carr, *The Glass Cage*.

c. Overlooking an unexpected result

   d. Failing to look for problems or issues

   e. Overly relying on technology to the extent that some core skills are lost

2. What is the follow-on or scale effect? Sure, only one small deviation might not be so important. But what if everyone were to do it? Now, there are potentially lots of small things going amiss. Few of them are the same or even similar. There are going to be extensive effects. Few systems can tolerate such a great amount of cumulative error or straining without faults.

3. Because many individuals are seemingly getting away with doing those small changes, it breeds the expectation that no one must be careful because everyone is doing it. Soon we legitimize the culture of ignoring, changing, or adding. All this makes it difficult to maintain protocols, procedures, and the expectation of being careful. In the end, it undermines much of the important infrastructure designed for safe, responsible, and productive manufacturing.

All in all, operators need to exercise more than ordinary care any time they are thinking about substituting anything unproven into their work. This is not about creativity. This is not about rote operating. So yes, operators are asked to bring their wealth of experiences and skills to the job. In doing so, they must be sure about what they are doing, have a reason for doing it, and give thoughtful consideration to how to do it and what might go wrong. This is precisely why procedures are so important. It is why you will want to develop operator competencies rather than just skills.

Automation Complacency

Keeping an operator engaged and aware is important where portions of operations are automated and other parts are not, and we expect the operator to make sure everything goes well. Yet it is human nature for individuals to do just the opposite. The simple act of turning over a task to technology encourages "tuning out." Operators assume that the automated part will go well, so they only need to attend to the parts that are done manually. A good illustration is the automatic spelling checker and correction in word processing software. We have all discovered nonsensical corrections that were done automatically. Some were amusing; some embarrassingly changed the entire meaning. A "fix" would be to ensure that all automatic corrections were flagged (by color change or other clear indication) until accepted by the writer.

In the control room, automation must be accompanied by clear indications of what the automation is doing and how well it is being done. This means providing the operator with meaningful and readily understood feedback not only on whether the task is proceeding normally, but also on what the task is doing, how far along it is, when it will be done, and where the process is expected to be at completion. Procedures need to be designed with all of this clearly in mind.

A simple feedback control loop provides an interesting illustration. First, the operator must know whether a variable is shown to him as a measurement only or whether automation manages it. We would rely on the graphical style guide for that. See Chapter 5, "The Human-Machine Interface," for examples. Second, when it is automated, the addition of an icon depicting the current position of the controlled variable would provide a good feel of how the control function is operating. A valve position indicator comes to mind. Broad experience has shown that a valve operating between 30% and 70% of range is normal. Outside the normal position suggests something needs looking after. Here we rely on the design functionality of the HMI graphics to keep the operator engaged and aware because both his manual operations and the automated ones are visible and at hand at the same time in the same framework.

Automation Bias

If automation complacency were not problematic enough, automation bias plays an even more damaging role. Where automation is actively working, the user tends to place too much confidence in it. So much so that it diminishes the operator's engagement in other surrounding tasks requiring attention and response. In effect, the operator gets lulled into a false sense of trust where he assumes that the automation will take care of everything. Unfortunately, that means that he fails to pay attention to other visible warning signs and indications surrounding the operation. For example, the risk intended by an alarm going off is downplayed because it is assumed that the automation will correct the situation. Consider another example taken from the widespread use of global positioning systems (GPS) for vehicles. It is not uncommon for a driver to be so trusting of the directions provided by the GPS that he completely misses seeing and understanding other traffic warnings. For example, a truck might crash into a low overpass despite several warning signs and sometimes flashing lights indicating the problem ahead.

Automation bias also describes the effects that automation has on the development and maintenance of skills that the automation replaces. The disastrous crash of Air France Flight 447 (Chapter 10, "Situation Management") illustrates this. The Airbus A330 possesses an unusually high degree of flight automation. Air France pilots were prone to utilize the automation extensively. They used it so much that their skills of situation awareness and manual response to flight operational needs were considerably diminished. When the autopilot failed, the flight crew was not able to identify critical dangers nor place the aircraft in a safe flight configuration. This is why certain industries are required to operate for limited times under more manual operator supervision so their operators maintain adequate skills. Refer to the example in Chapter 2, "The Enterprise," on the automated plant.

The Generation Effect

The *generation effect* explains how important it is to engage in active understanding. It is called the generation effect because it refers to the phenomenon that something is better remembered (and understood) if it is generated using one's own mind rather than simply being heard or read.[4] This would mean that while learning a procedure, for example, we need to understand why each step is done, why it is done in the order prescribed, and how to make sure it is working successfully. Moreover, the procedure would be explained so that it fits into a larger competency goal rather than working to master the rote performance of the steps.

An important message here is intended for the designers of tools and aids for operators. It involves providing analysis suggestions. It is better to suggest alternatives for the operator to consider rather than provide explicit analysis or directives (unless this specificity is required for safe operation). This way, the operator is encouraged to fully examine the situation and use experience and judgment to better understand it. After all, if the situation is rigidly clear from the outset, an operator might not be needed—the necessary action could have been automated.

7.6 Mental Models

A mental model might seem to be a bit too formal to describe what this section is about. But it will do the job. Simply stated, this discussion is about what each of us thinks about what we are doing and how the way we think affects what we do and how we do it. That is probably obvious. Less obvious is that our individual ways of thinking may make it more likely that we fail. Sometimes, when what we are doing is critical, that failure can be awful. Other times, it can be a bit embarrassing, or simply humorous. Here we examine some of the critical mental model issues that operators (and all of us) face. All are real. All can be controlled and managed if the enterprise designs for them.

The section develops with a discussion of several real incidents. None of this is abstract.

---

4 "Generation Effect," Wikipedia, last modified February 26, 2018, http://en.wikipedia.org/wiki/Generation_effect.

Expected Roles

Control rooms do not operate in isolation. Although they may be remote, manned by a single individual, and rarely supervised or questioned, they are an integral part of a plant or operation. There exist managers who conceive and provide guidance and wherewithal. There are engineers and technologists who design and maintain. There are suppliers and customers. There are operations team members who keep a watchful eye and a guiding hand.

Even within a single control room, at any point in time there may be several board operators, operator managers, supervisors, trainers, and other direct operations personnel. Each person has responsibilities and authorities. Each individual does what he thinks he is expected to do. The others count on each other that way. But what happens when individuals do not act according to their immediate responsibility but shape their actions based on what others expect of them, or base their actions on what they think others may think about what they are doing? For social settings, that is personal. In a control room, it invites disaster. Come aboard Asiana Flight 214.

Asiana Flight 214

Late on the morning of July 6, 2013, in clear weather and excellent visibility, a Boeing 777-200ER airplane owned and operated by Asiana Airlines and designated as international Flight 214 was making its final approach to land at San Francisco International Airport.[5] The flight originated from Incheon International Airport in South Korea. There were 291 passengers and a crew of 16 on board. At approximately 11:28 a.m., the plane crashed short of the runway after the tail section struck a seawall. Both engines and the tail section separated from the aircraft. The airplane eventually came to rest nearly a half mile from the first impact point. A fire resulted (caused by engine oil, not jet fuel) with damage but no explosion. There were 3 fatalities (one person was accidentally run over by a rescue truck after being thrown from the plane on impact) and 181 nonfatal injuries (see Figure 7-1).

Three captains and one first officer were on board the airplane. At the time of the crash, an experienced captain was acting as copilot and instructor (3,220 hours experience in a Boeing 777; sitting in the right seat); another captain acting as pilot was receiving initial training (43 hours experience in a Boeing 777; sitting in the left seat). The first officer was sitting in the cockpit jump seat (a spare, not normally occupied). The flight crew was Korean. At the time of the landing, the instrument landing system

---

5 "Asiana Airlines Flight 214," Wikipedia, last modified June 9, 2018, http://en.wikipedia.org/wiki/Asiana_Airlines_Flight_214.

Figure 7-1. Asiana Flight 214 July 2013 crash caused by the flight crew failing to collaborate regarding approach problems due to cultural customs overriding safe flight requirements.
Source: National Transportation Safety Board.

(ILS) of the airport was not functional, thus an ILS approach for landing was not possible. The Asiana flight crew was aware of this and had been trained to land without it.

Setting Up the Roles
At the time of the crash, the airplane was descending at too slow an airspeed and at too low an altitude. During final approach, cockpit voices were heard. The first officer announced "sink rate" as a notification warning that the airplane was descending too quickly for conditions. In the seconds before the crash, voices in the cockpit were heard to announce "go around" at two different times, but it was too late. At investigation, including flight data and cockpit voice recordings, it was ascertained that the "flying pilot" (left seat) felt the airplane was descending too quickly for conditions but deferred from taking any action, making statements, or asking questions that might appear to take seniority away from the "pilot in command" (right seat), even though as the flying pilot he was at the primary controls, had final authority, and should have taken independent action.

Takeaways
First, there is the cultural issue of failing to take charge. The pilot seated in the left seat of the airplane was flying the airplane. He was solely responsible and was required to act on his own recognizance. On Flight 214, he failed. He functionally deferred all

corrective action to his mentor sitting in the right seat. This was the copilot seat, regardless of the qualifications of the person. A responsible pilot would have taken action and modified it only if questioned by the mentor, or at least at a moment of uncertainty, asked for guidance. Sadly, this pilot waited for instructions, assuming that the current status was one that was understood and approved by the mentor (even if the mentor was acting as copilot). If that deference proves uncomfortable, so much better the lesson.

Second, the important role issues relating to this incident are defined and comprise a required protocol by the US Federal Aviation Administration (FAA) requirements for cockpit collaboration. Even where the "pilot in command" is actually at the primary controls, if anyone else in the cockpit makes an observation or conclusion that affects flight safety or integrity, the observer is obligated to announce it to the crew, and the command structure is obligated to understand and consider. There are no exceptions. The "sink rate" warning by the first officer in the jump seat should have triggered an instant reevaluation. For Flight 214 it did not.

Failure Avoidance

It is mandatory that an enterprise set standards for performance. Within reason, considerable leeway is granted for what those standards require. It is left up to custom and the culture of the enterprise to encourage compliance to the standards and manage situations of noncompliance. History and human nature strongly suggest that where a responsible balance between meeting standards and not meeting them is not achieved, disaster thrives. To clarify this message, when fear of failing is perceived to be more important than the full consequences of an actual failure, operational incidents will almost always result. These failures are often tragic. Consider the case of the Amagasaki rail crash.[6]

Amagasaki Rail Crash

At approximately 9:19 a.m. local time on April 25, 2005, a seven-car commuter train traveling at high speed around a pronounced curve derailed and struck an adjacent apartment building. The crash occurred on the JR West Fikuchiyama Line in Amagasaki, Hyogo, in the area of Osaka, Japan. The driver and 106 passengers were killed. Most of the deaths resulted from highly crushed passenger cars (see Figure 7-2). The cause of the crash was attributed to the train traveling at excessive speed for the track configuration. The single driver was experienced with this route and schedule. The maximum speed allowed in the curve was 70 km/h (43 mph). The actual train speed entering the curve was 116 km/h (72 mph). Engineering calculations determined that derailment would likely occur (with trains of the same configuration) at any speed above 106 km/h (66 mph).

---

6 "Amagasaki Rail Crash," Wikipedia, last modified May 21, 2018, http://en.wikipedia.org/wiki/Amagasaki_rail_crash.

426   Situation Management for Process Control

Figure 7-2. Aerial view of the April 2005 deadly Amagasaki, Japan, rail crash site caused by excessive speed due to the train operator trying to avoid excessively harsh punishment expected to come from minor operating delays on the trip.
Source: Reproduced with permission from Kyodo News.

The train driver passed a red light earlier in this run without stopping. This initiated an automatic train stop, which is a reportable event on his driving record. It caused a schedule delay. At an earlier stop on the same run, the train had overshot the pedestrian platform. Correcting for this added an additional 90-second delay. The acceptable schedule delay at the next destination station is less than 28 seconds (of a total trip travel time of 15 minutes). This requirement had been set due to platform sharing between trains and increased loads at passenger stations.

There were no automatic speed-measuring devices at the curve and no automatic external train-stop mechanisms. Just prior to the crash, the driver applied the service brakes instead of activating the emergency brake.

Setting Up the Fear
At the time of the crash, the driver had accumulated an 11-month unblemished record. Before, however, the young driver had faced harsh punishment: financial penalties and

months of an intentional demeaning punishment retraining program called *Nikkin Kyoiku*. The initiating event for the punishment had been his overshooting a station by 100 meters during his third week of employment.

> Drivers face financial penalties for lateness as well as being forced into harsh and humiliating "Nikkin Kyoiku" retraining programs. The final report concluded that the retraining system was the most probable cause of incident. This program consisted of [sic] violent verbal aggressions, forcing the employees to repent by writing extensive reports. Also, during these times, drivers were forced to perform minor tasks, particularly involving cleaning, instead of their normal jobs. Many see "Nikkin Kyoiku" not as a real retraining program, but as a draconian punishment and psychological torture.[7]

The Amagasaki investigation noted that it was likely that the driver, fearing a return to punishment, became so focused on the schedule that he failed to properly monitor track conditions, thereby operating the train at too high a speed in an attempt to make up the lost time. When the speed and curve were eventually noticed, the driver avoided using the emergency brake, as it would constitute another reportable breach. He had already accumulated two breaches on this single run.

Takeaway

The fear of punishment in this case was driven by its extreme nature. However, please consider that the issue here is not the nature or fairness of the punishment. The important point is that punishment fear, or even peer-pressure fears, can distract operators from their duties to the point that they are unable to focus on the job at hand. At this juncture, plant or operation activities are at risk for any eventuality that circumstance presents. The issues are both a failure to act when action is required and taking action inappropriate for the situation at hand.

Standards are necessary, and proper monitoring and enforcement are part of it. However, care must be taken to ensure that perspective is maintained at all times. This is a cultural matter to manage, not a standards issue.

Logic-Tight Compartments

Humans may not think consistently. All too often we find that in one part of our lives we carefully measure our options and deeply ponder which decision is the right one. We use logic and consideration and try to weigh all the necessary factors—but not always. Yet, in the stroke of a single moment, we are prepared to accept beliefs that logically we should have no business believing. There should be no reasonable way

---

7 "1,182 'Retraining Sessions' at JR West," *The Japan Times Online*, June 29, 2007, https://www.japantimes.co.jp/news/2005/07/23/national/1182-retraining-sessions-at-jr-west#.WyPZyFVKiM8.

that a thoughtful individual could honestly believe it. Any real belief in it would have to deny compelling evidence and distort logic. We have identified a *logic-tight compartment*.[8] These compartments are the "parents" of confirmation bias.

An amazingly clear example of such a compartment is the story of how the true cause for most stomach ulcers was discovered.[9] In 1982, Australian physicians Robin Warren and Barry Marshall first identified that the human stomach was host to a bacterium called *Helicobacter pylori (H. pylori)*. Moreover, this bacterium was shown to be present in the majority of people suffering from stomach ulcers. The physicians' patients responded remarkably well to antibiotic treatment. Not only did the medical community ignore the physicians' work for nearly 12 years, but they also ridiculed the physicians. The main argument against the physicians' findings was that the human stomach could not possibly be the site of any such bacterium due to its high acidic nature. It took the medical community until 1997 to fully embrace that most ulcers could be successfully treated with a common antibiotic.

Logic-tight compartments are so named because they act to close the door to anything different than expected. These compartments can be so strong that they can prevent all consideration of facts and events that in any way contradict current thinking and beliefs. Logic-tight compartments can be formed when a new scientific theory contradicts long-standing personal beliefs. They can result when an operator receives a new set of tools to use to do old tasks. Sometimes when a contradiction becomes evident, the dissonance can be so strong that normal response shuts down. Just when action is needed, the operator becomes paralyzed. All of this is not conjecture. In the face of compelling evidence, the failure to consider, failure to reassess, or failure to act is real and often is the root cause of mismanaged incidents and accidents.

Engineering design and management protocols need to anticipate such risks. First of all, the culture must reject misinformation. Critical judgments and effective action cannot result from attempts to resolve the wrong problem. Procedures must be designed from proven protocols, not anecdotes of a few anomalies of operator practices gone amiss. Incident investigators must carefully study objective reports to identify whether each near miss was caused by misstep, honest error, or logic-tight misconceptions. Take immediate action to correct the situation. Remember, a logic-tight

---

8   Michael Shermer, "The Mind's Compartments Create Conflicting Beliefs: How Our Modular Brains Lead Us to Deny and Distort Evidence," *Scientific American*, January 1, 2013, https://www.scientificamerican.com/article/the-minds-compartments-create-conflicting-beliefs/.
9   "History of Ulcer Diagnosis and Treatment," Centers for Disease Control and Prevention, last modified September 28, 2006, http://www.cdc.gov/ulcer/history.html.

compartment cannot be broken open by force. When coaching confronts a "compartment," carefully and respectfully find ways to gently but consistently open it. The most difficult type of learning is unlearning something and replacing it with a new behavior or even a new paradigm.

We respect the power of logical thinking. It generally is useful and beneficial. Trouble ensues when the logical or cogent evidence of the truth is inconsistent. On the one hand, the evidence suggests one position. On the other hand, the evidence points to a different position. Both positions cannot be true at the same time. We need to decide one way or the other. Faced with this dilemma, we are likely to pick the more agreeable truth. Most will pick the one that represents preconceived notions and what one wants to be true. This is the problem.

Surrogate Models

Surrogate models were first introduced in Chapter 6, "Situation Awareness and Assessment." Here we examine how they would be properly used. A surrogate model is a simpler or more familiar stand-in to help understand something that appears too complex and/or too hard to visualize or understand. A surrogate model is a substitution—something that works much like the original but is better known and more familiar. However, unless we are extremely careful, what we pick may not be a good surrogate. Rather than being useful to aid understanding, it acts as a filter or "red herring" that hinders understanding. For example, if the last shift experienced many problems with a certain electric motor driving a large, important compressor, and on your shift, you see laboratory values deviating, one reflex might be to focus on the compressor as the likely culprit. The compressor problem is selected for what might be going wrong in the unit with abnormal laboratory analysis values. Yet, a moment of thought about what the compressor does and what might do wrong would have quickly ruled it out for the issue of suspect lab values.

The message here is twofold. First, ask why you are considering each surrogate. Rather than settling on one, or working to decide which of several possibilities to consider, quickly list and test all for reasonableness. Here, the burden of proof is on "ruling in" not "ruling out." Each is out until demonstrated in. If any (or even all) are not clearly appropriate, do not use it. Second, as any surrogate is used, ensure that the ever-watchful eye remains on the actual process, not on the surrogate. Watch the lab values carefully and investigate the broader nature of those implications (bad samples, bad laboratory analyses, mistakes in reporting laboratory values versus actual laboratory results, piping leak, etc.). Give up the surrogate as soon as you begin to understand what might be the real problem or issue. Remember that the only purpose of the surrogate is to jump-start problem solving.

The Surrogate Model Test

Surrogate models can be helpful, but they can stymie progress in a big way if they prematurely or unduly narrow your thinking (get ready for *confirmation bias* in Section 7.9). Two simple test questions can be helpful. Look at both the surrogate (in your mind's eye) and what you can actually observe from the real situation and ask:

1. What are the real, specific similarities between what I see and what my surrogate depicts?

2. What are the real, specific differences between what I see and what my surrogate shows?

No matter how many similarities are found (remember that surrogates are always a supposition made in the absence of something more tangible), if any *real differences are present, it is quite likely that the surrogate is inappropriate* and should be immediately discarded.

The Deception of Two Reasons

The "two-reasons trap" is subtle and pervasive. Without being aware, we all too often use it. This trap appears to be subtlety implanted in places we depend on to be reasonable. Here is how it works. If there is only a single reason for something, our natural instinct is to ponder more. Yet when we see or find a second reason, any second reason, to either explain something or to base a decision on, we quickly lean toward accepting the decision. This seems logical. When there is a reason for something coming from one direction, and another reason that appears to be coming from a different direction, and both align, most of us tend to be satisfied that we have sufficient grounds for our decision. This would be true, except for an important problem. The two (or even more) reasons need not be anywhere close to the same weight or importance. Whereas the primary reason may be well founded and convincing, the second one can be trivial or insubstantial. Making matters worse, the second reason is often searched out knowing that one must find a "second" to ensure that what we want to do or believe is well founded. This puts a big crack in the notion of independent reasons. Sometimes even the first reason is shaky. Either way, the trap must be recognized and managed.

The lesson is that any belief must not be decided by counting reasons. It is the quality of each reason, not their number or order, that matters. There is an archaic English legal principle that counted witnesses for and against guilt.[10] The side with the most

---

[10] John H. Wigmore, "Required Numbers of Witnesses; A Brief History of the Numerical System in England," *Harvard Law Review* 15, no. 2 (1901): 83–108, https://www.jstor.org/stable/1323769?seq=2 - page_scan_tab_contents.

witnesses decided the case. Rather, any reason that is well founded and cogent would need its due process. Of course, even an important reason that cannot be confirmed or appears against the weight of the other evidence suggests caution. Your control room management tools will and must be designed to ensure all reasons are reasonable.

Remembering

We all think we are good at remembering. We think of something while doing one task and make a mental note to remember to get back to that thought when we leave the task for another. We examine something and observe several items that are useful. We see something interesting and intend to mention it to another at the first opportunity. We decide to remember; but few really do it well. This is not to suggest that most of us who cannot remember are somehow deficient or flawed. Our inability to readily remember is inherent in the way our minds work. For example, we are engaged in making observations that need to be remembered. We start a conscious effort to remember. The items may be thought of as a mental list, so we review that list in the mind a few times to "let it sink in." So far, so good. But let that mental review be interrupted by another brief but absorbing task, and most of the previous remembering will be lost. Add complexity to that situation, and quickly we revert to preconceived ideas and methods; we stop looking and start concluding and resolving.

The message is that "remembering" cannot be a skill to be relied on. It means that where one piece of information is to be compared to another, all of that information must be kept readily available for viewing at the same time in the same view. Thus, the text for the written procedure must be in view at the same time the operator takes the action step. It means that all task return points need to be readily seen for the operator to view and select them. A "task return point" could be the need for follow-up monitoring of an activity started earlier, an alarm that has not yet been handled, or a series of tasks to be done before shift end. Regardless of the nature of the tasks, all should be clearly but not invasively available for pickup by the operator without the need to remember what they are or how to find them.

Good Is Not Really Good Enough

We all try to be right. We know that no one is perfect. But how big is the difference between good enough and perfect? It turns out to be bigger than most of us think. A "rule" called the *80:20 Rule* illustrates this concept. It is also called the *Pareto Optimum*.[11] Figure 7-3 depicts this process.

---

11 F. John Reh, "Understanding Pareto's Principle – The 80-20 Rule," accessed August 8, 2018, https://www.thebalancecareers.com/paretos-principle-the-80-20-rule-2275148.

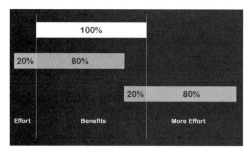

Figure 7-3. The 80:20 Rule (effort versus benefit view) illustrating the seductive nature of early benefits that seem beneficial yet are not enough for responsible work. While the first 20% of the effort yields about 80% of the benefits, often the real and lasting value is derived primarily from the last 20%. Gaining that requires an extra 80% of effort.

The 80:20 Rule goes like this: for any task, getting 80% of the reward (that doing a proper job will provide) takes only 20% of the effort (needed to do that proper job). As if that were not enticing enough, earning the final 20% of the reward requires 80% more effort. It is no wonder everyone would like good to be "good enough." Yet the reality of most things that we do in the control room is that they need to be done properly. In the control room, good is not good enough. There is too much at stake. At all times in the control room, there is much that needs doing that must be done as reasonably and as close to exactly right as possible. Control room management is about finding what is needed to do a proper job and doing all of it. Certainly, the efforts will be short of perfection. But the tasks demand well above "just good enough." Each enterprise will decide exactly how far it must go and build into its infrastructure the expectation and tools for everyone to do it that way, all the time.

7.7 Doubt

Let us get this off on the right foot. Doubt is an important tool. Overconfidence is one of the leading causes of mistakes. Doubt is a safeguard. Like any safeguard, we want to use it properly. No matter how extensive the training, no matter how important the job, no matter what the reason, if an individual does not believe something, and acting or not acting on that belief is required, proper consideration will not happen. We must ferret out doubts to keep them from being a barrier to good operation. Remember Piper Alpha (Chapter 1, "Getting Started"):

> The manager of the second offshore platform kept sending product to Piper even though he knew that there was a fire emergency on Piper, because he assumed that Piper would bring the situation back under control.[12]

---

12 Lord W. D. Cullen, *The Public Inquiry into the Piper Alpha Disaster,* Volumes 1–2 (London: Her Majesty's Stationery Office, 1990).

This lack of action contributed to the extent of the damage and probably added to the loss of life.

It has been said that human beings are hardwired to both have no doubts and believe nothing until proven. It seems like we cannot make up our minds. On the one hand, we want to be certain enough about conditions to not be constantly rethinking and reassessing everything. That requires much effort and is difficult to do. Sometimes the situation does not provide any time for that. On the other hand, we do not want to march headlong into a disaster that we should have seen coming but we avoided action until the last minute. So, we keep checking as we march even farther. When we can find a workable way to do something, we are able to balance "knowing" with "doing." The next sections delve even further into how doubt works. We will not pave a new road in psychology; we will just take a short look.

Possible Doubt

Make a list of all the possible reasons for something not being believable, and you have fulfilled the exercise. Sure, some of the reasons might be a bit of a stretch, but avoid the really far-fetched. If any reason fits the current situation, you have possible doubt. It is important that possible doubt concerns are not dismissed out of hand. One of the ways to keep this under control is to realize that weak signals (see Chapter 9, "Weak Signals") usually fall into this category.

Probable Doubt

Start with the list of possible doubts. From the list, pick the few doubts that are more likely to be true than not. They are termed probable doubts. Probable doubt is 51% certain. Prune away all doubts below a 51% likely-to-be-true line. We are looking for what now seems to be significant but that we cannot yet fully believe. What we have is a list of doubts that, if any might fit the current situation, we need to evaluate. In the control room, an operator with probable doubt who senses something being amiss must follow up. In general, these situations are abnormal ones. We are beyond weak signals; now something is likely amiss and needs discovery.

Reasonable Doubt

Reasonable doubt is the benchmark that captures both intelligence and street smarts. It is used in the United States as the legal test of criminal guilt "beyond any reasonable doubt." Although, to be truthful, many court juries thoroughly confuse reasonable doubt with shadow of a doubt—much to the loss of justice. In technical terms, reasonable doubt is 95% certain. There is a bit of room for being uncertain, but one feels that anything there would not be reasonable enough to consider.

Asking operators to only act (or to actively reject action) based on needing confirmation to be "beyond a reasonable doubt" is too high a requirement. There are no effective and timely mechanisms to enable timely operational decisions to be that confident.

Asking operators to only act (or to actively reject action) based on "beyond a reasonable doubt" is too high a requirement. There are no proper mechanisms to enable timely operational decisions to be that confident. As soon as probable doubts are evaluated and found to be well above the 51% point, they are reasonable enough. Each must be fully processed to decide to act or not. The result only must be credible. Any remaining uncertainty must be understood. This is a good place for collaboration and escalation. The constant and reliable safeguard is to always rely on "permission to operate" protocol (see Chapter 10, "Situation Management"). As you will see there, this is the written rule that keeps everything in the safe-operational frame.

Shadow of a Doubt

Remember that first list we made of all possible doubts? Some may have been far-fetched. Some may have even been very unlikely. Now go further and add to that list even more far-fetched or unlikely items (short of aliens and time travel). The result will be a long list, but some of the items could be doubtful. But doubtful is not enough. It is unreasonable to think that everything would be known beyond a shadow of any doubt. Actions that require knowledge beyond a shadow of a doubt are not useful actions. "Beyond a shadow of a doubt" is perceived as 99.9% certain. Try to take this extreme degree of certainty off the operator's table. No operational situation should be required to meet this standard.

Dealing with Uncertainty

Operators' decisions in the face of uncertainty or during unexpected situations are rarely crystal clear. There always is uncertainty. This uncertainty must be accounted for and acknowledged as part of the operating culture. This suggests the following consideration: design and practice open communications during operating events to encourage participants to fully express what they see and understand. These considerations include the following:

- What are the symptoms of the issue at hand?

- What is known and what is the "quality" of that knowledge (probability or certainty; take note of the *information fork* coming up in the next chapter)?

- What is not known but is essential to know, and why?

- What are the clearly perceived operational or other risks being faced?
- What are the reasonable choices, and why are they reasonable?
- What is the safest and most likely to be successfully executed plan?
- What is necessary to execute that plan?
- Do the operators have what is necessary to execute that plan?
- What is the backup plan in case that plan cannot be selected or successfully executed?

The tools and principles in this book are designed to guide us through the list above. Keeping the focus clearly on these points can help minimize guessing about relative importance and discourage jumping to conclusions caused largely by operators filling in missing items with conjecture. By the way, any time that an operator has a plan but no backup plan, it is cause for worry. This type of overcommitment to a single plan demonstrates loss of perspective.

Lingering Doubt

We have all had concerns or doubts that seem to keep lingering. Sometimes they are in the "back of our minds." They might go away for a bit, but they return again and again. We might have trouble "putting our finger on it." These doubts appear real; they appear vague and uncertain as well. They keep hanging around.

A lingering doubt is our subconscious telling us that something needs attention. Something must be decided, understood, examined more, or something else. Lingering doubts are our internal self's way of being exposed to a weak signal. Just when you might have thought that weak signals only belong to the control room, here they are again. If that lingering doubt is personal, so be it. Take it home and work it out. If it is related to the control room, shine some light on it and see if it can be worked using your situation awareness tools. If nothing useful seems to be there and the doubt goes away, that is fine. But if the doubt still seems to be lingering, it must be escalated. Doubt can be subtle and indirect. Doubt can also be useful and important. Work the process.

Managing "Truths"

When you are in a situation where two truths are conflicting, one of those "truths" needs more examination. Or neither is true. Look carefully.

Looking ahead to the discussion of alarm management in Chapter 8, "Awareness and Assessment Tools," you will see how the operator is encouraged to investigate all the possible causes first and to test each separate cause for validity and significance. A clear benefit is to be able to test truths in a structured and unbiased way. Many of the methods are laid out in that chapter. This is preparation for why they are so important.

7.8 How We Decide

We all make up our minds in personal ways. We weigh the pros and cons and compare risks and benefits for the short term and sometimes over the long. Or that is how we think we do it. In reality, we often decide in a quite different way. Sure, sometimes we touch most of the right deciding bases, but most times we do not. Fortunately, it does not matter much most of the time. But in the control room, everything matters. Every decision has the potential to push success a bit more or pull failure too much. How we make both ordinary and the most important decisions is based on largely unexpressed and unacknowledged internal balances between immediacy and the future.[13]

Short-Term versus Long-Term

Each of us is hardwired to value the short-term consequences more than the long-term ones. Instant gratification is a part of this, but to avoid going off on that tangent, we will focus on situation management.

Within a Shift

Think about a typical shift. Increasingly, we might be talking about a 12-hour shift, although 8-hour and 4-hour shifts are not uncommon. During the shift your operator has several different and often competing tasks and duties to perform. He must remain ever watchful for things amiss. He also is expected to attend to all the routine duties and responsibilities: some rather interesting and social, others boring and unchallenging. Everyone understands that all these activities and duties need proper attention and must be carefully and effectively completed. But that *understanding* is implicit. It is on the list, maybe even on an actual checklist.

What really happens during a typical hour is that the operator flows from one activity to another largely based on where his attention at the moment takes him. He might be filling out a requisition for the repair of a missing anti-slip tread appliqué on the step to a coupling inspection platform for a generating turbine. He might be reading the midshift laboratory reports on the current production quality. If these tasks are interesting, he might stay with them longer than needed. If they are uninteresting or

---

13 Jonah Lehrer, *How We Decide* (Boston: Houghton Mifflin Harcourt, 2009).

difficult, the operator will take the first opportunity to find something else to do, just for a moment, intending to get back to the earlier task soon. Short-term interests drive the schedule.

Along comes a real event, an alarm, or other important notification, and everything is diverted to it. That is as it should be. When the current situation settles down, typically the operator picks up another task, one that promises to be more interesting than the last one put aside for the moment. The tasks put aside are not ignored but they likely get less attention than others and are more likely to be performed perfunctorily.

This situation becomes problematic when time is managed poorly and important tasks get short shrift. Plants and operations depend on each day's work contributing to tomorrow's success. One of the important ways to ensure that is via continuous improvement activities. For an operator, this includes taking the time to request that a procedure be updated to include a step he found missing when he performed the procedure in the morning, noting something missing on a screen, or asking for more training or cross-training.

There is another more serious side to this. During the shift, what is going on always seems to take priority. Sometimes that priority is tremendously misplaced with expensive or tragic effects. For example, consider a pump that is run even though it is noisy. Stopping it will involve writing a report and likely shutting down operations for a few hours. The operator might think, "Let the next guy figure this out." Avoidance can be dangerous and expensive.

The takeaway is that explicit operator competency training and expectations must be set. There must be no confusion about what must be accomplished during the shift (identify easily postponed tasks and the tasks necessary for a successful shift) and the quality expected for each. This would be a good time to discourage multitasking. The initial benefit of "keeping both things on the list" might solve one problem by making other problems worse. This also is a good time to stand behind the enterprise's values regarding taking due care of equipment, people, and reputation. It is about setting the right culture, the right priorities. Do it safety. Do it for each task.

Over Many Shifts
Except for weekly or monthly reports and meetings, for the most part each shift looks like any other. We all like predictability. Although the weather might be different, the production and schedule might be different, or the shift might have swung from day to night or the other way round, shifts are more alike than unlike. It is easy

to put off training. It is easy to forget to make and use a personal improvement task list. We might not find enough time to do a walk-through in one part of the plant every week to look for things that seem amiss: missing covers, loose concrete, an odd odor or stain on the gravel, a noise that was not heard before. The bottom line is that if an enterprise needs operators to properly carry out long-term tasks within their sphere of responsibility, those tasks need to be explicitly incorporated into the culture and adequate time and energy allocated for training and task completion. Remember that focusing on the short term at the expense of the long term is a human trait. We are wired that way.

Loss Aversion

Another human trait that must be emphasized is called *loss aversion*. Being averse to loss means to strive so hard to avoid an immediate loss that we lose track of what is really important. Overreacting to prevent the loss hinders a careful examination of the real risks of that reaction. We all have a bit of this. Consider how hard it is to not speed up when you are running late driving to the airport. Going faster would reduce the risk of missing the flight. Missing the flight is a real concern—you are already adding up the inconvenience of changing hotel reservations and trying to reschedule a meeting. You are already feeling the pressure of looking silly or, worse yet, unprofessional. Hit the pause button. What you need to do is stop those thoughts and look at the real possibility of having an accident. You risk serious injury to yourself and others, significant financial loss, and sure destruction of any hope of schedule—to hopefully make a flight. Those who have suffered an accident in such a situation fully appreciate the folly of their decision to speed and wish they could have a "do over."

Not surprisingly, this same force is at work in the control room. Too large a percentage of accidents and incidents have loss aversion as a contributor. Operators who keep running knowing that something is amiss in the attempt to avoid a missed delivery do it. Operators who do not want to look silly in front of their peers or supervisors do it. Operators who take unnecessary risks with production equipment to avoid a shutdown do it. Even worse, other personnel in the enterprise seem to do it and get hailed as heroes until nature or the odds catch up.

To manage this, your enterprise must set clear limits to the exercise of operational risk. To make it work, the enterprise must reward every act that advances that culture. Even if it is shown that such an act might have been unnecessary after the fact, if it was prudent during the fact, it was the right thing to do. The enterprise would want the same decision to be made if the same situation were to recur.

Brazerman Auction

Loss aversion is not so easy to manage. It is always a trap. One of the most memorable illustrations is the *Brazerman auction*. The best discussion of this concept is by Ori and Rom Brafman in their book *Sway*.[14]

> On the first day of class, Professor Brazerman announces a game that seems innocuous enough. Waving a twenty-dollar bill in the air, he offers it up for auction.
>
> Everybody is free to bid; there are only two rules. The first is that bids are to be made in $1 increments. The second rule is a little trickier. The winner of the auction, of course, wins the bill [after paying the winning bid price]. But the runner-up must still honor his or her bid, while receiving nothing in return. In other words, this is a situation where second best finishes last.

Imagine that you are in this class. You can place a bid, up your bid as the bidding progresses, or hold your bid until the end. Once you bid, however, your bid cannot be withdrawn. What are you thinking? Not to worry, you will probably start off your bid as low as you can; two or three dollars if you are quick.

> Indeed, at the beginning of the auction, as people sniff out an opportunity to get a $20 bill for a bargain, the hands quickly shoot up, and the auction is officially under way. A flurry of bids follows. As Brazerman described it, "The pattern is always the same. The bidding starts out fast and furious until it reaches the $12 to $16 range."
>
> Everyone except the two highest bidders drops out of the auction .... Without realizing it [before] the two students with the highest bids get locked in .... Up until this point the students were looking to make a quick dollar; now neither one wants to be the sucker who paid good money for nothing .... They become committed to the strategy of playing not to lose.

Are you still in this game? Every moment you are not the highest bidder, you are not looking at the $20 prize but at your own bid. It is a loss directly from your wallet. You pay in full. You get nothing back for your bid. You see the only way to protect yourself is to change places and be the higher bidder. The bid passing $20 is not even noticed. It does not matter unless you are the loser. Sure, the winner can deduct the $20 from the cost of his bid. You are going to lose it all if you are not him.

> And so students continue bidding: $21, $22, $23, $50 [to shock the other bidder into submission], $100 [he didn't get the message], up to a record of $204 .... Regardless of who the bidders have been—college students or business executives attending a seminar—they are always swayed.
>
> The deeper the hole they dig themselves into, the more they continue to dig.

---

14 Ori Brafman and Rom Brafman, *Sway: The Irresistible Pull of Irrational Behavior* (New York: Crown Publishing Division of Random House, 2008).

If you are looking for a way out when you are one of the two still-in bidders, you will have to change the way you are looking at this. If you are the lower bid, you just stop. You will pay your bid and it is done. If you are the higher bid, you also stop. If the lower bidder ups the bid, you have already stopped. You have capped your loss. Another way out is to change the game from a competition to a cooperation. Both of you agree to stop where you are, split the $20, and pay off your bets. Each payment is $10 less than it would have been!

Anyone who decides that any hole over 2 ft is too deep will always stop. He can always fill in a 2-foot hole. He works this way. If gold, diamonds, or the missing pipeline is less than 2 ft down, he is good. He can dig more. Or he can stop if he thinks he should dig in another place.

Operators do not dig holes! Sure, they encounter them all the time. They always have "shovels," but they are not going to even reach for them. In the control room, operators are going to stop "digging" by the way you designed things. They will follow a clear, effective, and powerful protocol. It is called *permission to operate*. You will see it in Chapter 10, "Situation Management." Take a peek at Section 10.15 if you cannot wait. Do not forget to come back here afterward.

Sixth Sense

This topic under *how we decide* will lead us to a better understanding of an important tool to enhance situation awareness. John Lehrer provides wonderful motivation for this subconscious awareness through a story that takes place during the Gulf War. The following is a paraphrase from his story:

> As part of the campaign to liberate Kuwait, the Coalition had air and naval forces in place. In order to minimize civilian casualties, they were careful to engage the Iraqis outside of clearly urban areas. To do that, strategic air bombing was employed using aircraft-carrier-based planes. Britain provided naval air cover for the operations. Aboard the British ship were sophisticated radar and appropriate surface to air missiles. Monitoring the radar was a grueling and mostly monotonous job. Radar activity is only available to be monitored when the flying objects became "wet," that is, started flying over water. During one engagement, well into his 4-hour shift, one operator noticed a "blip" on the screen that appeared to be plausible but raised a level of concern that was unusual. Nothing to put a finger on—just a clear feeling of unease.
>
> This "blip" followed a usual pathway that returning airplanes have used to return to the American carrier. There was no transponder with the blip, but that was not entirely unusual. Outbound aircraft turned it off to avoid ground fire. Often pilots forgot to turn it on again once over water and on their way home. The blip speed was consistent with a returning aircraft. The particular radar equipment in use did not provide altitude. It

was well known that returning aircraft flew at around 3,000 feet but anti-ship missiles flew lower at 1,000 feet. As he watched, the blip lined up with the carrier and a nearby battleship. The situation was now 45 seconds into about a 60 second trip. If he launched a missile at a plane, two lives would be lost. If he did not launch and it was not a plane, a battleship and many of its crew would be lost. How to decide? At the last decision point he launched the missile—his gut feeling told him that it was not a returning airplane.

The blip really was a silkworm missile targeted at the battleship. He had made the right decision. Why? How? A careful analysis of all of the radar plots (everything these days is fully recorded and archived) turned up the reason why the radar operator had the "feeling" that it was not a plane. But it was only weeks later that it was found. It was extremely difficult to find the reason, but once found it was obvious. The radar equipment did not expressly indicate altitude. However, a plane flying at 3,000 feet became visible on the radar screen immediately as it crossed the land water boundary—at the first sweep. The silkworm missile, flying at 1,000 [feet] was so low that the radar did not register it until the third sweep—two sweeps after the land water boundary. This did not register consciously. But it certainly made a notable impression on the subconscious—the sixth sense.[15]

This situation management book does not suggest making primary life-and-death decisions based on anyone's sixth sense or hunch. But the power of subconscious evaluation is not to be lost. There is a place for it. If after employing available tools for becoming aware of a situation, and a conscious decision comes down to things that are equally probable with none standing out above the others, then is the time to ask Which one feels right? Which one appears to be pushing itself to the front of the rest? Still cannot decide? Escalate, collaborate, or delegate.

This discussion is not the end. At times, a decisive act puts the situation in another light. That difference is enough to suggest revisiting the evidence with another viewpoint. Sometimes, though not often, the evidence looks different and still reinforces the choice. On rare occasions, the revisit denies that choice but moves another to the front. Either way, we gain. Before we move on, let us look at how this relates to "hunches." There is a distinction to be made between sixth-sense choices and hunches. The hunch is either there or not. Hunches need more verification before we go with them. We will do just that in Chapter 9, "Weak Signals," when we go over the "sizes" of trouble indications.

7.9 Biases

Let's get this started with encouragement. Biases are real. We all have them. They are important, although they can be hard to locate and understand. Biases can and they will get in the way of good operation in a big way. But they can be recognized, described, and managed enough to make sense. A *bias* is an undue influence or

---

15 Lehrer, *How We Decide*.

predisposition. Please toss aside any social or personality aspects—they are not part of this discussion. This is about effects that are often subtle and largely unknown by the individual. One of the telling ways to observe where biases are present is to look for "highly trained, experienced people breaking their own rules."[16]

The problem with biases in the control room is that they usually get in our way as we do the job. They push or nudge us to think that something is more important or less important than it really is. They shove our thinking so that we shy away from one conclusion in undue favor of another. And in many other "below our own personal radar" ways, they cause our thinking to fail to be as clear as it must be. Nature does not deliberately hide problems and issues. We often do, although some are not obvious. If we are vigilant enough, observant enough, and questioning enough, information will be there to alert us to the problems. Our job as creators of effective situation management is to make sure that sufficient tools are at hand to do the operating job reasonably with few impediments. Understanding and managing biases are part of that job.

Let us take a look at the more prevalent and important biases.

Confirmation Bias

Confirmation bias is the tendency to confirm facts that agree with what we already believe.[17] Anything that is not what we expected is noted but set aside. Everything, well almost everything, that is consistent with or even advances our original beliefs is embraced and categorized as "see what I did tell you, it's true." We use this bias (unintentionally, for the most part) when information is gathered, remembered, or interpreted. Confirmation bias is strongest for emotionally charged issues and where deeply entrenched beliefs are involved. And we all have it to some extent.

Unfortunately, confirmation bias (also called *confirmatory bias* or *myside bias*) is busy working all the time, especially under stress. It is a permanent guest in the control room. Some important indicators are

- accepting and proceeding with the first thing that seems right,
- looking for small things that "confirm" that the present pathway is the right one,
- ignoring warning signs or cautions that suggest a rethinking of the situation, and
- failing to ensure that the current solution is actually working.

---

16 "Cognitive Bias Mitigation," Wikipedia, last modified May 29, 2018, https://en.wikipedia.org/wiki/Cognitive_bias_mitigation.
17 "Confirmation Bias," Wikipedia, last modified May 31, 2018, https://en.wikipedia.org/wiki/Confirmation_bias.

The basic underlying problem is that *once the operator thinks he has found the issue and starts to work it through, he rarely questions that decision*. Even when new information comes to his attention, it is almost always dismissed because it cannot be right. It is not clear why people act this way. Part of the explanation is likely related to the loss aversion discussed earlier. Explanations for the observed biases include wishful thinking and the limited human capacity to process new information. "Another explanation is that people show confirmation bias because they are weighing the costs of being wrong rather than investigating in a neutral, scientific way."[18] Clearly, this leads to unfortunate and often disastrous operating consequences.

You should take clear and effective steps to make sure confirmation bias is not allowed in your control room. Ensure that yours is a culture of acting right, not fast. Make sure that there is no reward for acting fast. Make sure that the operator knows this. Plan for acting right, train for it, and reward this behavior. The approach consists of four steps:

1. Provide effective operating protocols for situation awareness.//
2. Ensure that there will be enough time for situation assessment and remediation activities.
3. Provide sufficient operational and directional information to permit successful management of the abnormal situation.
4. Provide for safeguards in case operator intervention is not reasonable or not successful.

Confirmation bias often poses a danger when it blinds us to the magnitude or true extent of a problem. Once we mistakenly think that the problem is Problem A when it actually is Problem B, we act as if it is Problem A. If Problem A tends to be limited and unlikely to pose a great danger, the tools we use and solutions we pursue will not be appropriate for Problem B, which is extensive. No matter whether the two problems might appear to be alike, Problem B will not be addressed properly.

Remember, create a culture of waiting to form a belief until all reasonable information has been viewed and weighted.

Loss of Scale

Losing scale is like taking the first step into sand and sinking just a bit. Take another step and sink a bit more than you expect, but not too much. And then take a third step expecting the same response, only to discover too late that you are in quicksand. Walking into quicksand is not a linear experience! Any scale mismatch between the

---

18 "Confirmation Bias," Wikipedia.

Figure 7-4. Illustration of loss of scale showing a proper category of response (have a fire, then try to put it out) but one that is completely inadequate and dangerous for the true magnitude of the problem.

real scale and what we think it might be, will be damaging. This scale concern works both ways. Thinking that a problem or issue is much larger than it really is can be just as concerning as the other way around. When we mistakenly approach a problem that is actually much smaller than we imagine, it results in unnecessary delay or waste, as we think we need more planning and resources than is actually required. All of this takes time. Thinking that the problem is bigger than it actually is causes us to completely overreact. Even after its true size is understood, the earlier thinking has a biasing effect that often results in more time and effort spent at resolving, or using a completely inappropriate process or methodology, than the true issue requires (see Figure 7-4).

Illusion of Skill

No competent airline pilot, even one who has flown the same airplane for years, will do preflight checks without fastidiously using a checklist and referring to the procedures. Few craftsmen will take on a new task without stopping to consider what is needed and how it might be approached. There is an old saying that "the more you know, the less certain you get." With most wisdom comes caution. Unfortunately, caution does not always prevail. People, even some considered experts, often confuse the power of their own skill and experience. They use their *illusion of skill* as permission to cut corners, to be exempt from the folly of the uninformed or inexperienced, or to allow downright recklessness. This illusion can blind an individual from seeing or taking into account clear clues that were obvious to the nonexpert. If that were not enough, at times it is difficult for experts to not overclaim. Overclaiming means that they say or act like they know something when actually they do not. Using experts means that you will demand that what they say and what they do make sense. Always.

Continuation Bias

The tendency to continue with a chosen action once started is natural. The further along the progress, the more likely it will be continued. Even in the face of concerns or other indications that things might not be as they should be, we persist. This tendency is called *continuation bias*. We get caught up in continuation bias for a seemingly good reason: why lose all that investment (time, emotion, reputation, etc.)? Stopping or changing course usually means starting over. All the effort and resources already spent may end up in the box called "wasted."

A telling example of this situation is the lost ticket conundrum:

> You've lost a pair tickets to a very desired performance and refuse to pay again for replacement ones, thus missing the performance. Rather than wash away the first loss, it is emotionally added to the current cost of the performance. Completely forgotten is that the money paid for the now lost tickets will get you into nowhere. That the new cost is more than you would have considered had it been the first price, now holds decisive sway. To bring this example sharply into focus: Had the same individual simply lost the same amount of money in an unrelated way (say falling out of a pocket), unless available cash was very limited, the pair of tickets would have been purchased and the performance attended. In this situation the money lost wasn't connected to the tickets. Same loss. Different outcome.[19]

Managing continuation bias is important. The practices and technology for doing it are just about the same as those for confirmation bias.

Anchoring Bias

Anchoring bias is what happens with a first exposure to specific information. When we are working on making a decision or selecting a choice, we are heavily influenced by the first piece of knowledge or information we come across. This is unconscious. This is why sales personnel will first quote a very high price for a product only remotely related to the one you might be interested in and would clearly not purchase. We have been "anchored." Then when that same salesperson quotes a price for the item you are interested in, but the price is high, the tendency is to compare that price to the very high "anchor" one. Then we are much more likely to go along with it. Had the anchor price not been announced, but the same initial too-high price given for what you sought, it would have been questioned as being too high.

Anchoring is based on a natural human nature desire to please others. It is part of most cultures. There is nothing socially or personally wrong with that unless it becomes overly important. But there is a subtle effect in the control room that needs

---

19 Adapted from Daniel Kahneman, *Thinking, Fast and Slow* (New York: Farrar, Straus and Giroux, 2011).

exposure. Given a problem to solve or task to perform, almost without conscious knowledge, we often "build in" others' wishes. When faced with an important maintenance need, an operator defers making a repair request because he knows that the budget is running low and the technician will not like taking on another cost. When analyzing a problem, we use the current list of problems started by our supervisor rather than beginning fresh and maybe finding something unpopular or just different. This is the basis for situation awareness: explore, review, and understand the problem; work on solutions only after that understanding is verified.

Halo Effect

When we make conclusions about an individual's abilities or characteristics based on observed performance in one area, and automatically assume that it carries over to most other things that person can do, it is called the *halo effect*. There is nothing wrong with that if we are talking about respect. However, much can go wrong with that if we cede our own thinking and understanding to that person so that we act on his belief, or wish, or what we anticipate is his wish.

After-incident knowledge illustrates this idea. "I thought something was wrong, but Charlie, who has been on the board for 12 years, said it was okay." "I started to shut down the off-ramp due to excessive traffic backup, as suggested by one of the options in the revised operating protocols, but Jake said he never did it before and it worked out fine for him." Jake had received performance awards for the last 3 years.

Bandwagon Effect

The *bandwagon effect* is also referred to as the *herd mentality* and *safety in numbers*. The bandwagon effect acts like club membership. It works like this: "If the rest of the guys think this is cool, I do too. I'm part of it. I don't need to rethink this because I belong." The group identification is not necessarily formal. It can be just most of the people on the same shift. It differs from anchoring bias in that little original thought or effort is expended—what the group decided or thought is all that I need. It differs from the halo effect in that I do not spend any time worrying about how good the group is. I am a part of the group, and that is enough for me.

Before concluding this part on biases, it is important to draw a broad line between everyday biases and real psychological issues. We are not discussing individuals' personality disorders. This discussion is only about human foibles and tendencies.

Diffusion of Responsibility

When a problem arises in a group situation that all observe, each individual believes that someone else is going to take responsibility for it. The group need not be large, though the larger the group, the more likely it is that no one will act. There need not exist any organizational structure within the group that might suggest a natural responsible person. And the fact that no one appears to step up changes nothing. The *diffusion of responsibility effect*, also called the *bystander effect*, can prevent anyone from initiating action, with the result that no one does.[20]

Even in a control room setting where there should be a clear chain of responsibility, when individuals of senior organizational authority are present with those of junior or equal responsibility, the operator tends to wait and to defer. Remember the ill-fated flight of Asiana 214 earlier in the chapter. Where time to act is important, action is excessively delayed. Chapter 10, "Situation Management," offers useful ways to manage this problem. And it will need to be managed.

Post Hoc Ergo Propter Hoc

Causality is a pretty straightforward thing to think about. At its simplest, we observe or consider an event, action, or movement and begin to think that it will cause something to happen next. The thing it causes results from the thing that started it all off. When this is true, we call it a causal event. But we need to be careful. Not everything is so directly linked: we call it a one-to-one link if it is. Not much is one-to-one. Making it so when it is not can really throw things off. *Post hoc ergo propter hoc* means "after this, therefore because of this." The English translation does not explain enough. So let us look at two interesting illustrations.

> THE SETTING: At exactly 5:16 p.m. on the afternoon of November 9, 1965, an electrical power circuit breaker on a 230 kV transmission line near Ontario, Canada, malfunctioned and tripped. The power surges cascaded causing the rapid eventual blackout of the entire northeastern transmission network. This placed the entire province of Ontario in the dark as well as the states of New York, New Jersey, Delaware, Connecticut, Rhode Island, Massachusetts, Vermont, and New Hampshire without power for over 12 hours.
>
> POST HOC ERGO PROPTER HOC: The following stories were reported by various news outlets about how specific individuals found themselves to blame.
>
> BOY WITH A STICK: A terrified young lad ran home and confided to his mother that he was terribly sorry, because he was the cause of the power going out in his neighborhood. He did it. What he did was why the power failed. His story was that as he was walking home, he picked up a stick and carried it down the road. At some point he hefted it, drew it way back, and swung it as hard as he could against a utility pole as he walked by. At the very instant the stick hit the pole with a loud whack, the lights went out!
>
> MAN IN A BATHTUB: It was not his usual custom, but a man was taking a wonderfully nice bath to relax after a particularly vexing day at work. As the water cooled, the bath needed to come to an end. Rather than jump out, he decided to pull the plug and

---

20 Michael Shermer, *Skepticism 101: How to Think Like a Scientist* (Chantilly, VA: Great Courses, 2013).

enjoy what little of the bath might linger a bit longer. The tub stopper resisted. He pulled harder. The stopper did not budge. Then he pulled very hard and the stopper released its grip, and at precisely the same instant, the power went out! It was quite a while later until he was convinced that his pulling the plug did not cause the whole thing.

We know that this is not about sticks and tub plugs. It is about not associating events with causes that do not apply. However, when we see a thing happen and immediately blame what we just did as the cause, we are doing just that. We justify making this association because maybe the last time we did something similar, the same problem occurred. If we want to truly find an association, we need to work it out. Jumping to it may be a lot of blame work, but it is not the work we ought to do to find the real cause.

The "What Then" Question

A good way to manage first thoughts and early conclusions in order to check that there is not more to the situation, at times much more, is to ask the "what then" question. *What then* means if this is true, what are all the logical or reasonable things it might mean? For example, you see something on a laboratory report, read something in an article, or hear a politician promise something. In this example, you see that a certain blood chemistry reading of yours is a bit high. You compare it to the range of normal right next to it, and your reading is right near the edge. Maybe it is a bit over or under the range. Maybe it is barely inside. Not to worry, close is close enough, right? Well, yes. Being comfortably close to a normal boundary is not a problem. And there are a million reasons why that is fine and is as far as things need to be examined. But before we move on, let us ask the *what then question*: What if being on the edge actually might mean something? So, take a look. Find out what being on the edge might suggest. Ask someone or look it up. Unless some other thing or things are abnormal or suspicious, things can be just fine on the edge. But unless you look for those other things, you will not know. Always ask the *what then question* for situations that are (1) hard to believe; (2) too good to be true; (3) near the edge of normal, or maybe a bit over; or (4) appear okay now, but if they were not, it would mean that something really awful is happening or about to.

In summary, asking the *what then question* serves to avoid jumping to conclusions too quickly about what might really be going on—beyond what appears obvious. That is always a good idea.

7.10 Inattention Blindness

Inattention blindness is looking without seeing. It is also called *inattention bias*, which is part of the more general term of *selective attention*.[21] Inattention blindness can be a

---

21 Arien Mack and Irvin Rock, *Inattention Bias* (Cambridge, MA: MIT Press, Bradford Books Series in Cognitive Psychology, 2000).

Chapter 7 – Awareness and Assessment Pitfalls 449

Figure 7-5. Passing the basketball (still frame) setting up the situation of a foreground task requiring an outside viewer's full focus of attention.
Source: Figure provided by Daniel Simons.[22]

much more important issue than first meets the eye. To begin with, we all have this blind spot. And it really is a form of blindness. Inattention blindness is a failure to notice something unexpected when one's attention is focused tightly on something else.[23] That something else is not necessarily important. We often fail to see something that is obvious if we are not looking for it or are the least bit distracted. It is the side effect of something we do quite well. When we focus our attention, we filter out all the surrounding irrelevant and distracting content outside the focus. When our thoughts are working on something in the visual field, that focus will prevent us from seeing other even more obvious things elsewhere. We have to look in order to see.

A classic example is depicted in a short video that shows two groups of teens passing their basketball in an elevator lobby (see Figure 7-5).[24] Half are wearing white shirts and half are wearing black. The viewing audience is asked to count the number of times the ball is passed between the white-shirt team only. A "pass" is the ball leaving one person and being caught by the recipient. The players are all in motion. The action is continually moving in the visual field.

---

22 D. J. Simons, and C. F. Chabris (1999). Gorillas in Our Midst: Sustained Inattentional Blindness for Dynamic Events. *Perception*, 28, 1059–1074, www.theinvisiblegorilla.com
23 "Watch This!" *The Brain Games*, season 1, disc 1, games 7 and 8, directed by Jeremiah Crowell, released October 9, 2011 (National Geographic Channel, 2011), DVD.
24 Daniel Simons, "Selective Attention Test," *Surprising Studies of Visual Awareness, Vol. 1* (Champaign, IL: Viscog Productions, 1999), https://www.bing.com/videos/search?q=gorilla+ball+passing&&view=detail&mid=7 6C696714DFD83C1468A76C696714DFD83C1468A&&FORM=VRDGAR.

Figure 7-6. The gorilla, clearly in view, was missed by about half of the outside viewers of this basketball-passing exercise.
Source: Figure provided by Daniel Simons.[25]

During this ball passing, in the midground (behind some players and in front of others) and clearly visible, a person in a gorilla costume slowly passes through the scene (see Figure 7-6). At the end of the play, the audience is asked for their ball pass counts. As you might expect, the responses for the number of passes differ, although the majority are correct. But about half of the counting audience missed seeing the gorilla. They were astonished when they saw it during a replay after being told it was there.

The concern is that when we "focus" in a moving scene, we miss things. That is why investigators viewing surveillance recordings of an incident watch the same scene over and over again. Each time they try both to not focus so that everything is observed (amazingly difficult) and to focus on different things in turn. This can work for "instant replay" situations but is inefficient in the control room. There is no magic bullet for this problem. A reasonable approach is to develop a protocol of viewing the content of a display screen: first take to a quick overview of everything and then to take closer looks at what seems to need more attention. Two ways of looking at the same content will provide different information.

7.11 Partial Information

Humans are very good at working with limited information to understand a situation.[26] As soon as we can grasp a framework, what was initially vague and incomplete quickly snaps into clear focus. Even without a framework, the signal can jump

---

25 Simons & Chabris, "Gorillas in Our Midst."
26 "Watch This!" game 13.

JUMPING TO CONCLUSIONS

Figure 7-7. Partially obscured phrase that appears to be quite readable due to our ability to project meaning from incomplete causes. Often we are wrong, and without finding the truth, we act inappropriately.

out of the background. The most common example of this is hearing your name in a noisy, crowded room. Immediately, without any attention directed at listening to the blur of noise and sounds, you catch your name. This can be a useful trait. Most often it is not.

Boroditsky illustrates this concept with a simple word phrase test.[27]

Please examine the three-word phrase in Figure 7-7. Even though the black bar obscures more than half, we have little problem reading and understanding the phrase. The words appear to be the partially covered phrase "JUMPING TO CONCLUSIONS." How amazing is this? And there exist plenty of other examples of how adept we are at making sense out of what might at first appear too indistinct or fragmented. We make sense out of fragments by inducing from our experience and expectations. That can be a problem. We often conclude along a biased pathway. Many of our individual biases are shared widely with others. Fortunately or unfortunately, however you see it, this shared bias will foster a shared induction. Just because others decide to see what you think you see, it is proof only of a consistency of people seeing. It has nothing to do with the correctness of it all.

Look at Figure 7-7 again but imagine that the obscuring bar is removed. Here is the actual phrase: "IUMRING TQ GQNGIUSIQNS." In this case no harm, no foul. The actual phrase makes no sense. But what if it did make sense and it was a different sense than before? That is only part of the message here. The other part is that while observing obscured or partial information, it is human nature to jump to a conclusion that tries to explain what it really is. But the information is only what we think it might be. It is not necessarily what it is. The takeaway is that we must train decision makers to carefully confirm their assessments. With this thought in mind, we again call your attention to the process of weak signals. Looking ahead to Chapter 9, "Weak Signals," the clear steps of collecting the clues, considering the more serious problems they could suggest, and then looking for more evidence will not work if what we decide is wrong in the beginning. Looking for potential problems from sketchy clues works only if "looking" is not prejudged with what you want to see. Jumping to conclusions should not be done. This is a solid work process. It is one of many.

---

27 "Watch This!" game 13.

7.12 Myth of Multitasking

Setting the Stage

This is an important topic to understand. Please start this with caution. Multitasking might be used as a resource management tool for operators who find themselves overloaded. Overload is part of the equation of operations. There are few, if any, ways to control that part. However, this seemingly useful tool of multitasking (really juggling) is not at all effective. We are not capable of doing it well enough to rely on it.

Multitasking

Multitasking is the intentional act of working on more than one engaging activity at the same time or, simply put, trying to do two or more tasks simultaneously.[28] "At the same time" means that we interweave and overlap those activities so that it appears as if they are concurrent. These overlapping activities may be highly related or significantly different. Because driving a vehicle is such a common situation, let us use it for our brief illustration. You are driving (without any locational aids) and trying to find the right road to turn on next. Specifics that come to mind are: driving to your next appointment and talking on a mobile phone; or driving and trying to decide whether that meeting you just left means you are out of a job.

Is multitasking in your toolbox? Think you do it effectively? Maybe you do, but chances are you do not. Less than 3% of us multitask in an acceptable way—not good enough to be "effective," just "acceptable"! What does that mean? It means that when we try to multitask, we are really switch tasking. When we switch task, we stop paying enough attention to the task we just left and focus more on the one we just picked up. Even if we feel we are doing all tasks at the same time, our attention has shifted. One part we switch to; the other parts move along on autopilot. We are coasting with them. We are not paying enough attention to them to react properly. If nothing happens, we get away with it until we think we are done or something does happen. This going back and forth takes a toll.

The shifting process itself is much less seamless than you might think. Each interruption (read this as a shift, or anticipated shift) involves a certain amount of time, however brief. The time it takes to resume the earlier task is called "recovery time." Typical recovery times are 10 to 20 times longer than the time spent shifted. A 5-second quick intended look at something else would cost nearly a minute to fully refocus on the original thread. It might seem as if the return is much quicker, but the time it takes to fully refocus is longer. Refocus involves picking up the full thread,

---

28 "Watch This!" games 7 and 8.

not odd parts, such as the eyes seeing the road, but failing to register traffic conditions on the opposite side of the road or problems with the shoulder on your side. Research estimates that individuals who routinely engage in multitasking suffer a 20% to 25% overall *loss* in productivity.

To help explain this, understand what is going on inside our attention consciousness. We use different parts of our brain for those different types of activities. Mathematical and reasoning processing is done in one place (the parietal lobe). Memory tasks complete in an entirely different part (the frontal lobe). Vision and spatial processing tasks take place in yet another separate part (the occipital lobe). We are good at coordinating these parts for a consistent task (such as looking for a lost pair of glasses). We do not do so well looking for a lost pair of glasses and dressing for an important meeting. We are not good at using these parts for separate tasks. Either we forget a belt or wallet, or we take way too long looking for them. A missing wallet would create an interesting moment at the cash register, but a missed operational reading might spell disaster.

Let us cement the importance of this with a real-life illustration: Driving an automobile while engaged in almost any phone call creates the same level of impairment (to safe driving) as a blood alcohol level of 0.08%. That level is sufficient for a legal finding of driving while under the influence (of alcohol). While slurred words are not part of this, impaired visual and situational awareness are.

Multitasking should not be a permitted behavior in the control room. Control room management protocols need to be designed and followed as a series of threaded related tasks and activities. A general schedule must be followed. This schedule must be rich enough to provide operators with the confidence that between the alarm system and weak signal processing, operational risk will stay under control. The operator is then freed to pick up and follow other related tasks without any urgency or expectation of jumping back and forth on his own. At times of reflection, such as between thread tasks or upon returning to the board after picking up a coffee, a visual review of the human-machine interface (HMI) and alarm system should provide sufficient awareness orientation. This depends, of course, on the alarm system being proper and the HMI design being effective. Everything depends on your effective infrastructure being appropriately used.

Alarms and Multitasking
While we are thinking about the overhead of trying to multitask, let us take a quick look at how alarms insinuate inadvertently into your operator's real-time activities. Start with the plant or enterprise without any active alarms. Your operator is busy working

through his regular activities, paying attention to what they are and how they are to be done. His mind is occupied with the tasks at hand. Along comes an alarm. Now his whole attention is devoted to the new alarm activation. He must quickly assess the alarm and just as quickly park all his current, non-alarm activities. What if the things he was in the middle of before the alarm needed to be addressed more? Now the operator has to work on the alarm and at the same time wind up or park his other activities. Suppose those other activities are critical enough to need completion or movement to a predetermined proper status? What should he do? Doing both is not a proper option.

Normally, the alarm response time built into the alarm activation point (a full discussion is presented in Chapter 8, "Awareness and Assessment Tools") is sufficient to process the alarm. There is a bit of extra time, but probably not enough to do much else. Now we have a time resource problem. Here is a way out. If the alarm is active, all else must be dropped to work it. If the other stuff cannot be quickly handled (in a few seconds), and mishandling it will open up other issues, those items must be passed on to another individual or the operation rendered safe by shutting down or safe-parking it. Simple. Powerful. Uncomfortable? Sure, but because we really cannot do both things, we need to do the right thing. Running out of time to properly handle the alarm is not one of them.

Takeaway

"Multitasking is going to slow you down, increasing the chances of mistakes. Disruptions and interruptions are a bad deal from the standpoint of our ability to process information."[29] The basic understanding is that tasks that require even a modicum of concentration and understanding will need all our conscious attention. When we do not concentrate and understand, our performance slows down and declines in efficacy.[30] Design everything to follow threads; change threads only when important or necessary. Multitasking is not an acceptable working activity in a control room (or anywhere else, for that matter).

7.13 Personalities

Incidents and accidents usually have many causes. Each event usually can be traced to several failures of equipment, processes, people, and procedures. Here, we briefly open a small window into concerns identified in the literature. This material is intended to be used for design, procedures, and training, not for operator screening.

---

[29] David E. Meyer, Cognitive Scientist and Director of the Brain, Cognition and Action Laboratory, University of Michigan, Ann Arbor in Steve Lohr, "Slow Down, Brave Multitasker, and Don't Read This in Traffic," *New York Times*, March 25, 2007.
[30] "Slow Down, Brave Multitasker, and Don't Read This in Traffic."

Accident-Prone Behavior

Accident-prone personalities can be fostered to do harm by the environment within which they work.[31] Here are some behaviors (of operators as well as supervisors):

- **Impulsive behavior** – Acting without thinking, jumping to conclusions, and being the first one out the door to handle a problem, assuming instructions are for the other guy

- **Irritability** – Not only not enjoying the job, but also pushing back

- **Resistance** – Acting like it is "my way or the highway" ("I know what I'm doing" and "I don't need any help"), reflexively assuming one way is better than another way, "Rules are for everyone else"

Studies show such behavior does make a difference. Incident rates are five times higher for impulsive or resistant employees.[32] A foreman's resistance to questions or sticking to the rules leads to a noticeable increase in incidents by the crews working under him. This should not be overlooked or attributed to personality and set aside. It is that important.

The Quiet Ones

Operating a plant or other operation is a team effort. No complex system or activity can be so clearly understood and preplanned that it is successful without people actively engaged in its operation and coordinating with each other. Doing this creates openings for discovery of small problems or issues. Success depends on identifying anything unusual or operations that appear rough or strained. But identifying things only helps if the information is communicated effectively. Teamwork counts on team members speaking up, questioning, and discussing. That will not happen unless there is a culture that encourages and rewards speaking up. Success depends on *the quiet ones* doing their share.[33] This is both a cultural practice as well as a coaching goal. It is at the heart of *crew resource management* (see Chapter 10, "Situation Management"), an important collaboration activity that must occur to recognize and understand impending situations in which different members of a team might have different information or perceive the same information differently.

---

31 Paul Studebaker, "Reduce Incident Rates by Identifying Accident-Prone Personalities," *Sustainable Plant Today*, July 8, 2013, http://www.sustainableplant.com/2013/07/reduce-incident-rates-by-identifying-accident-prone-personalities/.
32 Studebaker, "Reduce Incident Rates."
33 Dirk Willard, "When It Comes to Plant Safety It's the Quiet Ones Who Are Dangerous," *Sustainable Plant Today*, November 7, 2011, http://www.sustainableplant.com/2011/11/when-it-comes-to-plant-safety-it-s-the-quiet-ones-who-are-dangerous/.

Situation Management Points

Situation management is built on safeguards and enablers that go a long way toward smoothing out this rough terrain. A large contributor to this problem is management's view of risk avoidance and lack of a receptive environment for feedback. Plants and operation centers that have a culture of pride of operation tend to be more fulfilling and rewarding to work in. Where job tools are sharp and effective, workers are more likely to trust them and use them effectively. Abnormal situations announced by alarms are designed to provide operators with both the direction and time to do a good job. Comprehensive training and follow-up produces confidence in operators and a willingness to trust the system to be on their side.

Near-miss reporting and follow-up encourage a climate of honest questioning and continuous improvement. Develop and nurture a no-fault reporting system for all irregularities and unsafe or improper activities or conduct. Build on this information through a well-integrated process of evaluation and learning the lessons that need to be learned.

7.14 Geography of Thought

Thinking about geography is interesting and fun. Travel to other places is always exciting. When we can, we do. But what if thinking were different in different places around the world? What if different people in considerably different longitudes had a different view of what is logical? What if they had a different view of how people see everything?[34] They do. Time zones and postal codes matter. We have all heard the expression "a clash of cultures." This is it. And that "clash" is not right versus wrong, where we think that the other guy has missed out on a truth or simple fact. Cultures do see facts somewhat differently. All are correct. They often conclude differently. "Understanding the thought processes of other cultures may very well turn out to be critical to the survival of all civilization."[35] Think of the world simplistically divided into two spheres of thought: East and West. These are just labels. The demarcation is approximate and there are gradations, of course. This will get us started.

Figure 7-8 suggests generally how each region might appear on a world map. You see the two regions East and West. The region in between is a combination, a Blended region. It includes a blend of East and West.

---

[34] Richard E. Nesbitt, *The Geography of Thought: How Asians and Westerners Think Differently . . . and Why* (New York: Free Press, 2003).
[35] *Providence Journal-Bulletin*, review of *The Geography of Thought* by Richard E. Nisbett.

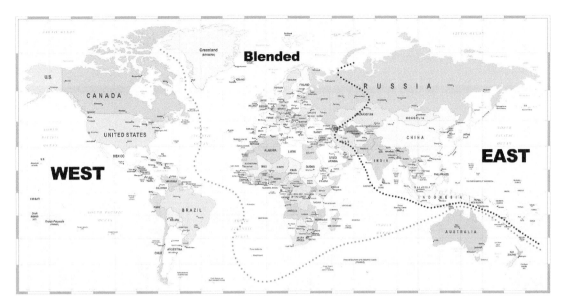

Figure 7-8. World map delineating East, West, and Blended cultural frameworks. These regions (marked by the magenta and green dotted lines) exhibit different norms with regard to their approaches to life and thought.

Norms and Conventions

Normal Visual Flow Directions

Let's start off with the basics. In the West, readers are acclimated to the natural flow of written and visual information as starting at the top, and moving from left to right and then to the bottom. If the visual frame (*display* in HMI nomenclature) is divided into subframes (*window* in an HMI display), each subframe is viewed with the same visual flow process. Move to the East and this natural flow is different. It starts at the top of the frame, but the natural flow begins at the right-hand side of the page. Also, instead of moving to the left, it moves down. On reaching the bottom, the viewer flow returns to the top and moves to the next position to the left. This is all well and good, until the subframes enter. Here a bit of conflict begins. The view process wants to continue downward, but the subframe wants the downward part to end so that the view returns to the top. Eventually, the subframe wins, but it is different. Figures 7-9 and 7-10 illustrate this effect.

There are assumptions "baked into" our ways of designing screens. The West expects the causal flow of information to be left to right and then back to the left, moving on an eventual pathway to the bottom. The top is considered to be the cause or the beginning (from a casual perspective) to what is below. The East expects the causal flow to begin at the right and move immediately down to the bottom. What is to the left is subordinate. The Middle East expects the causal flow from right to left and then

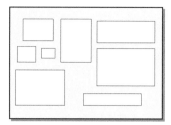

Figure 7-9. Illustrative page from an HMI screen showing the usual component layout structures of windows, formats, and elements.

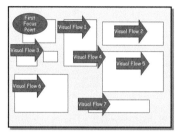 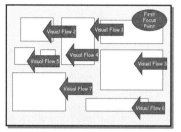 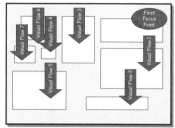

Figure 7-10. Eye flow on an HMI page (depicted in three windows, from left to right) illustrating different starting points and flow of view based on cultural convention. This suggests that layouts will need to guide the viewer's view. In the West it is left to right, top to bottom. In the Middle East it is right to left, top to bottom. In the East it is top to bottom, right to left.

back to the right, moving down for the next causal pathway. This is not in a rulebook, but it is the way different cultures view "natural." This means that screens designed for one culture do not necessarily flow in a causal way for other cultures. When this happens, viewers can be confused or misled.

The importance lies in what the information shown to the viewer is meant to convey. Recall that there are always two parties to this activity: the *designer* and the *viewer*. The designer has an objective in building the view—present needed information to the viewer. The viewer has the need to view this information—reaching (sometimes vital) conclusions about what it means and what, if anything, must be done about it. If the designer and the viewer are from different hemispheres, how can we expect things to have been designed the way the user expects? The short answer is that we must design in a way that the viewer will extract the necessary results in his customary manner. The work must reach a consensus of expectations and the correct methodology to achieve them. This HMI consensus about screen design just got much more important.

Illustration
Consider the following illustration adapted from an example making the rounds on Internet blogs. An enterprising soft drink marketer selling a branded cola decided to

Figure 7-11. Nonverbal marketing message whose entire meaning is dramatically different when the individual frames are viewed in order from left to right (the intended flow by a Western designer) or right to left (typical for the culture where it was targeted).

develop a succinct advertising message. Because the target market area was multilingual, a graphic advert was designed. The cartoon opens with a man lying on the sand in near utter exhaustion. The next frame shows the same man happily drinking from a bottle of cola. The final frame shows the man running off into the desert appearing vibrant and active. See Figure 7-11. We are conditioned to see exactly that story progression.

When the same message was shown to people in the Middle East, the advertisement was a complete flop. How could anyone think that people would want to drink cola that took a vibrant individual and then left him at death's door in the desert? Well, their flow of logic is right-to-left, not the other way round. Same pictures, same order, vastly different message.

Visual Perception

Sometimes the whole may not be a sum of its parts. Consider how we might view a nicely designed and stocked aquarium as depicted in Figure 7-12.

Figure 7-12. Aquarium showing fish and the environment of the tank consisting of the pebbled and planted bottom, added natural features of the bottom, and a rear tank picture to complete the view.

To the Eastern eye, the aquarium landscape is all about the harmony of the view: neutral gravel of varied size, the flowing grasses, and the blend with the rocky background. That eye finds the comfort of a scene in nature. The fish provide another element, not an essential ingredient. In fact, this scene would be enjoyed immensely without any fish at all. Simply the diversity of nature and the subtle undulations as it responds to the slight movement of water flow would be a cherished moment. Form and content are one.

A Western eye focuses immediately on the moving fish. To a significant extent, the grasses and background seem at first to get in the way of the view. They distract from the focus. It is all about the players and the court, metaphorically of course. Are the fish different? Which appeal more? The scene only becomes important as an interesting diversion or when a "story" can be developed that explains what is in the aqueous display.

Language Construction
Just as visual norms play an important part in design-versus-inference conflicts, inherent language construction differences can confuse as well. There is the larger class of language that builds meaningful words as a constructive process, as opposed to the other class of language that has a different word for (almost) everything. The thought construction (sentence word order) requires the reader to complete the entire sentence before the thought can be grasped. This resembles a whodunit. As the sentence unfolds, it might end with something good or useful or something damaging or dangerous. Last words can make this difference. These differences must be understood and managed to ensure that intent matches understanding. Sentence structure takes on even more importance in situations where there is a need for brevity or even abbreviation. It may take extraordinary care to make sure messages are effective.

Logic and Reason
General
Something that is logical is generally thought to be "reasonable and expected." A logical thing is so because it is "[n]ecessary and reasonable because of what has gone before."[36] Therefore, it is easy for anyone to hear something or think about something and consider it logical or not. Part of thinking about something as logical is that it does not have to be fully explained. Just the gist of it is enough. When it is part of a discussion, everyone is "on the same page," as the phrase goes. But what if your logical is not everyone's logical? Not only can nothing be taken for granted, it is also possible, even likely, that

---

36 David Bernard Guralnik, ed., *New World Dictionary of the American Language*, 2nd ed. (New York: Simon & Schuster, 1980).

one individual's expectation may be far different from another's. They may even be opposite, such that what makes sense to one person is nonsense to everyone else.

Such confusion would be uncomfortable and possibly embarrassing in a social setting. However, this confusion is frightening and dangerous in a life-or-death situation. Make no mistake; managing a hazardous operation can be life or death if done without proper regard to the requirements. Logic is not the same throughout the world. Consider the phrase, "When in Rome, do as the Romans do." Let us give the thought a little twist to reveal the point of this section: When thinking about what the Romans will do, you must think like a Roman. So how do you think?

In the West, the Greek philosopher Aristotle established the basis for logic or what is logical. Oversimplifying, if a certain truth implies a certain conclusion, and if a second conclusion based on the first is also true, then the first conclusion implies the second truth. For example, if Charlie is a horse and all horses have tails, then Charlie has a tail. And there are other constructions for getting at the truth through logical means. The value of these construction tools is their ability to use logic to "prove" something that is not obvious or was, until now, not experienced. More important, not only are the working principles shared, but also the use of logic is expected. We expect reasonable people to follow logic. For example, a wide-open shop door in an automobile parking area is not an authentic invitation for a driver to drive a vehicle into the shop.

Logic in the East has a far different heritage. What governs thought is *that which has been experienced by self or conveyed by others down the ages.* Anything that flies in the face of experience cannot be true. Moreover, the East has a radically different view of consistency and contradiction. Where contradiction in the West creates discomfort, contradiction in the East merely helps reinforce its realness. The world is full of contradictions and lacks a certain consistency that reflects its continuing change. It is our task to find ways to live in harmony within the understood contradictions. In this light, harmony is the goal.

Notably, because logic was not a pivotal part of its culture, China had no concept of science (not invented, used, or understood) throughout its entire early history. The need for harmony and respect for the present did not require reaching into the unknown for either answers or advancement. This is not to say that China's long history of careful observations and recording the lessons they tell did not leave the world with an enviable store of knowledge that has considerable relevance today.

Simplicity as Truth
As we learned in Chapter 1, "Getting Started," if there are two competing explanations for something useful, we tend to believe that the simpler one (that does not require

suspension of intellect, of course) is the more correct one. If a basic simple one does not work, then the next simplest one that does is the one to use.

Logic versus Fate
There is an honest expectation in many cultures that events will transpire the way they do no matter what. If a pipeline wants to rupture, it will do so despite any efforts at preventing it or managing it by people. If a chemical reaction wants to go exothermic, it will do so at a time of its own choosing. Therefore, it would not be logical for anyone to try and prevent these events from happening. The thinking goes:

> As an operator I am only the observer of fate. It is my lot to observe the good; it is also my lot to observe the not good. All is not up to me. It is out of my hands if it happens.

There is no place in control rooms and responsible enterprises for fatalism or predetermination. It is your task for effective situation management to ensure that any beliefs in fate by the operations team do not cause improper operations protocols and behavior. You will want to find culturally appropriate ways to manage this.

Nailing Down the Issue
Situation management builds on the expectation that truth will be consistent, that facts matter, and that proper management has a reasonable chance to affect outcomes. Inconsistencies require additional concern until fully resolved. These expectations are core to situation awareness and situation assessment. Even before you picked up this book, as an operator or engineer you were ever looking for something out of the ordinary, looking for things that did not "add up." Real, technical plants and operations act according to their physical reality. Stress corrosion does its action everywhere the same in the world. Heat imbalances portend a problem everywhere. In the control room it is essential that successful operating principles and beliefs be developed that are fully competent, understood, consistent, accepted, and practiced by operators, supervisors, and managers alike. Respect for culture is not against good operation; it simply asks for practices that can be understood and employed successfully.

Never mind geography that is a dozen time zones away. In the United States one easily finds enormous, often inexplicable, cultural differences within the same time zone. Take rural compared to urban environments. Take Northern compared to Southern cultures. More than accents differ. Moreover, as societies become more and more multicultural, the issue becomes closer to us all. As a systems designer, you understand the local personal culture with its diversity and must fully manage it in the control room. The same understanding and respect is required worldwide.

Individuality

One key to deciphering culture hinges on the different views of individuality. To the West, everything hinges on the individual. What are his rights? What are his goals? Standing out by being successful, admirable, or just plain funny is sought after. "Don't tread on me!" shouts the warning designed by Benjamin Franklin[37] placed on early American flags. To the East, everything hinges on getting along. Their higher good is realized by interdependence rather than independence. Cultural differences are real.

The point of all this is not to pick which view is better. It is to help both the East and the West to take particular care to ensure that what we ask our operators to do respects their underlying view of responsibility. The pursuit of successful operations must include effective and respectful incorporation of cultural norms into appropriate operations expectations and protocols without compromising safe and effective operations. Success or failure should not elevate or depress an individual's view of and respect for self.

Handling and Reporting Problems

There are a few universal cultural norms. One that is important here is the concern about avoiding embarrassment of self and others. There is an old yet valuable Middle East understanding that the surest sign of the onset of embarrassment is the mere appearance of any reddening to a person's face. Anyone who actually goes on to cause another's face to fully redden in public has committed a most serious affront. It is that important. In the West, avoiding embarrassment is referred to as *saving face*. We are all concerned about this. So, it is important to ensure that the handling and reporting of necessary things be done in a way that respects both the individual and the needs of the reporting.

When it comes to learning from mistakes or near-mistakes, it is clear that the first thing we must know is that an actual mistake happened or almost happened.

> Without proper accounting, experts say, airlines can't learn from mistakes and regulators can't properly assess safety risks. Even smaller incidents—like component failures and near misses on the runway—are key bellwethers for major crashes. Left undocumented, botched procedures are left to grow endemic.[38]

Before the FAA instituted no-fault and anonymous reporting of airline issues in 1979, a large percentage were unreported. Since then, over a million anonymous tips have been reported, with 4565 incident reports being written to share with the industry. However, the rest of the world is slow to adopt similar reporting standards and mechanisms. Therefore, a much larger number of potential lessons to be learned go unnoticed.

---

[37] The slogan is accompanied by a coiled rattlesnake, as depicted on the Gadsden flag during the American Revolution.
[38] Daniel Stacey, "In Asia's Skies, Mistakes Go Unreported," *Wall Street Journal*, July 14, 2015.

This is especially prominent in jurisdictions that impose extremely harsh penalties for errors. Recall the Amagasaki rail disaster discussed earlier in this chapter. It was caused by an operator reacting in apparent terror of being called to answer for slight operational irregularities. Reporting systems, training, and collaboration all need to pay careful attention to overcoming residual cultural norms if they are going to be effective. Without effective ones, the entire operation of an enterprise is at unacceptable risk.

Three Postal Codes

The West and East distinctions were used to clearly lay out geographical differences. They are real. In fact, they are so real that cultures that lie between these two anchor points exhibit merging (looking much like an averaging) of cultural norms. You will not find anything as exact as a proportional culture, but we do see clear blending. This suggests paying even more care to cultural expectations and roles. Doing so might seem like reinventing the United Nations. Take heart, nothing that extreme is intended. Just understand that there are cultural differences that are rarely neither explicit nor expressed yet require understanding, respect, and accommodation. The accommodation referred to here is that which ensures operators are able to adequately and properly perform the needed activities with both personal dignity and enterprise operational certainty.

7.15 Institutional Culture versus Individual Responsibility

So far, the discussion in this chapter has worked to expose the unintended consequences of being human and of living in diverse cultures. The final pitfall cannot take its pedigree from so noble a base. This pitfall is the failure of institutions to align their culture with the appropriate degree of operational responsibility required for a sustainable enterprise. Almost every major industrial disaster has its root cause in a failed institutional culture. There is no way to excuse it. There is no way to suggest that it might somehow be an offspring of the unintentional. Failure is not actually intentional, but by failing to expect, look for, and accommodate them, it is certainly not responsible. Without enterprise responsibility they cannot have operator responsibility.

Polarity

The term *polarity* comes from a situation that appears to have two apparently necessary requirements that are both needed but appear to be "poles apart."

> A paradox or *polarity* is a pair of interdependent goals that need each other over time to create and sustain success . . . . However, when an organization leverages them both as a system, it is better able to achieve goals.[39]

---

39 Rosa Antonia Carrillo and Neil Samuels, "Safety Conversations: Catching Drift and Weak Signals," in *Professional Safety* (Park Ridge, IL: American Society of Safety Engineers, 2015), 22–32.

All operating situations that appear to be polar will expose the plant to undue risk for incidents. Examples include the following:

- Operate this plant to not miss a scheduled delivery, and make sure that there are no reportable environmental releases.

- Make sure all equipment is properly maintained, but ensure that all plants operate at full capacity so that the current market shortages can be capitalized on.

No amount of polarity can be allowed to exist if the enterprise is to operate in a safe, responsible, and productive manner. Management must explicitly set down the rules and guidelines that effectively eliminate any need for operators (or their supervisors) to ever confuse the business objectives with the responsible citizen requirements. Slogans like "safety first" or "any environmental release is one release too many" are not the answer. Management must set out specifics of the poles and clearly indicate which governs and how operators are to make decisions. This must be consistent in operating procedures, incident investigations, and performance evaluations.

Alignment Failure

A failed culture is one that has not adequately set operational protocols for its activities to require and enable its personnel and equipment to be capable of maintaining safe, reliable, and environmentally responsible activity. A failed culture places operational and ethical barriers on its personnel that pit responsible operational decisions at odds with enterprise needs and practices. Few operators are strong enough to make the hard decisions of the moment necessary to challenge or overrule corporate expectations (explicit or implicit) and do the safe and responsible thing. Sure, in hindsight, no enterprise will attempt to justify such a conflict or even to suggest that it exists. The enterprise will point to published policies that laud safety or expose its proud record of environmental compliance. Scratch below that surface and other realities come to light. Unless those with operational responsibility know for sure that every level of the enterprise will back them up when they consider and/or make operational decisions they expect will render appropriate operation, no enterprise will be safe or responsible. This support for safety must be integral to the culture we seek.

Historical Incidents

Offenders are both the large and the small. Let us examine a few historical incidents of note.

Challenger Space Shuttle

Perhaps the most famous (or infamous) example of cultural failure was the catastrophic explosion aboard the space shuttle Challenger with the loss of all hands seconds after launch.[40] At 11:38:00.010 Eastern Standard Time (EST) on Tuesday, January 28, 1986, the US National Aeronautics and Space Administration (NASA) launched space shuttle flight (the Challenger) STS 51-L from the Kennedy Space Center at Cape Canaveral, Florida. Seventy-three seconds later, all signals were lost and the aircraft was observed to have exploded in flight.

The proximate cause of the explosion was the failure of the large-diameter right booster aft O-ring that was intended to seal propellant inside of the solid rocket boosters used to provide additional launch thrust. At 58.788 seconds into the launch this failure permitted contained fuel to leak out of the right booster, which promptly ignited. The failure sequence included the thermal destruction of mounts for the booster that led to other structural damage and the eventual rupture of the hydrogen tank, resulting in the destructive explosion.

The disaster can be traced to two clear institutional failures: (1) the prior-to-launch engineering failures that led to the launching of a spacecraft that was not airworthy, and (2) the prelaunch command and control failure by Launch Control to properly evaluate the actual risk to flight safety once it had been brought to their attention. Each will be discussed briefly.

Prior-to-Launch Technology Management Failures
Nine years earlier, shuttle engineers noted and documented serious concerns about the ability of the O-rings to properly perform during launch. From October 1977 through February 1979 there was official written documentation of those concerns. In November 1980, the O-ring joints were officially classed as critical. At postflight evaluation of two launches in February 1984 and again in January 1985, erosion of the O-rings was observed, indicating they lacked the ability to adequately seal. In July 1985, a full two years before the Challenger disaster, the engineering contractor, Morton Thiokol (MT), officially expressed concern to NASA. Yet no substantive design changes were ever made.

In fact, for shuttle launches beginning in July 1985 and continuing up until the fatal Challenger event, launch constraints were routinely placed relating to the O-ring problems. The Solid Rocket Booster technology manager regularly waived each constraint, thus resulting in launches without restriction. Both NASA and MT appeared to process issues and concerns about O-ring problems, yet each exhibited

---

40 "Report of the Presidential Commission on the Space Shuttle Challenger Accident" (Washington, DC: Presidential Commission on the Space Shuttle Challenger Accident, June 6, 1986) http://www.dtic.mil/dtic/tr/fulltext/u2/a171402.pdf

a systematic common failure to effect any remediation. *They accepted that each close call that did not cause an actual flight failure to be a validation of their decision not to take the matter seriously.*

Prelaunch Command and Control Management Failures

No information concerning launch constraints and the six previous waivers were made known to the management of Flight 51-L at the time of the Flight Readiness Review. It was known, however, that there was a significant possibility for O-ring failure in cold weather launches. At 2:30 p.m. EST on January 27, 1986, an engineer at Morton Thiokol raised a serious warning about low temperature at a Thiokol internal meeting. These concerns were raised with NASA with the MT recommendation not to fly if the predicted low temperature was present. Under intense management pressure from NASA, MT reversed its no-fly recommendation and the launch was allowed. It should be explicitly noted that engineering specialties from neither NASA nor MT participated in that last-minute reversal. They launched the shuttle. It exploded 73 seconds later.

Summary

This is an unfortunate but clear illustration that no matter how robust the design process may be, no matter how proceduralized the review process may be, and no matter how fully the current situation is evaluated, unless management follows their own rules, it is all for naught. And this process must permit the essential full consideration of minority concerns.

BP and Transocean Deepwater Horizon

This incident was carefully laid out in Chapter 1, "Getting Started."[41] Here we will observe the specific institutional conflicts between management and operations effectively saying, "Do what I say and not what I do." At approximately 10:00 p.m. CDT on April 20, 2010, in the Gulf of Mexico 48 miles off the US coast near the state of Louisiana, there was a fire on board the Transocean *Deepwater Horizon* drilling ship with a resulting loss of life of 11 individuals and the injury of 16. The drilling ship sank the next day. This caused the nation's second largest man-made environmental disaster (after the 1930s Central Plains dust bowl). At the time of the incident, Transocean had the worst quality record of deepwater drillers for two years running. It had merged with GlobalSantaFe Corporation. This led to a significant reduction in US-based engineering and design staff. Two years before that, it had had one of the best. BP has had

---

41 Fred H. Bartlet Jr., "Macondo: The Gulf of Mexico Gulf Oil Disaster," Report to the President of the United States (Washington, DC: National Commission on the BP Deepwater Horizon Oil Spill and Offshore Drilling, 2011).

a colored record punctuated by the 2005 Texas City disaster and then the *Deepwater Horizon* disaster.

Operation Prior to the Incident

On the basis of misleading declarations regarding probability of incident and state of emergency preparedness, BP was not required to file an environmental impact statement, nor have special remediation plans or ability at hand during operations. Transocean, prior to the incident, made modifications to the blowout preventer (BOP), the primary safety device that it knew increased the likelihood of failure. It was the failure of this device to properly operate that enabled the massive quantity of escaped oil. BP was also warned that the specific metal casing used for drilling might collapse under the high pressure, yet it was used. In the months prior to the disaster, there were unusual problems with sudden gas releases, pipe falling into the well, and several failures of the BOP. No formal operational assessment was done nor changes made. The work was days behind schedule, and management had been complaining.

During the Incident

"There were several serious warning signs [on the *Deepwater Horizon*] in the hours just prior to the explosion."[42] Instrumentation indicated that gas bubbling was occurring, which suggested an impending blowout. Against prevailing judgment and protest, the heavy sealing drilling mud was replaced with seawater (which is much lighter). But the protest was weak because it was understood that workers could get fired for raising safety concerns. Important operational alarms were turned off (ostensibly due to their annoyance factor).

Postscript

At review, "A House Energy and Commerce Committee statement in June 2010 noted that in a number of cases leading up to the explosion, BP appears to have chosen riskier procedures to save time or money, sometimes against the advice of its staff or contractors."[43] Moreover, "On July 22, Sky News reported that in a survey commissioned by Transocean, workers on *Deepwater Horizon* raised concerns 'about poor equipment reliability, which they believed was a result of drilling priorities taking precedence over maintenance.' In the survey, carried out in March 2010, 'less than half of the workers interviewed said they felt they could report actions leading to a potentially 'risky' situation without any fear of reprisal . . . many workers entered fake data

---

42 Henry Fountain and Tom Seller Jr., "Panel Suggests Signs of Trouble Before Rig Explosion," *New York Times*, May 25, 2015, http://www.nytimes.com/2010/05/26/us/26rig.html?src=me&ref=us.
43 "BP Engineer Called Doomed Rig a 'Nightmare Well,'" *CBS News*, June 14, 2010, http://www.cbsnews.com/stories/2010/06/14/national/main6581586.shtml.

to try to circumvent the system. As a result, the company's perception of safety on the rig was distorted, the report concluded.'"[44]

With flawed institutional auditing and the apparent lack of attention to safe and reliable operation, operators and others charged with responsible operation will be unable to carry out their duties effectively.

7.16 Close

What should you make of all of the pitfalls and their potential effects on operation? Everything here is real. And much is documented by experience. Anything can keep your operator from doing an effective job. Everything allowed to progress unmanaged in the control room will lead to problems. These problems could and should have been avoided. The takeaway from all of this is why situation management must be what it is. Effective design will use the technology and practices that you construct and utilize in the control room to minimize unwanted effects. The bottom line is that you will need to build practices that require the operator to *always look for what he does not expect* with an open mind and an inquiring attitude. Here are a few of the ways to do so:

1. During the shift, the operator must view the plant or operation by asking the question, "What might be going wrong that I have so far missed entirely?"

2. During the shift, the operator must examine all the collective clear indications of things being normal and ask, "Which of the indications of normal might be incorrect?"

3. As new information comes to the operator's attention, the operator must test the new information and prove that it does not require a modification and/or rethinking of the rest of the information around that particular operation or activity.

4. At shift change, the relieved operator must orient the relieving operator with only demonstrated facts and conditions. He should skip communicating entirely any conclusions about normal or abnormal until the end so that it is clear to both that those conditions must be proved independently by the relieving operator (see number 1).

---

44 "Gulf Of Mexico Oil Disaster: Transocean Reports Highlight Workers' Concerns Over Deepwater Horizon," *Sky News*, accessed July 23, 2010, http://news.sky.com/skynews/Home/Business/Gulf-Of-Mexico-Oil-Disaster-Transocean-Reports-Highlight-Workers-Concerns-Over-Deepwater-Horizon/Article/201007415669165?lpos=Business_Third_Home_Page_Article_Teaser_Region__5&lid=ARTICLE_15669165_Gulf_Of_.

5. At shift change, the relieving operator must prove that there are no abnormal or important situations that he is unaware of and that might be missing from the shift handover exercise.

Even well-trained and alert operators make errors. The unfortunate few are significant. Where the process or operation is forgiving, the consequences of error are usually a near miss, perhaps a bit of embarrassment, and always a relief. Where operational "forgiveness" is not there, impactful incidents result. Where operational limitations are fully challenged, catastrophes result. Remember, this is about well-trained and alert operators. Now you know that much more is going on in the control room that affect operators. Now you understand that carefully designed and implemented procedures and policies for operation must be in place and used by everyone, all the time. *Those procedures and policies must specifically take into account the naturally occurring cognitive foibles and limitations of the human operator.* This book has been conceived and designed with just that intent in mind. You are invited to take full advantage.

7.17 Further Reading

Endsley, Mica R., Betty Bolté, and Debra G. Jones. *Designing for Situation Awareness: An Approach to User-Centered Design*. Boca Raton, FL: Taylor & Francis, 2003.

Seconds from Disaster, season 6, episode 7, "Runaway Train," aired December 10, 2012.

8

Awareness and Assessment Tools

Intuition will tell the thinking mind where to look next.

Dr. Jonas Salk (Physician; Discoverer of the Polio Vaccine)

When you find yourself on the side of the majority, you should pause and reflect.

Mark Twain (Author; Samuel Langhorne Clemens)

Operators must ever look for problems, issues, irregularities, and anything else that might be useful to point out operational concerns that need knowing. This task is an enormous responsibility. Think about it. Your operator is the next to last line for the enterprise's ability to manage serious operational challenges. Anything missed or handled ineffectively leaves the enterprise's sole means of protection entirely up to robust and safe design, the safety shutdown systems, and physical protections and containment. When carefully and responsibly designed and maintained, they generally do the job. But it is at a cost. Emergency shutdowns, automatic shutdowns, and containments are not pretty. They do not do their job gently. And they will not necessarily position the enterprise for a comfortable restart, if restart is even in the cards.

Operators are therefore tasked to use their competency to observe and collect clues of impending problems and concerns while they could be manageable short of other harsh, protective systems. The operator's job is to recognize the obvious and also search out and bring to the surface the hidden, obscured, or subtle goings on of their operation as early as reasonable. To be useful, *situation awareness* requires the suspension of any preconceived assumptions about problems and outcomes. That is the job of *situation assessment*. That comes later on, not here, not now. We want operators to see what is lurking about first, *without any attempt to assign any meaning to what they find*. Just look. Collect the clues. Then in situation assessment we will work out what it means.

8.1 Key Concepts

Operators Are Not Born with Situation Awareness Genes	The capability to find out or figure out what is going on inside the process or operation must be provided for by the enterprise. There is no reasonable way that any individual can be expected to keep on top of subtle operational threats without competency training and purpose-built tools.
Knowledge Fork	All information, data, procedures, and operating requirements fit on one of the three tines of the fork: known, unclear, or assumed. Nothing but knowns can be used. Throw away the rest unless it can be fully verified as known.
Role of the Alarm System	Each alarm activation is a situation in need of management. Alarms are designed to provide the last clear opportunity for operator intervention to resolve an abnormal situation before it challenges design limits or safety systems, if present.
Abnormal Situations	All abnormal situations that need operator attention must be either alarmed or identifiable by the operator using other tools in the operator tool kit (not left up to chance).
Role of Messages	Alerts, notifications, and messages provide predetermined notice (unprejudiced—not necessarily good or bad, just information) of states of operation, risks and irregularities, and specific tasks to consider doing.
Shift Assessments	The shift-handover process can provide a powerful protocol and tool for in-shift assessments of operational status.

8.2 Introduction

Situation awareness does not happen just because our operators need it. Carefully crafted tools, protocols, and competencies must aid it. It is a vitally important skill for operators to have. The better they are at understanding and managing data and information, the better the job is done. Most of their capability is usually built around experience, availability of useful procedures, intuition, and no small measure of luck. Luck, in that bad things that happened during the shift were manageable. Luck, in that unmanageable situations did not occur on the shift. This chapter takes the operator's responsibility for knowing what is going on away from luck, ad hoc, and "just keep an ever watchful eye" to the level of a supportive technology. This is a competency that we now know much more about how to provide. The technology addresses *what* to provide and *how* to provide it for a properly effective capability. The tools for that job are in this chapter. In Chapter 6, "Situation Awareness and Assessment," we went over how operators develop the ability to understand and work with situation awareness. Now you will see the working power of the tools. By understanding and being able to design and implement them, you can provide an important measure of capability to the operator for *situation management*.

Knowledge Fork

We want to be able to bring information, opinions, and observations into play. Let us work on making sure all are clear, accurate, and verified. The *knowledge fork* helps ensure that each bit and piece brought to the table is dependable (see Figure 8-1). Every observation, every bit of data, and every other item that is put on the table must be clear. As each bit of information (including observations as well as procedures that appear to be relevant) comes in, use the tines of the fork and classify it as known, unclear, or assumed. Make sure that everything is classified. Use written notes, annotation, and sticky notes on documents or screenshots to track it. This can be done as each bit comes in or after a few are in. Remember that the later it is done, the easier it is for assumptions and unclear parts to lose their proper category and just fall into the basket of knowns. Only known information can be relied on. The rest must be ignored until properly verified.

The knowledge fork categorization must be used to clarify all information used by operators. It needs a prominent place in the tool kit! Operators will use it to pick up everything.

Awareness and Assessment Situation

The job of your operator is to successfully operate the plant or operation. Success means that operations are in conformance with the requirements. During the assigned shift or operations window, operators will utilize skills, employ available tools, and rely on established protocols and equipment. Operators must be aware of anything

Figure 8-1. Knowledge fork with the three "tines" describing how to qualify every bit of information available and intended to be used for problem identification and remediation (if needed). These classifications ensure that decisions are made from only known information. Anything unclear or assumed must be verified to the level of "known" to be used.

going on that represents a threat or potential threat. They need to take appropriate remedial actions against those threats. Without awareness, no operator would know that anything might need assessment and management. So, what does awareness need to look for?

> Only a few incidents started with the sudden failure of a major component. Most started with a flaw in a minor component, an instrument that was out of order or not believed, a poor procedure, or a failure to follow procedures or good engineering practice.[1]

A cornerstone of situation awareness is the alarm system. A proper alarm system aids awareness. Alarms are used to identify all problems requiring operator intervention that result from all abnormal situations that enterprise designers and operations experts identify. This critical understanding for alarm system design leads to a high level of alarm system effectiveness. Figure 8-2 depicts the general situation. The desired state for the plant to be is *process normal*. When things go wrong enough, some part of the process

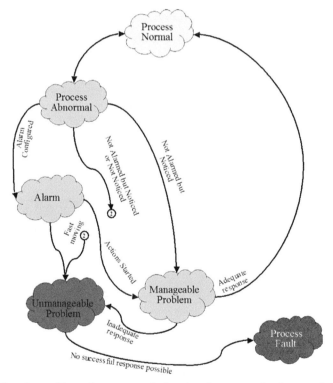

Figure 8-2. Operational conditions from normal showing the current primary structure in the control room. Most conventional operator monitoring relies on either alarms or serendipitously coming across the other problems before alarms. This chapter adds other tools to this.

---

1 Trevor Kletz, *What Went Wrong? Case Histories of Process Plant Disasters*, 3rd ed. (Houston, TX: Gulf Publishing Company, 1994).

becomes *process abnormal*. If alarms are configured for these abnormal conditions, an alarm will activate. If the problem is manageable and the operator works the situation successfully, the situation will return to process normal. On the other hand, if the situation is an *unmanageable problem*, the operator will not be able to prevent a *process fault*.

If alarms are not configured for the process abnormal condition, but the operator notices that there is a situation that needs attention, one of several outcomes could occur. If it is a *manageable problem* and it is done correctly, the process normal will be restored. If it is done incorrectly, or if the process abnormal situation is entirely missed by the operator, the situation will be an unmanageable problem and a process fault will result.

How will the operator find all of the ways that threats might impact a plant or operation? What can operators use to find them? Besides waiting for alarms to notify of threats, operators have little else in the way of specific tools and other defined resources. In an attempt to fill this gap, many planned ways have evolved. A partial listing includes the following:

- Carefully designed HMI that is closely monitored
- Clear and complete operating procedures and protocols
- Routine scheduled monitoring of important operating parameters and conditions
- Periodic monitoring of operations looking for problems or concerns
- Using materials and energy balances
- Preparing for and conducting shift handovers
- Conducting training in looking for problems, issues, and potential concerns

All of these are rather general. With the possible exception of the balances, they do not much relate to any specific minute-by-minute ways an operator would be able to identify potential problems. This means that for most operating situations not covered by alarms, the operator might not have enough help.

Figure 8-3 collects all these situations within the "cloud" labeled "process is not normal but not yet abnormal." For the moment, focus on the top four clouds: *process normal, process abnormal, alarm,* and *process is not normal but not yet abnormal*. We have already gone over the pathway from process normal to process abnormal to alarm. The

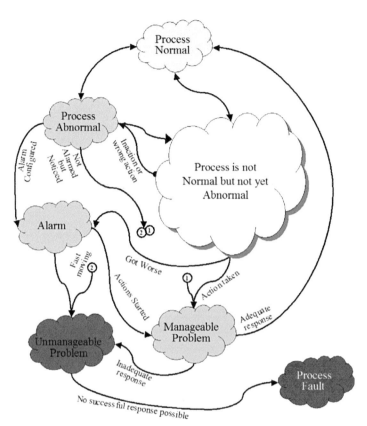

Figure 8-3. Operational conditions from normal showing an extended awareness structure in the control room. In addition to alarms, the process of identifying and evaluating off-normal situations is added to the kit.

rest of the problems the operator must deal with are contained in the cloud *process is not normal but not yet abnormal*. There are four states that the plant can move to:

1. Return to process normal.
2. Move to process abnormal.
3. Go directly into alarm.
4. Be judged a manageable problem (for the operator to begin to manage).

The first three states constitute the usual operator situation. They will have adequate procedures, and the operator is trained to handle them. We will now work on extending the operator's skills by introducing situation awareness for the operational state where their plant or operation is not normal but not yet abnormal. This chapter provides many clear and useful tools in those four situations. We will start with the alarm system.

8.3 Alarm System

A proper alarm system is a cornerstone for responsible operation of industrial plants. It enables operators to be notified of a subset of abnormal situations in time to make the difference. Providing alarms for all abnormal situations needing operator intervention is part of the process design and operation requirements. These abnormal situations should be specifically identified during the process design, implementation, and operational design activities. Figure 8-4 shows how alarms fit in. Note that operators also need a way to identify those other abnormal situations not covered by alarms. This chapter and the next are designed to find those. Finding and managing them is a core requirement for proper situation management.

An alarm is an announcement to the operator initiated by a process variable (or measurement) passing a defined limit as it approaches an undesirable or unsafe value. The announcement includes audible sounds, visual indications (e.g., flashing lights and text, background or text color changes, and other graphic or pictorial changes), and messages. The announced problem requires operator action. An alarm is a construction by which an aspect of manufacturing operation is identified and configured in a binary way to be either *in alarm* or *cleared* (i.e., not in alarm). The condition of *in alarm* is passed to an operator via intrusive sounds and notices placed on video display units or other devices to gain attention. The operator can manage these sounds and notices only via specific *silence the alarm* or *acknowledge the alarm* actions using the existing, planned infrastructure of the alarm platform. Usually, this alarm platform is an integral part of the process control system (PCS) infrastructure.

Alarm Fundamentals

Successful alarm systems are built on four fundamentals (Figure 8-5). Proper alarm design is a straightforward engineering process. Everything you know about

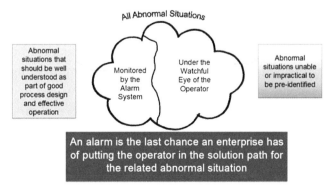

Figure 8-4. Alarms for abnormal situation identification are the primary responsibility of the process and equipment designers to build. All the other abnormal situations are left up to the operator to find, somehow.

Figure 8-5. Alarm design fundamentals. Every configured alarm must satisfy all four. A configured alarm that does not satisfy all four should not be an alarm (unless a special situation demands it and everyone accepts it as an exception). Anything that is not already an alarm, but satisfies all four fundamentals, should be an alarm.

engineering, system design, and project execution remains useful. Intuition is going to be useful. Moreover, while there are some who believe that alarm improvement is a complicated, delicate, and unforgiving process, it is certainly not. Yes, experience can be useful. If it is available, embrace it. But do not let anything get in the way of relying on these four fundamental requirements for an alarm—every alarm. Together they form the foundation of everything we need to know about alarms. They govern all successful alarm system designs. Using them should resolve almost every simple and usually every difficult decision you will face in understanding and designing new alarm systems that work.

1. Every alarm requires timely operator action, and that action must be a necessary one.

2. Every alarm activation must occur in time to permit the operator to successfully remedy the situation, if that remedy is at all a reasonable outcome (given the realities of the situation).

3. Adequate information must be provided to the operator for how to work the alarm.

4. Only alarm important conditions/situations.

Anatomy of an Alarm

Figure 8-6 represents a single high (process measurement) alarm configuration situation. The variable alarmed has a range limit between engineering units LOEU and HIEU. In this example, the measurement is a flow rate measured in cubic feet per minute (CFM) with an LOEU of 0.0 and an HIEU of 1,000. A normal range for the flow rate is illustrated between 425 and 800 CFM (as shown by the gray rectangle). The production operation would be abnormal if this flow rate exceeded the 925 CFM High Trouble Point (as shown by the red dashed line). If the flow appeared to be able to exceed 925 CFM, the enterprise has decided that the operator should intervene before it did to make sure that it did not. It was determined that if a process alarm were set to activate

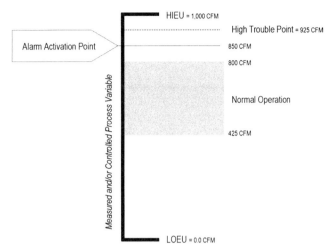

Figure 8-6. Anatomy of a configured alarm. The normal range of flow is between 425 and 800 CFM. If this flow increases to 925 CFM (High Trouble Point), it causes production problems. To ensure that the operator has enough time to intervene to keep this from happening, an alarm will activate at 850 CFM.

at 850 CFM, the operator should have enough time to make the appropriate evaluation and take the proper actions to keep the flow rate below 925 CFM. The alarm activation point must always be between the normal operating range and the trouble point. If it is not, an alarm cannot be used to manage this abnormal operating situation.

This section will provide an introduction to and overview of the basics of a proper alarm system and the process for making it a reality in your enterprise, called *alarm management*. Please accept that while alarm management might be discussed and presented in a limited way here, it is a careful discipline with important work processes and activities. Please refer to the reference book *Alarm Management for Process Control* for a complete comprehensive coverage together with detailed procedures and practices you can use.[2]

Alarm Philosophy

The first thing is to decide the overall alarm management plan. We call this the *alarm philosophy*. The philosophy is the design basis for the improvement of an enterprise alarm system used by operators responsible for maintaining appropriate production. It contains the foundational decisions needed for a working alarm design for alarm improvement programs. It is the fundamental design guide for evaluating, designing, and improving site alarm systems. While the basic principles are well known, each enterprise may place different importance on the various aspects. In addition,

---

2 Douglas H. Rothenberg, *Alarm Management for Process Control—A Best Practice Guide for Design, Implementation, and Use of Industrial Alarm Systems*, 2nd ed. (New York: Momentum Press, 2018).

each enterprise may have special requirements and objectives that only a customized philosophy can adequately cover. All key enterprise personnel will develop the plan together. They will have been empowered to provide the recommendations to management. All site personnel, all contractors, and all consultants will rely on the alarm management plan. Incident investigations will use it.

Note that all detailed specifics for implementing the philosophy will be located in supplementary documents that must comply with the philosophy in all respects. This would include the activity of changing the alarm system parameters in the control system (distributed control system—DCS, supervisory control and data acquisition—SCADA, programmable logic controller—PLC, etc.), the redesign and modification of HMI graphics, the redesign and modification of operating procedures, the redesign and modification of training, and the myriad of other aspects necessary for carrying out the new alarm system design.

Alarms

Unless every abnormal situation will be manageable without operator intervention by the existing production and controls equipment, a plant cannot be safe without a properly designed and maintained alarm system.[3] Alarms are the formal design tool to direct the operator's attention to a problem that requires attention. It is the plant or operation's last chance to put the operator into the solution path for the specific problem before the rest of the safety management systems are challenged. Make no mistake, if the operator is not able to bring things back sufficiently, the only remaining lines of defense are the safety shutdown systems, the physical design strength of the processing equipment to withstand a seriously abnormal mode of operation, the physical separation of the equipment from personnel and environmentally sensitive areas, and the ultimate dangerous nature of the plant or operation.

Alarm Management

Alarm system design is governed by four critical design fundamentals. To reiterate, every alarm must:

1. Require an explicit operator action (no action needed, no alarm should be configured)

2. Activate in time for the operator to respond appropriately (if not possible, no alarm should be configured for this abnormal situation)

---

3 Tyron Vardy, "Can Luck Cover for a Lack of Alarm Management?" *Plant Engineering*, August 26, 2014, https://www.plantengineering.com/single-article/can-luck-cover-for-a-lack-of-alarm-management.

3. Have clear and effective procedures for operators to use in order to remedy the abnormal situation indicated by the alarm activating

4. Be important enough to require the operator to cease all other duties (except perhaps working other alarms) to attend to this one

Alarm management is all about the understanding, design, implementation, and operation of an effective alarm system capability for production plant operators. Alarm management is the design and implementation process for the entire redesign of the portion of the process control system capability that is used to alert operators to conditions where an alarm is needed. The minute-by-minute managing of an alarm is but one small part of this much more encompassing technology. The full process includes the following:

1. Benchmark analysis of present alarm system performance, including its impact on production, safety, and the environment

2. Development of a philosophy governing the operation of the enterprise sufficient to specify a design basis for the required alarm system and supporting plant infrastructure

3. Selection of which variables to alarm

4. Setting of alarm limits

5. Setting of alarm priorities

6. Determination of recommended operator actions (alarm response sheet)

7. Design of advanced techniques to facilitate improved alarm performance

8. Addition of plant condition monitors and decision support tools

9. Incorporation of new alarm system design back into the plant infrastructure

10. Continual audit, assessment, and modification for improvement

Alarm management is a process. When successfully done, alarm management will result in a fully functioning alarm system suitable to meet production requirements to better realize enterprise goals. Not only does it require a complete redesign of the alarm system itself, it will also need the entire rest of the production infrastructure to be supportive. This includes the human-machine interface (HMI), operating procedures, training, incident investigation, equipment maintenance, and management policy.

Alarm Rationalization

A properly operating alarm system provides situation awareness, assessment, and management to all abnormal situations known in advance (by designers, through operating experience, or by other means of knowing beforehand). Rationalization is the work process for alarm design. It is the process and technology by which each needed alarm is selected and the configuration design and supporting information for the task is built. In this section, we will learn the objectives of alarm rationalization and the steps to complete a rationalization. Rationalization is a structured process. It includes deciding which points to alarm (including calculated variables), determining the alarm activation point, setting the priority, and all other remaining alarm response information including potential causes, ways of confirming the existence of the abnormal situation, appropriate operator responses, and likely consequences of error. Specifically, alarm rationalization is the work to complete items 3 through 7 in the earlier task list.

The objective of alarm rationalization is to create an alarm configuration with the correct number and configuration of alarms. When properly and carefully done, the new configuration should result in significantly fewer alarm activations. Additionally, those alarms that do activate will be important and provide useful operator guidance. There is only one recommended basic approach to rationalization: start from zero (no alarms, build them up as required).

"Starting from zero" begins with a blank slate (or blank sheet of paper; the "white sheet," if you will) by initially assuming that the entire enterprise has no configured alarms whatsoever. The enterprise is first divided into its primal set of smaller components, both for ease of understanding and to facilitate the design process. This is why we spent a lot of time in Chapter 2, "The Enterprise," on decomposition. Here is where we put that to good use. There are two categories of smaller components: small ones called *key repeated elements* (e.g., a pump, a compressor, an on-ramp to a highway) and large ones called *key subsystems* (e.g., a pumping station, a distillation tower, a line segment in a transportation pipeline). For each key repeated element, an analysis is conducted to decide the minimum number and type of alarms needed to manage it properly. This method is currently the best practice as it significantly bolsters the ability to not miss important abnormal operational situations requiring alarms. The result is to make sure that each configured alarm will be understandable, prioritized, relevant, unique, and timely. For most plants, a successful rationalization results in a significant reduction in the number of configured alarms—on the order of five- to tenfold, which results in a reduction in alarm activations during operation. It is not unusual to completely eliminate alarm activations during normal operations, to have very few alarms during abnormal operations, and to have a manageable number of alarms during upsets.

Key subsystems are built up by using as many of the key repeated elements as needed and supplementing with what else is needed. At this point, you will have a substantial list of enterprise pieces framed out. Use each piece as the starting point for all other similar ones throughout the enterprise. As you use each, make any necessary changes and modifications to match the part you are working on. There are two approaches to working through the plant equipment and infrastructure to ensure that all aspects are covered:

1. The method of flows

2. The method of elements

Refer to Figure 8-7 for the illustration of how this works for *flows* and to Figure 8-8 for *elements*. Note that in the figures, flows are identified as F-#, pressures as P-#, temperatures as T-#, analyzer values as A-#, and so on. The green arrows identify the major workflow path. As one part is complete, the next part is begun following the green arrow direction. The method of flows is used by picking a major flow of whatever is being done in the enterprise: raw materials, pipeline contents, electrical power, and the like. Once one flow is done, pick the next one and so on until you are done. To use the method of elements, pick a large component in the enterprise and work your way completely around it until everything is covered. Then pick the next one and next one until you are done.

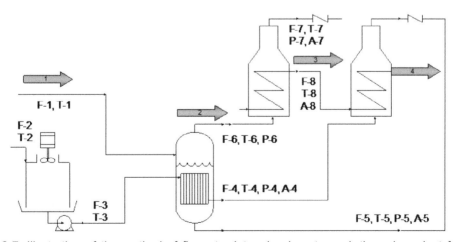

Figure 8-7. Illustration of the method of flows to determine how to work through a plant for alarm rationalization. This method is best for significant flows where understanding what is happening before aids understanding of what follows.

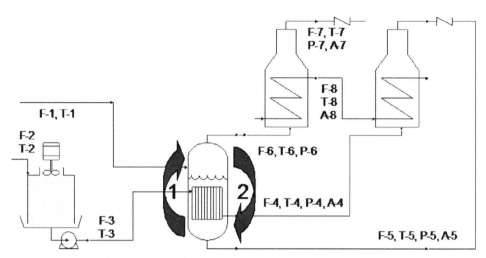

Figure 8-8. Illustration of the method of elements to determine how to work through a plant for alarm rationalization. This method is best where understanding what is happening fully at each element is most efficient.

Each method is designed to build on the working experiences gained by working on the parts done just before. As one examines the preceding work, it will be necessary to look a bit at what is around where you are working. It makes sense because you have already had a look around at things close, so why not move to one of the things you looked at a bit before? Those things you have looked at are easily noted from entries in the alarm response sheet (see the next section), so picking the next thing can be a simple choice.

Choose one method or the other, or a combination of both ways to work your way through the alarm design from each abnormal situation to the next until the design is complete. Best practice is to use the method of elements to select an element and rationalize it fully. Then select the next element by using the method of flows. Proceed this way through the entire operator area.

Rationalization begins with a review of the alarm fundamentals, examines the alarm philosophy to identify the specific work process to be used, and identifies the key performance measures of success. The steps are as follows (starting from zero approach):

1. Identify all "required" alarms and properly rationalize them. (Note: A "required" alarm is any alarm that must be configured due to statutory requirement, enterprise standards, HAZOP review, or other explicitly identified methodology that supersedes the generally accepted practice of alarm design under the requirement of special need.)

2. Identify all important abnormal situations that the process activity might exhibit.

3. Select the most appropriate alarm(s) to notify the operator of each abnormal situation.

4. Build the complete alarm response sheet for each alarm.

5. Determine the alarm priority based on predetermined risk management protocol.

6. Determine the appropriate alarm activation point (alarm set point) to ensure enough time for the operator to be able to respond.

7. Review the need for advanced alarm controls and propose options for implementation.

8. Continue the process until all abnormal situations requiring operator intervention have been covered.

Alarm Response Sheet

The alarm response sheet documents all the operator knowledge surrounding the proper handling of an alarm once it has activated. Here is what a typical one looks like (Figure 8-9).

The details of the sheet will be explained below. Note that the text in parentheses is from the fields on the alarm response sheet to show you what is what.

Header Information
The header block contains all the descriptor information and most of the configuration information. The first item is the Tag ID (FI-2009). Next, the point descriptor, which is the actual wording, is used for identifying the tag in most text fields within the PCS (PB Feedwater Flow). There is a place for general Comments (blank in our example). The nominal control Set point is shown (195) and Units (KLbHr). The Alarm Priority is shown (High). Finally, it includes the location of the point within the PCS point architecture, Operating Group (12), and Process Area (PB).

Configuration Data
This section contains the range values for the analog signal and the alarm activation points for all configured alarms. In our example, the tag FI-2009 has a single alarm configured, a high flow (High Alarm) with an Alarm Point (200). The upper (275) and lower range limits (0) are also provided. Examine the remainder of the sheet next.

```
         Tag ID:  FI-2009                    Alarm Priority:  High
Point Descriptor:  PB Feed Water Flow         Alarm Status:  Enabled
       Comments:                            Operating Group:  12
       Setpoint:  195   Units:KLbHr           Process Area:  PB
     High Alarm:  200   (0 to 275 range)
```

Abnormal Situation:
1. High steam consumption
2. Unusual boiler operation

Causes:
1. Failure of LVC/LIC-2011
2. Plugged flow element
3. Incorrect steam flow measurement
4. Instrument failure

Confirmatory Actions:
1. Check steam drum level LIC-2011
2. Check strip charts, check "periscope"
3. Check BFW header pressure FI-2078/2077, DEA Pressure PI-2098/2099
4. Check BFW pumps
5. Check flow control valve FCV-2009

Consequences of Not Acting:
1. Boiler could overheat and seriously damage tubes
2. Insufficient steam flow, high press.

Automatic Actions:
None

Manual Corrective Actions:
1. If steam drum level and steam pressure are high, put feedwater valve in manual and begin to close it.
2. If steam drum level is low, manually increase opening of feedwater valve

Advanced Alarm Considerations:
1. Disable alarm when the boiler is shut down
2. Enable alarm when boiler is started up

Safety-Related or Testing Requirements:
HaZop 12-4453; Yearly testing required

Figure 8-9. Example alarm response sheet. This is a fully described document that an operator will use each time an alarm activates. It includes all the information and steps to manage the specific alarm. Each alarm in the operator area will have its own alarm response sheet.

Abnormal Situation

This section contains the basis for deciding that the high flow should be alarmed. This alarm has been determined to be the best indication for two abnormal situations: (1. High steam consumption) or (2. Unusual boiler operation).

Causes

This section documents all reasonable causes that would likely lead to the alarm activating—that is, would cause a high-priority, high-flow alarm for FI-2009 to activate. In our example, there are the following four: (1. Failure of LVC/LIC-2011), (2. Plugged flow element), (3. Incorrect steam flow measurement), and (4. Instrument failure). Any one of these failures, if not corrected, would potentially lead to the high-flow alarm activating. Interview operators, engineers, technicians, and all others intimately familiar with this part of the enterprise, and compile this list by reviewing operating procedures and all incident reports. Incident reports examined include those relating to the primary concern, FI-2009, as well as all the other "cause" instrument issues or failures.

Here and in the next parts of the alarm response information, you can observe how the seemingly long list of background and operational documentation items needed for rationalization now becomes important and useful. It forms an important part of the backbone of operator knowledge.

Confirmatory Actions

Before we ask the operator to work this alarm, we must take the additional step of ensuring that the alarm is a true representation of a physical situation gone abnormal. The fact that our alarm has activated does not mean, for sure, that the physical flow (boiler feedwater) measured by instrument (called FI-2009) is actually flowing at too high a value. A transmitter might have failed, but the actual flow remains correct.

There are five confirmatory actions in our example (1. Check steam drum level LIC-2011), (2. Check strip charts, check "periscope") (strip charts refer to an existing analog pen recorder; periscope refers to a special level detector installed on the process), (3. Check BFW [boiler feedwater] header pressure FI-2078/2077, DEA [deaerator] Pressure PI-2098/2099), (4. Check BFW pumps), and (5. Check flow control valve FCV-2009). Any one or a combination of items from this list would confirm that our actual flow is too high. If none are out of the ordinary, the operator should consider that the flow is probably okay, but some instrument might be in error.

Yes, we do understand that one of the "rules" for responsible operating is "trust your instruments." Nothing will be said here to contradict that guiding principle. However, trust is not always good if blind. Experience has shown that a brief time spent confirming the alarm condition can prevent a great deal of wrong focus in situation management. Moreover, the simple process of checking and confirming will often uncover many of the direct causes for the alarm. No time would have been ill spent.

Consequences of Not Acting

Consequences are important to bring out here inasmuch as they play a large part in determining the priority. Alarm priority is later on. Just as important, providing the consequences here enhances the impact of the issue for the operator and assists the operator in understanding the underlying process operation. Not acting is a dual concern. First, we expect the operator to see this alarm after activation and start to work the issue. Of course, if there are other alarms of higher priority, then those will come first. This alarm is a high priority. Unless there are emergency ones activating or there are several other high-priority ones at the same time, our operator is expected to direct his efforts here. Second, even if our operator is quickly focused on this alarm, the consequences are meant to convey what will likely happen if efforts to "fix" the problem are not appropriate or not successful in time.

Returning to our example, there are two consequences: (1. Boiler could overheat and seriously damage tubes), and (2. Insufficient steam flow, high press. [pressure]).

Automatic Actions

Alarm activations are sometimes used (on rare occasion) as digital switches to automatically initiate a process operation. Sometimes they are used to automate a blowdown or start an overlevel protection pump or other odd equipment. On the other hand, there are times when the alarm condition happens so quickly that there will likely never be enough time for the operator to fully react and the process recover solely as a result of operator actions. In these cases, an alarm event might set in motion automatic actions to control the situation with immediate follow-up by the operator to make other needed actions. Those automated actions are documented here.

For our example, there are no (None) automatic actions.

Manual Corrective Actions

Manual corrective actions are the bread and butter of the operator's handling of alarm conditions. Often all that is needed to start the process heading back toward where it should be is to put the problem controller in manual. This will stabilize the process if process or controller variability was the likely cause of the alarm. Slight control adjustments may also be made. Other times, another related part of the process would be manually changed. At times, our operator may direct that the position or alignment of certain manual valves in the field be modified. Still other times, our operator might shut down one part or another in response to broken or malfunctioning equipment.

In our example, there are two possible recommended manual actions: (1. If steam drum level and steam pressure are [both] too high, put the feedwater valve in manual and begin to close it [thus diminishing the flow until it is back into an acceptable region]); and (2. If steam drum level is low, manually increase opening of feedwater valve [to add more feedwater]).

Advanced Alarm Considerations

Here is where any alarm activation or other controls on the alarm are identified. Remember that a foundational principle of alarming is that all alarm activations must be actioned by the operator. If the equipment is in a state where the alarm activates and nothing must be done, then the alarm must not activate. In our example: (1. Disable alarm when the boiler is shut down) and (2. Enable alarm when the boiler is started up).

Safety-Related Testing Requirements

Some alarms are used to recognize and manage more serious operating conditions. Many of these might not be so apparent on the surface. Consequently, our alarm might have been added due to a specialized hazard condition. Alternatively, it might be present as part of a safety warning system but does not require a safety integrity level (SIL) rating of its own. Moreover, the instrument itself might be prone to certain degradations that must be checked. Testing may be a legal or contractual requirement. In any event, this is where any of those unusual conditions are documented.

For our example this alarm is part of a HAZOP and requires testing (HAZOP 12-4453; Yearly testing required).

Example Online Alarm Response Sheet

It is all well and good to have collected this information. Its true value, however, becomes realized when it can be made available to the operator when needed. Some plants choose to integrate links to the alarm response sheets. Figure 8-10 illustrates what this might look like as an online operator support tool. The figure contains mock information, so please do not examine the actual text too closely for content.

A control loop faceplate is depicted on the left of the figure. The button at the lower right (with the check mark) is the link to check the alarm response information. The right portion of the figure is the information that it is linked to. There is a box that depicts all alarms for the selected tag. Eight alarms are depicted; they represent all possible alarms for this tag. However, in actual practice, you will see only one or very rarely two used. The two dialogue boxes at the bottom contain the response information as well as additional point descriptors and other configuration information. The example was prepared for use on an Emerson DeltaV DCS system. It links through a standard configuration parameter "ALM_Help."

Process Trouble Point

Before we decide where the operator action should be initiated (the alarm activation point; see next discussion), we will need to specify the portion of the operating region we want the operator to protect. Alarms are not designed to warn the operator that

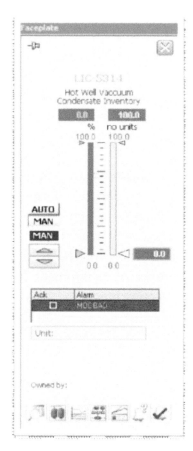 

Figure 8-10. Example online alarm response sheet. This view depicts a formatted view of the information in the alarm response sheet in Figure 8-9. It is designed to be accessed by the operator on the HMI to guide the alarm response activity. This one includes access to the loop controller in the event the operator must make adjustments there. Otherwise, the proper response adjustments will be made elsewhere.
Source: Reproduced with permission from Ergon Refining.

the process is becoming abnormal. Alarms are designed to call for operator actions to protect an abnormal situation from escalating to the point of causing harm or damage. Figure 8-11 illustrates the operational situation for alarms. The general operating regions are shown to include the Operating Target, Good Operating Region, Safe Operating Region, Upset Operating Region, Dangerous Region, and Damaging Region. Alarms (depicted on the figure by the red alarm bell icons) will be designed to activate within sufficient time for good operator action (if such action is possible) to prevent the process from crossing into the Dangerous Region. The approach for doing that will be described later on.

We need all the abnormal situations that can lead to serious problems to have been identified in advance and appropriate safeguards and alarms designed. Our job, our pursuit, our intent, is to pre-identify them and design for them as best we can.

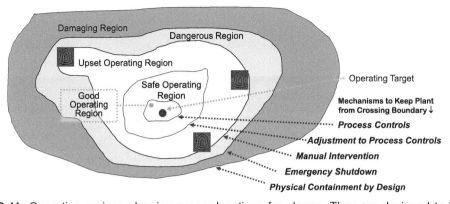

Figure 8-11. Operating regions showing proper locations for alarms. They are designed to help the operator see abnormal operation early enough to respond. Depending on how far from normal the problems arise, the response will involve increasingly more intervention. At the farthest distances, the operator is expected to move to a safe state instead of trying to keep operating.

Alarm Activation Point

One critical requirement is that the alarm activate in enough time for the operator to be able to manage the alarm (if management is at all reasonable in the specific circumstances). We will now examine how to determine appropriate alarm activation point settings. The proper setting of alarm activation points will be one of the more valuable parts of the knowledge that alarm management has to provide. These procedures will enable you to determine alarm activation points that ensure that the timing of alarm activations will have a reasonable chance to assist the operator to maintain production integrity. In Chapter 1, "Getting Started," we learned how important it is to be able to give the operator enough time to respond to a process upset. Sufficient time is necessary for the operator to detect a problem, understand its nature and extent, decide what to do about the problem, implement the changes decided on, and wait for the process to return to a satisfactory status. We call this *fault tolerance time* or *process safety time*. Figure 8-12 repeats an earlier figure from Chapter 1.

The bottom-line message is that in order for the operator to have a chance at managing process upsets, we must arrange for the notification to the operator of abnormality sufficiently early to be able to correct things. For that to happen, the Time to Manage Fault must be less than the Process Safety Time. If this is so, then the process should have sufficient time to recover. We use this concept to determine the alarm activation point—that is, alarm activation points are set not near the level of unsatisfactory behavior but early enough to provide enough time to manage the upset.[4]

---

4 Robert Weibel and Douglas Rothenberg, Alarm management system, US Patent 7,936,259 B1, filed December 17, 2008, and issued May 3, 2011.

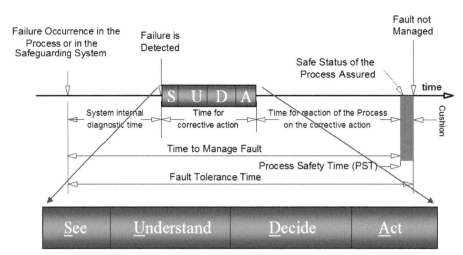

Figure 8-12. Process safety time, fault tolerance time, and SUDA (see, understand, decide, and act) time illustrating the basic concept that alarm management is time management. Alarms are designed to activate early enough to provide the operator response time needed to successfully manage that specific abnormal situation.

The actual time needed for all of this is the time it will take the operator to complete SUDA and the time it will take the process to respond enough to avoid reaching the unwanted fault. However, just barely having enough time would be too close. We add a cushion time (the region in green) to account for variability and other intangibles. The determination of the proper time will require specific information about the process and alarm response. The technical challenge here is to convert needed time into alarm settable values for the control system (e.g., change time into measurement value).

We discuss this via the following illustration. Consider the data and calculations shown in Figure 8-13. We show a temperature variable. The plant has determined that most temperature variables can change about 6°F per minute for fast ones, about 2°F per minute for normal ones, and about 0.5°F per minute for slow ones. Also, in general, it takes operators about 1 minute to manage easy loops and about 4 minutes to manage hard loops. Time to Manage includes SUDA and response times (basically, the effective process "dead time" during which a change in operational command appears to have no effect; after which the change appears). The enterprise philosophy has set a cushion at 1 minute (you can use either time or percentage of total time).

For this particular temperature, the process will get into trouble at or above 85°F. If the temperature loop were "fast" and "easy" to manage, the proper alarm activation point would be 73°F.

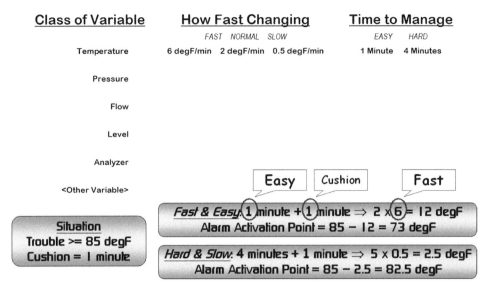

Figure 8-13. Example alarm activation point calculations for two hypothetical temperature alarms: one that is easy for the operator to manage, but the temperature changes quickly; and the other that is harder to manage (due to perhaps a more complex process activity) but slower changing. The dynamics and difficulty are specific engineering and operational characteristics. The actual computations are quite simple.

$$\begin{pmatrix} \text{Alarm} \\ \text{Activation} \\ \text{Point} \end{pmatrix} = \begin{pmatrix} \text{Process} \\ \text{Trouble} \\ \text{Point} \end{pmatrix} - \left\{ \left[\begin{pmatrix} \text{Time} \\ \text{to} \\ \text{Manage} \end{pmatrix} + \text{Cushion} \right] \times \begin{pmatrix} \text{Process} \\ \text{Change} \\ \text{Rate} \end{pmatrix} \right\} \qquad (8\text{-}1)$$

Equation 8-1. Formula for calculation of the alarm activation point to position the alarm to activate in sufficient time for the operator to reasonably act to remedy the underlying abnormal situation.

$$73 = 85 - [(1+1) \times 6] \qquad (8\text{-}2)$$

Equation 8-2. Formula for calculation of the alarm activation point for example situation where the trouble point is 85°F, the time to manage is 1 minute, the cushion is also 1 minute, and the process variable is changing at about 6°F per minute.

If the temperature loop were to be different, say slow and hard to manage, then the alarm activation point should be set at 82.5°F (in practice, you would probably use 82°F). While not shown in the figure, if the temperature loop were to be hard to manage and fast, the proper alarm activation point would be 55°F.

Being able to set a proper alarm activation point is central to alarm management. If one cannot be found, an alarm cannot be used to request the operator to manage that situation. No matter how important the situation is, no matter how much one wants the operator in the solution path, if the alarm cannot be activated in time for the operator to do his work, some other means must be found to manage the situation.

Alarm Priority

The specific procedure to set alarm priority is specified in the alarm philosophy. While there is much hand-wringing and wordsmithing around alarm priority, it is a really simple (but very powerful) process. The priority of any alarm is used to assign the proper level of operational risk to the abnormal situation indicated by that alarm activating. Alarms at the same priority level are assumed to represent an equivalent risk to the enterprise. This means that operators would work alarms in priority order, from highest to lowest until all available time is used. All alarms not worked will most likely mean that the underlying problem would result in unwanted things happening. By using priority to determine the work order, those unwanted things should be the ones at the lower end of risk.

Before we look into how priority is determined, we need to settle a potentially confusing nomenclature problem. We call some abnormal situations *critical*. We call other abnormal situations *emergency*. We call some threat levels *emergency*. The same words are used for different abnormal situation descriptors and at the same time used for alarm priority level descriptors. All of this can get confusing fast. The best way to handle this potential for confusion is to simply think of words like *critical*, *high*, and *emergency*, as just labels. As labels we will use them in more than one place: abnormal situations, operating procedures, alarms, and elsewhere. No one need spend any time or effort to keep the cross-uses consistent. A critical abnormal situation will not have a critical priority alarm configured for it, unless it actually qualifies for a critical priority alarm and there is an alarm priority level with the label *critical*. This bit of confusion is just part of how everyone uses labels. We are not going to resolve it here. Sorry.

Getting back to alarms, the current practice in setting alarm priority is to start with the existing approved enterprise risk management protocols and definitions. Forget about any of the probability or likelihood of occurrences (when you get the alarm, it really did happen!). Next, adjust those descriptions to use for alarms. Your risk management definitions are going to have several types of risk (called *impacts*) along the lines of safety, financial loss, and environmental exposure. For each type of risk, several levels (called *severities*) will be assigned for alarm use. They are usually arrayed in a matrix form. An example with guiding definitions for the *safety* consequence is shown in Table 8-1. Next, you will decide on definitions for *environmental* consequences; *financial* consequences, if you have those; and any other types you have to round out your entire set of consequences. Using the words in Table 8-1 as an example of *safety* you come up with how important each table cell is. Relative numerical weights for all consequences are then determined (done in the alarm philosophy you built, of course). The sample results are shown in Table 8-2. You can see how this all works in the alarm priority example later in this chapter. Here we lay out the necessary parts to get it started.

Consequence	Severity Level	Working Definition
Safety	None	Negligible
Safety	Low	Minor injury to one or more; perhaps requiring first aid but no lost time
Safety	Medium	Injury to one or more; possibility for limited lost time; all return to work in existing capacity
Safety	High	Serious injury to one or more; likely lost time; possible disability furlough
Safety	Critical	Very serious, usually permanent injury to one or more; possibility one or more deaths

Table 8-1. Alarm priority for safety *consequences* and *severities* definitions only. The table is read row by row. There are additional consequences that will be identified and explained in a similar manner (not shown); see Table 8-2 for their labels.

Each alarm is evaluated for its safety, environmental, and financial aspects. This means asking the alarm rationalization team what level of severity each of the consequences would have if this alarm were entirely missed (or managed improperly, or took too long to get to, or anything else that prevented good alarm management). Once the three consequence values (the appropriate ones for the specific alarm being worked on for safety, environmental, and financial) have been chosen, look up the numerical number for each from Table 8-2 (that were set by the alarm philosophy team much earlier), add them up and get a score.

Before we translate this "score" to an actual alarm priority, we first ask if there is something important about attending to this particular alarm from a time point of view. Yes, if we get to it now, the alarm activation point should provide sufficient time. But suppose this alarm's priority was almost on the fence between two priority levels. If the alarm needed to be attended to quickly, then might it be a good idea to bump it a bit to see if it moved into a higher level. Or if it did not need quick attention, might it be useful to work on it later than another that was quicker? All of this is computed into priority by using the urgency multiplier shown in Table 8-3. To use it, take the

Severity→ ↓Consequence	None	Low	Medium	High	Critical
Safety	0	25	75	100	300
Environmental	0	75	100	150	250
Financial	0	25	50	75	150

Table 8-2. Alarm priority consequences and severities values. The rationalization team uses a table like this but will use the word *descriptions* (Table 8-1 etc.) for each cell to guide them to determine the priority of a given alarm. In this table, the relative weights of each consequence shown for each severity level are shown that are used in the numerical computations for a candidate priority assignment.

Time Available (in minutes for effective action)	Multiplier (multiplies consequence-severity combined raw score)
> 30	0.9
> 10 but ≤ 30	1.0
> 3 but ≤ 10	1.2
1 ≤ 3	1.4

Table 8-3. Alarm urgency multipliers used to adjust the alarm priority numerical calculations to take into account how quickly an alarmed variable will actually get to the trouble point. This will slightly increase or decrease the priority for alarms that have a numerical value quite close to the deciding value between priorities (Table 8-4).

score from Table 8-2 and multiply it by the proper multiplier from Table 8-3 to get the new score.

Now take the new score and use Table 8-4 to determine the proper priority. This work process is straightforward and consistent. It does require the rationalization team to be particularly careful in making the specific judgments that go into the procedure. And like any procedure, once a result is obtained, it is always prudent to take a step back and ask whether or not the end result is proper. Sometimes, this process produces a priority that for some other reason is better set at a different value. Set it at the different value and comment why.

Alarm Priority Example

The alarm priority determination might at first appear to be quite complex and somewhat subjective. In actuality, it is neither. Experience has found that during alarm rationalization, there is a remarkable degree of consistency among the working individuals assessments and in the final results. As an example of the actual work process, consider a low-temperature alarm (on anything you might be thinking about in your

Priority	Breakpoint Value
Emergency	From 500 and above
High	From 350 up to 499
Medium	From 250 up to 349
Low	From 100 up to 249
[Might not be an alarm?]	0 up to 99

Table 8-4. Alarm priority from alarm score. Once the total score of the alarm is computed (Tables 8-2 and 8-3), this table is used to determine the recommended alarm priority. Notice that if the alarm score is very low, this is an indication that the candidate alarm is not important enough to actually be an alarm. Remember the four fundamental requirements for alarms?

particular plant or operation). Remember how we determined that we needed this specific alarm: this low-temperature alarm has been identified as the best indicator for the abnormal situation that requires operator action. Using the definitions from Table 8-1, the rationalization team decides that the safety consequence is *high*, the environmental consequence is *low*, and the financial consequence is *medium*. They did this by referencing the defining words from Table 8-1. Next, they looked up the corresponding scores from Table 8-2 and found 100 + 75 + 50 for a combined total of 225. They also evaluate the urgency (Table 8-3) as *high*. This means that the basic score of 225 should be multiplied by 1.2. This results in a new score of 270. A score of 270 maps to a recommended priority of *medium* (Table 8-4). As a reminder, the rationalization team should take a brief step back and ask, "Is *medium* the right priority for what we know about this abnormal situation?" If so, this alarm priority is set. If not, they would debate the merits, assign a better priority, and document their decision.

Alarm Rationalization Step-by-Step

First and foremost, make sure the team doing the alarm rationalization is experienced and together have sufficient knowledge of the entire plant and its operation. In addition to operators, you will need equipment specialists, operation support engineers, and safety and financial experts. At the end of the day, the enterprise will end up with a fuller and more grounded understanding of all operations. Let us review the best practices.

Before Starting
Identify all "required" alarms. Required alarms are those that, regardless of any other basic alarm management guidelines, must be present due to legal requirements, statutory requirements, or enterprise demands. All will be properly rationalized (step 1 through step 5), even if some or all would not have otherwise been selected to be alarms according to good alarm management practices.

Step 1
Identify all individual key components of a plant: for example, pumps, blowers, compressors, heat exchangers, reactors, crushers, separators, storage systems, and utility systems. This list should contain more than 90% of the parts of the plant. If appropriate, the list is further subdivided into large ones, small ones, steam-driven or electrical drives, and such.

Step 2
For each individual key component (one at a time), define its purpose or job. For example, the job of a heat exchanger is to move thermal energy between a process fluid and an energy fluid. Thoughtfully identify them all.

Step 3
Now, taking each key component, ask what process condition, operational condition, or issue can get in the way of the selected key component doing its job? For the heat exchanger example, only two things can get in the way: not enough thermal driving force between the process fluid and the energy fluid, and not enough flow through the exchanger (generally, an unacceptably high resistance to fluid flow in the tubes for a shell and tube exchanger) to enable sufficient energy transfer.

Step 4
Taking each condition or issue above one-by-one, ask: What is the best thing to alarm so that the operator will know there is a problem? For our heat exchanger example, it would be a too-low temperature difference between the two fluids, and it would be a too-high pressure difference across the tubes.

Step 5
For each alarm identified, produce the full set of rationalization and configuration information: causes, confirmatory actions, manual corrective actions, consequences, advanced alarm controls, testing, urgency, priority, alarm activation point, and any other needed documentation.

Step 6
Repeat steps 2 through 5 for all the other key components of the plant or enterprise. Now you should have a fully rationalized alarm set for each of the key components. These will be used as starting-point templates for those key components everywhere they occur in your plant or enterprise.

Step 7
Work through your entire plant by picking up the appropriate alarm template set for the key components as the starting point. Then "adjust" the template for what might be differences between the template and what exactly is there. Where no template exists, work it out using steps 1–5 for each.

Finally
Take a step back and look for anything missing. Examine the entire set of old existing alarms to see if anything in the new alarm set might have been missed or needs a revisit. You are looking for the odd missing abnormal situation. This is not a back door for bringing lots of old alarms back.

Alarm Rationalization Guides and Tools
Teams do the hard work of rationalization. Understand the process. ANSI/ISA-18.2, API RP-1167, EEMUA 191, and *Alarm Management for Process Control* are very helpful

references and guides. Using a purpose-built tool will facilitate the rationalization process, link up with the alarm configuration databases of the control systems, and directly export to online documentation files for alarm response.

Alarm Metrics

Alarm management practitioners often make a fuss over alarm operating measurements. While measuring alarm performance and their nuances may seem like a good idea, this should be approached with caution. The only measurements recommended relate to specific operational issues or alarm design issues. Specifically, make measurements to (1) properly design and maintain alarms and (2) determine operator loading. Any time taken up by doing more is better spent on something else. There are two areas of measurement: configuration and performance.

Alarm Configuration Metrics

This might be hard to believe, but we really do not care how the alarm system is designed so long as it follows the four fundamentals and the working alarm system does not overload the operator. As true as the previous statement is, the basic problem remains: it will not be possible to determine how the new alarm system will work until it is done and used. This means that all the alarm work must be done before anyone knows that the alarm system will do the job. No enterprise would want to design that way on speculation. Therefore, first ensure that all alarms meet all four critical design fundamentals (see the "Alarm Management" section earlier in this chapter). Now all we need to do is to work toward not going overboard with configuring alarms. While the "starting from zero" approach should do that job for you, it is also necessary to watch the growth of the number of alarms. Only process alarms are considered here for evaluation. Alarms for security, extreme hazard detection (facility fire and gas), and control system integrity are not managed here. The general design factors are as follows:

- Configure about one alarm per control loop.
  - *A control loop is considered to contain a measurement, a desired set point, a feedback calculation mechanism, and a control output action. For most process industries, each loop usually has a control valve.*

- Configure about one alarm for every two analog measurements (not a part of a control loop).

- Configure about one alarm for every five digital measurements (level switch, pressure switch, etc.)

- Configure alarm priority.
  - About 80% of the alarms at low priority (the lowest risk level)
  - About 15% of the alarms at medium priority
  - About 5% of the alarms at high priority
  - Less than 1% of the alarms at critical priority (the highest risk level)

At completion, the total number of configured alarms (of all types and priorities) per operator should be between 900 and 1200 for each one individual sitting in the operator chair. If any operating area has more than one operator sitting in the chair, the above numbers are increased proportionally. If the above numbers are exceeded, it suggests that the alarm load on the operators will likely be too much and/or the operational risks are not properly understood or applied.

Alarm Performance Metrics

Here we are providing ways to check to determine if the actual (as opposed to potential, as we did earlier) alarm system performance might be overloading the operator. However, regardless of these measurements and values, the only determination that has any value is the one that properly identifies true operator load. The following list is for guidance and recommends the target number of alarm activations per the categories below:

- Average number of 6 per hour (at a mix of priorities; see priority mix below)
- Short-term average of 12 per hour (for an hour or so)
- Low-only alarm priority: 10 per hour
- Medium-only alarm priority: 2 per hour
- High-only alarm priority: 1 alarm every hour and a half
- Critical-only alarm priority: almost never (thus they do not enter into loading estimates)

Table 8-5 illustrates a few example situations of alarm activations per 8 hours for a mix of three priorities. Other shift durations work proportionally. You can see how this works.

Please keep in mind the operator loading topic just before this. The ability of the operator to have enough time to handle alarms is the only issue here. It assumes that

Situation	Low Priority	Medium Priority	High Priority
Example 1	38	5	0
Example 2	40	8	0
Example 3	33	2	1
Example 4	16	7	2

Table 8-5. Maximum operator 8-hour alarm loading for a mix of priorities. Read each line across. An example would be that 33 low-priority, 2 medium-priority, and just 1 high-priority alarms activating during an 8-hour shift would fully occupy that operator in that shift. The plant in this example used only three alarm priority levels (not recommended).

the operator base load (see Chapter 3, "Operators") is nominal: between 25% and 35% of time. This base load includes all the general and normal operator activities needed to manage a usual shift without any abnormal situation events.

Operator Alarm Loading

Once an alarm system is implemented and all configured alarms follow the four fundamental requirements, if the operator alarm loading is too high, all remedies must be done outside of the alarm system. These changes may include adding more operator time, adding advanced controls and automation, improving existing controls, improving instrumentation, and modifying process design or equipment.

Alarm management, once you have a properly designed and working alarm system, reduces to the simple but important issue of time management. Alarm management is time management. Alarm design technology is geared to ensure that abnormal situations are announced to the operator in enough time for him to be successful. The better the job of providing tools and training and other support for the operator to work the alarm, the less time needed and the better the chances of success. You saw alarm metrics in the previous section. The design ones are intended to maximize the chances of the alarm system that you are building (or have built and are examining) to work within the time capabilities of your operator. The performance ones are designed to measure how well you did just that. Remember, the operator can only work on alarms until he runs out of time. If there are too many, regardless of whether or not the operator selected the ones to work in order of risk, the ones not selected will not be addressed at all. This reduces to a simple responsibility: system designers must expend sufficient effort to ensure that operators have the time (and resources) available to them. Figure 8-14 illustrates a hypothetical hour of operator time.

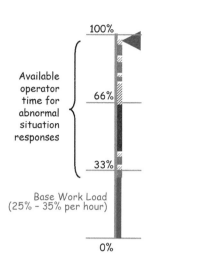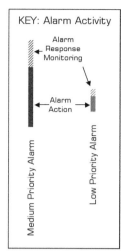

Figure 8-14. Illustration of operator alarm loading for a hypothetical 1 hour. Shown is the base load (in blue) that is taken up by planned and regular duties (reports, regular monitoring, shift changes, etc.). Added to that is the time needed to respond to five low-priority alarms and one medium-priority alarm. All this activity leaves just a small amount of time left over (above the magenta pointer at the top).

The figure depicts a single hour of time. Shown are the base load and the time taken up to manage six alarms. There are one medium-priority and five low-priority ones. This is the most that the operator will have time for. This is shown in the guidelines examples in Table 8-5 (row two interpreted for 1 hour). Each alarm is represented by a colored bar. Each bar has two time portions: (1) the actual time needed for the alarm to be directly managed, as solid bar [SUDA plus enough response time to avoid the consequences]; and (2) the time needed for the operator to keep an eye on things to make sure the consequences avoided stay that way and nothing untoward else happens from this situation, dashed bar.

If alarms occur too often for operators to have enough time to manage them, it could be possible to defer a part of the base load activities until the alarm load lessens. Each enterprise would need to establish plans for how this would work. On the other hand, if the actual base load of operators was higher than the assumed normal, the alarm performance metrics will need proportional reduction. A simple example would be if the base load were to double. Now it is up from 35% to 70%. Now all alarm loading targets need to be reduced to half. New values would be:

1. Average number of 3 per hour (at a mix of priorities; see priority mix below)
2. Short-term average of 6 per hour (for an hour or so)
3. Low-only alarm priority: 5 per hour
4. Medium-only alarm priority: 1 per hour

5. High-only alarm priority: 1 alarm every 3 hours

6. Critical-only alarm priority: almost never (so they do not enter into loading estimates)

Contribution of Alarms to Situation Management

Alarms provide the frontline tools and technology for situation management. *Proper alarm management is required for all abnormal situations that responsible and prudent enterprises are capable of identifying as part of good design and operation.* Period. Extensive effort is required to ensure that the alarm system identifies all reasonable abnormal situations that operator action is required to manage. Each alarm so designed will be accompanied by a fully identified, detailed procedure for management. Remember to use the knowledge fork to always make sure all information is known.

All the rest of the abnormal situations that arise must be found using the skill and intuition of the operator coupled with the rest of the technology and tools in this chapter and the next one.

8.4 Operator Ownership Transfer at Shift Change

A shift change is one of the more vulnerable operating situations. It is also one of the most valuable. At no other time when operations may not be upset, is it essential to fully understand what is going on. This understanding must be totally shared with another individual. Chapter 3, "Operators," covered the detailed requirements for shift change, including: report status, show up ready, understand and own, and short joint operation. If you have skipped that, you might want to read that chapter first so this will make better sense. Here we develop a tool for shift change. The topic is important enough to revisit from Chapter 3. This section reviews the process for the arriving operator to take proper ownership of the chair. This important step, when coupled with the extra step of "getting in the harness and pulling," cements the operator firmly in role and function. We are discussing this topic here because ownership is such an important situation assessment and awareness tool. It also provides a natural segue into the later topic on collaboration, which involves similar activities and goals. The handover requirement from Chapter 3 is our starting point:

Requirement: The leaving operator is responsible for transferring all shift handover information to the arriving operator. The arriving operator is responsible for taking full ownership and control of the operating chair.

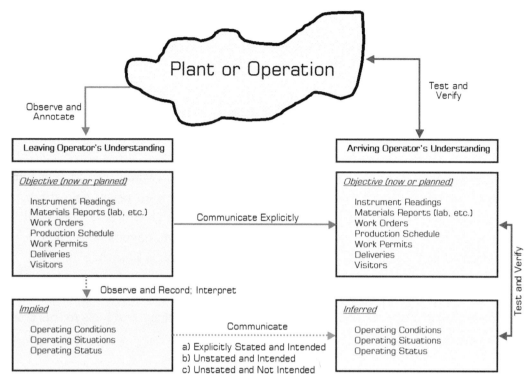

Figure 8-15. Information and decision flow for shift handover. The leaving operator gets everything ready and then shares it with the arriving operator. The arriving operator must fully understand what is being communicated as well as test and verify that it is true.

Begin at the point that the leaving operator has everything needed to brief and the arriving operator is present, fit for duty, and ready for the brief. Follow the process depicted by Figure 8-15.

The solid green arrows (→) indicate information directly taken from the plant or operation. This information flows into the depicted green boxes (□) in the figure. Information includes instrument readings, laboratory values, alarms and alarm history, trends, work orders (ongoing, just completed, scheduled to begin soon, etc.), deliveries, testing, and all manner of other activities that are part of plant or operation life. This information is usually collated into a specified shift handover (or turnover) report and will be used directly in the shift-change briefing.

Transferring Information

Observations gleaned during the shift, and the information compiled by the leaving operator to suggest current operating conditions and situations will go into the shift report. The dashed green arrows are used to depict material based (partially or totally) on conclusions and suppositions. The two classes of shift information—*objective* information (direct observations and recorded data) and *implied* (or *inferred*) information—are shown in their own boxes. This separation is important. The objective information would come directly from readings and other material that required it to be recorded without modification or alteration of any means. Differences from what actually exists in the plant or operation must be due to errors, such as instrument malfunction, report writing errors (noting a lab value in a wrong location of the report), transcribing errors, and intentional deception. Implied information is that which an individual conjectures as being relevant by use of personal supposition, conclusion, or conception. It may be anchored on or suggested from objective information. But such pedigree is not proof. This is an important and useful distinction.

Receiving Information

Moving our attention to the yellow boxes ( ) in Figure 8-15, we see a very similar structure for the arriving operator as for the relieved one. There is the objective information box and separate inferred information box. Because the objective information flows between the two operators, it is presumed to be accurately transferred. *Transferred* means read or seen on a screen or paper or other report. The word *inferred* is used to reflect that regardless of what might have been intended by the leaving operator (who does the implying), the arriving operator (doing the inferring) takes it in through his own thoughts and expectations. All this shift information is now in the hands of the arriving operator (all the yellow boxes in the figure). The stage is set for the next activity of verification.

Verifying Information and Taking Ownership

At this step, the arriving operator has all the information. Two important events need to happen: (1) he must verify that what he has is the best evidence for the truth, and (2) he must understand it and take ownership. Verification is simple to say but can be a bit involved to actually do. It is best done in two steps. Note that these steps are shown as solid red arrows (→) on the figure. The first step is to verify that the objective information is correct. The operator will examine it for reasonableness. He will spot-check a representative sample. He may make historical comparisons. In short, he will utilize tools in his toolbox of competencies that do this. See the red arrow up and back from the Plant or Operation. Once the objective data is verified, it is time for the operator to turn his attention to what it means. See the red arrow between the Objective and Inferred yellow boxes.

Now the arriving operator is asking whether the objective data and the implied data are consistent. The questions are: Is the process actually operating like I think it is? Does my thinking about where the plant is operating match the objective data (that I have verified to be correct, so far)? If so, then he reasonably knows where the plant is and what to watch out for and manage. He can take ownership of the operational chair (shift). If the data does not match, then the arriving operator does not know to any reasonable level of certainty what is going on. Work with the leaving operator. If sufficient agreement and claim cannot be reached, it is time for escalation and "permission to operate" protocols to kick in.

In-Shift Handover Emulation

This shift handover and ownership process may be useful for other activities in addition to changing operators. If you get it in place, you might as well exploit it. Make it into a reusable operating tool. Remember how hard it is for the operator to really know what is going on during the shift? Well, how about using a streamlined version of the shift handover process during an actual shift? Each time one is done, the operator will feel more confident that he understands what the process or operation is doing and whether anything needs special attention. Doing this periodically during the shift would also facilitate preparing the end-of-shift report, because a lot of the information would have already been sniffed out and processed. The section "Situation Management Flow of Operator Activities" in Chapter 9, "Weak Signals," lays out a model for incorporating these during-the-shift evaluations. The bottom line of this book is providing operators with natural, straightforward, powerful, and readily useable tools to do the job well. This is one of them. Consider adding in-shift evaluations as part of the standard shift activity.

8.5 Alerts, Messages, and Notifications

There are events that are not (and should not be) alarms. They are part of the safe conversations (discussed in Chapter 9, "Weak Signals"). They provide information to the operator about events, status indications, and anything else useful, but do not necessarily require timely action. We can notify operators via automatic equipment (control systems, computers, etc.) or by use of less structured hand-held devices (pagers, smartphones, etc.). For the sake of simplicity, let us refer to alerts, messages, and notifications collectively as *notifications*.[5] Notifications are purposely passed on to the operator. A notification can take any appropriate sensory form: text on a display screen, hard copy, sound, visual, olfactory, and touch (vibration, etc.). These forms can occur either singly or in combination. Here we focus on the traditional text and visual ones. Notifications need to be in keeping with the nature of the situation as well as the location and status of the person (operator) for whom the notification is intended.

Notifications as Weak Signals

A notification is information (in any of the earlier mentioned forms) that is not an alarm. By definition, an *alarm* is a situation that requires explicit action by the operator

---

5 Rothenberg, *Alarm Management for Process Control*.

in order to have an activity continue properly under his direction. Notifications may or may not lead to actions. If they do, they do not by their nature require any action to be timely. The question is posed: Are any (or most, or all) notifications candidates for weak signals? The answer is yes. This means that once the operator gets one, he can use the tools from the next chapter to work them out. For now, let us work out how to construct and get them to the operator. First of all, a notification is something that is intentionally made available to the operator to convey something of interest that he would benefit from knowing. Notifications are preconstructed; that is, there is a specific condition for each that is set up in the control system logic or some external monitor that is passing the notification to the HMI for the operator. This structure is important. We will get to it later in this section.

Event Notification
Notification of event – Equipment status identifying: stopped, running, bypassed, and such for individual items or compact systems. Examples: FCV-23401 closed, Pipeline Segment 4400 out of service, truck in loading rack 12, control room outside door open.

It is suggested that all event notifications not intended to be weak signals are possibly unnecessary. Their status can be indicated via appropriate HMI changes only. Of course, they would be logged in the event log.

Action Notification
Notification of action – Examples: add 2 lb of catalyst 4402, turn off supplementary heater, start sample process, control room pressurization needs restoring.

Action notifications are not intended to be weak signals. However, an action notification that reminds the operator that he is late for an action would actually be configured as an alarm if being late for the action puts the operation at risk.

Situation Notification
Notification of situation – Examples: pig launcher is blocked in, traffic loading is approaching peak travel period, 2 hours until early shift handover, Tank 4 fill valve has been open more than 5 hours. More examples include events informing operators about where an automated process is in relation to conditions, production requirements, and so on; early notifications about what system, process events, or potentials are likely to take place in the near future; the inverse of alarm conditions, for example, the clear absence of any unusual situation(s) can itself be a condition that would be nice to know; at the current production rate, the required delivery will be met and the transportation costs are within proper limits.

Any situation notification not intended to be a weak signal should be either reconsidered to be a weak signal or rethought. By their nature, situation notifications are intended to advise the operator to something somewhat out of ordinary that the operator might need to consider.

Properties of Notifications

Notifications can be single entities or they can be continuous cumulative entities that depict a current situation dependent on a history. Although not requiring action by the person notified, they usually have a significant degree of temporal importance. Therefore, their timely delivery is essential. Notifications may or may not require receipt acknowledgment, although this type are expected to be very few. Receipt acknowledgment should depend on the specialized and individual nature and requirements of the particular notification.

Notifications in Combination

Clearly, notifications have intrinsic value as single entities. And that is their primary usage. Notifications, per se, do not require direct action; however, this is not to say that a chain of notifications would not point to a situation that would eventually require action. Such action would usually be in advance of any alarm. In this way notifications can have remarkable additional value when taken in combination. Such combinations are termed a *story*. Therefore, there is a need to have notifications contain within themselves enough structure to permit single ones to be put together into any number of collective stories.

If a combination of notifications creates a story that points to a potential problem or specific concern, then it would be included within the notification. For example, "The temperature of Bath 1 and Bath 2 are both trending down (Potential of low fuel gas pressure or heat quality to their shared burners)" where the part in parenthesis indicates a weak signal forward projection; you will see more about how it works in Chapter 9, "Weak Signals."

Other examples might be to retrieve collective information about past activities that would be useful for the following:

- To trace the performance of tools and/or usage of equipment and systems. This could be useful as an engineering or operations aid.

- To suggest better plans for infrequent operations using experiences of earlier ones.

- To identify production anomalies. This means that if a notification is issued to suggest that something is amiss, another notification must be issued when that situation has gone away.

Compared to alarms, the active lifetime of a notification would appear to be long. The term *active lifetime* implies retrievability as well as continued linking to other notices and conditions to build potential stories. Thus storage, retrieval, and story generation capabilities must be in place for a suitable time period.

Sorted Notifications

Examples would be a topic-based report such as "All notifications on the third shift on a certain date," or "All notifications relating to environmental emission weaknesses in our production of product X."

General Design and Implementation Guidelines

You are already using notifications. In the past, you probably used part of the alarm system to deliver them. We now know that alarms must require timely operator action and are only given when sufficient time and adequate advice will accompany them. Therefore, notifications cannot be intermixed with alarms. They must be delivered some other way. A notifications system would be a natural solution. Unfortunately, very few controls operating platforms (PCSs) include this technology. In order to have it, you will need to either build it or purchase it. Once implemented, it must seamlessly integrate into the operational infrastructure.

Notifications are a tool. Like any tool, they need to be crafted and maintained. Crafting means that all text will be arranged into messages in a predictable manner using words and abbreviations that mean exactly the same thing to all. Message content must follow carefully constructed forms. For example, it is desired to let an operator know that someone has identified a partial blockage in a standby compressor kickback line. An *ordinary message* might look like this: There might be a partial blockage in the kickback line for the injection gas compressor. A *structured message* would look like this: Standby Compressor C-43A Partial Blockage potential in kickback line (area 17, segment 4, line 334). Note that the structured message begins with the equipment (unambiguously identified), proceeds next to the problem or core of the message content, and ends with the relationship's (problem) relative location. All notifications should be complete enough that the operator will know exactly what the message is and precisely to where in his operation it pertains.

Finally, the deployment of this tool must be managed so that its use is neither excessive nor sparse. Excessive use will demote all messages to the category of noise, that is, "not worth any effort." Too little use will mean that few will remember to use it; and when used, few will understand what it means and much will be left uncommunicated.

Formal Notifications Delivered by the Controls Platform

Notifications can be delivered to the operator in several ways. They may appear on supplementary displays that are not part of the DCS, SCADA, or other formal controls system. This form of delivery uses a separate external display, not a part of the control platform suite, but in the operator area. This provision generally requires a separate notifications construction and delivery engine. Where this real estate is shared, visibility assurance management practices need to be in place. They might be delivered via hand-held devices or other special purpose devices.

They might also be delivered somehow using the control system. Notifications that must be delivered directly via the controls platform infrastructure present problems to systems designers. Specific HMI tools are needed to generate the notification. If the controls platform is generally closed to externally linked processors (in the sense of a computer being a processor), these messages will need to be produced, routed, and presented by the use of existing platform infrastructure. Very few controls platforms have this capability. A common workaround is to "borrow" unused areas of the existing alarm system (e.g., a priority not used for alarms) and create shadow alarms that are delivered to a replicated alarm display different from actual alarms. This will require careful management to ensure that they cannot ever be confused with real alarms. It will also require the ability to remove notifications no longer needed, as well as the capability to sort them in useful ways. Because these will not utilize the traditional alarm display route, the designer must arrange for the appropriate location on screens within existing display locations.

Informal Notifications Provided by Email, SMS, and Other Means

The general tools for email and SMS (short message service; e.g., text messages to mobile electronic equipment built to receive them) are well designed by the service providers. But that design does not incorporate sufficient requirements for form, content, or security of delivery. By security, we mean that all notifications issued get delivered intact to the intended recipient in a timely manner. Both the sending and the receiving parties must know about any delivery issues in a timely and unambiguous manner. Both the sending and the receiving parties need to have the same understanding of the context and content.

Security of delivery is only one of the critical aspects. There must be a methodology for the person being notified to be reasonably sure that he sees those notifications. In this age of texting, it is all too common for the more useful messages to be obscured or even lost in the sea of other ones. Most portable electronics are lacking in the capability of secure delivery. Consequently, it is imperative that notifications delivered by this method be understood to be less secure. This extends to situations where there are multiple people being notified. Protocols must be in place to ensure that all recipients will understand whether they are an owner of the notification or a bystander.

Notifications and Logs

Understanding weak signals has led to clarity about notifications. At this point, we have clarified which notifications are meant for the operator directly and which are not. *Directly* means provided to the operator without the operator explicitly looking for them. This would mean notifications indicated on the HMI as a status and/or appearing in a notification window on the HMI. They inform the operator because we expect the operator to process them in the here and now. *Here and now* means at the time the operator comes across it. Some are important enough to place in dedicated notifications windows. Some are important only when the operational aspects relating to them are in focus (a particular operating screen is in view). Notifications are placed where they make sense for the operator to know.

There are a number of events and other information that need to be captured and available, but not in the "here and now." The operator will not need to ponder or observe them. They are meant for logs only. They may be consulted for preparation of shift handover reports. They may be useful for event investigation. They also may be useful for statistical data collection regarding equipment performance, operator performance, or other uses.

8.6 Putting It All Together

Good operation critically relies on all operators having proper situation awareness at all times. *Awareness* is defined as perception of and cognitive reaction to a condition or event.[6] Awareness is the state or ability to perceive, to feel, and/or to be conscious of events, objects, or sensory patterns. With this level of consciousness, an observer (operator) can confirm data without necessarily implying understanding. Make no mistake, understanding is important, but it comes later. This awareness can be the process of observing; actively seeking; or just a feel, often referred to as our *sixth sense*. Awareness provides the "data" for the operator.

This chapter contains important concepts and useful practices and procedures. The reason all material has been carefully explained is to give operators effective ways of determining what is going on in time for it to be of good use. Simple. Powerful. Necessary.

---

6 "Awareness," Wikipedia, last modified April 15, 2018, https://en.wikipedia.org/wiki/Awareness.

You are now ready to fully appreciate situation awareness requirements:

1. Determine the minimum essential information to provide the operator that will enable a reasonably confident assessment of whether or not things are within normal as well as whether there is a reasonable expectation that normal will continue. "Provide the operator" means that this information is readily discernable from the HMI displays used for process management or other formal parts of the operational infrastructure. Once the situation is identified or suspected, follow-up confirmation would be direct and discerning.

2. When *normal* is not the current situation, be able to determine how abnormal the situation is, where the likely abnormal situation(s) is located, and how likely the situation will either become less abnormal or escalate.

3. If remediation is in progress, determine how effective it is in restoring proper operation.

Your goal is to understand how to provide proper situation awareness capability to your operators and their line supervision. Yes, supervision is a vital partner to operations. Ask yourself: Why do we have a supervision infrastructure? What role will it play and how will it aid and ensure effective operations? Having read and understood your options, you are ready to ensure that your operational kit is ready as well. Build or modify the operators' tools to provide what they need. Pick and choose from what you understand here that will ensure adequate situation awareness. Make sure that you have a "critical mass" capability, not just a few interesting possibilities. Build it into the controls and operator interfaces. Include it in the operating procedures and operators' training. Be sure to assess its effectiveness through operator evaluation and incident and near-miss investigations.

8.7 Making Situation Awareness Happen

Now that you have made it all the way through this chapter, you might be asking yourself some important questions:

- How do I know whether this stuff will actually do the job my operators need?
- It all seems understandable and makes sense, so how can I manage to get all this engineered and implemented?
- Is all this straightforward enough for ordinary operations and technology people to use?

- How do I communicate and sell this to my management?
- How do I evaluate whether or not it is working after we use it for a while?

Let us take a look at each question one at a time.

Capable of Doing the Job

There is nothing contained in this chapter that is not already well understood and in use throughout industry and other operations. Sure, some tools work better than others; some are more expensive than others. All will perform capably and effectively when properly designed, implemented, and used. The test is not one of capability, but of need. Does your enterprise have operational exposures that can be closed by adding specific situation awareness tools and capability? If the first answer is "probably not," I suggest that you go back and reconsider. Often the enterprise that feels most strongly that this is not needed, actually needs it more. Awareness of situation awareness needs is itself a problem that must be overcome. Please do not wait until after a serious incident.

Design and Implementation

Maybe this is not as big a job as it first seems. Starting from scratch, it is an uphill climb. But surely, your enterprise is not starting from the beginning. There are parts of these tools already either in procedures or in general practice by your operators. Now that you have a better understanding of their purpose and how they might be better designed and improved, you will benefit from cleaning things and sharpening them. Consider working this task into next year's plan, if you need to spread it out. Otherwise, the longer the delay, the longer the unnecessary stress on the operator and risk of bad things happening.

Usability

How can situation awareness not be useful and usable? And remember, this technology is not replacing anything.

Selling Management

What would any operator, engineer, technician, or manager give to be able to convince senior management of the essential need for situation awareness? There is no magic here, unfortunately. Operators need reliable, effective, and useful tools to see what their plants are doing. Managers are the ones who ensure that such tools are provided. Managers should be prepared to see the benefits and accept that responsible stewardship requires enough of these tools for the operator to feel comfortable and competent in his job. They must make it happen. It is part of their job.

Auditing

Auditing is employed to provide useful and useable feedback on how your operator's use of situation awareness technology is working. It provides information on what is useful and what is not. It should provide important guidance for you to make it better. There are two kinds of audits to consider: performance and process.

Auditing the Performance

Performance audits look at outcomes. One looks at the number and severity of abnormal situations properly managed before and after incorporating an effective situation awareness program. In addition to numbers, one should make sure to gather operators' opinions as to which tools and techniques worked well and which did not. Ask for specifics and relate their experiences to actual annotations made in operator logs and shift handovers. An excellent way to make this happen is to incorporate this audit as an integral part of incident and near-miss reporting. Try to build it into training and certifications. And make sure that all of the audits are fully reviewed, analyzed, and acted on.

Auditing the Process

Process audits look at the technology itself. Here are some key points to consider:

- How easily did operators recognize when the tools and procedures should apply to the current situation?

- How effectively did operators employ the tools and procedures for situation awareness during the everyday shifts?

- How effectively did operators employ the tools and procedures for situation awareness during serious situations?

- Were the tools used as designed?

- How often were the tools on-the-fly modified (for purpose, or for no reason at all)?

- How often did operators simply skip using the tools and procedures? For what reason? For how long? Why?

Remember that you are not really looking for numbers and statistics. You are seeking to understand how to improve your operators' ability to see abnormal situations early. You need the kind of feedback that points to strengths and weaknesses in their gaining and achieving proper situation awareness throughout their working shifts.

8.8 Close

Much of the technology in this chapter has been around for quite a while. In one industry or application or another, it has been used successfully for years. It is largely proven and well understood. This, together with the new power of weak signals in Chapter 9, "Weak Signals," gives operators vital tools and processes for doing their job. The power of this book is having the best practices and ideas explained well and in one place. Respect the underlying problems and issues they address. They are for you to see and understand. Try them on in your mind. The ball is in your court.

8.9 Further Reading

ANSI/ISA-18.2-2016. *Management of Alarm Systems for the Process Industries.* Research Triangle Park, NC: ISA (International Society of Automation).

API RP 1167. *Pipeline SCADA Alarm Management.* 1st ed. Washington, DC: API (American Petroleum Institute), 2010.

Blevins, Terrence L., Gregory K. McMillan, Willy K. Wojsznis, and Michael W. Brown. *Advanced Control Unleashed.* Research Triangle Park, NC: ISA (International Society of Automation), 2003.

Endsley, Mica R., Betty Bolté, and Debra G. Jones. *Designing for Situation Awareness: An Approach to User-Centered Design.* Boca Raton, FL: Taylor & Francis, 2003.

EEMUA (Engineering Equipment Materials Users' Association). *Alarm Systems—A Guide to Design, Management and Procurement.* EEMUA Publication No. 191. London: EEMUA, 2007.

Hollifield, Bill R., and Eddie Habibi. *Alarm Management: A Comprehensive Guide.* 2nd ed. Research Triangle Park, NC: ISA (International Society of Automation), 2011.

Shermer, M. *The Believing Brain.* New York: St. Martin's Griffin. 2011.

9
Weak Signals

All things are easy to understand once they are discovered. The task is to discover them.

Galileo Galilei (Astronomer; Italy, 1564–1642)

In solving a problem of this sort, the grand thing is to be able to reason backwards.

Sir Arthur Conan Doyle (Author, creator of Sherlock Holmes)

When people stumble onto the truth they usually pick themselves up and hurry about their business.

Sir Winston Churchill (British Statesman)

Weak signals is a very new concept. It is new to the control room setting. It is new to our thinking in any situation. Properly understood and competently conducted, weak signals can provide a valuable and sharp tool for operators to identify situations that otherwise might have gone unnoticed and thus unappreciated. Certainly, operators have lots of other ways to identify things going amiss and potentially likely to cause trouble. Earlier chapters have covered many. Weak signals is another tool in the box. It is a very sharp one. You should find it extremely useful. Besides the obvious benefit of helping operators make sense of small indications of trouble or concern, the process works extremely well in getting around preconceived notions arising from biases (Chapter 7, "Awareness and Assessment Pitfalls"). It also makes sense of how the human-machine interface (HMI) screens are designed (Chapter 5, "The Human Machine Interface") to help find both large and small problems. We are now positioned to take full advantage of situation awareness.

Operators have the critical job to be on the lookout for problems in the making and find them early enough to prevent bad things from happening. These things can be hard to see. Operators need to be able to pick up early on operational clues and any suspicions about something not looking quite right. Qualified operators with proper training and experience are imperative. A purposeful, effective control room design and strong personnel access protocols are expected. High-quality supervision is essential. Success relies on a comprehensive human-machine interface (HMI) designed and built to provide an open window into the process. The HMI puts operators in the best position to focus their talents and actions for watching, evaluating, and managing. Using it, operators will sense things amiss and work out what might be going wrong. HMI design topics have been covered in earlier chapters. Now we are going to use them. This chapter lays out a powerful tool to help all operators find subtle plant operational issues and problems as they develop. But, as powerful as it is, and as useful as it is, please keep perspective. Without careful attention to all the earlier material in this book, even the best tools and talented operators cannot do well enough jobs. The single purpose of weak signals is to identify potential operating problems. *Weak signals* is not a tool used to diagnose. It is not used to remedy. It is simply used to identify and confirm.

Abnormal situations come in two parts: the ones that are alarmed and the ones that are not. Those that are well understood, are alarmed. Those that are not alarmed are the ones the operator must find pretty much on his own. Figure 9-1 illustrates this situation. Alarm system designers are responsible for ensuring that all abnormal situations they can uncover during equipment and process design, during the development of operating procedures and training, and during the operational safety analyses, including HAZOP and preoperation safety review (PREOP, also referred to as

Figure 9-1. Abnormal situations are divided into the portion that competent designers must identify during the equipment and operations design process, including the HAZOP and process hazard analyses and other applicable design and operational safeguards. For the former, alarms are designed. All the rest are left up to operators to find using effective monitoring during operation. HMIs, plant operation guides, protocols, and procedures facilitate this task.

pre-startup safety review—PSSR), have proper alarms configured for them. The operator must discover all the remaining abnormal situations by his own effort. It is part of the job. Chapter 8, "Awareness and Assessment Tools," discussed the infrastructure tools operators rely on. Chapter 4, "High-Performance Control Rooms and Operation Centers," and Chapter 5, "The Human-Machine Interface," added to the building up of operators' support mechanisms to help. This chapter adds a new *situation awareness* tool called *weak signals* to the kit.

Weak signals are observable; suggestive but not definitive; and when examined properly, lead to the discovery of abnormal situations early enough to be valuable. Operators using this tool have a powerful skill to assist them. It helps operators focus on suspicions, clues, hunches, nagging issues, and the rest of those subtle nuances that might rightly be early (sometimes very early) indicators of things going amiss. It allows operators to separate them from noise. Weak signals as a tool moves the operator's capabilities from being useful to being effective and reliable. Its single purpose is to aid operators to find operational issues early. It is a true situation awareness tool. Using it, operators can more easily and effectively evaluate early irregularities and other puzzling concerns that they come across to determine which might be meaningful and which are likely not. The use of weak signals can be a game changer. Every operator who understands how to use it and uses it well can have a better chance of maintaining good operation. Every enterprise that successfully incorporates weak signals into its control room infrastructure can take advantage of the increased ability of operators to find potential operational problems and issues.

This is a tool that you will retrofit into your existing control room operations. It fits seamlessly into shift changes and periodic "walk-throughs" during the shift, and it can be used as needed when troublesome things seem to pop up. The chapter lays out the foundation and the work process. Operators who understand and incorporate weak signals into their shift activities will have a powerful methodology for increasing situation awareness. Be assured that there is nothing weak about using weak signals. It is a powerful capability. The term *weak* implies that subtle changes or cues provide relevant clues. These clues offer important avenues for assessment. But they first must be identified. According to Paul Schoemaker and George Day, a weak signal is:

> [a] seemingly random or disconnected piece of information that at first appears to be background noise but can be recognized as part of a significant pattern by viewing it through a different frame or connecting it with other pieces of information.[1]

---

1 Paul J. H. Schoemaker and George S. Day, "How to Make Sense of Weak Signals," *MIT Sloan Management Review* 50, no. 3 (2009), http://sloanreview.mit.edu/article/how-to-make-sense-of-weak-signals/.

This chapter is about how to identify and assess weak signals. Weak signal management has a single purpose: early identification of potential abnormal situations. Weak signals is a situation awareness tool. Once identified and confirmed, the rest of *situation management* takes over. It uses all the existing plant infrastructure and operations management tools. This work process builds seamlessly onto the technology and methodology of Chapter 8, "Awareness and Assessment Tools," and into the processes and work procedures you will find later on in Chapter 10, "Situation Management." Together, operators have a way to find the out-of-normal and abnormal, hopefully in time to make the difference. Weak signal management provides the ability to take something not well formed and provide confirmation (or disconfirmation) of its likely existence.

Weak signals have an interesting history. The term was coined by H. Igor Ansoff in 1975.[2] Ansoff recognized explicitly that almost no significant event emerges without some warning, although those warning signs can be extremely subtle and often indistinct. For the next 20 years, weak signals were an obscure label of something to look for. Bryan S. Coffman revisited them to extend and add structure to the concept in his MG Taylor monograph series in 1997 titled "Weak Signal Research."[3] Here was the beginning of a useful process. Some of the earliest uses were to make predictions about what new technology innovation might become a commercial success. Specifically, what might be the next big innovation to capture the market as a new product? The difficulty is that a real innovation does not build on something already there. The next innovative thing is not a continuation of an existing thing. It is an entirely new form. When people are asked what they would like to see in new things, they only see more wonderful versions of what is already there. So, to make the leap of insight, we must look for clues about what people do (as opposed to what they use to do it). We must ask what barriers might be in the way of doing it better. The clues, if any, are the weak signals.

This chapter lays out this tool for operators. Operators will use this tool differently. They will be looking for the next problem that they did not otherwise see coming. Besides the obvious utility of alarms, finding other potential problems is rather difficult. It is not so easy to be able to "keep an eye out" for things that could be amiss. The alarm system covers a really important part—everything that could be pre-identified for the enterprise design and operational requirements. But there is a lot left over that must be found. Weak signals fills a large part that gap. It provides an important tool

---

2 H. Igor Ansoff, "Managing Strategic Surprise by Response to Weak Signals," *California Management Review* 18, no. 2 (1975): 21–33.
3 Bryan S. Coffman, *Weak Signal Research—Part I: Introduction* (Louisville, KY: MG Taylor Corporation, 1997).

for operator effectiveness. We will explain the technology and process. There is a lot to it. This discussion has been carefully laid out for you and the many supporting side threads examined. Try to keep this in perspective. The concept is simple. It is straightforward to use. Operators will actually use weak signals around 5% of their on-shift time. The rest of situation management is about the remaining 95%. Together they provide a game changer for your operator.

Before we get started, let us clarify how weak signals are to be used. There is a popular notion that we should call hunches, intuitions, clues, and the like weak signals. As if calling them that will make a difference and somehow elevate the situation. It does not. The term *weak signals* is used here to mean both a dependable way for observing them *and* a principled plan for using them to identify true implications. It is this dual activity that promises to be the value. Let us see how it works.

9.1 Key Concepts

Weak Signals	Weak signals are the foundation of a new technology that changes the operators' role from "hunting around hoping to find the problems" to an organized process for identifying likely ones.
	Weak signal analysis forms the core for simple yet powerful tools to expose abnormal process situations to the operator's view. Every off-normal condition is a weak signal.
	Weak signals (1) are observable; (2) are suggestive but not definitive; and (3) when examined properly, could lead to the discovery of an important abnormal situation early enough to be valuable.
Strong Knowledge	Weak signals are only useful in situations in which underlying knowledge of the function and inner workings of the process being managed is strong and used.
Solid Foundation Infrastructure	Effective weak signal management can build only on an existing solid foundation. The quality of the entire enterprise is the single highest influence on the operator's ability to find, understand, and manage operational threats.
Short-Term Strategic Tool	Weak signals is a short-term strategic tool providing a tactical look-ahead of likely developing abnormal situations, thus permitting operators to fully focus on the current task of maintaining regulation without blurring the need to find developing problems.
Do Not Escalate	Weak signals do not usually become stronger with time or worsening of the underlying off-normal situation; though it often becomes easier to backward-project them to look for evidence the longer they persist.
Confirmation and Disconfirmation	Weak signals evidence may exist as *confirming* (there is a real abnormal situation) or *disconfirming* (a real abnormal situation is unlikely here); both types must be sought and evaluated to guard against operator biases (Chapter 7, "Awareness and Assessment Pitfalls").
	All confirmations must be clear, not vague.

Primary Situation Awareness Tool	Weak signals are the "front line" of situation awareness. To be useful, control rooms must be designed for their effective detection.
Weak Signals Question	"What is wrong with the current operation and my assessment of it that I may not be aware of now?"
Some Not Alarmed	Fast-evolving *transformative weak signals* can progress directly to consequences. If those situations could be anticipated in advance, alarms would have been designed for them. It is important, therefore, to identify and attempt to process transformative weak signals as soon as possible.
Alarms Plus Weak Signals	All process abnormalities worth knowing about that are not covered by appropriately configured alarms need to be available for exposure by weak signals. This is an *affirmative requirement* for HMI design, reporting protocols, and operating procedures.
	If they are neither, either the alarm system or weak signals need more work. This gap suggests a design failure in the understanding of process and operational risk.
"Soft" Observations	Weak signals can also arise from general observations of the process being managed or other sources surrounding that process.
Bottom Line	Weak signals is a powerful information noise processing algorithm.

9.2 Introduction

This chapter is about strengthening your operator's ability to figure out what might be going wrong. This is a daunting responsibility. It is difficult for an operator to be constantly on the lookout for anything and everything that might be amiss in his plant or operation. It is a task made all the more difficult because most of the time things are okay. Success depends on years of experience and on-the-job skills that vary considerably from operator to operator. Even for the same operator it depends on how the day is going. Your best operator is not on every shift. Even your best operator has limitations. Weak signals augment alarms to expose much that might be subtle, indistinct, elusive, insidious, indistinct, fuzzy, or otherwise inconspicuous to the operator working hard to achieve adequate situation awareness. Not too long ago, alarm management was not useful enough, and it often interfered with the operator assessing situations. Early alarm management was done intuitively and incrementally. This has all changed. Alarm systems now have a strong and powerful role in the management of abnormal situations. We understand their basic principles and know how to apply them. And when we do so, they work remarkably well. Weak signals is an important aid for finding the rest of what might be going wrong.

A Word to the Reader

Weak signals is a new topic. It takes some getting used to. This chapter will walk you through it. Mastering weak signals will allow you to relate what might be happening in the plant or operation with the other activities of operating in a control room.

To do this, the chapter is organized in layers. The layers begin at a concept level and progress deeper into the details and options. It is suggested that the reader read the entire chapter fully. This way, as important techniques and provisions are introduced and discussed, preliminary ideas can build into an understanding. As you read, it will become clear why this book carefully prepared the control room for getting ready to help this to work. The operating discipline of shift change both assists the leaving operator to make sense of his shift and at the same time orients the arriving operator to what to expect. The HMI is carefully designed to convey the critical functionality of the plant in a way that exposes as much of the irregularities (all those weak signals) as a good understanding of the plant will permit. Being able to understand and manage uniquely human thought and planning limitations keeps the operator better focused on finding out what might be amiss without getting in his way of keeping an open mind as he gathers confirming or disconfirming evidence for it.

This book could have been written almost the same had we not known about weak signals. In fact, this book was initially conceived without them. The entire frame of the book had been laid out without them. It was clear that for operators to do a good job, we needed them to be well trained with an in-depth understanding of the plant and equipment. We needed properly designed plants with strong management systems. We needed effective maintenance programs. We needed control rooms designed to facilitate the long shifts and the varied work going on inside them. We needed the HMI to fully engage the operator and allow him to fully open the windows to the process and provide the handles to adjust as needed to keep things on course. We needed a collaboration and operational charter as the tools of last resort. All of this puts situation management on as firm a footing as our best practices and experiences could provide. And that is where it was for a while.

Add weak signals to the mix and not much seems to change. But in reality, everything has changed. They give a clearheaded operator who is not fatigued and is looking at his HMI, the ability to see those subtle things that suggest something is amiss. And once they are seen, he has a process to evaluate their importance and significance. This is an extraordinary tool for finding abnormal operations early. Weak signals do not take the place of everything else. Rather, they build on everything else. Meaningful use of weak signals depends on those other things!

Before we look into this remarkable topic, I offer some advice. As you read how the concept of weak signals works and how the operator fits into its functionality, many of you are going to get a feeling that to do it well will be extremely taxing. Alarms are one thing. They sound when needed, gain the operator's attention, and provide most, if not all, the necessary guidance to manage the situation. Weak signals do not do any of that. They are subtle. They need to be observed or discovered by intuitively finding them or observing them during dedicated looking-for-them activities. The more conscientious the operator, the more potential weak signals have to cause anxiety and concern. As an operator, please accept the guidance that looking for weak signals must be a part-time activity. It is not intended to occupy all the time left over. You know and understand the main activities of your job. You know how to do them. Please feel comfortable following that role. Along the way, if you happen across something that seems not to look right, of course, do the weak signals thing. At planned times during the shift, actively look for them. Do a careful review as you prepare for shift handover. During shift handovers, do a review with the other operator and follow through them together. When you do all this, please try your best to be comfortable and do not spend more time looking for or worrying about them!

9.3 Weak Signals

Our mental model of danger can be misleading. We would like danger to be announced clearly and loudly, like in the movies: the right music, the right change in camera angle and movement, and all the other appearances of something sinister that prepare us for impending shock and doom. Movies are not reality. We cannot depend on any such clarity in real life. Nature is not much given to music or announcements. When problems do come, they all too often quickly sweep in the very events they portend. That is not at all comfortable. Or they come slowly. So slowly that even if we are looking, it would be nearly impossible to see them. And if we did see something, anything, what we saw would not necessarily mean that much. We need better ways to be warned and advised in real life.

Being able to see and understand barely visible subtle indicators is one of the powerful understandings that have come out of recent careful reviews of tragic and impactful industrial incidents. This is the idea behind weak signals. They are an important tool for operators. By understanding how to recognize and process them, they provide a strong structure to deliver on the operator's situation awareness needs. Weak signal management is a compact way for collating "seemingly insignificant" notices. It drives an answer to the operator's dilemma: How do I exploit what I sense may be going on but cannot seem to put my finger on? It leads to a methodology to catch, collate, and recognize incidents. It provides a means for the operator to identify root problems early for active consideration. And it integrates seamlessly and naturally into all other aspects of operations.

So how do weak signals actually fit into the picture? Let us imagine operational conditions are as depicted in Figure 9-2 (taken from Chapter 8, "Awareness and Assessment Tools").

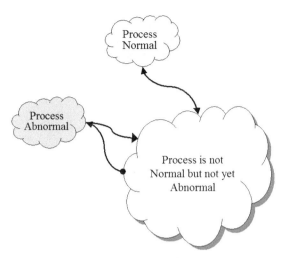

Figure 9-2. Focus of operational conditions moving away from normal to not normal. Note that problems that cannot or did not get managed eventually may lead to faults.

Focus on the situation of *process is not normal but not yet abnormal*. It is this situation that brings weak signals into the control room. Here is where operation is at the curves and edges of our experience. The better the plant is operating, the harder it is to look here. Most everything will appear just fine. The longer the shift, the more difficult it is to stay focused and alert. The more complex the plant, the subtler the abnormalities can appear to be. Besides the traditional alarm system, there is not much else in the way of good tools and workable processes. Finding out what might be going amiss and working out whether it actually is can be daunting. At the present time, we just ask the operator to keep an eye on things and look for what does not seem to be okay. It is a simple expectation. But it is not easy to do. Sure, some unusual things can be spotted as the operator makes his rounds of the screens. A lot cannot. Weak signals is a methodology for finding what might be amiss. It is a new sharp operator support tool for the kit.

Let us look inside the *process is not normal but not yet abnormal* cloud. Figure 9-3 details how weak signals link to normal operating situations gone off-normal. Weak signals could either be too weak to really mean anything—*immature*. Or they might be weak yet real indicators of something going progressively astray—*parametric*. Or they might be weak but important indicators of something ready to dramatically evolve—*transformative*. All need to be recognized. Each is handled differently.

Keep in mind that this is only the core weak signal part of things. We will fit in the other, more direct parts later after we understand this part. The individual white clouds in the figure show the inner structure. The weak signals come from processes that move from normal to off-normal. Off-normal comes from normal but is not normal

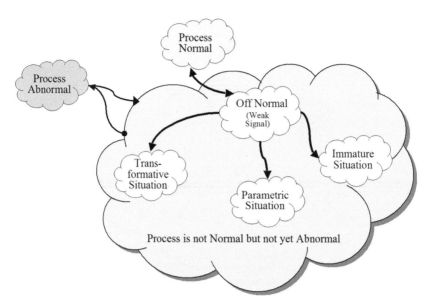

Figure 9-3. Weak signals as off-normal operations showing the three possibilities. Each possibility is important, so the sorting of the weak signal into its proper category determines how it would be treated.

anymore. You are probably familiar with the terms *normal* and *abnormal*. The term *off-normal* may be new. It represents that situational region between normal operation (with its variations and movements) and abnormal operation where it is evident that something is amiss. In off-normal, our process is no longer normal, but it is not yet abnormal. This is a gray area where the unusual may remain unremarkable or just as likely move to take a turn for the worse. We need to know both. In off-normal, one or more weak signals can appear.

Weak Signals Announce

We have an intuitive understanding about what a weak signal is:

> A "weak signal" is something which, in and of itself, may be relatively insignificant and does not [necessarily] justify action. . . . Sometimes however; weak signals are early indications that something is indeed wrong.[4]

It is this dual potential that requires understanding. You will understand how to construct a working methodology to decide whether that weak signal needs more attention or if it is just noise. By the time it becomes clear to the operator that he is facing a serious problem that demands his best (at times, heroic) efforts to resolve, successful remediation may already be beyond reach. By the time the call of danger is

---

4   International Association of Oil & Gas Producers (OGP), "Cognitive Issues Associated with Process Safety and Environmental Incidents," Report No. 460 (London: OGP, July 2012).

heard, the problem may have reached beyond the tipping point.[5] The best way out is to know earlier. The only way to know earlier is to plan and implement effective ways to see the weak signals earlier.

Perhaps nowhere else in popular literature is the subtlety of weak signals so interestingly expressed as by Isaac Asimov in his book *The End of Eternity*.[6] The story is pure science fiction. I shall not retell it. However, the situation is everything. What follows is a hypothetical synopsis taken from the spirit of Asimov's writing:

> Somewhere in the far distant future, a small and responsible group of individuals discover an amazing source of almost infinite energy. They are determined to put this to the benefit of all civilization. Using this invention, they develop a successful method of time travel. Realizing that this is such a massive responsibility, they determine to use it for beneficial results. One of their first "discoveries" is that the future is not a single thread. The evolution of the future is not at all deterministic (i.e., it does not flow from one determined place to the next determined place and so on according to a causal flow). "Mistakes" get made, so that one future that would be beneficial to society and may have occurred on one thread, might not occur on the other thread that the society's future may actually follow. The group forms a society called *Eternity*. Their job is to uncover all such "mistakes" and go back in time and correct them. The cardinal rule for correction is that they must make "the minimum necessary change." It must be imperceptible to all that any change was made. As a consequence, the places in time that those imperceptible changes must be made are only those where they have (almost) no bias. That is, the change could be chosen pretty much either way. No single way would have a clear upper hand.
>
> Let us propose a purely made-up situation (not from Asimov, but from simple imagination in the theme of Eternity). Our hero is Dr. Jonas Salk, the inventor of the first vaccine against polio. We look into the past. He was attending a conference in 1948 (again, purely our invention here) at which a paper was to be presented about some foundation or another in virology by a respected researcher. Salk heard this paper, got off on the wrong track, and never invented his famous vaccine. And lest you think that society is off the hook, we offer that Albert Sabin (who developed the oral polio vaccine) contracted an illness for which he was mistreated by a trusted physician, who was distracted by his own child's suffering with polio—polio that would not have been contracted had the Salk vaccine been available. Sabin was so impaired after his illness that he never worked again. Okay, you get the picture.
>
> Enter Eternity. The decision was made to find a way to prevent Salk from hearing that (wrong track) paper. Salk's car was made to stall on the way from his hotel to the conference. In the few minutes it took to get it started again and for him to be on his way, he missed the opening parts of the paper. He never heard (via his own ears and own

---

5   Malcolm Gladwell, *The Tipping Point* (London: Little, Brown and Company, 2000).
6   Isaac Asimov, *The End of Eternity* (New York: Doubleday, 1955).

cognitive processes) the offending principle. He did not get distracted and so went on to make history. A stalled automobile was an imperceptible change. Stalling would provide a nuisance, of course, but raise no concern beyond the actual inconvenience.

There is a subtle but powerful message. When we understand what it is, our apparent ability to easily differentiate important decisions from ordinary ones might be a bit less dependable than we think. Yes, many important ones are clearly recognized. They firmly, sometimes relentlessly so, announce themselves, weigh heavily on our consciousness, and often continue to bother after being made. They stand out. In the end, they are easy decisions to make. "Easy" in that it is reasonably clear what the probable costs and likely benefits might be. We can give both sides due thought. We know what we are deciding.

On the other hand, if those decisions are for situations that do not seem to matter which choice is selected because the situation could easily go either way, much more real consideration is needed. Alternatives that appear so close as to not matter, often matter the most. At times after selecting one choice, we would surely like to have selected the other one instead.

Finding Weak Signals

Weak signals are everywhere. They can appear even if nothing much is amiss. Hopefully, they appear when something is going on and might be a problem. Real plants vary their activity as raw materials change, weather changes, equipment ages, valves stick a little, pressures vary, and all manner of other things move a bit this way or that. Mostly this is the stuff of normal. Yes, we use control systems to try and keep things steady. And we try to operate with as much predictability and "sameness" as before to keep the variations small. However, variations, and their cousin disturbances, seem to be everywhere. Many will be obvious. The less obvious, the ones we actually do look for, can take a lot of time and effort to find. After finding them, each will need to be looked into to discover what it might mean. This all sounds promising. Seeing everything might at first appear to be a gift, but on practical reflection is really a never-ending task. Clearly, this genie needs understanding.

Consider two ways to do this. First, you will want to be receptive to weak signals that come to be noticed during the normal course of doing what operators do, such as checking on production or operating values, watching laboratory reports, or preparing or validating work orders and permits. Second, we suggest that specific times during the shift be reserved for making weak signal rounds of the plant or enterprise. Chapter 10, "Situation Management," covers useful ways to incorporate weak signal search into the regular flow of operator responsibilities and activities.

Weak Signals for Situation Management

Once you perceive a weak signal and understand it, a whole host of other signals may become visible.[7]

Now that you have been introduced to weak signals, let us see how they fit into the situation management picture. Figure 9-4 shows the whole picture. The diagram is a relational one. The weak signals part is depicted by the white clouds. This is a very busy figure. You have seen parts of it earlier. I included this expanded version of it here for one reason: it covers every operational situation an operator sees. Each pathway (follow any arrows) is a real-life situation mapped out. At every juncture, the operator comes up against a state that he must be able to handle. Adequate situation

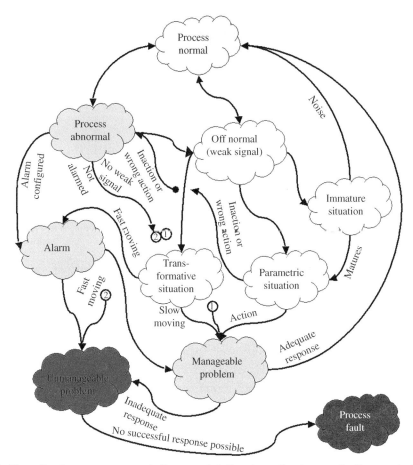

Figure 9-4. The situation management diagram detailing how the types of off-normal situations are sorted. Each situation would be handled using different methods and tools.

---

7 Coffman, *Weak Signal Research*.

management should provide tools and processes to make success happen, if it can at all. Let us follow along from the operator's point of view.

Start with the process *normal* (the top green cloud of the figure). The process can stay there (what it has been designed to do and hopefully will do). It can go *off-normal* and back to *normal* (*normal* and *off-normal* are quite close, and crossing a boundary does not necessarily mean that things will stay there) or remain *off-normal*.

Off-normal has three categories (shown on the figure as the other three white clouds). A weak signal may be *immature, parametric,* or *transformative*. If it is *immature*, the operator must decide whether it is noise or something more. The operator will normally discontinue considering it if it is noise. If it is not *immature*, then it can be *parametric*. This problem should be *manageable*. He may then decide to intervene directly if he considers it necessary. Otherwise, he can wait to see if it will become *abnormal* and will alarm. This can be problematic unless he knows that an alarm is configured for this specific (underlying) problem. If no alarm is configured and the problem is ignored, it is likely that the abnormal situation will escalate to an *unmanageable problem*. Incident reports are full of this scenario.

If the weak signal is a *transformative* one, and it was identified promptly, it may still be *manageable*. Promptly and properly managed, things should return to *normal*. If not properly managed or it was not noticed early enough, it will become *unmanageable*. All *unmanageable* problems not adequately protected by process safety systems will lead to a *fault*.

On the other hand, the process can become *abnormal* directly from *normal*, which is often the case where something goes sufficiently wrong for it not to return without assistance. If an alarm has been properly configured, it will *alarm*. If not, there is an alarm design fault, and all bets for a good outcome are up to luck or the operator acting proactively. An *alarm* alerting the operator or a proactive operator action may be able to manage the problem properly. However, even with an alarm, if the correct actions are not or cannot be taken, we have an *unmanageable* problem. The outcome is now up to the process safety system to save.

This covers the procession of operational states. We now use the same diagram to define operator situational responsibilities. The dotted gray lines in Figure 9-5 delineate the regions' boundaries. This figure has a lot of moving parts. There are a great deal of situations and connections. Please slow down and take a good look. Everything in this book has a place on this figure. See how the responsibilities, tools, and discussion flow around it. It should help you put it all into perspective.

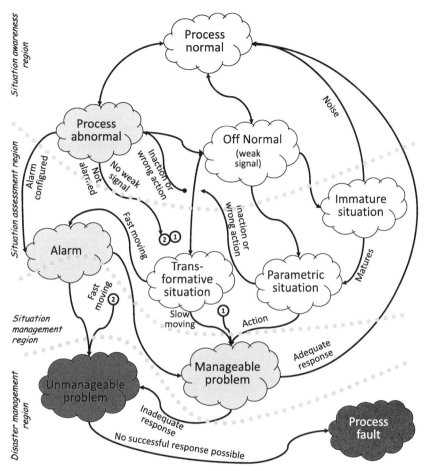

Figure 9-5. Situation management diagram showing operational regions (at left border edge) that identify what general activity is going on for each. The differences between abnormal situations (most likely leading to an eventual alarm activation) and the not-so-normal (off-normal) situations requiring special investigation to observe and consider managing are identified. Note that the operator does not ordinarily do the disaster management.

The Regions

- **Situation awareness region** – The process is *normal*, *off-normal*, or *abnormal*.

- **Situation assessment region** – The process has an *activated alarm*, or the weak signal is *immature*, *parametric*, or *transformative*.

- **Situation management region** – The process is *manageable*. This entire book is devoted to providing ways and tools to prevent *faults*.

- **Disaster management region** – The process is *unmanagable* and will lead to a *fault*.

Disaster management is another story indeed and not part of this book. It should be a part of yours.

Categories of Weak Signals

Weak signals have an important role to play. They can form the core of effective situation awareness. Weak signals bring a new clarity and technology. Clarity comes from clearly defining the boundary between situation awareness and situation assessment. Clarity comes by the way we organize the various types of weak signals leading to approaches for useful situation assessment. Technology comes because they provide a clean methodology to examine the weak signals to determine whether they suggest problems or not. Weak signals come in one of three types: *immature signals*, *parametric signals*, and *transformative signals*. Refer back to Figure 9-5. Each is explained next.

Immature Weak Signals

Immature weak signals are what natural variations are made of. They could be normal variations in energy or raw materials. They could be control loops that are working but could be improved with a bit of analysis and effort. They could be operators tinkering a bit too much. The latter is often due to operators' limited understanding of process dynamics or just the human nature to fidget. It is all a matter of degree. Just as in real life, noise is noise, or it can turn into something more. If it rises out of the noise box, then something is really afoot. Deciding when the noise is more than noise is divided into two parts. First, there is the part of making immature weak signals visible. This can be done with context icons and dashboards. Second, there is the need to see whether they are immature weak signals or something else. If an immature weak signal matures, it should mature into a parametric weak signal. That would be reasonable because it is rare for something so significant to suddenly rise out of noise. It is rare enough for us to count on. Otherwise everything ultimately can be too important to miss. If that were so, operators will be faced with an impossibly heavy load.

If an immature weak signal does not mature, it is considered noise. In other words, if it is not readily clear that an immature weak signal is noise, then it is assumed to have matured. It will be considered as a parametric weak signal. *Maturity* is what we are able to deduce from looking at that signal in different ways. Maturity is a classification based on a reasonably short evaluation. This is not to say that once an evaluation classifies one as noise, an immature weak signal is forever noise. Different operational conditions can easily result in that particular weak signal being noise no longer. Until then, it is considered to be noise. The operator looks for awareness of situation elsewhere.

Parametric Weak Signals

These are the weak signals that are close to or have crossed the boundary of customary and usual. Their detection requires good understanding of process operation and the careful setting of the boundary between noise and what is potentially important. These signals represent process conditions (usually individual variables, but not always) that are not normal and likely to change in ways that have a proportional impact. Small, slow changes of weak signals should have small and slow changes in operational vulnerability. Faster or larger changes of weak signals suggest faster or larger changes in operational vulnerability. At this point, those situations should represent manageable problems. Depending on the enterprise operating policies and the available operator resources, early action would be suggested. Such action is good operation protocol. That is why there is an operator there. These situations underpin the operation of the alarm system. The alarm system exists to ensure that no weak signal that can mature to cause problems is allowed to cause problems if these problems can be avoided or managed by operator action. The other alternative is to wait until the situation actually becomes abnormal. An eventual alarm is our best hoped-for result.

Transformative Weak Signals
As in the case of parametric signals, these are the weak signals that are close to or have crossed the boundary of customary and usual. Like before, their detection requires a good understanding of process operation and the setting of the boundary between noise and potentially important. These signals represent process conditions (usually individual variables, but not always) that are not normal but likely to change in ways that have a *dramatic and unpredictable impact*. Small, slow changes in them are expected to cause significant changes in operational vulnerability. Faster larger changes are expected to expose the enterprise to potentially serious operational vulnerability. Those situations should represent manageable problems. If these weak signals were limited, then timely actions would prevent the situation from becoming more serious. If these weak signals are faster moving, actions are unlikely to prevent alarms. Our weak signal is weak no longer. Once in alarm, the alarm response procedures become active.

Weak Signal Concepts
Weak signals have always been with us. They are part of all information. They are the stuff that data contain. They are ever-present. And they can be visible to those who are sensitive to them. They can be that just-barely-different-sounding noise coming from your automobile's engine. They are differences you may or may not hear, but a trained mechanic will pick up immediately. They can be the fleeting look in someone's face as he or she tells a story. We all know about gamblers who are adept at "tells," which to them broadcast truth or not. But we just now started calling them by the name *weak signals*. Naming is useful, of course, but the value is the way we are able to build an

understanding using names. This understanding has led to our ability to exploit them for real benefit. There are two points to bring up:

1. **Weak signal importance** – A weak signal can either announce a problem in maintaining the status quo or portend a serious event in the making. It is an operational necessity to identify which is which.

2. **Dominance** – "No weak signal ever rises to dominance by itself."[8] It is the co-occurrence of the weak signal and the potential exposure to risk that can lead to serious operational situations.

What Weak Signals Look Like

Let us dive in and see what they look like and where the operator can find them. Weak signals appear in plant or enterprise operating observations that seem slight, small, or otherwise nearly imperceptible. Some may be noticed by the operator but not considered useful to follow up. Of course, the operator cannot be expected to follow up on every hint or suggestion that something might be amiss. So there must be a balance here. But we do expect him to follow up on for the ones that need it. Weak signals are a way to find out. The illustrations that follow are examples of what the operator may encounter on the control board, in the electronic controls system, or simply while in the field attending to the physical unit. Where individuals work jointly, it is important that they communicate the weak signals and work together to process what they mean. They process weak signals as a team.

Weak Signals from General Observations

We start with a lot of "soft" weak signals. Later on we can get much more "technical." There are two columns in our examples illustrated below: *Observation* (the weak signal candidates) and *Possible Meaning* (something important [or not?] that might be causing the weak signal). There is a lot of information under the heading of *Possible Meaning*. This is purposeful. While the observation may seem innocuous at first glance and more than likely passed off for more pressing demands on the operator's time, when one takes the time to actually consider what might be behind the observation, it becomes quite evident that some suggest real operational issues including existing problems or developing problems. This is what weak signals do. This is why we label them *weak signals*. They usually deserve more examination. Please note that the examples that follow are not designed to be complete or all-inclusive. They are intended

---

8 Coffman, *Weak Signal Research*.

only to give you an introduction into how it might work. While attending to duties, our operator observes the following:

Observation (*weak signal*)	Possible Meaning (*forward-extrapolation*)
A control valve position is near the ends of its range.	The process is operating in a situation that is near or outside design expectations.
	The valve itself is not working properly.
	The measuring instrument is not operating properly.
	The control loop is not working properly.
	The process design is faulty.
A feedback controller causes lots of action (valve movements).	The control loop is working just as expected.
	There are possible tuning problems with the control loop.
	There is possibly too much noise in the measurement. • Excessive process noise • Interfering electrical signals • Missing or misadjusted deadbands
	There is potentially a sticking valve or leak in the pneumatic tubing.
Manufacturing or production efficiencies are changing more than 5%.	The process is operating in a situation that is near or outside design expectations.
	The process design is weak or faulty.
	A measuring instrument is not operating properly.
	The efficiency calculation is not functioning properly.
	The process is operating according to design, but operational choices are affecting the efficiency (either improving when improvement was not being sought or worsening).
Additives usage is changing more than 5%.	The process is operating in a situation that is near or outside design expectations.
	The process design is weak or faulty.
	A measuring instrument is not operating properly.
	The additive composition is changed or changing.
	The additive addition calculation is not functioning properly.
	The process is operating according to design, but operational choices are affecting the usage of additives more than expected.
Normal variations of process measurements seem to have changed.	The process measurement system is operating unreliably.
	There are faults in telemetry or data flow.
	Some aspects of the operation have changed beyond normal variable limits or expectations.
	Operator interface faults or failures are not receiving updated information, or are receiving updated information but failing to display it properly.

Observation (*weak signal*)	Possible Meaning (*forward-extrapolation*)
Warning indicators are present for situations that are considered normal.	The process is operating in a situation that is near or outside design expectations.
	There are instrument faults or installation errors.
	The process design is weak or faulty.
	A measuring instrument is not operating properly.
	The actual operating situation is abnormal, but the operator is misreading the situation.
Laboratory reports are late. and/or Laboratory values are missing.	Reports have been misdirected (either electronically or otherwise).
	There is an unusually high demand on laboratory resources (equipment and/or personnel).
	There are difficulties in processing laboratory samples due to laboratory equipment issues.
	There are difficulties in processing laboratory samples due to the unusual nature of the samples.
	Administrative or other personnel problems or issues exist at the laboratory.
	The laboratory protocol is weak.
	Laboratory priorities are not aligned with the manufacturing requirements.
Personnel are inside the plant operating area who are not "checked in."	Access control protocols for operating area access are weak or not enforced.
	Unauthorized or unknown maintenance or cleanup activities are in progress.
	Unauthorized or unknown operations activities are in progress.
There are open electrical boxes in the field.	Unauthorized or unknown maintenance activities are in progress unless they are in a known maintenance area.
	Housekeeping is poor or maintenance and inspection protocols are weak.
	Physical damage is preventing proper closure.
There are leaks on the ground or elsewhere.	Leaks representing spills or events are unrelated to operational problems or issues and thus represent sloppy work and/or poor housekeeping.
	Leaks represent actual weakening and/or failure of designed containment of operating equipment.
	Leaks representing equipment operated outside of the design limits are causing temporary or permanent containment breach.
There are changes in noise or smell in the unit.	Changes represent events unrelated to operational problems or issues.
	Chance observations reflect transient operating conditions that are not indicative of any operational or equipment issue.
	Changes representing equipment operated outside of the design limits are causing temporary or permanent containment breach or damage to the mechanical integrity of equipment.
	Changes represent actual weakening and/or failure of the designed containment or stresses to the mechanical integrity of equipment.

Let us pick one for a closer look: *warning indicators are present for situations that are considered normal*. This situation has been an important cause of operators being unable to properly see potential problems. It has prevented them from recognizing the early warning indicators of problems in the making. The operator was thinking that the process was in a known situation and considered it okay. Because he considered it okay, he dismissed those subtle warning indicators as not relevant. "They do not apply," he thought. In reality, some small part of the process was actually operating a bit too close to or even outside the design limits. Had he followed through and examined the warning indicators "for cause," the departure from the normal operating situation could be seen. Subtle, sure. Important, of course.

On the other hand, these situations could also be discovered through careful training or by following effective procedures. So weak signals, per se, would not be the only way to deal with this. However, training and procedures have an inherent limitation: detailed training and specific procedures must be in place for each unique situation. This means lots and lots of specific procedures will be required—too many to make this an effective way to do things. What weak signals provides is a general protocol with powerful tools that cover almost all the situations. Rather than dealing with a large number of special situations, operators have a tool with sufficient generality to be broadly useful and effective.

Weak Signals from Operating Observations
Operating observations are the things the operator sees as he operates. They are what he is looking for by watching the controls and displays, reading the laboratory reports, discussing events and concerns with field operators, identifying during shift handovers, and monitoring all the other activities in the control room. The rest of this chapter is designed to assist these activities.

Examples of Weak Signals
Because the idea of weak signals is still new, perhaps a tragic historical example might open a door enough to understand and appreciate them. First, let us look at a scene from the movie *The China Syndrome* and next at the Texaco Milford Haven event.

China Syndrome
> *The China Syndrome* is a 1979 American thriller film that tells the story of a television reporter and her cameraman who discover safety cover-ups at a nuclear power plant. It stars Jane Fonda, Jack Lemmon, and Michael Douglas, with Douglas also serving as the film's producer.
>
> The film was released on March 16, 1979, 12 days before the Three Mile Island nuclear accident in Dauphin County, Pennsylvania, which gave the film's subject matter an unexpected prescience. Coincidentally, in one scene, physicist Dr. Elliott Lowell

Figure 9-6. The vibrating cup of coffee from the movie *The China Syndrome*. These subtle vibrations (weak signals) were observed during an operational change. The true cause was faulty construction of the nuclear power plant that was covered up by falsified welding inspections. The main subplot was the search for the meaning of the vibrations, which eventually led to the discovery of the fault.

Source: Reproduced with permission from Columbia Pictures.

(Donald Hotton) says that the China Syndrome would render "an area the size of Pennsylvania" permanently uninhabitable.

While visiting the (fictional) Ventana nuclear power plant outside Los Angeles, television news reporter Kimberly Wells (Fonda), her cameraman Richard Adams (Douglas), and their soundman Hector Salas witness the plant going through an emergency shutdown (SCRAM). Shift Supervisor Jack Godell (Lemmon) *notices an unusual vibration while grabbing his cup of coffee which he had set down.* He then finds that a gauge is misreading and that the coolant is dangerously low (he thought it was overflowing). The crew manages to bring the reactor under control and can be seen celebrating and expressing relief.[9]

The cause of the turbine shutdown was excess vibrations detected during a mild earthquake. This shutdown is a planned form of equipment protection. However, the unusual vibration afterward of a coffee cup placed on a printer table was not (Figure 9-6). It was during his search for the cause of these barely noticeable extra vibrations (weak signals) that the shift supervisor uncovered faulty construction and falsified welding inspections. Left uncorrected, it could result in a catastrophic reactor failure.

Texaco Milford Haven (Pembroke)

Background

In the early hours of the morning of July 24, 1994, at a Texaco refinery in Milford Haven, England, lightning struck the refinery, causing problems with the vacuum unit, the alkylation unit, the Butamer, and the fluid catalytic converter (FCC). Immediately after

---

[9] "The China Syndrome," Wikipedia, last modified May 2, 2018, https://en.wikipedia.org/wiki/The_China_Syndrome.

the strike, several units shut down due to lack of utility power. A production control valve closed (most observers think that it moved to its failure position) and a unit started to fill with liquid hydrocarbon. That control valve showed an erroneous state of being "open" on a process display graphic.[10]

In response to the hydrocarbon buildup, as designed for safety protection of the vessel, the overpressure relief valve "popped" three times. The escaping liquid entered the relief system, eventually ending up in the knockout drum. The knockout drum eventually overfilled (in part due to an earlier modification that had not been properly assessed at the design time nor appreciated during the upset). The overfilled knockout drum then spilled liquid into a relief line that was not designed to contain such a flow. The resultant failure released about 20 metric tons of hydrocarbons into the operating plant. The vapor cloud produced by this release eventually ignited about 110 meters away from the rupture, causing a major explosion equivalent to 4 metric tons of high explosive. The explosion caused $80 million damage and injured 26 people (thankfully, none seriously).[11]

The Incident
The initial cause was the stuck valve on the deethanizer (a fractional distillation column used to separate unwanted material from liquefied petroleum gas). This caused the level problems on the debutanizer (used to separate butane from natural gas), the trips to the compressor system, and the liquid levels on the towers. Would you class the flows and pressure on the stuck valve line as weak signals? They did not have an alarm, but no or reduced flow caused the problem on the previous unit, which was generating lots of alarms.

- **Knock-out (KO) pot** – If the operators had seen the alarm on the KO pot (a type of liquid-vapor separator), would they have been able to prevent the disaster? If so, we have a case of poor alarm management, poor operational practices, and/or poor HMI design. However, an effective weak signal identification and management policy might have been able to provide second chances. This is not the primary role for weak signals, but a possibility nonetheless.

- **The stuck valve** – The stuck valve causing flow and level problems (debutanizer; tower levels?) would be a weak signal case; the trip to the compressor could be another, although it should have raised a clear "red flag."

---
10 Mike Gray, T. S. Cook, and James Bridges, *The China Syndrome*, directed by James Bridges (1979; Hollywood, CA: Columbia Pictures).
11 Health and Safety Executive, *The Explosion and Fires at the Texaco Refinery, Milford Haven, 24 July 1994* (Sudbury, Suffolk, UK: Health and Safety Executive, 1994).

- **Alarms on the previous unit** – Here we have a case of poor alarm management and poor operational practices. However, an effective weak signal identification and management policy might have been able to provide second chances. This is not the primary role for weak signals, but again a possibility nonetheless.

Hunches and Intuition Might Be Weak Signals

Weak signals are made-to-order for honest hunches. Many a hunch or other intuitive suggestion that something might be amiss, something seems not quite right, or something is missing, is just the thing that could be investigated. Operators are encouraged to follow up on all of them *only* using the structured work process for weak signals. Although, if an operator is having trouble moving from a vague feeling that something is amiss to finding weak signals, you will want to wait until "Active Looking for Weak Signals" in Section 9.7 for good ways to look for them. This is the power of the approach. But before things get too carried away, let us recommend a few boundaries for all of this. The first thing to understand is that all of this is not dependent on hunches or other forms of intuition. That is a side benefit. Operators now have a sharp new tool in the kit. You do not want operators to be so preoccupied with looking for every oddity or concern that processing them becomes the most important thing they do. Sure, the hunches can be useful and every one found could be processed. But the operator should not make finding them always at the top of the list. Doing that can overload and distract the operator from the core task of keeping on top of the big picture. You will need discipline. Keep in mind that the value of weak signals is not in finding as many of them as you can. The value is the methodology for sorting them into those that do not matter and those that do. Those that do matter should be looked into.

In Chapter 10, "Situation Management," we build the process of looking for and evaluating all questionable operating items (looking for weak signals) into the regular flow of the working schedule for the shift. This provides the timing and methodology to give the "looking" robustness. It has the benefit of relieving the operator from having another tool that seems like a burden more than help. Doing it this way keeps the focus on the main responsibility of proper operations. The power of situation management is that it provides both the tools and a usable framework to get them working together.

Expectations Will Interfere with Weak Signals

Weak signals and expectations do not mix well. Weak signals are not meant to be viewed or sorted in any probability order—"most probable" or "most likely" are not what to look for or assess now. Weak signals are not something that must be figured out now. They go on the list for later. They are used to lead a discovery process. You will want to work this tool in an orderly fashion. Avoid looking directly for an explanation of what it means to match a specific weak signal to a conclusion—this will almost always lead to confirmation bias errors or worse. Please remember the material of Chapter 7, "Awareness and Assessment Pitfalls." To use weak signals properly, we take everything in proper working order and trust the working process. No shortcuts, please. What comes to mind to underscore this is the working process used by the mythical character of Sherlock Holmes created by Sir Author Conan Doyle. It is the perfect thought model for what we seek:

> When you have eliminated the impossible, whatever remains, however improbable, must be the truth.[12]

Anything that appears to be a bit unusual is grist for Holmes. We ask the operator to do the same. We ask operators to first *observe*. Only *after* all the "clues" are noted does Holmes's deduction engine start working. For operators, we prime the pump of detection when designers build those clue producers. They are the stuff of Chapter 5, "The Human-Machine Interface." We get to explain their important use in this chapter. Are you beginning to get the sense that all of this is meant to build and fit together to help you provide something important to the control room? This is the power of *situation management*. Stay with me on this.

9.4 Building and Displaying Weak Signals

Weak signals are always present. But not all weak signals are visible. We will need to change this for our operators in the control room. Our task then is twofold. First, we must ensure that we build and use enough good tools to expose the needed weak signals to examination. Any weak signals will need to be available for detection when the operator looks. For if they are not exposed in this way, how can anyone find anything so well camouflaged? Second, we need to provide the technology for the operator to be able to recognize what he sees. When weak signals are exposed to discovery, the operator must be confident that he can and will discover them and confirm them (or disconfirm, of course). Here are the tools. Both the *expose* part and the *discovery* part are built into them.

Characteristics of Weak Signals

By now you should be ready to consider that weak signals are an important tool for situation awareness. You will want to know how they work and how the operator can use them to keep on top of his production plant or operation status. Here are some

---

12 Sir Arthur Ignatius Conan Doyle, *A Study in Scarlet* (London: Ward Lock & Company, 1886).

useful ways you can design and implement them. Let us begin with a review of what we want a weak signal to do. All weak signals need to:

- Be based on something real, something easily related to something real, or something felt that could not be readily dismissed

- Depict a reasonable difference between "seemingly normal" and "seemingly off-normal"

- Look normal when things really are normal

- Indicate likely off-normal when there is a reasonable likelihood that things can be really off-normal

- "Accumulate" in a reasonably repeatable way as off-normal situations add more weak signal clues for the situation that was related to the same or closely related earlier off-normal one

- Have the ability to be forward-extrapolated

- Be backward-projectable to useful opportunities for investigation

Weak signals can be derived, built, or exposed from a variety of places. These places are discussed next.

Weak Signals from Direct Measurements and Observations

This category is about as straightforward and direct as this concept gets. Examples include tank levels (near top, near bottom, moving too fast or too slow), valve positions (near closed, near open, or seemingly unrelated to process normal positions), average message length, vehicular traffic flow, on- and off-ramp vehicle flow values, and more. These are conveniently depicted by icons for ready understanding and shown on operator displays of appropriate levels.

Process overview displays are useful to aid the operator in finding weak signals. These displays are designed to provide the primary top level for situation awareness. The overview contains all the components or units in the operator's area of responsibility. At this level, all active alarms will be noted (although not individually depicted, of course). Any identified operational threat will be clearly depicted. The full set of key variables will be clearly displayed with current values shown and compared against good operating targets. The pattern of values for the key variables is intended to convey a rich understanding of the global operational condition of the entire unit or plant. And this pattern would suggest to the operator that something deeper might need looking in to. Rather than exposing weak signals directly, the overview would usually be suggestive of whether or not a search for problems or issues in the plant or operation might prove productive.

Weak Signals on Operating Displays
This is the usual way we expect operators to find the "needle" in the proverbial haystack. The word *needle* is used on purpose. Anyone who has taken shifts in the "chair" knows full well how hard it is to find out what is truly happening. Each operator has his routine. That routine is part good experience, strongly colored by any bad experiences (so the operator is extra watchful of a few certain situations), and mixed in with some hints of superstition and ritual. Oh yes, operators are not immune to the importance of good luck. This book is about finding ways to replace luck with success.

The way operating displays are designed and the information they contain assist operators to see what is going on and find anything that suggests problems. Here is the inherent conflict. On the one hand, the operator would want to use the displays to confirm his suspicion that everything is okay. When he senses no problems or issues, finding confirmation of nothing wrong would be comfortable. This would free him to shift (temporarily, of course) to other duties and tasks he must complete. Confirmation bias really does have a seductive pull. On the other hand, if our operator is keeping an open mind, he will want to search for everything that might suggest trouble. Look for weak signals.

All of this needs perspective. Operating displays must be appropriately designed. That design would include the use of process-variable icons, operational plausibility models, energy and material balances, gauges and dashboards, and more. These are discussed later. Note that it is up to training and expectation to ensure that displays are used properly. Figure 9-7 shows the current operating view (before invoking the comparison function).

Depicted is a tower top feed from a stirred tank reactor, a bottoms feed from a separator, a product storage tank, and a top solids separator and tank. All key operating variables are shown via icons (graphic rectangles and ovals with embedded value bars). Refer back to Chapter 5 for how key variables are identified. There are horizontal bars for flows, vertical bars for levels, vertical ovals for pressures, and small horizontal bars for valve positions. This plant is entirely fictional to illustrate a concept. Those who are well versed in process design are kindly asked to avoid trying to figure out what is being made and just jump in to see what this is illustrating. So far, a couple of valves appear to be quite near an end of travel and a level appears to be quite high. Each unusual one is a candidate for weak signals. The next figure is designed to assist operators to observe how those early indications (potential weak signals) are changing over a short time history.

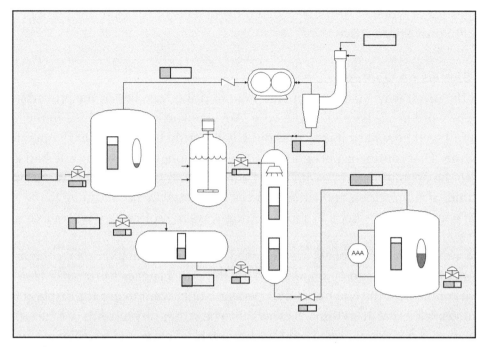

Figure 9-7. Current operation display with icons for key operating parameters shown. The key variables are depicted by the graphic rectangles and ovals with embedded value bars. This encourages the viewer to readily see what appears to be off-normal or abnormal in context and in relationship to other key parameters.

Weak Signals from Comparison Operating Displays

Comparison operating displays are a useful new tool. The idea is simple. To build a comparison operating display, start with any of the existing ones. Each display will contain information and values depicting the current operating situation. Flows, temperatures, levels, pressures, compositions, and all the rest will be present and as up-to-date as the infrastructure will allow. To build a comparison display, start with the display like the one shown in Figure 9-7. To that display, add content to compare what is current now with what was current at some convenient time in the past. Useful times might include an hour ago, just before the last shift change, or 24 hours ago. Each enterprise can determine what works best. Here is how to show the information:

- For all values that have changed less than an indifference value (which is a way of saying the change really does not mean much of anything), show the current value.

- For any values that have changed more than a given threshold of concern, show the current value and prior reference value indication (e.g., a previously determined short-term average). Call this a measurement change indication.

In Figure 9-8, each measurement shows up with a *measurement change* indication. Each blue bar on the measurement or position icon shows the value in the past. Each

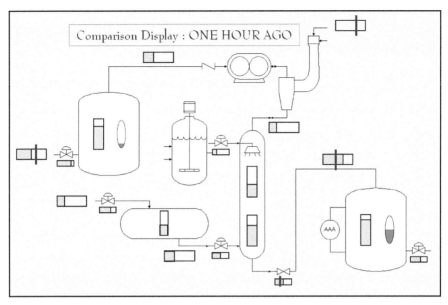

Figure 9-8. Current operation display with icons for key operating parameters shown. This view is a comparison operation display highlighting all variables that have recently changed significantly (blue bars). The gray section shows all current values. Blue bars indicate 1-hour-ago values.

unusual difference is a likely candidate for a weak signal. Four variables in the figure have changed enough to warrant indication. Note that your eye is directed to "catch" those changes. Investigate them all. Without a blue bar, the current measurement has not changed much during the measurement change window. What is shown represents what is current. This illustrates where weak signals can be an integral part of a screen display.

Lest we miss an important benefit from Figure 9-8, all the current values of the key operating values are useful. Even if the current value is not very different from what it may have been earlier, the operator will view it and observe whether or not it seems right. The use of comparison indicators is an aid to call operator attention. Such indicators are not meant to be a filter that classifies all the rest as being just fine.

Weak Signals from Indirect Measurements or Observations

Indirect measurements or observations refer to calculations or algorithmic inferences made from a combination of measurements or deeper analysis of a single measurement. The general categories include the following:

- Mass and energy balances
- Operational plausibility values
- Difficulty of control measures

- Statistical process control information
- Baseline operational measures
- Instrument condition monitoring
- Equipment reliability indicators
- Key performance indicators

Each is discussed below.

Mass and Energy Balances

Balances have long been staple tools for engineers in the design and analysis of plants. They have now become important operational tools. They facilitate detection of problems that otherwise can be quite difficult to observe. When balances do not close (approach zero), operators suspect a problem, and most of the time they know where to look. An imbalance is an important way to "suspect" a problem. Generally, the places to look are fairly limited and ordinary: for example, a leaking valve or one inadvertently left closed when it should have been open. Balances are such an important tool that they are often formally recommended following major incidents and accidents. For example, Recommended Item #4 of the Milford Haven HSE Report specifically mentions them as a needed requirement.[13]

By their nature, balances are a computational tool to suggest subtle problems with the operation of a plant in very specific ways. The two widely used are mass (or material) balances and energy balances. They work due to fundamental principles of physics: neither mass nor energy can be created or destroyed. Yes, we all know about Einstein, but the processes he refers to are beyond any we are considering. Balances can be computed instantaneously or over a given time frame. The longer the time frame (within reason, of course) the better the quality of the balance. Pick a time frame that provides results you feel can be trusted.

The way balances are used is straightforward. The first step is to draw an imaginary boundary around the physical system you are concerned about. This usually means the boundary is drawn around an important portion of the operating plant. Figure 9-9 depicts a plant with such a boundary shown by the dotted magenta line. The only places where we will need to know (by computing using simple procedures) values are the places where any energy or material crosses the balance boundary. Notice that only four heavy lines (these are pipes) cross the dotted magenta line.

---

13 Health and Safety Executive, *Texaco Refinery*.

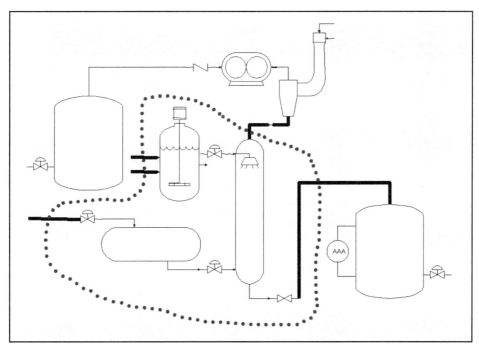

Figure 9-9. Plant with area selected (magenta dotted line) for calculation of a balance. All changes in the sum total of what is being examined must be in balance. Anything out of balance means that something is missing, suggesting a problem that needs further examination.

Let us begin with a mass or material balance. First, measure all the material flowing into the balance area. Next, measure all the material flowing out of the balance area. Next, subtract the outflow from the inflow. Finally, account for any increase or decrease in accumulation of material inside the boundary. What remains is the imbalance. The closer it is to zero, the less likely that material is getting lost or mysteriously increased within the balance area. If the measurement time frame is long enough to rule out normal variations, then any imbalance suggests that what is going on inside the balance area is not normal. The greater the imbalance and longer the measurement time, the more likely the *not normal* is really *abnormal*. Before *abnormal*, it would have been *not normal*. *Not normal* is a weak signal.

Showing Imbalances to the Operator
Figure 9-10 illustrates a gauge to indicate material balance with the key for the symbols and labels. This gauge shows a normal-appearing situation. It is judged so because the short-term imbalance is just a bit low and the accumulated imbalance is just a bit high. All of this suggests measurement noise and natural variation, as opposed to an actual problem.

A "right-click" option on the icon brings up a data box with numerical values as shown in Figure 9-11.

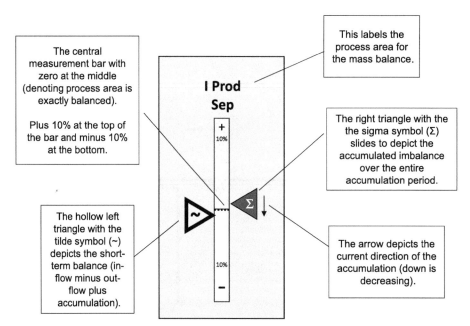

Figure 9-10. Mass balance gauge showing how it looks and its symbol key defining each visible characteristic. This provides a quick, at-a-glance overview including reference target, current status, and current direction of change. The symbol key is not a part of the operator display.

Figure 9-12 shows how this gauge looks when placed on the graphic (upper right). The mass balance is depicted as oversize on this illustration just to show clarity. In actual use, the rest of the content on the screen will determine its size and location. From this graphic, the operator can see at a glance how the unit balance is performing

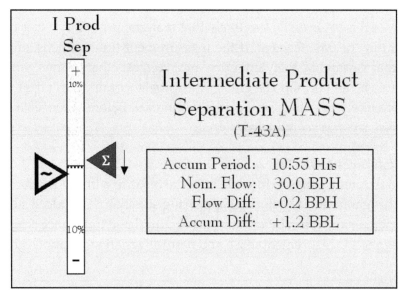

Figure 9-11. Mass balance gauge showing additional "right-click" information providing useful numerical values.

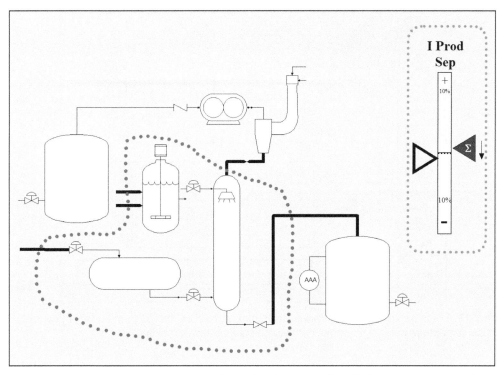

Figure 9-12. Plant with area denoted (dotted line) for depiction of the balance and the actual balance icon shown at the upper right. Note that there is nothing notable with the balance closure by the way the icon appears.

without any further investigation. Notice that the balance area actually shown to an operator is clarified by a gray dotted box instead of the earlier magenta color. The magenta was to show what area is being discussed for this book.

Figure 9-13 illustrates a gauge measuring material balance that suggests an unmonitored flow into the area might be present. This is suggested (as a weak signal) by the measured imbalance being quite negative (about 8%) and the accumulated inflow also being negative (about 3%). This means that over the 3-hour balance time period, about 3% more has seemed to disappear. Now the operator knows he must look. Note the light red short-term triangle, light red long-term triangle, and the light red warning box are added to the gauge information.

Energy balance is handled the same way as material balance, except instead of measuring material, we measure energy (usually heat). *As an aside, this example subsystem uses a gas-fired reboiler. It is under very good temperature control. That means that the control system adjusts the gas firing to properly maintain the correct process temperature. That temperature is normal.* Figure 9-14 depicts a gauge illustrating our situation. Figure 9-15 shows the same gauge with the additional dialogue information.

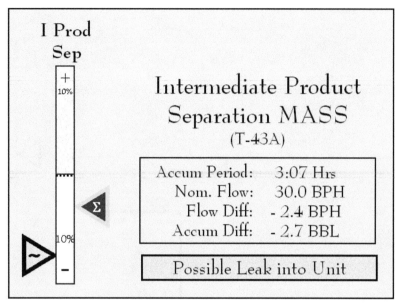

Figure 9-13. Mass balance gauge showing additional "right-click" information and a notice suggesting a possible leak (meaning that the mass balance calculation did not "close").

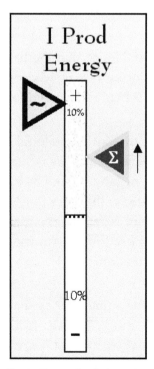

Figure 9-14. Energy balance gauge illustrating potential combustion problem identified by an energy balance calculation that is not "closing."

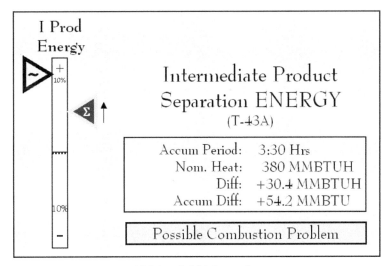

Figure 9-15. Figure 9-14 with additional "right-click" information.

Here we see the short-term energy imbalance shows about 8% more coming in than going out or stored. The 3.5-hour cumulative imbalance shows 5% more. This is a significant amount. The way this is calculated is to measure the gas flow rate to the burner and then, knowing the heat content, multiply to get the energy. We compute the energy going out in the traditional way. Check for any energy being stored within the furnace in the traditional way. The temperature of the outflow is not climbing (remember, the temperature controller is working just fine). If the stored energy is not changing, this leaves the energy balance at question. It could be in error due to an instrument problem, due to a gas efficiency problem (it somehow has 5% less heat energy), or because the combustion process has become less efficient. Any way you look at it, this is worth looking into.

Operational Plausibility Values

An operational plausibility model is a straightforward idea once you get the picture. Basically, we identify a well-understood part of a plant and then provide realistic values for all key variables there. Any variable that is outside its realistic value is cause for attention and should be examined further. The way we compute the realistic values is generally not highly sophisticated and does not involve significant computational resources. This is why departures are not immediately alarmed. And this is why we prefer to classify them first as weak signals. Doing so will invoke a nice, evaluative process leading to additional attention if warranted.

If the pilots of Air France Flight 447 (see Chapter 10) actually used their plausibility model values for thrust, as they are required to do in all of their training, the

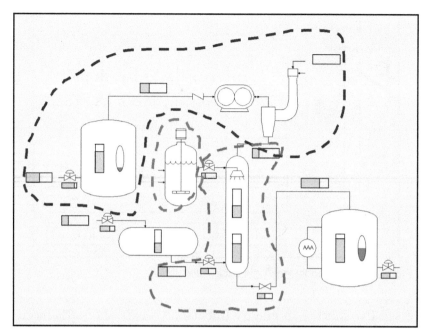

Figure 9-16. Plant showing identification of operational plausibility component boundaries (dashed lines). Calculations are made for the portion of the plant within each area to determine whether or not each seems to be operating properly.

plane would have never crashed into the Atlantic with all lives lost on June 1, 2009.[14] It is well known to pilots of the Airbus 330-203 (and all modern jet passenger planes, for that matter) that there are "safe settings" for engine thrust and control surface positions that will ensure that the plane will continue to fly adequately. And in fact, unless the pilot is performing unusual activities, it will maintain altitude with little departure. If there is a reasonable knowledge of altitude, small adjustments can be made (which are well within the "safe settings") that will correct and maintain proper altitude. Unfortunately, the Flight 447 pilots were so overwhelmed by the erroneous airspeed that they failed to even consider the idea of whether their settings were in the "safe" region. In their mistaken belief that they were flying too fast, they departed significantly from those settings and made operational changes that were completely unreasonable for flight. They had a required operational plausibility model available as part of their training, but they failed to use it. To be clear, there was no discussion or indication in the cockpit that they even thought about the proper protocol.

Let us get back to the control room. Figure 9-16 illustrates an example of picking parts of a plant for this procedure.

---

14 "Air France Flight 447," Wikipedia, last modified June 8, 2018, https://en.wikipedia.org/wiki/Air_France_Flight_447.

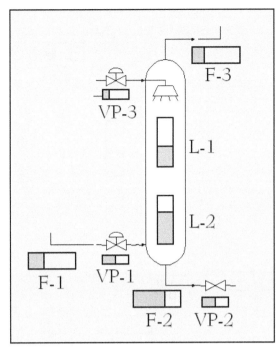

Figure 9-17. Part of a plant (within a selected operational plausibility boundary) showing selected operational plausibility components that are used for observation and any calculations.

This example plant has been divided into three major subsystems. There is an airborne particle collection subsystem identified by the dashed blue boundary line. There is a stirred tank reactor identified by the dashed green boundary line. And there is a separation tower subsystem identified by the orange dashed boundary line. For our example, select the tower contained within the dashed orange boundary line. This is shown in Figure 9-17.

We build an operational plausibility model for the tower using basic chemical and mechanical engineering tools. The model predicts that only the following variables are important: flow 1 (F-1); flow 2 (F-2); flow 3 (F-3); tower level 1 (L-1); tower level 2 (L-2); and the three valve positions, VP-1, VP-2, and VP-3. From the way this tower is operated, it is known that for each production rate level, the set of key variables should be within a small range of known values. Our operational plausibility model becomes a table lookup of the set of values for each specific production rate. Any reading that is different from the expected value by an unreasonable amount announces an off-normal situation. The graphic shown to the operator will display those values. See Figure 9-18.

Figure 9-18 shows what this model might look like if it detected problems. There are only two variables that are not as expected. The violet bars in F-3 and L-2 show the expected values for these two variables. Neither of the current values for F-3 and L-2

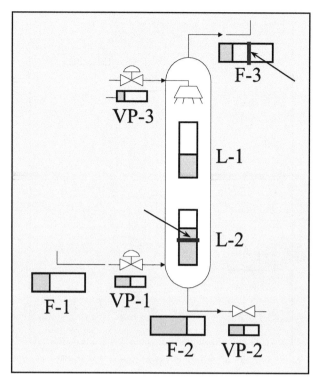

Figure 9-18. Operational plausibility model showing comparison values (larger purple bars at arrows) that single out ones different from current values (black bars).

is abnormal in and of itself. Therefore, they would not have been in alarm, had alarms been configured for them. By exposing them as weak signals, we notify the operator to have a look. If they appear to be useful weak signals, he processes them. If not, he might keep a watchful eye out just to make sure.

Difficulty of Control Measures

This is not about looking at the controller error (set point minus measurement) but at how hard the control loop is working to keep the controller error within bounds. One of the important aspects of a feedback control loop is that its design functionality will almost always keep the controller error low, until it cannot. When it cannot, it is always due to a lack of ability to move the manipulated entity anymore. The controller simply runs out of resources. A simple illustration would be valve position. If it is near the end of range, then the control loop is running out of maneuvering room to keep the measured variable at the control point. Until the control loop actually "runs out of maneuvering room," it will keep the error as low as it does when there is a lot of maneuvering room. This means that the actual measurement will be as close to proper as the control loop has been designed to keep it. Looking at the error will not provide any advance clue that trouble may be brewing until it actually boils over. Examples include the following:

- **Level control** – Keep track of the surge capacity remaining in the tank or vessel.

- **Vehicular traffic flow** – Keep track of how close the flow is to the carrying capacity of the motorway, not the average speed or inter-vehicle spacing.

- **Message routes** – Keep track of the capacity and availability of alternate routes, not the actual traffic load on the selected route.

Statistical Process Control Information

Statistical process control (SPC) is a highly developed and extremely useful methodology to identify unwanted and unusual operation that can be qualified (by appropriate measurements and analysis) and continually monitored. Its effectiveness is the ability to separate unusual variability (called *special sources of variation*) from normal variability (called *common sources of variation*). So long as the observed variability (e.g., temperature measurements and opacity measurements) is within the range of normal, nothing need be done. In fact, nothing more can be done inasmuch as the range of normal variability is just about as good as the particular process can perform. Attempts to do better only make things worse. This is perhaps one of the few cases where the advice "If it ain't broke, don't fix it" is probably correct. This normally expected region establishes the control range comprised of a lower control limit and an upper control limit. If the procedure detects variability outside the control range, it strongly suggests that the process is now outside where it should be. This variation is due to special causes. They are "special" in that they are due to something different from normal or expected goings on. Now you know that you should be looking for something unusual.

Figure 9-19 depicts the framework for an SPC gauge. The actual temperature range of interest is from a low of 2°C (Celsius) to 11°C, shown as the minimum to maximum frame line. The desired temperature is 6°C, so the control loop control set point (SP) is set at that value. Based on product blending ability, the best variability range is from 4.5°C to 6.8°C, shown by the wide gray bar. Because this temperature is quite critical, perhaps there is a significant economic loss if it goes out of range, there may be alarms at both the low and high ends of the range (not shown here). There was an SPC study for this process, and a lower control limit of 3°C (left "L") and an upper control limit of 8°C (right reverse "L") were determined. Therefore, the range between 3°C and 8°C is within the normal range of process noise (measurement noise, control valve positional repeatability, etc.). By convention for gauges, no actual numerical values are shown.

Looking at this situation through the eyes of weak signals, the following gauge situations might be useful. Each identifies a weak signal.

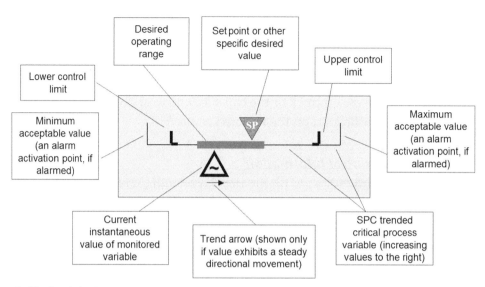

Figure 9-19. Statistical process control gauge and its symbol key defining each visible characteristic. This provides a quick, at-a-glance overview including reference target, current status, and current direction of change.

In Figure 9-20, observe the following:

- The temperature controller set point value (inverted gray triangle with "SP," usually determined by the operator) is outside the desired operating range.

- The desired operating range is comfortably within the upper and lower SPC control limits (the black left "L" and right reverse "L" symbols) but has failed to properly respond to the controller target.

- The current value of the monitored (and intended for control by a control loop) value (see lower triangle) is at the center of the desired operating range (see the thick horizontal gray bar above the triangle). The current operating target value is steady (no arrow below the lower triangle).

The situation in Figure 9-20 strongly indicates an operational mistake or controls malfunction. It is easily recognized from the gauge. A process set point is outside the

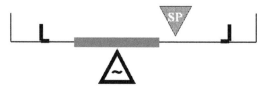

Figure 9-20. Statistical process control gauge showing the process set point outside the desired operating range. This suggests an operating target was set in error either by an operator or by a cascade controller miscalculation.

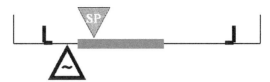

Figure 9-21. Statistical process control gauge showing the current value is steady but outside of the desired operating range. This suggests that the process is unable to meet the desired value and has been there for some time.

Figure 9-22. Statistical process control gauge showing the current value at the edge of the desired range and getting worse.

desired operating range, but the observed value of the process variable is inside the desired operating range and steady. Either the controller set point change was just made, or it was made a while ago but is not being properly acted on by the control system.

The situation illustrated by Figure 9-21 indicates an operational issue likely related to quality. It is easily recognized from the gauge. Note that it is steady, getting neither better nor worse.

The situation shown in Figure 9-22 suggests an operational issue likely related to continuing and continuous degradation. Again, this is a situation that is easily recognizable from the gauge.

The importance of the several gauge examples reviewed here is the ability of the operator to use them to quickly assess the situation within the frame of what is desired and what actually is present. It should be clear that something is amiss when they differ. And from where and how much they differ, the first line of examination is suggested.

Baseline Operational Measures

Baselines are expected values. Generally, they have ranges. But they are static and not associated with other measurements or values. So, for a measurement or other variable to be at or within a baseline would mean that it is reasonable or expected. Most important process measurements and other values will have baselines. Most others will not, although a particularly bad experience will usually result in an operator having some pet ones, which generally are not useful ones, to be candid. Having baselines can be useful to suggest off-normal, but only when the values are objectively set.

Baselines are related to but quite different from operational plausibility values. Operational plausibility values are calculated based on the current operational situation using current measurements and a method of relating them in a way that estimates or approximates what other values should normally be. So, it is entirely possible for all measurements to be within their expected baseline values but the plant to be operating at a place that is unexpected and not normal (or even abnormal).

Instrument Condition Monitoring

This discussion introduces condition-monitoring technology. The goal in bringing it up here is for completeness in the coverage of useful and available situation awareness tools. Depending on the criticality of your enterprise or the harshness of your processing conditions on instruments, you might want to further investigate some of these tools. Before we get into the details, there is an important distinction to be made. There are two very broad categories of how this technology might be used.

The first category is *clear notice of malfunction*. The instrument, valve, or other device has failed and the device itself is telling the operator this situation. If the controls configuration permits and the designer thought about it, there will be an alarm (if action is required) or other direct notification. A simple example is a transmitter that suddenly goes out of range. Situation awareness will be obvious. Get to the management part of this problem directly. These situations are not discussed here.

The second category is *vague or perhaps indistinct clues* that something may be amiss. It is for this category that the tools for situation awareness may be of benefit. Therefore, a brief discussion of the general areas follows. It is left up to the enterprise engineering and operations policies and practices to decide what is useful and how to develop a functional capability for using it.

Basic Sensor Validation

Some of you may have seen what a sticking valve might look like. You have observed firsthand what a biased field transmitter might do. Many have experienced these problems on occasion but may have missed some of them. These problems may have been missed at times when we did not explicitly move the valve but noticed an unusual response. We likely missed where the control loop seemed quite happy with where things were, not knowing that some intermediate variable was not being adjusted because it was sort of close to where it needed to be, but was actually where the sticking sensor was keeping things. Unless special techniques are used, or added effort is expended for this purpose, there is not much that can be provided to the operator that would be helpful—that is, unless a few special techniques are employed for the operator.

Specific sensor validation procedures can be quite effective. There are several general concepts.

- One technique is to periodically measure the statistical variability in the sensor readings. Once a base is established, any relevant departures usually point to something changed. Typical causes may be control loop degradation, process or raw material changes, and actual sensor operational changes. By identifying something unusual, it should be a straightforward engineering activity to find out the cause.

- Another technique relies on a drifting mass or energy balance to identify potential problems.

- Still a third technique relies on exploiting alternate measurements that can be used to either compute or estimate the variable under measurement by the primary sensor. Temperature profile measurements are one relatively direct way. The operating plausibility models discussed earlier is another.

- Other ways may involve combining several flow measurements whose piping arrangements require them to be consistent. Any inconsistency points to a problem or issue.

There are others. Important measurements, for example, would be excellent candidates for employing appropriate validation measures.

Smart Field Devices
Be they Fieldbus, HART, or other protocols, smart digital devices have changed the playing field for industrial instrumentation and control. Not only are they better at doing their jobs, they also can dramatically reduce wiring costs and the need for additional separate instrumentation. Some smart valves are proficient at providing effective flow measurement (and therefore control) due to built-in sensing. They also contain a powerful set of diagnostics and health monitoring capabilities that are easy to use and extremely effective. It is now possible to almost eliminate physical inspection and bench testing in order to determine fitness.

There is an important systems design challenge regarding how to use their broad diagnostic power: what to provide the operator and what to provide engineering and maintenance. How this decision is made can provide extremely useful information to both teams, overload both teams, or overload one team and put the other out of the loop. The recommendation is to provide the full set of diagnostics to engineering and maintenance, but provide operations with notifications only when the issue impacts operations (or in the case of known failures, alarms). The operator will mostly use these notifications as indications of issues to which he must devote more than ordinary attention. They are not usually weak signals unless several are suspect at the same time, suggesting something more systemic.

Equipment Reliability Indicators

Equipment reliability information can be an extremely rich and useful source of early indications of possible trouble brewing. Good reliability tracking protocols can be full of clues to be examined and evaluated. We will not go into much detail here because your program should be well known to you and the related support personnel. Rather, let us take a quick look at the general forms of information you may wish to track and use. This is not just about equipment problems, although they may be real enough and require attention. This is about using equipment problems as leading indicators of operational problems. There are three categories of interest: baseline, differences, and changes.

1. **Baseline** – This area includes overall plant or enterprise reliability information (or statistics, as some might call it) and compares it to expected similar industry or enterprise results. During comparison, any of your areas that are notably outside what peers are experiencing should suggest corresponding problems or issues for further evaluation. This evaluation must be broad. Under consideration would be the maintenance practices themselves, equipment selection as-built or replacement, and operational stresses. Operational stresses may be due to equipment being used for different purposes than they were designed for, or operation too close to or outside proper limits. Because this information would reference specific plant or enterprise areas or equipment, operating conditions should be readily examinable. Note that this category of situation awareness has a broader or wider time frame than an operational shift. Consequently, this evaluation may be shared with engineering. Nonetheless, it is an operational responsibility, and operations must be responsible for ensuring it is done and properly utilized.

2. **Differences** – This refers to baseline information but is focused more carefully on differences that may reflect operational anomalies. This means that if a plant's baseline reliability and maintenance experiences are generally comparable to industry or enterprise expectations, but certain parts of the plant differ

from what is generally expected, even if these differences do not change over time, these differences would suggest examining them for something more. Again, the time frames for differences are much wider than an operational shift.

3. **Changes** – When reliability information changes (either gradually or suddenly) from one value to another, it suggests developing problems. Where those changes are unintentional or unexplainable, they usually point to deeper issues. These would be examined to see what the changes actually mean. These time frames may be within an operational shift or well beyond. Appropriate changes need be examined in the time frame where they arise.

Key Performance Indicators

Key performance indicators (KPIs) are performance measurements that organizations use to assess, analyze, and track operations or enterprises. These measurements are commonly used to evaluate success in relation to agreed-on and tracked goals and objectives.[15] They are widely utilized because they are simple yet clear indicators of important aspects of an enterprise. Unfortunately, most KPIs are not going to be of much use for situation awareness due to the time frame of their calculation. Situation awareness for operators must be timely. Many KPIs are designed to be long-term indications (certainly well longer than a few operational shifts).

The first step in using KPIs for situational awareness is to be certain that the ones in use are responsive and powerful enough to provide either clear operational information of something going wrong, or are amenable to weak signal analysis. For example, suppose that a plant is using energy consumption per unit of production as a KPI. Four hours into the night shift, the consumption makes a clear upward movement. If there were one prime user of energy, this would suggest examining that activity for something that has changed and needs attention. The operator would examine that activity to narrow down the issue. Continuing with our example, say that most of the energy is electrical and is being used by a very large compressor. Our operator can quickly and effectively follow an electrical consumption change checklist and see what might be going wrong. Some items on that list might be malfunctioning transformers, overheating motor windings, power factor management gone awry, or a misapplied safety brake designed to quickly slow down an unloaded compressor. If, on the other hand, there are many users of the energy and it is of differing forms (e.g., natural gas and electric power), then weak signals would be a more likely tool.

---

15 "Seven Common KPIs for Production Monitoring" (white paper, York, PA: Red Lion, 2013).

The takeaway is to use KPIs that do the following:

- Track performance aspects of the operation that have the clear potential to show normal from abnormal or usual from unusual

- Indicate clearly enough to be sharp and distinct on their surface

- Have a change time frame well within a shift or be incorporated in a shift-to-shift monitoring program that is integrated with shift operation and shift handover

- Suggest clear avenues of investigation, either direct mapping or via weak signals that fall clearly within the operator's span of responsibility, when they deviate from the expected values or ranges

Weak Signals from Trend Plots

Trends by their nature are ideal for exposing weak signals. An upward trend jumps out to a viewer where it may be quite well hidden by text data. Likewise, a pattern of "blips" is obvious from a trend, as is cycling (with the periodicity being easy to see and qualify as to amplitude and cycle rate). This is not to say that it is a good idea to throw a lot of trend plots on display screens and ask the operator to keep an eye out for things that look suspicious. Rather, it is a suggestive trend included within an operating screen that artfully tips the balance to a potentially useful weak signal opportunity. This is all about developing a reliable way to get at weak signals that are going to be useful.

Another example is the use of sparkline graphs as extra information. Sparklines can be revealed (e.g., by "hovering" or a "right-click").[16] Figure 9-23 shows what that might look like (the rectangular trend at the extreme center left with the green normal bar).

The flow into the tank at the left shows that the current flow is reduced from that of an hour ago (blue vertical bar). The addition of the pop-up sparkline should be all that is needed to see if this reduction is proper or unusual. In this case, it appears to be proper and therefore of no consequence. Yes, there is a "blip," but most of the time the variable falls within the expected values as framed by the horizontal green bar. On the other hand, if the sparkline together with the comparison operating display values do not seem to appropriately match, then this would be a weak signal worth investigating.

---

16 "Sparkline," Wikipedia, last modified May 26, 2018, http://en.wikipedia.org/wiki/Sparkline.

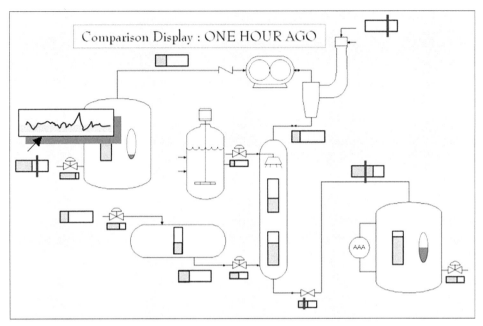

Figure 9-23. Sparkline pop-up added to a comparison operating display to provide a mini history for the affected key variable. Each comparison variable would have a sparkline available for the operator to select if desired.

Another example could be illustrated by examining two related trends. When viewed together, anything unusual should be able to be identified. Figure 9-24 illustrates this. For most of the trends, both variables appear to be generally tracking each other. However, during the last two or so time periods, this relationship is completely reversed. Knowing what the variables mean and observing such a reverse could be just the thing that needs further investigation.

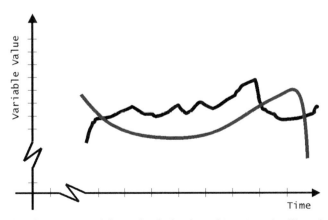

Figure 9-24. Weak signals suggested from the behavior of two trends. The observation is that the trends appear to be different in the beginning, begin following each other in the middle, and differ at the ends. Because both trends are expected to agree, this difference is now readily observable.

The Role of Icons, Dials, Gauges, and Dashboards

Icons, dials, gauges, and dashboards (IDGDBs) are the visible front lines for weak signals. Their design should include as many weak signals and other awareness information as can be identified by reasonable means. Operators will use IDGDBs to help find potential problems. Remember, all nonnormal situations that are not contained in some form of exposing medium, like these, will be difficult for any operator to find. Please understand that the examples shown in this book are provided to illustrate the tools and concepts. They are not necessarily to be considered as the only way or even a best way.

Recall from Chapter 5, "The Human Machine Interface," that part of the essence of IDGDBs is that they provide clear indications of current values and normally expected values, where the normal ones are not close enough. Do this to enhance weak signal detection. As an example, we look at Figure 9-25, which is taken from a similar one in Chapter 5 (see Figure 5-30). The blue circular region (not actually a part of the IDGDBs) points out a weak signal. For this case, unless an adjustment is made to the rate, the delivery will not make the needed delivery (termed *nomination*). However, an adjustment should be able to resolve the problem. In addition to making the appropriate change, the operator should investigate why any change was needed. This may rise in importance if other adjustments were needed in the past. See the discussion later about "flags." This is another good place to distinguish between off-normal and abnormal. Because it is natural to make the needed minor adequate adjustments, it would not be alarmed; therefore, it would not be considered abnormal. Yet it is something that should be watched and perhaps adjusted.

Another example is illustrated by Figure 9-26.

Also recall this figure from Chapter 5. This is a particularly good example of a way that weak signals can be displayed. Each of the 36 process values (vertical bars on

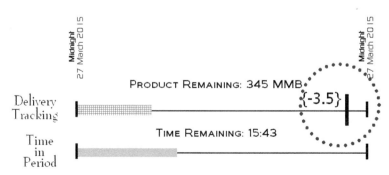

Figure 9-25. Nominations gauge illustrating a weak signal (annotated within the blue dotted circle that is actually not a part of the gauge but shown to identify the weak signal). If absolutely nothing changes, unlikely but possible over the next 15 hours and 43 minutes, the delivery will be under by 3.5 MMB.

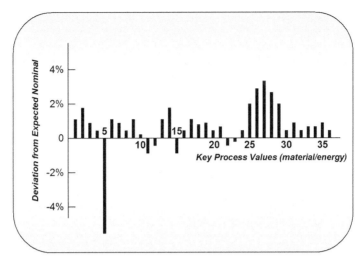

Figure 9-26. Deviation diagram (repeated from Chapter 5, *"The Human Machine Interface"*) showing an overview of an entire operator area. Depicted are 36 critical operating values and their departure from expected value. The arrangement is the order of the significant movement of product from area entry (at the left) to area exit (at the right).

the figure) represents an important measurement. None are, or should be, in alarm. Yet many appear to be headed for trouble down the road. Trouble might be brewing at variable 5 (greater than 4% below nominal expected value). However, because variables 4 and 6 are adjacent in the plant to variable 5, yet show no difference from what is expected, it would be quite likely to presume that whatever is going on at variable 5 is unique to it. A SCADA, PLC, or DCS error or instrument problem would come to mind. Trouble also is suggested around variables 25 through 29. All appear to have departed enough from their nominal expected values. This suggests a closer look. Because a group of variables appear to be involved, this case might well represent an actual process problem in that plant area. In any event, the operator is advised of these off-normal areas and should likely investigate them for cause. They show up because they appear out of the ordinary in this form of information display.

Intuition and "Raw" Information

Intuition is an interesting thing. We all have it. Not all the time, and likely almost never when we really need it in our personal lives. Nevertheless, we do have it. It is mostly a useful thing. But here in the control room for the purposes of situation awareness, it must be put aside. The reason is quite simple and fundamental. Your operator is, by the nature of being in control, proffered lots of information—information from sensors, analyzers, and the like. You want it that way. You want him to see everything that is going on. You do this because you expect him to be able to find what is right and what is not right with his plant or enterprise. You want detection, examination, and then careful analysis for cause. You want him to find what is really there, but nothing more. You want him to "look," but not before it is appropriate, and "conclude" the meaning of what is seen. To summarize: the operator should look for root causes and—then and only then—substantiate what they might be by finding supporting (or denying) evidence.

The operator will need to use the alarm system as intended. The operator will need to use the weak signals tools as they are designed. These two capabilities, the alarm system and the weak signals tools, used together will form the backbone of detecting abnormal operating situations in time to manage them.

9.5 Models for Weak Signal Analysis

A process model is an artificial thing. It is man-made. Engineers build process models and try to use them to aid in the activities of design, evaluation, operational recommendations, and all related activities that can best be done before something costly is actually built and tried out. To build a model, someone (well, lots of someones) examines the real process and develops something that works a lot like the real thing to be used in its place. Because a model is not real, when we use it in our minds and something happens that breaks it, we are not breaking anything real. It does not cost anything. We do not actually hurt anyone. No one will be deprived of a high-quality product. We have to wait long enough for a real thing to do its thing, but we can often push a model to work much faster. Depending on how accurate and how complete it is, we can use our model to help us make decisions about what the real process might do when subjected to the same situation as our model. Models are also used to explain and train.

To do weak signal analysis, operators must have and be able to use good mental models of their plants. They get their mental models from their training and experience in operating. We can test their knowledge during competency evaluations. All of weak signal analysis depends on operators having a very good understanding of how the real thing works and how it can get into trouble. He will use that understanding to "suppose" what the real thing might be doing when he sees anything of concern. He will then use that understanding to suggest where to look for confirming (or disconfirming) evidence. And he will use that understanding to evaluate what he finds for the evidence.

9.6 Weak Signal Management

Now we have an idea of what weak signals look like. It is time to use them. Weak signal management is the way. It does not take much imagination to see that lots of weak signals could be easy to find anywhere one looks. They need to be easy to find.

But we do not want to find more than are there. This is a skill, not an algorithm. This means that the HMI tools and views need to be carefully designed and implemented so that they adequately convey proper clues. It is a major reason for style guides and consistent design. We are going to use weak signals to uncover what might be going amiss. Start with the operator identifying something as a weak signal. He must be able to decide which type it is. Is it immature and therefore either noise or possibly something else leading eventually to parametric? Is it parametric suggesting early action or watchful attention? Or is it transformative suggesting immediate intervention would be a good idea? Here is how that decision is made.

The Work Process

The approach is actually quite simple to express. Coffman first suggested the approach, and it was addressed more recently by Rosa Antonia Carrillo and Neal Samuels.[17] What we will do here goes way beyond their earlier scope and work. We use it as a sharp tool. Start with the requirement to have an effective way for the operator to see weak signals. Remember all of those icons, dials, gauges, and dashboards (IDGDBs)? They convey a lot. There are other suggestive aids as well, but that is later on. Now let us take a look at how the approach works. This is an approach that uses a new working methodology. Do not worry if this brief introduction does not sound complete enough. Each step will be carefully explained soon. The five-step procedure is introduced and described next.

For each identified weak signal, perform the following steps:

- **Step 1.** *Suppose* the weak signal is an indicator of something much, much more abnormal. To "suppose," you are going to use your rich process model knowledge.

  Take that small indication of something that concerns you and suppose that it is a clue of something important that is affecting your operational area.

- **Step 2.** Identify in your mind (using that good process model in your head) what is the *worst that might happen* if this specific weak signal is actually a true early warning.

  Remember that you are looking for the significant thing (quite bad, really unwanted) that might be happening if it is not fixed or worked on. The less-than-important ones would not really show up early as weak signals. Even if they did, there might be many and they might not need early attention.

---

17 Coffman, *Weak Signal Research*; Rosa Antonia Carrillo and Neal Samuels, "Safety Conversations—Catching Drift and Weak Signals," *Professional Safety* (January 2015): 22–32.

If nothing can be identified, then the weak signal (for the moment) is classified as immature *(meaning that it would be considered noise and not (at this time) important. Let it go for now.*

- **Step 3.** For each worst thing that might happen, *suppose all potential causes* (possible things that can go wrong) for it to happen.

  In steps 1 and 2, you have noticed something unusual. You have an idea of the worst it might cause if you do not work on it. Now you want to look for all the ways something can go wrong that would lead to that bad thing happening. Just make the list in your mind.

- **Step 4.** For each potential cause, look for anything real (back in the plant) that would *confirm* (or *disconfirm*) each supposed thing that was wrong.

  Now use the list from above and look for specific evidence to prove or disprove whether or not each possibility exists. If there is no confirmation, then the weak signal that started this whole thing is classified (for the moment) as immature. *Let it go for now.*

- **Step 5.** For each confirmed weak signal, you know you have the indicated real problem.

  Confirmed *means that the weak signal was noticed, forward-extrapolated to a worst bad thing, backward-projected to look for evidence now, and enough evidence was found. So the original forward-extrapolated problem is very probably quite real. It will need attention.*

Figure 9-27 diagrams the activity in the previous list. Begin at the *off-normal* cloud at the top left. As each weak signal is noticed, one at a time, the operator works through an analysis for each. There are only two possible results. Either there will be evidence enough to confirm that what is suspected as abnormal really is and the situation needs operator attention, or there is not enough evidence to confirm or disconfirm the off-normal situation, and so it could either wait for later review or just be dropped.

Why does this actually work? It works for two powerful reasons. First, it is a direct way to identify early potential problems for investigation (the forward-extrapolation part). Second, it is a clear process for investigating each of the potential problems (the backward-projection part). Backward-projection is a way to find any likely underlying symptoms for the problem. We started with only a single weak signal that could be the tip of an unknown problem or could be nothing. The forward-extrapolations are the way to find out which, if any, problem it was "leaking" out from. Backward-projecting provides a way to look for more evidence of the problem. If enough evidence is found, it is used to confirm that something is really going wrong. Before,

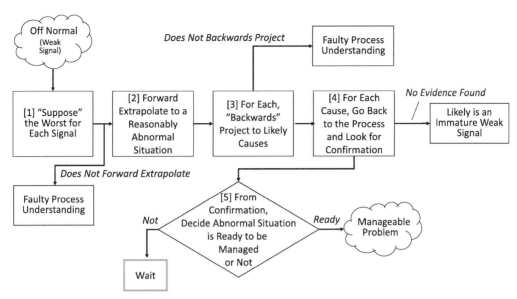

Figure 9-27. Weak-signal flowchart. This is a step-by-step work process. Follow the progression.

without a weak signal suggesting something might be amiss, the operator was not even aware that something might be going wrong. Problems were there, but he did not know to look for them. Now he knows. "Looking" this way, the operator will have specific things to look for and, if there, find. He is using weak signals as the precursors to the disaster he "supposed" might be happening. He is looking for all indications (yes, they can be other weak signals) that might or might not have been specifically suggested by icons, dials, dashboards, or other observation. He is looking for a parametric weak signal or a transformative weak signal. If he finds neither, he most likely has an immature weak signal. With proper evidence, he can act or wait, as he deems appropriate.

Let us take this carefully step-by-step in more detail.

Step 1: Identification

At this step we are only interested in *identifying* observable weak signals of interest. This step is only to find them. No attempt should be made to explain them at this step. The operator will not be doing exhaustive searches for all potential weak signals. That would be time-consuming and frustrating. Most often, nothing will be there. Operators have much to do besides going on an extensive problem hunt with absolutely no indication that there is or might be a problem. What really must happen is the operator "visiting" each area of his operation on a dependable and regular schedule. He may observe things on existing displays as icons, gauges, and dashboards. Perhaps

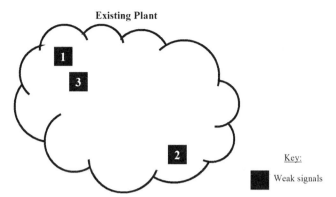

Figure 9-28. Depiction of weak signals 1, 2, and 3. The operator identified them by either looking for them or just coming across off-normal indications for each. This is the first step in weak signal management. Be careful not to get ahead of the process.

it is observed as part of his periodic "rounds." Perhaps he just happened to be looking at something that caught his eye. Whatever the initiator of focus, he is there. While he is there, he will notice weak signals. He should make a list of all them. These are the ones that need to be examined. Eventually, all the weak signals he finds will make it onto that list. All entries on the list will be examined. At this point, none are. It must wait for the next step.

Figure 9-28 illustrates the identification process. At the time an operator was looking at the plant or just looking around, weak signals 1, 2, and 3 were observed. Weak signals 1 and 3 appear closer together. This suggests that they were observed coming from the same or closely related part of the plant. Weak signal 2 appears to be coming from a different part of the plant. But this is just our superficial use of a visual illustration.

Step 2: Forward-Extrapolation

At this point, you, as the operator, have already observed one thing (or more) that does not appear to be quite right. Your goal now is to find out what it might mean. If it is important, then you will want to consider doing something about it. Back to what you see. You observed things by looking at the actual plant or enterprise data or information. From that list-of-the-moment weak signals, take each one in turn. Ask the hypothetical: What could be going on if that specific weak signal were announcing its worst? When you ask that question, you are now thinking using the model in your mind of what the plant is. This is a jump. Look into your model of it and open those doors to your imagination, not fancifully, but reasonably. This is a reasonable moment for some "gloom and doom" predictions, but do not go overboard here. You are just asking, what can this weak signal be suggesting that is bad? *Bad* means that it

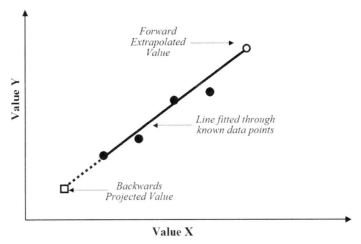

Figure 9-29. Arithmetic linear example of forward-extrapolation (solid black dots connected by a line to the open black dot at upper right) and backward-projection (dashed extension of black line to the open square at the lower left).

is something that an operator really should know about and attend to. Remember that only the early suggestions of significant problems are useful. Note them all. Usually one or two will come out of this activity. If you get more, that is okay too. To do this well, you will need a rich understanding of the process or operation you are managing. Forward-extrapolation uses specific knowledge of what you are operating.

Figure 9-29 illustrates a simple linear arithmetic example with both forward-extrapolation and backward-projection. Our known data comes from the four solid black dots. Extrapolate to the next value of time, and we find a black open circle. The open circle is where we expect things to end up a certain time in the future if things moved forward linearly. We do it linearly only because our model is linear. When we backward-project (the dashed line) to before any black dots, we find a black open square. Imagine for the moment that all the black dots have come from the open square at some earlier time. Use this idea to see how it might work as we work with weak signals instead of dots on a graph. Figure 9-30 illustrates this.

The cloud on the left represents the plant. It is from there that any weak signals and other possible indications of trouble can come. They are not necessarily a part of any weak signal group. A grouping may be present but not necessarily seen at this moment because no one is looking for it. Nor should they. Another reason might be that they are visible somewhere else in the process, but not here. Or they were not visible at all, unless they were explicitly being looked for. But who knows to look? Looking for everything is not a reasonable activity. Eventually, they could all be found. This work process is designed to assist the operator doing just that.

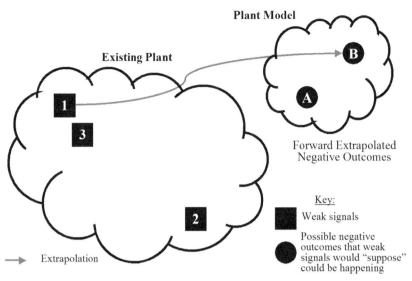

Figure 9-30. Forward-extrapolation illustration for a single weak signal (1) to the likely worst situation (B).

Returning to our example, weak signal 1, weak signal 2, and weak signal 3 are the ones the operator sees for the moment. The cloud on the right represents all likely forward-extrapolations from the weak signals on the left to the "supposed" worst problems that could be happening. In our example, weak signal 1 extrapolated to possible worst problem B. We are now done with the first forward-extrapolation (1 to B). It is not shown on the figure, but weak signal 2 extrapolates to possible worst problem A. That is why 2 and A appear on the figure. Weak signal 3 has not been examined yet. Repeat this procedure for each observed weak signal in turn. At the end of this activity, the operator will have several pairs of weak signals matched to possible worst problem(s). He will have one pair for each weak signal. If a weak signal extrapolates to more than one possible worst problem, add this "pair" to the list. In this case, that pair would have the same weak signal as the other but a different possible worst problem.

If any specific weak signal does not have a forward-extrapolation, keep an eye on it and seek collaboration or escalation. It suggests that one of two things is likely happening. Either the forward-extrapolation was not done carefully enough (due to inadequate experience, carelessness, etc.), or there is poor understanding of the particular process or operation being managed. Remember, this case is not about finding something actually amiss in the plant. This weak signal may be so weak that it does not mean anything at this moment. However, other times it might. When an operator is not able to "suppose" anything that he could suggest to be considered abnormal, it would be problematic.

Special Case of More Than One Weak Signal Extrapolating to the Same Possible Worst Problem

On the other hand, if more than one weak signal extrapolates to the same or a very similar possible worst problem, we have a special case. In this case, there is enough evidence to begin to presume that the twice-reached possible worst problem might well be true. For this case, do the backward-projection now. If you easily find evidence confirming both weak signals, then start working on that possible worst problem. If evidence is not so obvious, go back to carefully working the weak signals process. Getting two weak signals to extrapolate to the same problem is likely due to a real problem existing.

Special Case of a Weak Signal Extrapolating to a Seriously Possible Worst Problem

There is another important nuance to this otherwise straightforward, step-by-step procedure. In the extraordinary case that any weak signal clearly extrapolates to a really, bad possible worst problem, immediate attention to verify and manage that outcome will take priority over all other weak signals of the moment. Separate out this weak signal and proceed directly to step 3, step 4, and step 5. Be even more disciplined as you do this to ensure that you do not permit the perceived seriousness to shortcut your careful proof. You are likely working on a transformative weak signal. This is only one way to find them. Stay objective.

Step 3: Backward-Projection

Backward-projection is the second most important part of working with weak signals. This is because we must have a very good understanding of the plant or enterprise. It must be good enough so that these extrapolations and projections are done expertly. We have forward-extrapolated each weak signal to identify as many as possible of the potential candidates for the possible worst problem(s). Now we want to know where our "supposed" worst problems could have come from. Remember, we forward-extrapolate to find all supposed worst problems. Now we want to turn time around to find the possible source(s) for each. It is important to note that you will be looking for *all* early indicators. It is entirely reasonable to be able to do this. If a worst problem was actually in progress, then the evidence for this should not be hidden from the view of someone actually looking for it. *Backward-projection* means that we are actually looking in the plant for that evidence (or lack of evidence). Backward-projection requires specific, detailed knowledge of what you are operating:

> … big failures usually have simple causes while marginal failures usually have complex causes. If the product [being produced by the plant] bears no resemblance to design, look for something simple, like a leak of water into the plant. If the product is

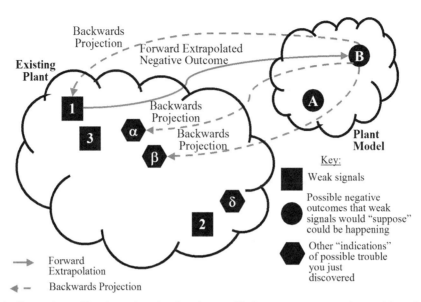

Figure 9-31. Illustration of backward-projection from a likely worst outcome for problem B to look for specific confirming evidence. While that evidence is shown in the figure (as δ, α, and β), let us wait until the next step to find them.

slightly below specification, the cause may be hard to find. Look for something that has changed, even if there is no obvious connection between the change and fault.[18]

Let us see how this works. Figure 9-31 illustrates backward-projection for our weak signal illustration. It depicts this only for problem B for simplicity. Potential negative outcome problem B, if it exists (remember, we think that it does from observing weak signal 1; now we look to see if it really does), should correspond to indication α and indication β being eventually found to be true. What we are saying is that if evidence confirming both indication α and indication β can be observed, then it is likely that problem B is an actual problem now. At this point, no attempts to prove indication α and indication β are present; we just put them on our list. Just for completeness, we would next do the same for problem A, but let us not interrupt the flow for now.

The particular weak signal you extrapolated *forward* may or may not be on the *backward-projection* list. Look for evidence as explained below:

If the backward-projection leads only to the original weak signal and it alone, without any other evidence of something amiss, this weak signal is most likely immature or noise. Unless the forward-extrapolated problem is a serious one, keep it on your list

---

18 M. Gans, D. Kohan, and B. Palmer, *Chemical Engineering Progress* 87, no. 4 (1992) in Trevor Kletz, *What Went Wrong? Case Histories of Process Plant Disasters*, 4th ed. (Houston, TX: Gulf Publishing Company, 1999), 350.

for the next time you do your "rounds." If it is serious, seek assistance. See also the topic "Persistent Weak Signals" in Section 9.7.

If the backward-projection does not include the same weak signal that was used for forward-extrapolation, there is cause for concern. It is always possible that the forward-extrapolation may have led you to an improbable result. Or you made an error. Carefully assess how the forward-extrapolation was done to ensure that what you extrapolated to is a reasonable worst problem. If you reach the same forward-extrapolated result(s), you must find other potential source clues for problem A when you backward-project. If other source clues are there and you find them, things are still working. Go with what you find. If they are not there, you have made an error—the process understanding is too weak or the work process is flawed. Seek collaboration or escalate.

Remember, at this step you are only making lists. Please do not look for evidence! It is too early. Starting early is going to open you to jumping to conclusions. Do all of this carefully. Now you are ready for the next step.

Step 4: Evidence for Confirmation

Now look for the evidence. Each separate backwards-projected item from the original weak signal must provide additional evidence to factor into a decision as to whether there is a real problem that needs attention. Evidence cannot be weak—it must be reasonably clear. It does not need to be in-your-face obvious, though it can be. It must be strong enough that if you came across it some other way, once you saw it you felt that you had a problem brewing. And the problem you had would have been the one you would have backward-projected from had you been using weak signals to find it.

Refer again to Figure 9-31. We look for confirmation (and disconfirmation) to ensure that all backwards-projected evidence you find (indication α and indication β in our example) is clearly present. *Clear* means you can see the actual confirmation for (or discomfirmation) for α and β that anyone would agree to be more highly suggestive of the truth. By asking what you observe to be clear, the weak signal analysis process avoids jumping to an unsubstantiated conclusion. To drive this important point home, it is part of the process to get a "hunch" that something might be going wrong; it is not okay to decide that the hunch is right by finding more hunches that cannot be confirmed.

Evidence (or clarity) comes in two forms: *confirming* and *disconfirming*. *Confirming* evidence is finding believable information reasonably suggesting that *a* problem one is thinking about might actually be *the* problem. *Disconfirming* evidence is actively finding believable information that points against the abnormal situation actually existing. We want to be as objective as we can when we look for problems. Looking for both confirming and disconfirming is one of the best ways to guard against many of those preconceived notions and biases carefully explained in Chapter 7, "Awareness and Assessment Pitfalls."

We are now ready to illustrate the process of backward-projection. This is where the power and utility come out. It is straightforward to actually do a backward-projection. Here are the steps:

1. From the selected potential problems (e.g., B), pick one of the backward projection indications (e.g., α) to prove true or not true.

2. Look for everything to suggest that the backward-projection item (α) is true or is not true.

    a. *If this indication exists,* (e.g., α) did I actually find convincing evidence that it is true?

    b. *If this indication does not exist,* (e.g., α) did I not find convincing evidence that it is true?

3. Look for all potential evidence from the candidates in both the confirming and disconfirming categories above. Remember that you are working to either prove evidence for it (e.g., α) is present or that it is not. Keep in mind that you are looking for the specific α to be in a condition that if it were, B is most likely happening (and needs to be managed).

    a. If you find one or more pieces of evidence that are as abnormal as they might need to be for the worst thing to be going on, each confirms that a worst thing is probably happening.

    b. If you find one or more pieces of disconfirming evidence present, it suggests that the abnormal situation is not happening.

4. If you find no confirmations, then presume that the weak signal is immature and for the moment does not strongly enough suggest a problem or issue. Keep it in mind for later (or let it go).

5. If you find confirming evidence and no convincing disconfirming evidence, then conclude that the potential problem is actually present and get to work resolving it.

6. If you find both confirming evidence and convincingly disconfirming evidence, either

   a. cautiously work out an approach to resolve the original abnormal situation, taking care to ensure that all remediation efforts recognize that the original abnormal situation might not be present or be present to a lesser degree; or,

   b. proceed expeditiously to a second cycle weak signal analysis (see Section 9.7).

      i. If the second cycle weak signal analysis provides no confirming evidence, conclude that the original weak signal is immature.

      ii. If the second cycle weak signal analysis provides both confirming evidence and disconfirming evidence, proceed using step 6a above.

Degrees of Evidence

This is a good time to visit the reality of what you might find when you (backward-project) to look for evidence. This is about how believable any evidence found might be to you. When you looked (backward-projected), you might have found a clear indication that the *supposed worst situation* was actually true. What you found was enough for you and others to easily conclude that the supposed worst situation is what is going on right now. No doubt. No ambiguity. In fact, had anyone seen or observed those indications from the outset, it would have been clear what was happening. That is one of the possibilities. It worked out this way because the weak signal that started this off really was an abnormality that "leaked" out of the actual worst situation. You noticed it. And you had a really good understanding of the process or plant that it was a part of.

This is not the only way it might have turned out. That weak signal you observed might not have been as clearly connected to the actual worst situation. Or your understanding of the process or plant might not have been strong enough. Or you might have been distracted or misled by other factors. Or the weak signal you observed was clearly connected to the actual worst situation but was in its earliest stages of going wrong; or other factors in the plant were working effectively enough to temporarily accommodate the abnormal operation. In these cases, you might find evidence that was not so strong. Refer to Figure 9-32. What you decide is tempered by the strength. However, it must be a decision.

We do a backward-projection looking for confirming and disconfirming evidence of an abnormal situation existing and needing recognition. What you find can be strong (clear and evident) or sufficient (reasonably convincing). On the other hand,

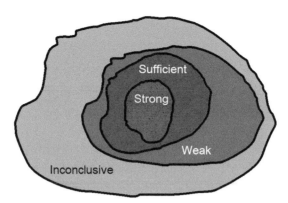

Figure 9-32. The backward-projections are done to look for confirming and disconfirming evidence of an abnormal situation existing and needing recognition. That evidence can be strong (clear and evident), sufficient (reasonably convincing), weak (maybe good enough, but probably not), or inconclusive (pretty much useless one way or another). Either way, it is on the table.

what you find might be only weak to cause you to feel uncertain but not so much that you can leave it alone entirely. Or what you find is pretty much useless (inconclusive) so that you feel either that you might have done something not quite right or that there is nothing there and you are imagining the whole thing. All of this should reinforce the need to actively look. You need to find believable evidence. Care must be exercised so that only the weak signals are the weak part. Evidence cannot be weak. Confirmation must be convincing.

No Evidence Found

If no evidence is found (inconclusive, or nothing), it is likely that no real problem exists. Each backward-projection suggests an item of evidence to look for. An example of "evidence" is a specific flow rate that is trending higher than normal, but is not abnormal yet or abnormal but not observed until the operator was looking in the right place. Another example is a temperature profile with a more pronounced bend where only a slight one should be there. You get the idea. These are reasonably definite things that, when the operator expressly looks at them, he can see that either something appears a bit off or it is really not there.

Only Weak Evidence Found

What if only weak evidence is found? What can you do when all the evidence from doing a backward-projection seems itself to be weak? Yes, you did find things that did not seem to be where you expected them to be. And they were enough out of the usual to raise suspicions, but not strong enough to confirm the problem they were backward-projected from. What now? Do you have anything of value? The answer is yes. In fact, what you have is not evidence; it is a new set of weak signals to work with. Work them just as before. We call this a two-cycle weak signal process (see Section 9.7). So work them. If nothing clear comes from doing that, stop working this as weak signals. Something might be going on, but you will need another way to find it. Collaboration probably is not going to be enough. Escalate the situation.

Backward-Projection Illustration

Referring to Figure 9-33, we will discuss a few situations. There are two forward-extrapolation pairs: 1 to B and 2 to A. Start with the first forward-extrapolation pair. The backward-projection of possible worst problem B leads to all the following potential problems:

- Potential expected problem indication α

- Potential expected problem indication β

- Weak signal 1 (Good—it tends to confirm the forward-extrapolation.)

The list above is just that, a list. By this it is meant that the list can answer the question: When possible worst problem B is true, we should be able to find evidence for one or more things being abnormal or amiss. The operator makes this list using

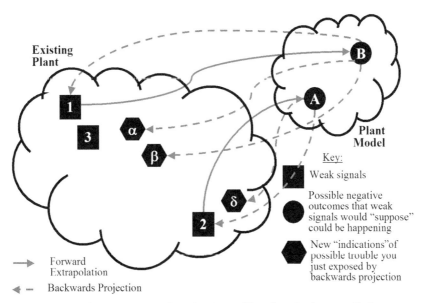

Figure 9-33. Illustration of two backward-projections. The first is from a likely worst outcome B. Projecting B backwards, we find specific confirming evidence 1 (the first original weak signal) and newly exposed α and β. Finding confirming evidence is sufficient to make a determination that (negative) outcome B is likely to be happening now. A second example is when we backward-project A, we find confirming evidence 2 (the other original weak signal) and newly exposed δ. Again, this is sufficient to make a determination that (negative) outcome 1 is also happening now. Because we found different confirming evidence for the two, it is unlikely that they are related.

his understanding of the process he is managing. Let us say that worst problem B is a tank emptying too fast. No evidence is looked for or used to make this list of problems. Only now will the operator look for evidence related to confirming or disconfirming the problem. Here is how he does it. For each "potential expected problem" in the bulleted list earlier, look for both confirming evidence and disconfirming evidence. We will look for all the evidence. Say problem indication β is one of the potential problems we find when we first look for confirming evidence. Let us also say that indication β represents a misaligned piping path between two parts of the plant. What follows below is the activity list the operator would use to examine each potential problem indication to look for *both* confirming evidence and disconfirming evidence.

Confirming Evidence

- Can the operator find one (or more) manual valves that appear to be in the wrong position?
- Can the operator find one or more flow measurements that would indicate that the flow has been incorrectly aligned?
- Can the operator find one or more pressure measurements that would indicate that the flow has been incorrectly aligned?
- Can the operator find one or more temperature measurements that would indicate that the flow has been incorrectly aligned?
- Can the operator observe (or does the field operator see) any evidence of leaks?
- Can the operator observe any evidence of unusual accumulations or losses (e.g., in tanks or in other processing vessels) that could have been connected in error?

Disconfirming Evidence

- The operator was not able to find any manual valves in the wrong position.
- The operator was not able to find any unusual flow measurements.
- The operator was not able to find any unusual pressure measurements.
- The operator was not able to find any inappropriate or unusual temperature readings.
- The operator was not able to find any evidence of leaks.
- The operator was not able to find any evidence of unusual accumulations or losses.

The Weight of the Evidence
In this illustration, it is easy to understand that some confirming evidence can be conclusive. The first bullet in the confirming evidence list (a manual valve out of position) is a clear indication that there is a flow path problem—or is it? Actually, for it to be *the* problem, the valve in the wrong position would need to permit the flow to be misdirected. After all, the valve might lead to a section of piping that is blinded off. Or it may be one of two valves that bypass a piece of equipment, but the other one may be in the right position (shut, if shut is the proper state). This is straightforward. It does require detailed familiarity with the process and equipment (and ready access to documentation that can provide the needed information). But that is as it should be. Other confirming evidence may not be conclusive but suggestive. Follow the threads. Gather all the evidence.

It is just as possible that there is not enough confirming evidence to persuade the operator that the potential problem (B in our example) really exists. If this is the case, then table it and move on to the next one. It is also possible that there is about enough confirming evidence to put the possibility of the potential problem "on the fence" of the operator's belief. In this case, if there is little to no disconfirming evidence and care is taken with working the solution to the potential problem, the operator can decide to do that. "Care being taken" is pointed out to ensure that the operator is more watchful in this case as he works to resolve the situation. Remember that working to correct a wrong problem can be worse than not working the right problem.

Finally, if this same situation of inconclusive or too little confirming evidence also has reasonable disconfirming evidence, the operator might ask for a second opinion or just table the investigation until the next planned round of weak signal activity for the shift. On the other hand, if it represents a significant potential threat, the responsible course of action would be to escalate it using the established protocols and procedures for doing so. Clearly, something might be going on. An inability to quickly resolve it should not cause it to be passed over.

Step 5: Resolution
Abnormal and off-normal situations identified by the weak signal process are meant for the operator to follow up on. The first and most vital step is to confirm the plausibility of the problem. Even though finding a weak signal identified the situation, it was forward-extrapolated to a potential problem or problems and backward-projected which found more evidence, make sure that the found evidence is reasonable and sufficient to confirm the likely existence of that problem unfolding. This is the reasoning behind ensuring that both confirming and disconfirming evidence is sought.

Given sufficient evidence, the operator will follow existing procedures to remedy the found problem. Remember, weak signals are about finding what is wrong. The rest is business as usual. Please note that the entire activity of remedy is outside of weak signals technology. Remedy will be guided by the totality of operator training, plant procedures, and other preexisting infrastructure built for that express purpose. It is the responsibility of management to ensure that this capability exists, is fully exercised, and is properly utilized.

Never Assume the Problem

The first step in weak signal processing is to ask, "What is the worst reasonable thing that just this weak signal might suggest could be going on?" That question is very different from "What do you think might be wrong?" That question is too broad. Even thinking about it this way invites preconceived notions to jump headlong into this. "What do you think might be wrong?" is way too susceptible to all pitfalls that mislead.

"Humans tend to judge too quickly when presented with ambiguous data; we have to work extra hard to consider less familiar scenarios.

Whenever multiple pieces of evidence point in opposite directions, or when crucial information is missing, our minds naturally shape the facts to fit our preconceptions."[19]

A misled operator leads to missing or mishandled important problems. The weak signal work process is designed to overcome these very natural biases and pitfalls that get in the way. Shortcutting is not a good idea.

Recapping the Steps

Let us review what this means. Operators are using weak signal management to find early abnormal situations in progress. This tool starts with looking for suspected off-normal things (the weak signals). In our example, the operator found three. Then he worked each one in turn. For each one, he "supposed" all the possible worst problems (by forward extrapolation) that it would point to. All weak signals are forward-extrapolated before any are backward-projected. We build those pairs. Now we have things to examine. Before we only had weak signals to work with. Now we have potential possible worst problems to look for. This should enable the operator to either confirm that there is a problem or conclude that there is not one. Remember, these are only potential possible worst problems at this point. The operator must either find sufficient

---

19 Schoemaker, Paul J. H., and George S. Day, "How to Make Sense of Weak Signals," *MIT Sloan Management Review*, 50, no. 3, (Spring 2009), http://sloanreview.mit.edu/article/how-to-make-sense-of-weak-signals/.

evidence for them (by looking for evidence at the backwards projected locations), confirming each problem, or not and conclude that the potential worst problem (found by forward extrapolating the weak signal) is not happening or not likely to happen.

Here is how it works. Refer again back to Figure 9-33. Examine each of the current (potential) worst problems (A and B, there could be more) that were forward-extrapolated to from all the visible weak signals (1, 2, and 3).

1. If more than one weak signal forward-extrapolated to the same problem (A or B), you already have enough evidence to know that the problem is the one you found by forward-extrapolation. Get to work.

2. For each of the rest of the weak signals that forward-extrapolate to different potential worst problems (other circles in the figure), take each one at a time and look for its presence in the actual plant by doing the backward-projection. For each indicator (hexagons in the figure), the operator should know where to look. The operator will look for real things that are abnormal, which would be there if the problem exists (may be clearly present, may be just building, or may not be there at all). For example, potential problem indicator β could be tank level LIC-2433 on the low side. Look for a reading too low for LIC-2433. Potential indicator α would be a laboratory reading for the pH on the high side. Get the current lab values and the recent history and see what is going on there. Go through all the other potential problem indicators one at a time that were backward-projected using this same method. Even though the previous illustration discusses steps for confirming evidence, remember to look for disconfirming evidence as well.

3. If several of the potential problem indicators can be confirmed (the hexagons in the figure), then you have now validated that weak signal 1 likely indicates a real problem: B. You know what that problem is. It is the one you forward-extrapolated to when you "supposed." Work problem B appropriately.

4. If any weak signal forward-extrapolates to a single potential problem and that potential problem backward-projects to only the weak signal from which it was extrapolated from, it suggests an inconclusive situation. It is likely an immature weak signal. Keep a close eye on it.

5. If only one potential problem indicator (the hexagons) can be confirmed in addition to the weak signal that started this cycle, let us call it indicator φ; start all over with that one as the "weak signal." "Suppose" all the worst for it. If you only get the same potential problem that the weak signal forward-extrapolated to, you are done. That one is probably real. You will have to decide whether this

problem is something to be worked on (it is now a possible parametric weak signal), watched (you think it is an immature weak signal), or ignored as noise (you think that it is not enough of a weak signal to be meaningful at all).

Refer to the "Special Case of a Weak Signal Mapped to a Specific Problem" section later in this chapter for more details.

6. If no potential problem indicators (the hexagons) can be found and confirmed, you are done. The weak signal you forward-extrapolated from to a potential problem has no indicators except for the initial weak signal. You have shown that it is most likely an immature weak signal. Decide whether to watch it to see if it matures or conclude that it is noise.

All of this is specifically designed to take all weak signals and see if there is something more to each one. The methodology of forward-extrapolation and backward-projection is a powerful "tool of the trade." Follow the process to where it leads. If the operator is not able to find confirmatory evidence for a weak signal representing any discernable situation, it remains immature with the most likely result that it is nothing more than noise. If it was confirmed without sufficient disconfirmation, then the operator knows that he has a likely problem, and what it is. He can move forward to the business of situation assessment and more. The situation awareness function has done its job for now.

Please do not get so carried away that you even come close to just reactively looping from extrapolated potential outcomes to potential problem indicators. This is not a mechanical process to be pushed or forced until it produces definitive proof. It is a tool. Use it consistently, but thoughtfully.

If any specific weak signal does not have a backward-projection, something is amiss. A weak signal that forward-extrapolates to a suspected problem must backward-project to itself. Either the backward-projection was not done carefully enough (because the operator does not have enough experience or has a poor understanding of the particular process or operation being managed) or the weak signal is being misinterpreted.

All of this can appear complicated and way too involved to be useful. If that is your first impression, pause. Remember that this is a new skill. There are lots of moving parts. But they do fit together. And the fit is really good. Any part that seems complicated or unnecessary likely seems so because either it is so new that it has not had enough time to sink in, or you might need to spend a bit more time with it to clear up the wrinkles or misunderstandings. The bottom line is that weak signals can fill in many of the gaps between alarms and luck. You know a lot about how to do alarms; make sure you do it. Luck, on the other hand, is not something to depend on in real life.

Classifying Weak Signals—A Review

The work product of backward-projection—to get more information and then confirm or disconfirm that a real problem exists—either promotes the weak signal to a real potential problem or relegates it to an immature weak signal (unimportant at the moment). The topic of classification for weak signals is discussed next. Classification is important only in that it leads to appropriate operator action.

Classifying a Weak Signal as Immature

An immature weak signal is a temporary home. We have an intuitive inclination to see normal, not nonnormal. Suspend expectation in favor of skepticism. The task is to decide whether the weak signal is noise or something of value. If a ready explanation is not present, it is noise. If it is noise, then move on. If you find something not normal, then decide if it is enough of something to be a *parametric weak signal*. If a ready explanation proposes various alternatives, the signal is confirmed to be an immature weak signal and not noise. It stays on the watch list. Keep a ready eye on it.

Classifying a Weak Signal as Parametric

If the result of the backward-projection of a weak signal leads to plausible evidence that is suggestive of things going wrong in a direct but not rapidly changing way, we have a parametric weak signal. It is likely a good indicator of something going wrong, but it is doing so in a manner that should be understandable and progressive. Intervention should be planned and scheduled. That intervention can be started now, if resources are available. It may be started later if the situation suggests. It may be deferred until an alarm is activated (not recommended), if resources are scarce. At the point an alarm is activated, the operator follows the usual clear course of action for alarm response as documented in the enterprise alarm response requirements.

Classifying a Weak Signal as Transformative

If the result of the backward-projection of a weak signal leads to plausible evidence that is suggestive of things going wrong in a way that is likely to change dramatically or with outcomes that are likely to be extensive, we have a transformative weak signal. Every effort should be directed to preventing the situation from reaching the "tipping point."[20] *Prompt action, in advance of alarming, is indicated.* For some situations, it may be the only opportunity for managing the problem. This is a major benefit of weak signals.

---

20 Gladwell, *The Tipping Point.*

Summary of Extrapolation and Projection

Utilizing weak signals as a tool depends on being good at extrapolation and projection. If the extrapolation part is not done so well, when you go to look for evidence (projections) you will be looking for the wrong thing or looking in the wrong places. If the projection part is not done well, you also will be looking for the wrong thing or in the wrong places. Either way, all those weak signals you manage to find are not going to get you anything.

If you find yourself in certain situations, there are ways to understand what to do next.

A Single Weak Signal Continues to Extrapolate to a Single Insignificant Implication

When you backward-project to look for confirmation, you can find only inconclusive evidence. Do a second cycle extrapolation and then projection, or (if you are really suspicious) a third cycle if you still find inconclusive evidence on the second one. We consider this because either the extrapolation might have been imperfect or the projection might have been imperfect or both. If you keep finding nothing convincing, you have probably found noise. Move on. If it worries you, just keep an eye out the next time you look for weak signals.

Several Weak Signals Found and Extrapolate to Several Significant Implications

This is the situation where several weak signals are found and extrapolate to several significant implications. When you backward-project each one looking for confirmation, no useful confirming evidence can be found. You have weak signals that appear to lead nowhere. These cannot be relegated to noise. Escalate this situation or treat it like weak signal flags, discussed later on in this section.

Different Weak Signals over Time Extrapolate to the Same Significant Implication

This is the situation where different weak signals over time extrapolate to the same significant implication. When you backward-project each one looking for confirmation, each time you look, no useful confirming evidence can be found. You have a persistent significant implication that cannot be confirmed. Something is working hard to let you know that it might need attention. Escalate this situation or treat it like weak signal flags, discussed later on in this section.

Weak Signal Management: Before and After

You have seen how weak signal management works. But before we wrap this up, let us clarify what may have been in the back of your mind: What is the difference between formal weak signal management and what we might have been doing before without it? What is the difference now with how we use hunches and the rest? Well, there are differences. And the differences are what allow weak signals to work. The first difference is that we design the control room and the HMI specifically to expose as many of the essential off-normal concerns as reasonable. The second difference is that we do the forward-extrapolations with context intent, not by reflex. Our desire is that seeing weak signals everywhere does not overwhelm operators, or that every one they see seems to be overly significant.

Design for Weak Signals
The basics of control room design in Chapter 4, "High Performance Control Rooms and Operation Centers," clarify the physical separation between control room operation and most of the rest of the goings-on at a plant or enterprise. This reduces unnecessary outside distraction that increases the likelihood of missing things or exaggerating things beyond their proper values. The principles of HMI design (Chapter 5, "The Human-Machine Interface") delivers sufficient related information in context so that things that are off-normal or a bit unusual are more easily observed—and once observed, can be put in context of the larger control goals and situations. Following good weak signal management will help to reign in our natural human tendencies of sometimes seeing the peripheral yet missing the obvious. At the same time, when we do find something of note, we do the examination in a way that increases the likelihood of finding what we need to.

Proper Forward-Extrapolation
Being able to find what we need to find is an important part of weak signal management. By collecting the observed weak signals as a group and then forward-extrapolating them collectively, we avoid following (possibly hidden) biases in the forward-extrapolation, which reduces the tendency to overextrapolate. Remember that the forward-extrapolation work process is to imagine and then look for what the worst reasonable thing the weak signal, if truly a harbinger of trouble, might be trying to tell us. The important part of the forward-extrapolation is the requirement for looking for "reasonable" but impactful problems. We are discouraged from looking for the spectacular or Armageddon ones unless they are the only ones. We use our deep understanding of the plant or enterprise to guide us.

9.7 Digging into Weak Signals

Now that the full concept and work process has been laid out, there is more to assist with understanding and use. They provide additional ideas about how to use weak signals. By now you are already thinking that using them can be useful, but as you think more about it, you might see some sticky parts. Let us visit them.

Trouble Indicators Come in Sizes

This entire chapter is about small indicators of trouble. So, let us firm up exactly what *small* means. We will work with all signs of trouble. But any that are *not* small will be different. Let us identify the small ones. Here are the categories for four different types of small indications: category I is *Not Visible*, category II is *Barely Visible*, category III is *Visible but not Apparent*, and category IV is *Evident and Distinct*.

Not Visible Indicators (Category I)

Category I not visible indicators are just that. No one can actually see them. They are things like hunches, feelings, and indistinct or vague concerns. These aspects were introduced earlier through the telling of a captivating "war story" in Chapter 7, "Awareness and Assessment Pitfalls," under the topic of sixth sense. They emanate from the back of your mind. You do not see anything. Nothing tangible causes them to come to you. When you try to focus on where they might be coming from, you can see reminders or thoughts from something else. Or you are reminded by nothing at all. There was nothing to notice. You just sensed it. Something did not feel right. Or it felt too right. Everything was vague. It lingered.

Barely Visible Indicators (Category II)

Barely visible, category II, means that you actually observe something out of the ordinary. You thought you heard something. You thought you saw something. Or you actually did hear or see something, but it was vague or indistinct. Something is probably there. Nothing is clear. A valve is a more out of its expected position than you thought it would be. Something just does not look right. Something you see is off. These indicators have a starting point in something tangible. You notice. You have a place to put a finger but are not sure it means anything.

Visible but Not Apparent Indicators (Category III)

At this level, category III, you know specific things. Maybe a lab report is missing. You see a closed valve that should have been open. You read a message and understand what it is saying but not yet what it means. You see a reading that is definitely out of normal even though you may have doubts about its accuracy. You know that something, or a few things, are not where they should be. What you do not know is why. And you do not know what it means.

Evident and Distinct Indicators (Category IV)

You see enough at category IV to be confident that something is amiss. You know you have a problem, a situation, or an opportunity. You are not totally sure but know this one is not going away. You need to nail it down.

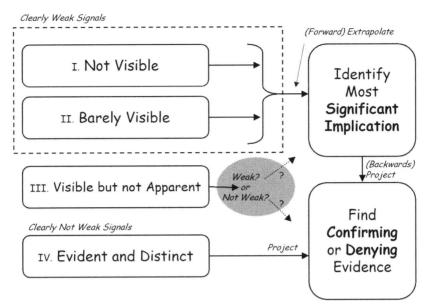

Figure 9-34. The four types or "sizes" of signals observable by operators and how they are used to find abnormal situations needing attention. The first two, Not Visible and Barely Visible, are weak signals. The fourth, Evident and Distinct, is not. The weak ones are extrapolated and then projected to find evidence. The last one needs only to be projected to look for evidence. But is Visible but not Apparent weak or not weak? That classification will determine how we handle it.

Tying Up Loose Ends

We have categorized the levels of the indications we observe. Doing this is important to be able to decide which process to use to make sense of them. Every type of indication requires processing. The single objective is to confirm there is sufficient evidence to reasonably know what the indication means. Weak signals involves forward projection to potential significant implication(s) and then backward projection to find confirming or disconfirming evidence. For the not weak signals, we already know what the significant implications might be. All we need to do for this category is (backwards) project to look for confirming or disconfirming evidence. See Figure 9-34.

But what about those indications that are in between weak and not weak? That category is labeled Visible but not Apparent. Visible but not Apparent means that we see something quite clearly that suggests abnormal or clearly moving toward abnormal. What is not clear is what situation is likely to develop from what we see. Because we really do not yet know enough about what we are seeing, looking for evidence (to confirm or disconfirm) cannot get a proper start. We would just be guessing where to look and what to look for. In order not to guess, we will treat this type of indication as if it could actually be a weak signal. This situation looks like Figure 9-35.

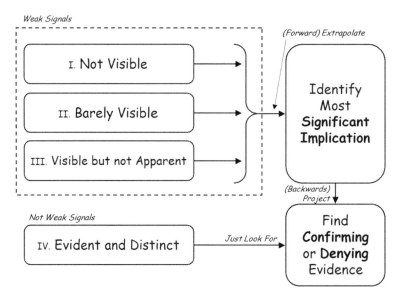

Figure 9-35. The four "sizes" of signals observable by operators and how they are used to find abnormal situations needing attention. The third category, Visible but not Apparent, is now linked to the first two, Not Visible and Barely Visible, and are all handled as weak signals. The fourth category, Evident and Distinct, is not. The weak ones are extrapolated and then projected to find evidence. The last is only projected to look for evidence.

This process shown in Figure 9-35 works all the time, except when it does not. There is an interesting case where taking the signals in category IV and looking for evidence produces no evidence, or maybe just a splinter of evidence is found. There is definitely not enough there to convince anyone. What does this mean? Well, it means something important. And finding out what that is will reaffirm the power of weak signals. Evidence not found strongly suggests that the situation we thought we had from the category IV signal is probably not the right situation. What we thought was evident and distinct is still visible just the same. There was no doubt about it being real and there. But when we looked for proof, we could find none. So now we have a twist. It is our conclusion that needs revising. Something else is the problem. Something threw this off. Some foible in our thinking got in the way (remember Chapter 7, "Awareness and Assessment Pitfalls"). To get back on track, think like a weak signal. Open it back up and now "suppose" and do the (forward) projection you skipped. We skipped it before. Now we cannot. We will now do it with an open mind. Find any? Good, backward-project each to look for evidence. Did not find any evidence, or couldn't think of any significant implications? Get help. Something is really wrong and it is hiding deep.

We should visit one other aspect before moving on. By now it must have occurred to you that the first category, Not Visible, does not seem to fit into the weak signals work process. That work process can only start with "suppose the worst (or best) of what a noticed weak signal might be." But we have nothing to suppose. There is nothing to see. There is nothing at all visible. It is all in our mind. It is feelings or hunches or whatever. Everything is vague. Well, no way that is going to work. We need something tangible, even if it is just the slightest thing. Only when we have something tangible can we work the process. Fortunately, there is a really good way to get our start. To do that you will need to actually look for something. You will actively look for it. And you will look deeper than what is on the surface as discussed next.

Summing Up
Before we go to the next topic, please remember the bottom line of weak signal analysis. Weak signal analysis only provides a believable hypothesis that a situation can be present. It can only confirm knowledge about the current state of things. It provides just enough certainty. You know that this may not be full proof; it is just enough to reasonably believe that something is true enough to work on. The actual working on the problem is out of scope here. Weak signals stops just before that. It stops as soon as you have a good, confirmed idea of what's potentially going on now that you should look into as problematic.

Actively Looking for Weak Signals
One of the hardest ways to use weak signals is to let them get under your skin. Yes, they are always coming up. Everywhere you look, you might see them. If you were to work on every one, it could easily be all you do forever. This will not do. We need to figure out a workable approach. The only reasonable way out is not to look for weak signals all the time. If they show up when you are not looking and are not clearly of concern, put them aside for later. This means you will figure out a way to manage when you actively look for them. The next chapter (Section 10.12) provides specific recommendations for when to look for them during preplanned parts of the shift. Sure, operators must always be open to recognizing them when they pop up. But, again, this must be a tool, not an entire shift's work. So let us assume that your operator has a good workable plan to time manage the "looking" on a schedule. This looking will review the available operating status information: HMI screens, laboratory reports, repair orders, permits, production schedules, visitors, and all the rest. Operators are looking at what the system designer planned for. It will be done systematically.

But what about those small indications that are not readily visible here? What about the ones that seem invisible? Here are some useful ways to actively look when your operator must. A good way is to pick "filters" that he can use to keep everything from looking like that "nail" that you have a great "hammer" for. Remember, weak signals should represent abnormal situations that are early in their progression (to something bad or worse) but hard to see. Here are a few good ways to look for them (from Amy Webb[21]). The ways are *contradictions, inflections,* and *extremes*. Each one is a good working filter. When anything fits into one, a weak signal is probably there. Remember, having a weak signal does not mean that you have a problem. Every one must be confirmed. This is just a good way to find some that might have been missed or overlooked.

Contradictions

Contradictions are looking for things that are familiar but for some reason do not relate to each other in the same way we come to expect them to. It is their current relationship rather than their individual activity that is suspect. They are not behaving the way we expect. That is unusual. We want to find them so we can take a better look. A contradiction is:

> [w]hen two or more things succeed or fail simultaneously [or very close to the same time], though usually they would track in opposite directions [one fail where the other succeed; or the other way around]; or when things track in the same direction, though typically the reverse would be true.[22]

Another contradictory situation is when two or more distinct things that are not thought of as being connected now become connected (by seeing them seemingly to be related), or when things that are normally connected now appear to not be. Examples are delayed deliveries of supplies with an observed increase in process upsets. Who would relate them in a causal way? But now they seem to be. Another one is seeing a curtailment of overtime by management resulting in a noticeable increase in late arrival of operators. Yet another is the issuance of more detailed and informative operating procedures that now seem to cause more alarms to activate. All can be advance hints of problems brewing.

Inflections

An *inflection* is a noticeable change. It might be the slowing down of something that was changing quite normally. It might be a movement or change in something that was usually steady. Either way, you notice something unexpected. It could be a new union contract or a pending work action. Or you notice a change that on first thought should not affect you, but then you realize it might. Examples might be a curtailment of production by another part of your plant but in a part that does not usually affect you, a significant news event, or talk of a corporate takeover of your company. Surely, some have a hard time affecting what operators do in the control room. Others might not be so disconnected. Here we are looking for the unusual that would not be directly

---

21 Amy Webb, *The Signals Are Talking: Why Today's Fringe Is Tomorrow's Mainstream* (Philadelphia: Public Affairs, 2016).
22 Webb, *The Signals Are Talking.*

linked to variables on a process graphic or in routine laboratory analysis reports. Those you are already looking for. Inflections are new.

Extremes

Operation is entirely normal. Nothing seems amiss. Everything appears within the accepted operating envelope. But a few things are near the edge of normal. Not over the edge, but quite near. There are several important reasons for concern when this happens. First, the operator must find out why the process is in this place. It is not outside of design. The plant may have been there before. Now that it is there again, we need to know what is going on. If it needs fixing, the operator must know about it first. Weak signals, just to remind us, is only about knowing.

Second, even if nothing is wrong now, and you know it for sure, if down the road something comes up, the operator has not "left himself an out" as needed from the principles of defensive operating (see Chapter 1, "Getting Started"). There is less room for error. There might be much less time to figure things out. The operator's freedom to select what to do and how to do it has fewer options. You need to find out why you are at the edge and move away if you can. If operations cannot be moved, then you need to bring more resources close by just in case a changing situation needs it. Operators in this situation cannot "go quietly into the shift" to think about a related literary thought. Preparations need to be made.

Special Case of a Weak Signal Mapped to a Specific Problem

Refer back to Figure 9-33. Consider a situation that forward-extrapolated from weak signal 3 to problem E (not shown on Figure 9-33). The operator examined problem E and found it to be completely and quite apparently true without any weak signal management. Pause. Remember that the benefit of weak signal management is to avoid bias and keep from jumping to conclusions that are not warranted. However, in this specific situation, once E was proposed, it was clear and apparent that without any intervention or luck to change things, worst outcome E would happen. There is no need to backward-project. It is clear from just looking. Direct management of this situation is what you will do.

Here is how it works:

- if a weak signal points directly and clearly to a unique problem,
- if there are no other reasonable ways needed to confirm that that problem is actually developing,
- if managing the problem now will not unduly put other operations at risk,

- if the operator knows how to manage that problem, and
- if that problem can be managed directly—do it.

This often is the only way to manage natural disasters in the making. Remember the transformative category of weak signals? In these cases, the points of intervention are so few and the power of progression is so enormous that reacting at a weak point of nature's plan may be all we have. But make sure that your operator is not working unduly under any biases!

There is another useful way weak signals can directly point to an underlying problem. If more than one weak signal forward-extrapolated to the same problem, this is a sufficient basis for concluding that that problem actually exists. Yes, it might be of additional use to backward-project that problem to see if more evidence exists. That evidence would mean that you found one or more of the hexagonals to which the original weak signal plus any other weak signals that are backward-projected to. However, you already have enough evidence to know that the problem is the one you found by forward-extrapolation. Get to work.

Example of Acting Directly

We have all likely encountered a simplified situation in which the above situation is true. Consider that you are driving an automobile in the winter. The weather is a bit dodgy with temperatures hovering around freezing. There are off-again and on-again snow flurries. Visibility is decreasing. All in all, this is not a pleasant drive. But you are in control of your vehicle and progressing nicely along your chosen route. Then you begin to sense that your vehicle's traction might be slipping ever so slightly. Not all the time, not even when you can predict it might happen, but you sense it. This is your weak signal. The worst that might happen is you will lose control and crash. Without any further need to examine the road more carefully or better determine the correct outside temperature, you know that you should immediately slow down to a more manageable speed, avoid any sudden changes in power and direction, and avoid using the brakes.

You may have over-adapted. So what? You are safe. Yes, you will likely arrive at your destination later than planned. Yes, you will be paying closer attention to all aspects of driving, and it will be more than a bit tedious. Unless you fail in vigilance or get unlucky, you should arrive in one piece.

Prove True or Prove False

Each weak signal is considered an indicator of a real problem. It is what they are. Weak signal management is designed to get the operator as close as reasonable to the evidence that there is a problem. Its thrust is to find evidence of the problem being present. The operator is looking for the problem being true or for the problem to be false. This is important. The task at hand is to take a weak signal and determine what it means. Either it is a small indication of something going wrong, or it is not anything important or useful. The operator must look for true first. If not found to be true, look for false. If the operator is not able to find enough evidence for either, weak signal management calls this an immature weak signal. The operator can conclude that it is noise. Or if it is still suspicious, mark it for a watchful eye.

Weak Signals Observed by Experts

Some weak signals relate to matters for which you are expert. Just the briefest of a hint is all it takes to put a meaning or inference to the front of your mind. You know what it means. You know what to do about it. This is yours. Your experience is valuable. Yet is it *always* right? Probabilities aside, it is not the times that you are right that matter. It is the single time that you are not. This is a situation that tests your discipline, not your expertise. It will be your discipline that makes this work. It will be your expertise that makes the work successful.

For each situation that your expert knowledge thinks is a "slam dunk" into an explanation, pause and work it as a formal weak signal activity. Do the forward-extrapolations. Do the backward-projections and look, really look, for proper evidence. You should be able to prove true or prove false for every one. If you cannot, then it is time for a closer appraisal of what your expertise might be really telling you. You definitely must not be trying to prove to yourself (or anyone else) that your expertise is in question. There is always room to learn more. In fact, it is the things that we know and understand the best that we can learn from the most. Work the process. Learn.

Collaboration and Consensus

So far, weak signal management appears to be an activity that an individual operator would use. Each operator on shift looks for weak signals, processes any, and moves on. That is a usual situation. However, there are important opportunities here for collaboration and consensus. These opportunities enrich the activity.

The first place we find weak signals collaboration useful is for training activities. This is a natural fit; there are many variants. The training might be done with a supervisor or trainer looking over the shoulder of the operator and both actively looking for and processing weak signals. The training might be just hypothetical or "tabletop." Or it may be at the control console of a plant. Coaching includes locating weak signals, forward-extrapolating them, and backward-projecting them. This is the perfect time to identify weaknesses in basic process knowledge. It is useful as well to expose weaknesses in how to do weak signal management. Agreeing on the weak signals, the forward-extrapolation, and the backward-projection outcomes are integral parts. The particular type of consensus we are suggesting is agreement step-by-step. That is, study, examine, and agree on what weak signals are present. Examine and agree on how each would forward-extrapolate and to what. Backward-project and look for and agree on all evidence found or not found. Of course, if a problem were confirmed, then cooperatively handling it could be part of the active training, but we are getting beyond the scope of weak signals and this discussion.

The second place for collaboration is during the actual process of working the weak signals during the shift. Operators would seek out other operators or supervisors for specific concerns. The focus is to gain additional certainty that the individual weak signal management steps are being worked properly. The individuals will exercise care to ensure that the cooperation progresses one step at a time, not jump from a weak signal to a probable underlying problem.

The third place for collaboration is at shift changes. Of all the opportunities, this should be formally integrated into the shift process. A weak signal review will be performed at all shift changes. This means that every shift operator will formally conduct at least two of these per shift: one coming on shift with the relieved operator and the other going off shift with the relieving operator. To reduce bias, the relieving operator leads the review. The shift handover should continue until there is consensus.

Weak Signals Do Not Escalate

Weak signals generally do not escalate individually. This means that each one will not usually become stronger with a worsening of the underlying situation. They can, but not reliably enough to depend on it. Rather, expect the number of weak signals to increase with a worsening of the underlying situation. This is important to understand. It means "do not expect a weak signal to become less weak (become stronger) as time progresses, even though the underlying cause (possible abnormal situation in the making) may become worse." Remember what weak signals are: they are the stuff that is "leaking out" of a potential abnormal situation. They are not so much a part of the "main event" but emit from it like the rippling wake of a boat. Because they are often peripheral, there is no reason to expect their strength (or value) to be in any way proportional to the seriousness of the main event. Their importance is that they "leak." We need to find them. They are valuable because we have a way to use them once found! Most often a worsening situation creates more visible weak signals. So finding and tracking them in a careful way is going to be useful. And to be most useful, we need to find out what they mean before it gets too far along to effectively respond to.

Before the Tipping Point

We want your operator to find what is going on early enough to be able to evaluate and intervene as needed. It is why you are building this weak signals tool for him. Remember that *early enough* means before the situation gets too abnormal to be properly managed—in the words of Malcolm Gladwell,[23] before the tipping point. Once the situation tips, it usually will not respond to anything preventative. The best we can do is react and try to reduce the bad impacts. Before the tipping point, expect category I signals to stay that way. Hunches and vague feelings stick around. A few category II weak signals may emerge. They do not replace the category I ones. They are in addition. Likewise, there may be one or two category III ones showing up. These too are in addition. Because all these are weak signals, we should not expect any increase in category II or category III as time grows. No weak signal has its visible moment necessarily related to any other one. They appear as a result of the underlying situation emerging as it develops. They pop out when the real situation has a leak of what is going on.

This is true most of the time but not all. There is one situation where it is not. If a particular weak signal is, by some unusual aspect of design or circumstance, the indication of a single prominent aspect going amiss, it will escalate. For example, consider a special tank that is heated and pressure controlled, and all inlet and outlet flows tightly managed. It could be that something happens with the temperature management inside the tank that causes the temperature to rise. The first weak signal is the temperature appears to be at the high edge of normal. This temperature continues to rise. This weak signal grows from category II to III. But nothing else appears amiss. The pressure is spot-on, the flows are close to exactly what is expected, and the mixed product is analyzing out pretty much as it should. What is happening? Not for sure, because we are not looking at a detailed design of this tank, but suppose that it includes really good pressure controls, carefully designed flow ratio controls, and the like. This means that even though the temperature is getting way out of normal, the other controls are still able to keep up. Until they cannot. But for now they do. Those other weak signals that might leak are not leaking. Yes, a savvy watcher of this situation would know that keeping track of the value of most manipulated variables that are under feedback management generally does not provide as much critical information as where the controller is in its range of operation. If valve positions were readily available to the watcher, these would start to emerge as weak signals. Now you see why HMI design is so important. But in our case, they are not easy to find and see. When the compensatory management fails, the entire tank contents gets spoiled. We crossed the tipping point.

---

23 Gladwell, *The Tipping Point*.

After the Tipping Point

After the situation tips, weak signals take a secondary role. Before the tip, their purpose was to provide advance notice of something going amiss. Once the situation crosses the line, there is little subtle about the weak signals. To be sure, there are some that move from weak into alarm condition. Some may be really close to activating a safety interlock. Yet, weak signals still have value. Rather than being the primary means of looking for potential problems, the weak signals take on an added new secondary role. Weak signals now become useful for the following:

- They may confirm or contradict the primary evaluation of the abnormal situation now being faced.

- They may confirm or contradict the effectiveness of the remediation and other situation management efforts now underway.

- They may suggest another abnormal situation, in addition to the primary visible one, that requires recognition and possible management.

In this case, expect the number of weak signals to significantly increase. Even though your operator is busy working the primary event, ensure that someone is watching out for the rest of the process. This leads naturally into the next topic.

Weak Signals as Flags

Weak signals come and go. After all, they are weak. Part of being "weak" means they tend to show up early. *Early* means that whatever may be influencing them to appear is just as likely to cause them to go away. This movement between normal and not normal is fluid. However, there may come a time when the accumulated real-time operational knowledge from weak signals gets suggestive enough to push the real operating situation into a stronger case for being abnormal. This does not make it a strong signal. Strong signals are alarms and other clearly abnormal situations. We are not going to define another class here. It is enough to say that the number of weak signals growing suggests a more urgent examination and decision—no longer are we in the realm of waiting and classifying. Here we have good reason to think that something might be amiss. We need to find out what. We need to find out whether the weak signals are growing in a related way or just numerically. Watching for and understanding accumulating flags is a useful operational practice.

Most of us are familiar with the use of flags in sports. In American football, a flag on the field means that a referee has identified a violation of the rules of play. When one is identified, a flag is tossed onto the field. A single flag in the field is enough. The game progresses only after the situation has been resolved. However there is the dramatic event in a game where several flags are tossed out at the same time. That means many things went wrong and were observed. Back to the control room. For *weak signals*, more than a single flag also may be useful. Each weak signal can be thought of as a single flag. During an operation, if the number of active weak signals starts to increase (generally, we use three or four as the number), the operator should assume that something unusual might likely be happening. This situation is particularly important when the operator feels that nothing much is wrong. The more flags, the more likely something really is going amiss. This can be especially important when responding to an alarm. It might suggest that a wrong or ineffective approach is being used. It might suggest that another important problem needs attention.

When the number of visible weak signals grows, suddenly each individual weak signal takes on an entirely new importance—so much so that we expect the operator to pause in his work and directly begin to ask the following questions (in the order listed):

1. Do the flags appear to be related?

    a. If *yes*, the "relationship" should suggest a new problem directly. Work on it now.

    b. If *no*, the "relationship," if any, is unknown, but a problem is clearly present. Ensure that the more general problem is found now. Escalate if needed.

2. Is the problem being worked the actual problem?

    a. If *yes*, either the solution is inadequate or more problems exist that need to be worked on.

        i. If no more problems are found, then the working solution must be modified or changed to fit the situation (this usually requires escalation or invoking permission to operate).

        ii. If more problems are found, identify them and work on them.

    b. If *no*, stop work on the wrong problem and look for more than one concurrent new problem that is not worked on yet; find them and work them. Collaborate if needed.

3. Is the solution the operator thought would work, still working for that problem?

    a. If *yes*, then other problems are becoming important; get help or find them and work on them concurrently.

b. If *no*:

   i. Ensure that the operator is working the solution properly.

   ii. If the problem is getting worse much faster than expected or more problems are arising:

   1. Reassess to make sure the problem being worked is still the problem.

   2. Reassess to make sure the solution being worked is the correct solution for the supposed problem.

4. Are the indicators for success (or failure) still what the operator was expecting?

   a. If *yes*, then other problems are becoming important; get help or find them and work on them concurrently.

   b. If *no*, it is likely that the wrong problem is being worked or the wrong remediation efforts are being used; reassess the plan and revise it as needed; otherwise escalate.

Asking these questions should get you well on the way to counter the problematic aspects of both confirmation bias and continuation bias. The answers to these questions should be sufficient to reenergize the solution activity along more successful lines or to suggest that the problem might not be resolvable. Better start thinking about escalation, permission to operate, and other operational safeguards.

Notice that there are a lot of steps in this process. No one can be expected to remember them and follow them step-by-step. However, they are important. Working the list this way ensures a careful analysis. One of the ways of working this is to have at the ready a work-process sheet listing the steps and pull it out and work it down. Another way is to have someone nearby (possibly an extemporaneous collaborator) reading the list and checking it off as you go along. Use one or the other or something else just as reliable.

Two-Cycle Weak Signal Analysis

There is an important and potentially powerful variant to how weak signal analysis may seem to not find any underlying problems when done as described in the work process, yet still be quite useful. Take the case of weak signal 1 (Figure 9-33) forward-extrapolated to potential problem B. As before, problem B backward-projects to indications α and β, and weak signal 1 (the original weak signal that started this off). Suppose that when the candidate indications were examined, none appeared to be problematic or indicative enough to suggest something is wrong. Confirmation (or disconfirmation) for neither α nor β could be found. The weak signal is too weak. So, according to weak signals, the initial one, number 1, must be just noise. But you are troubled. You feel that something is amiss, but you cannot find it. In this case, consider performing a second weak signal analysis.

For the second cycle, start from those indications (Figure 9-33, indications α and β) that resulted from the first backward-projection. These were too weak to be confirmatory on the first round. Thus, problem B may not exist because it cannot be readily confirmed. Now think of indications α and β as new weak signals. You did not see them at first, but now imagine that you do. Here is how it works:

1. Take each new weak signal (problem indication α and then β) and one at a time, forward-extrapolate each. Assume that one of them forward-extrapolates to a new problem: say problem G (not shown). Okay, now we have something more to work with.

2. Backward-project problem G to get, for example, problem indication β. Problem indication β was the same as you got from backward-projecting problem B. No matter—and this is where the power of the second cycle comes in. While you did not find anything abnormal enough with indication β with respect to problem B, it might just be abnormal enough for problem G—abnormal enough to confirm it. Problem G is different. For example, problem B could be a faulty pump, and indication β was a somewhat low pressure in the discharge line. When you checked that pressure, it was not enough to point to a pump problem. But problem G (that you got from the second cycle) could be a downstream leak; a somewhat low pressure in the same discharge line might just be the clue you needed. Now you would look for it. If it was just low enough, you found that there probably is a leak. Summarizing, when problem B is backward-projected, the result can be a different indication than before from the first cycle where problem B was backward-projected, or it can be one of the same ones. Remember, the same initial facts can mean different things and have different importance depending on what specifically you are looking for.

3. Investigate each just as you would in steps 1 through 5 of the standard work process for weak signals. If nothing comes from the second cycle, common wisdom suggests that a third cycle would be out of the question. However, and this is an important *however*, if you still remain troubled, it may be just the time to escalate or consider permission to operate protocols.

Accentuate the Negative, Eliminate the Positive

There is an old song that asks one to look at the bright side: accentuate the positive; eliminate the negative.[24] Sure, we all want to look at life by cherishing the good things. That is a wonderful outlook. But weak signals is not about an outlook. It is a power tool. Its power comes from the inherent ability to ferret out any little things that suggest something is amiss. This is why we always forward-extrapolate to find the worst (bigger) things that they suggest. For it is when everything seems to be going right that finding what might not be is important. Safe and effective process operations are not a voting matter. There are no benefits from having a preponderance of evidence pointing to everything being okay. What is really needed is to know that safe and effective operations are being achieved beyond any reasonable doubt. Each doubt is a weak signal to be processed. When all have been processed and nothing untoward still remains, then they have served that purpose.

Weak Signals and Checklists

Checklists can be extremely useful tools to reduce the chances that nothing important for a task is left out or done in the wrong order. Their worth has been proven in everything from aircraft cockpits (pre-flight, in-flight emergencies, post-flight) to hospital surgical operating rooms. Each item on the list has a description and a box to check off. The order of the list is the order in which the items must be performed. Instructions are given to clarify what a check mark (√) means or under what circumstances a check mark must be skipped. This specificity would suggest that there would be no room for weak signals. Such is not the case. Each item on the checklist appears as a binary (yes or no) choice. Real life usually is not always that sharp and clear. As an illustration, consider this line item on a checklist:

☐ **Turn on the auxiliary pump and make sure the discharge pressure is over 42 psig**

The line is checked after the pump is turned on, and the discharge pressure is observed to exceed 42 on the gauge. But suppose that the pump was observed to be slower starting this time than it is normally—or that the pressure was above 42 but just barely, or moving around from 48 to 70 where such movement is unusual. Even for activities as formalized and straightforward as a structured checklist, there is ample room for weak signals to be used to suggest conditions that might be off-normal and useful to look into for possible problems. This is what weak signals can do.

---

24 Johnny Mercer and Harold Arlen, "Accentuate the Positive," song performed with the Pied Pipers (1945), 3:30, https://www.youtube.com/watch?v = f3jdbFOidds.

Persistent Weak Signals

What if a weak signal or a few, either related or not, seem to hang around for a while? You see that each is definitely weak—a suspicion, not definite—yet noticeably there, and there for a period of time that has been longer than just some transient thing. The problem here is that when each one was "processed" earlier using the weak signal tool, nothing came of it. At that time, each was judged immature and would have been relegated to either noise or put on an informal watch list. What now? How should we use this situation to our advantage? A weak signal that hangs around yet is too weak to confirm can suggest one of three things:

1. Just part of the regular variations real plants might exhibit during normal operation. While the magnitude or noticeability can be at the edges of normal, nothing clear is visible nor can it be found by looking.

2. The early beginning of a situation or problem, but too early to know whether or not it might fix itself or need later attention.

3. A situation or problem that is progressing and likely to be important, but not necessarily fast enough moving to be clear now. While it is too early to know for sure, this case needs clarification to make sure it is recognized and work started to ensure proper understanding and/or management.

The weak signals tool suggests that persistent weak signals belong to the third alternative above until ruled out. It is important to make sure that any persistent weak signal is not anything important. This means from this point on, once a weak signal is classified as likely persistent, you, as the operator, must go through the activity to make sure that trouble is not lurking that must be dealt with. So how do you do it? Well, for sure, just taking another look to see if forward-extrapolation and backward-projection will turn up anything different than the first time is not enough. Sure, do it again. But if nothing shows up, do not force it. Now is a good time for collaboration and/or escalation. Both should be done without you leading the thinking. Let the collaborator or escalator lead the activity. Keep as much independence as you can here.

After all of this, if the weak signal is still persistent without anyone being able to provide a sufficient explanation, it is time to work out a way to get to the bottom of the situation. Involve engineering or call in experts. Leave the explaining to them for now. Keep an eye on the signal while they are doing it.

Weak Signals That (Seem to) Lead Nowhere

Weak signals can be easy to find. A well-designed HMI that purposefully contains a wealth of information to guide operators to find them is important. However, gaps in

HMI design or procedures or operator experience can easily provide misleading weak signals or get in the way of deciding what they mean. So far in this chapter, you have seen the recommended method of working weak signals to see if they are meaningful. You do that using the process of forward-extrapolation (supposing the "worst") and backward-projection (to find any evidence for the "worst" being likely). If evidence is found that is convincing enough, you work on the problem (the "worst" thing or things you found evidence for). If not enough evidence or no evidence at all can be found, the weak signals procedure suggests that that the weak signal is immature. *Immature* means that it is either noise or a possible real problem, but it is too early know whether it will be one. Either way, you do not worry about it anymore. If the underlying problem is going to get worse, eventually you should be able to find evidence for the same weak signal. All of this is to say that the weak signals tool works as it should.

On the other hand, some weak signals never seem to lead anywhere. Periodically, when you look for and find the current crop of weak signals every time you process those too-weak ones, they never seem to lead anywhere. What does this mean? Are you doomed to just keep cycling them about or should you consider something else? The short answer is that something can be really afoot in your plant. Weak signals are sort of whispering a message, but they are not going to lead anywhere by themselves. There are several suggestions for what to do next. It is important that they be pursued.

Escalate or Collaborate
The two-cycle approach brought up earlier suggests that you consider escalation or collaboration. Basically, consult with others to see what you might find together. And that should be an option in this situation. Consider it.

Review Your Model
What model? Your model! The only way you are able to look at information about the plant (on the HMI, etc.) and find potential weak signals is because you, as the operator, understand what the plant is all about. You are able to determine what seems to be okay and what seems to be not. The not okay things that are not alarms going off are most likely weak signals. The way you evaluate what seems okay and what does not is by using your understanding of the plant. Your understanding of the plant is actually your (mental or conceptual or supposed) model. The better your training and background, and the better your model and the more effectively you use it, the more likely you are to determine what the weak signals indicate.

Weak signals that appear to lead nowhere might be caused by

- weaknesses in your model;
- weaknesses in your understanding of the model;

- weaknesses in the design of the HMI in that not enough information that is necessary for understanding is visible, or it is visible but too obscure to be useful enough; or

- weaknesses in your ability to work the weak signals tool carefully enough.

Operations management should have an ongoing program to identify those weak signals that do not seem to go away. Each should be examined for cause to find the underlying issues. This activity is an engineering and supervisory one; please do not just leave it to operators to figure out.

Looking Too Hard
Weak signals can a powerful tool. Early in the process of learning about it and working it into an operator's tool kit, it is not too much of a stretch to see how it might become too important. It is easy to confuse *useful* with *use it often*. This is one of the important reasons for processing weak signals at planned for times during a shift (see Chapter 10, "Situation Management," Section 10.11). Try to pay careful attention to make sure the tool is not being overused.

Accept the Noise
Noise is intrinsic to your plant and will never go away completely. Sometimes weak signals that somehow seem to stick to the noise will not go away either. After all else is ruled out, it might be useful to just label them as noise and train around them. However, they must be revisited periodically to make sure nothing has changed. It is also a good time to consider the procedure of looking for "disconfirming" evidence that rules out a weak signal (see "Step 4: Evidence" in Section 9.6 and "Prove True or Prove False" in Section 9.7) from being important.

Weak Signals among Strong Signals
Strong signals are not subtle. They are evident and obvious. They are clear in their presence. What they might mean may not be obvious, but they cannot be missed. Most times we see them and work out what they mean and deal with them. But there can be a sinister side. It all has to do with our nature to "oil the squeaking hinge" and think the job is done. For the operator, his "oil" is his proper attention. Where he places his attention is important. And here is where things can get problematic. Whenever weak signals exist scattered around strong signals, the weak ones can get ignored. Either they are not seen, or if seen, they are considered to be insignificant. This is part of our human nature, but that does not make it a good idea. Weak signals among strong ones just might be the exact time that they may be the most important.

Even before the strong signals are factored in or even worked on, it should be part of the operator's protocol to look for any weak ones. It may be those weak signals that suggest something different may be actually going on. Process the weak signals to see where they lead. If they reinforce the strong ones, good. If they are ancillary to the strong ones, then bookmark them for later attention. But if they contradict or muddy the waters around the strong ones, be careful. This just might be the cautionary "red flag" that calls for very careful attention to be paid.

To summarize, the search for weak signals (no matter how brief) must be an integral part of recognizing important operational situations. Finding out what they mean must be part of managing every important situation. It is already built into alarm management where the operator looks for confirmatory actions. Build it into situation management.

Actively Looking for Weak Signals

There is a powerful yet seemingly small part of weak signal management yet to tell. It is the requirement to "actively seek out" as opposed to "simply observe." This means that the operator would be specifically looking for weak signals as an active, ongoing part of the job. This might be best illustrated by digressing into how we deal with "things that go bump in the night."[25]

Things That Go Bump in the Night

Let us create a fictitious situation in the spirit of classic children's prayer literature. This message is instructive.

> You are alone at home reading the newspaper a bit later in the night than is your custom. Everything is quiet: house, basement, garage, and yard are all settled in. Then you hear, or think you hear, a noise. Everything stops. You hold the paper still. You take a quick look around. You turn up your "listening" to as high as it can go. You listen more. You listen for another noise. You are listening for confirmation. You listen and listen, and then, hearing nothing more, you breathe again, go back to the paper, and slide the memory of the "noise that was not" into the background. In the morning you discover a bicycle missing from your garage.

What you did not know, and did not figure out while you were reading your paper late into the night, is that a seasoned burglar was lurking about behind your garage. He happened to dislodge an empty tin soup can from your recycling bucket. He caused the sound. He also knows that 999,999 people out of a million (of course, we made

---

25 "The Phrase Finder: Things that go bump in the night," Phrase Finder, accessed January 20, 2014, https://www.phrases.org.uk/meanings/378900.html; *The Cornish and West Country Litany*, recorded (in written form) prayer, 1926.

this number up; it is intended to be very small), on hearing an unusual sound, wait to hear more. When they do not, they suppose that the hearing of the sound was either a mistake or nothing much to worry about. The burglar knows this and simply waits you out. You decide it is nothing. He then has a clear go at whatever he wanted to see, do, take, or destroy.

On the other hand, a seasoned clandestine spy, reading the same newspaper in the same house at the same time, knows that he must prove with certainty that no one was lurking. *He must prove that what he thought he heard was unimportant.* He will initiate deliberate activity to verify the security of the house and grounds. He will stealthily egress and, in ways known only to his trade, search and investigate. He will find anything that needs finding. Please take a lesson from an expert.

Critical Task
You should be able to find abnormal things as part of careful observing. Weak signals must be observed. Confirmation must be sought. So, when you find one, begin the *active* activity of processing it. Remember, weak signals normally do not escalate. This means you *must* seek out the confirmatory clues (what backward-projection implies) to indicate things going astray. This does not mean to go back and look again at the thing that raised your interest and led you to conclude that you had a weak signal. That would be like waiting for another sound. It means that you need to look for different things. Those different things become known as you extrapolate forward and then backward-project. Go after them. Weak signal management is a structured and active technology. The "other shoe" may have already dropped; you are tasked with finding it! You do it by using effective weak signal processing.

9.8 Weak Signals Might Not Persist Very Long

Keep in mind that a weak signal is subtle. It is by its nature vague and indistinct. For the most part, a weak signal is off-normal but not yet abnormal. This is important. There is nothing obvious or definite about weak signals. They are termed *weak* because any underlying nonnormal, off-normal, or (rarely) abnormal condition that may be present is not that easy to see. But it should be observable and then examined using weak signal tools. This is not to say that the underlying situation could not be dire and cause imminent disaster and havoc. It means that whatever that situation might be, it might be hard to see. So, yes, a weak signal may progress to abnormal as one of its life-cycle pathways. But even in the extreme where things may be poised for a most terrible event, what might be visible is only a slight indication. This is not unusual. But it is not so unusual that there is not an important lesson to be understood.

Weak signals sometimes have a limited life span. Afterwards, the weak signal may fade because the subtle problem has resolved itself. Or, the problem changed a bit so it now may appear somewhere else as a different indication. Any useful evidence to examine (via a backward-projection) would be gone. On the other hand, any persisting underlying abnormal situation may progress to an early manageable problem or to an alarm. To be useful, each weak signal must be evaluated in a timely manner to rule it in or out as an indicator of a likely abnormal operating condition. Any operating protocols that do not include timely evaluation will not be effective enough. Let us take a closer look.

Weak Signal Life Cycle

Weak signals usually goes through a natural cycle. The cycle steps and the alternatives in each step are listed here:

1. Off-normal process situation exists.

2. Off-normal process observable conditions (number of weak signals that could represent the off-normal situation)

    a. stay pretty much the same,

    b. slowly increase,

    c. quickly increase,

    d. slowly decrease, or

    e. fade away.

3. Weak signal becomes stale or outdated.

    a. Weak signal is missed.

    b. Weak signal attention is delayed too much.

    c. A process alarm activates, or bad things actually happen but no alarm is configured or is activated (if it was configured).

    d. Evidence at the places that one would backward-project to does not persist for a long enough time, or evidence is there but it is too diffuse or ambiguous to be of any dependable use.

Stale weak signals are of little use. To avoid this case, operators are encouraged to regularly look for them and process them in a timely manner. But all of this attention and care must be kept reasonable. It would be most unfortunate if we were to replace the old situation of not being able to find subtle problems with a new situation of spending too much time worrying about and working on weak signals. It is a tool, not a life's work.

9.9 Other Weak-Signal-Type Extrapolations

You now know how weak signal management works. Its power is the way it helps operators look for potential problems and issues in a way that offers a good procedure to reduce our natural inclination to jump to conclusions and other potentially biased actions. However, it may come as a surprise to know that this process of picking up on a "clue" about a problem and using that clue to hone in on realistic potential problems or issues is not new. Although it is not as formal as what we do in weak signals, near-hit investigations, what-ifs, HAZOP, and even alarm management do a form of extrapolation and projection. Next time you work on any of these, consider using some of the formalism of weak signal management to do those forward-extrapolations and backward-projections.

Near Hits (Near Miss)

Yes, there was no operational problem that needed management. Something was there that prevented the abnormal situation or condition from causing harm. But you were close. The fact of being this close can be exploited for immediate operational benefit. Alarms, when they are properly designed and managed, reduce the exposure to hits—and to near hits as well. The special part about weak signals and their role in near hits is that the better the operator is at managing weak signals, the more likely a hit can be avoided. A hit that is avoided by proper use of protocols and procedures is not a near hit. It is a job well done. Only where something failed but the enterprise still managed to dodge the impact do we have a near hit.

See also Section 9.14 for the big picture of tactical versus strategic weak signals. What follows is a specific treatment for near hits.

The Tactical Part

The term *tactical part* means that there is a use for weak signals in the here and now of operation. As soon as a near hit is recognized, the operator will focus attention on the part of the plant immediately surrounding the area involved. This immediate area might be geographically close (physically in the same vicinity) or operationally close (at a distance but connected by pipes, close energy coupling, etc.). The weak signal can be a process variable, control action (manual or otherwise), or something else that almost caused trouble. Trouble was avoided because it was a near hit and not a straight-on hit. But there just might be other trouble lurking close by. The near-hit part was just the first to show up. Look for more.

Here is how it works. Use the specific thing that was directly involved in the near hit—a flow too high, a pressure too low, or a valve open when it was supposed to be closed—as the weak signal. Forward-extrapolate it to *any other* worst situation that it might affect. By "any other" we are looking for things other than the situation that was avoided, only because that one never happened. The rest of the weak signals work process you already know will be used.

The Strategic Part

Weak signal tool design can also directly benefit from incorporating the knowledge gained from analyzing near hits. A near hit could have been caused by something "out of the blue" that presented itself without warning. If that was truly the case, you are going to rely on whatever is already in place for safeguards. However, the causes of many near hits could have been observed if we knew what to look for and how to look. If the causes can be observed as they build into problems, either an alarm might be configured if it fits into the plant's alarm system design, or the appropriate weak signal(s) might be built and incorporated into the HMI and/or operational protocols. Alarms have precedence. That is, if an added alarm will alert the operator of an impending problem that needs attention, then it is the choice. But alarms have special requirements. If the situation does not meet those criteria, then an alarm cannot be used. Weak signals are used instead. If weak signals can be predictive enough to provide early notification, both might be employed.

What-If and HAZOP

We will not turn this into a primer of either what-if or HAZOP analysis. This message is that they are only of value if the investigator has a depth of understanding in the processes and plants being checked. The usual way these investigations start is with something happening that is definitely not a weak signal. Then they look for the impact of the event to find out what it means for damage or other bad things. What weak signal analysis can tell us is that it might be a good idea to take the looking for the bad things to a second round if it is not already being done. This results in assuming actual damage from the event and asking, Well, if this happens, what else might happen after this happens? The extrapolations idea helps both to look more carefully for what else might happen and to look for clues to confirm or disconfirm the extent of the current examination. Remember, these examinations are done in our mind. Adding things to consider helps to ensure a broader approach and understanding.

Root Cause Analysis

Root cause analysis is what everyone wants to do after a problem occurs. Finding the root cause makes sure that we understand what happened. And it gets to the heart of finding ways to ensure that it and similar problems do not get a chance to happen again. The additional useful problems are found during the backward-projections. We already know what happened, but we only know one way that caused it. We need to find all of them. One of the causes we find will be the root one. Weak signals can help by reexamining the situation that happened. But now we are looking for any other small indications of trouble brewing. Look for weak signals. If any are found, process them to find any other potential root causes that might have been missed before.

Alarm Rationalization

Alarm response information relies heavily on the concept of a backward-projection. The entire job of finding confirmatory actions (see Chapter 8, "Awareness and Assessment Tools") requires knowledge of what can be found from backward-projections. Think of looking for the confirmatory actions as a formalized exercise in backward-projection adapted for alarms. Alarm management also relies on forward-extrapolation to determine the "consequences" used to set alarm priority. So, your tool for weak signal management can be useful across the board. Things more and more are fitting together nicely. The more you understand this, the more your operational tools have in common.

9.10 Weak Signal Templates?

Most plants and operations are composed of many individual components linked and interconnected in ways necessary to do their jobs. When we examine the individual components, we often find common ones that appear in several different places. Examples are motors, pumps, reactors, entry ramps, or individual aircraft in the air. All the entry ramps, for example, are not exactly the same, but they are similar enough that much of the knowledge about their inner workings and how to manage some of them would be close enough to serve as templates for understanding and controlling others. We have seen earlier in Chapter 8 ("Awareness and Assessment Tools") how using templates for alarm management is a way to achieve significant efficiency and consistency of design. The alarm management term is *key repeated elements* and the larger groupings of *key repeated subsystems*. To what extent can weak signal analysis take advantage of those common components? Would there be any benefit from prethinking what individual weak signals might present in each of the "repetitive" components?

The answer to this will test our understanding of how weak signals are intended to work. And the answer is probably not. There might be a slight benefit from the operator having a pre-worked-out set of weak signals to monitor for pumps and on-ramps. However, those pre-worked-out weak signals need to be carefully set out with all reasonable forward-extrapolations, all likely backward-projections, and all useful conditions to look for. This would take a large amount of engineering. Documenting everything would be a large task. And being able to access what the operator requires at the moment of need would be challenging. We are talking about large checklists and lots of dedicated supporting HMI graphics—probably not a good idea.

9.11 Retrospective Weak Signals Case Study

A way to illustrate weak signals in action is during actual process operation. In this section we will take a minute-by-minute retrospective look at operating activities for a significant industrial incident: BP Texas City.

HMI displays are able to show any desired screens containing the needed data. However, very few enterprises archive all the data all the time so that it can be available for post-event viewing and auditing. A great deal of, but not all the data is archived to a "plant historian." This points to a considerable difficulty to illustrate how a weak signal analysis might have worked for an event for some time after that event has passed. What follows is a weak signal illustration using the Texas City disaster. It should give the idea. Please do not read too much specificity beyond these general illustrations. We will look at the disaster and try to imagine how weak signals could fit in. To be sure, no such weak signal process was in place or used. This framing using weak signals is done only to illustrate how it might have worked. Remember, we will be looking for events or situations that (1) are observable, (2) are suggestive but not definitive, and (3) when examined properly could lead to the discovery of an important abnormal situation early enough to be valuable. Each one is a weak signal.

Texas City

The March 23, 2005 disaster at the BP Texas City refinery was introduced in Chapter 1, "Getting Started."[26] If you turn back or recall, there were eight steps in the disaster chain. If any one step had been properly managed or accounted for, the loss of life and property would not have happened or would have been considerably less. Examining the disaster chain failures is important to get at the bottom of fundamental enterprise deficiencies. That is after the fact. What about minute-by-minute during the event? Managing these is a key responsibility of the operator.

---

26 John Mogford, "Fatal Accident Investigation Report—Isomerization Unit Explosion Interim Report for Texas City, Texas, USA, ACCuSafe" (BP Internal Report, May 12, 2005), http://sunnyday.mit.edu/16.863/Texas-City-interim.pdf; BP US Refineries Independent Safety Review Panel, "The Report of The BP US Refineries Independent Safety Review Panel" (Washington, DC, January 2007).

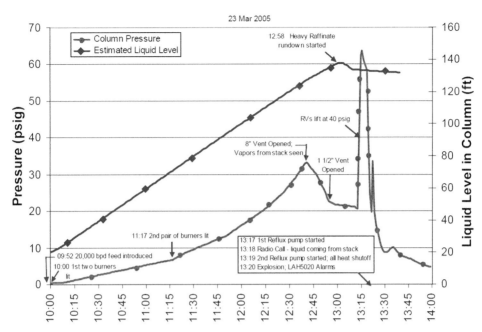

Figure 9-36. Texas City splitter pressure and level over time. Even at this level, things are not going as they should.

A most important observation is: this incident was unfolding (unraveling might be a more appropriate description) over a very long time frame as it refers to a shift—11 hours. Lots of "red flags" should have been visible. Red flags are an important part of the definition of weak signals. Figure 9-36 tracks the splitter pressures and temperatures over the time period of interest.[27] Raffinate (the part of a liquid remaining after its more soluble components have been extracted by a solvent) "boils" at between 143°F and 309°F.

Figure 9-37 shows a process flow diagram of the relevant refinery units.[28] The proximate cause of the disaster was the Raffinate Splitter (E-1101) fully filling with flammable liquid that overflowed liquid into the Blow Down Drum (F-20). This liquid then exhausted out the top of the drum into the atmosphere and almost immediately caused the explosion and fire.

Figure 9-38 tracks the splitter pressures and liquid levels over the same time period.[29] Note that even a casual inspection of these results should have been cause for concern.

---

27 Mogford, "Fatal Accident Investigation Report."
28 Mogford, "Fatal Accident Investigation Report."
29 Mogford, "Fatal Accident Investigation Report."

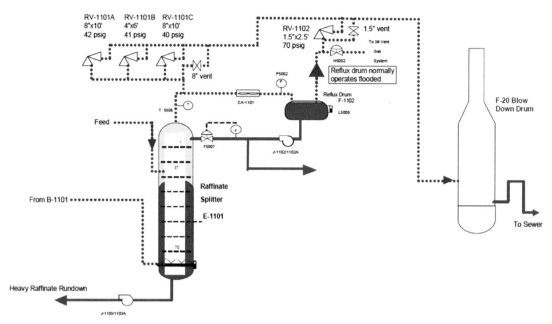

Figure 9-37. Texas City splitter process flow diagram depicting the specific process equipment in the operator area where the problem occurred. The excessive level was in the Raffinate Splitter (E-1101). The liquid that eventually exploded was expelled from the Blow Down Drum (F-20).

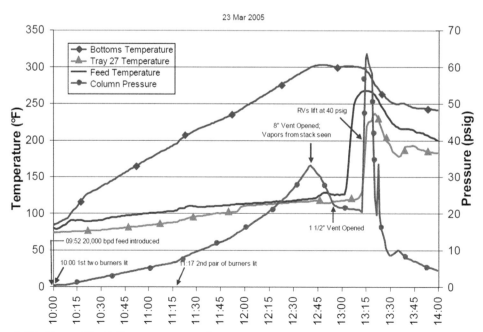

Figure 9-38. Texas City Splitter pressure and temperature over time. Refer to the list in Table 9-1 for the candidate weak signals that can be observed from this graph.

This example will identify as many of the weak signals as the documented actively supports.[30] Forward-extrapolations (to suspected worst-case problem) are done. Backward-projections (to look for evidence of the worst-case problem) will be attempted, but the reader should understand that little supporting evidence has been documented. Thus, we will rely on the greater understanding of what actually happened—and therefore what data would in all likelihood be readily available to the operator. See Table 9-1. Green shaded items were noticed by the operators and checked. Yellow shaded items were not noticed or were noticed but so superficially processed as to render that work ineffective. *While alarms are ordinarily not considered weak signals, they are considered here only because the alarm system was so poorly performing as to be largely ignored by sitting operators.* Pink shaded items depict the disaster event. The table information is, of course, for readers of this book; none are contained in any original reporting or previous examinations by others.

9.12 Relationship between Weak Signals and Alarms

This section covers the importance of the ability of weak signals to inform the operator and how this complements the alarm system alerting the operator. *All situation awareness not announced by alarm activations must be gained from serendipity observation or weak signals analyses.* All three activities together form a strong structural framework for all situation awareness and assessment. Figure 9-39 is a graphical illustration of the relationships and regions.

A critical responsibility of the alarm management process is to identify all abnormal situations that can be reasonably thought of using a thorough examination of the plant or enterprise and all of its constituent components. If the operator is not expected to manage them, they must be managed with explicit automatic controls or other controls or equipment safeguards. If the operator is needed to manage them, appropriate alarms must be identified for each one. They are depicted on the figure as the red alarm bell symbol. They are configured to protect all abnormal situations in the orange region. All the rest of the off-normal or abnormal situations (the yellow region) must be identified and handled some other way. Weak signals are depicted on the figure as the white clouds in the yellow region. For the rest, the red region, things can go very wrong with the operator having no structured way to find them unless alarms, weak signals, an attentive operator, or luck does the job.

---

30 Mogford, "Fatal Accident Investigation Report."

Table 9-1. Retrospective weak signal analysis of 2005 Texas City disaster. Note that weak signals were not in use in 2005; therefore, this is all hypothetical and shown for after-the-fact illustration.

Weak Signal	Forward-Extrapolation	Possible Backward-Projection	Comments and Conclusions
0300 h: High-level bottoms alarm activated.	Tower overfilled.	Verify other level measurements are valid. Check other level alarms; redundant level alarm not active.	No other independent level measurement devices being present should have required use of indirect level measurements—not done. Failure to follow through and resolve existing level alarm conflicts.
0941 h: LCV-5100 stroked properly; moved to closed, but flow indicated 3–4.7k bpd.	Leaking valve.	Heat exchange between this flow and feed bottoms flow indicated no exchange.	Valve not leaking; probably a flowmeter calibration problem.
0950 h: Liquid flow into splitter increased without any corresponding outflow at all. (Unnoticed by operators.)	Splitter inventory is increasing when it should not be indicating a clear operational problem.	Find the missing inventory or Find the malfunctioning metering or measuring devices.	Failed to notice weak signal.
1117 h: Reflux drum (F-1102) showed no level despite suitable temperature in the reboiler. The lack of level in the reflux drum was confirmed by field.	Splitter (distillation column) not working due to lack of vapor or lack of condensation at upper trays.	Recheck splitter levels by: a. Independently verifying level measurements b. Use tray temperatures to infer presence of liquid.	NOT DONE Note: Investigation showed that the temperature profile at the time indicated a liquid level up to tray 27. Also tray 33 temperature was a match to feed temperature.
1240 h: High splitter pressure alarm activated. Pressure control valve did not operate. Manual venting of blowdown drum done and showed vapor venting.	Splitter was operating incorrectly, and because the usual controls were not working, something significant must be wrong.	Completely reassess the operating state of the splitter.	Not done.
1300 h: Operators opened outflow valve (reason unclear), but this valve should have been open for a significant amount of time. Operators failed to monitor the effects of waste heat recovery on the inflow temperature to the tower.	The liquid inventory inside the splitter is very out of normal or appropriate. The feed inlet temperature is out of control.	Find out where the inventory is. Find out where the energy is going.	Not done.
1314 h: High-pressure relief valves activated.	Tower pressure is out of control.	None needed—immediate shutdown and evacuation.	Not done.
1315 h: Explosion	n/a	n/a	

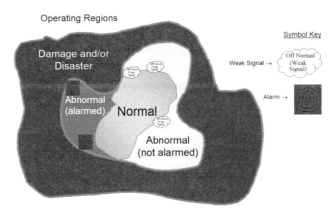

Figure 9-39. Relationship of abnormal situations and how alarms and weak signals are used to keep the operator informed. Alarms are configured for all known abnormal situations requiring operator action and identified at plant design. The rest of the abnormal situations may or may not require operator action but could not be identified at the design phase of the plant. These are left up to operator skill and diligence to find them; weak signals are a tool to help do that.

Situation Awareness Depends on Both Alarms and Weak Signals

When the alarm system is properly designed, implemented, and operated (here is where the power of alarm management is really demonstrated), then all process abnormalities worth knowing about that are not in the alarm system need to be observable as weak signals. If they are not, then there is undue risk exposure. Perhaps the alarm system design missed the potential problem. Or there was a missing weak signal. It is entirely possible that HAZOP and the other operational risk assessments that were done missed or ignored this exposure. Once again, we come face-to-face with the importance of each individual component tool for safe production management. When each component shoulders the responsibility for which it was designed, the other components can rely on it and are positioned to do their part.

> Progressive (as opposed to sudden) abnormal situations that lead to an eventual alarm if they progress far enough, should be visible as weak signals beforehand.

There is another important side of the relationship between weak signals and alarms. Abnormal situations that often (not necessarily most often, but that too) lead to alarm activations if uncorrected should be visible before alarm activation on the HMI as either weak signals or strong signals. This usually means that each abnormal situation as it develops would be visible as one or more weak signals. Weak signals do not always lead to alarms. However, all progressive (as opposed to sudden) abnormal situations that would lead to an eventual alarm if they progressed far enough, should be visible beforehand as weak signals.

Few Weak Signals Lead to Alarms

There is yet another important point to be made about weak signals and alarms. *Most weak signals that are useful will probably not have alarms configured for the underlying abnormal situations.* Remember, alarms should be configured for those abnormal situations where operator intervention is needed and should be identified in advance by engineers, operators, and designers. When the underlying abnormal situation is not exposed by something that is alarmed for appropriate operator action, there will be no alarm. For most of these, weak signals may be the only useable advance notice. Many disasters announce themselves by weak signals, and only weak signals, prior to crossing the tipping point. Weak signals are that important!

Precedence of Operator Activity

Alarms always take precedence over weak signals. Always! Weak signals belong to situation awareness. When things settle down and the operator has time and opportunity to get back into looking for things that might be going wrong, he gets back into the situation awareness business. When he does, weak signals can be an effective process to use.

There is an exception to the alarm precedence. Recall the earlier discussion about weak signal flags in Section 9.4. When a critical number of related flags accumulate (e.g., three or more), even during the alarm response process, it would be advisable to do a quick aside and have a look at the likely meaning of these flags. Accumulating flags usually indicate that something might really be going amiss. That something might be a problem with the current management plan for the alarm. Or it could be different or unrelated. A quick reconsideration might be a good plan before continuing.

Weak Signals from Alarm Activations

We close this discussion about alarms and weak signals with a useful weak signal application. Differing from the earlier applications in this chapter, this is a design one. Actually, think of it as a *strategic* weak signal tool. Its purpose is to uncover situations that may offer weak signals as early warning, possibly early enough to avoid the alarms activating. They are uncovered by *retrospectively* examining alarm activations.

We revisit the status of an alarm system after a proper rationalization (implementation and deployment) has been completed. Our alarm system was (properly) designed to provide indications of problems that you have decided the operator must act to fix. No longer will nuisance alarms distract. No longer will alarms that are appropriate in one operational situation be permitted to activate in a different operational situation where they are not required. In short, any alarm that activates signals a real problem in need of resolution. This means that behind every alarm activation there was trouble. There was a specific, defined problem with production, raw materials, energy management, or something else going wrong.

Alarm activations should regularly be examined after the fact on a scheduled basis. Take those alarm activations and identify the root cause for each problem. For each root cause, determine if it makes sense to identify that cause early. If it makes sense, it will usually show up when the process is off-normal but before abnormal. Weak signals would be the way to do this so the operator has a way to see them. He will look for them as part of the usual activities of situation awareness. Ensure that they would be visible to the operator.

9.13 Relationship between Weak Signals and Critical Variables

Recall the "loss of view" situation introduced in Chapter 5, "The Human Machine Interface." Operators were in a control room when all information on all HMI displays went blank—scary and disconcerting. Remember the takeaway from that? The lesson was to recognize that there is a set of values or readings from each process area that should provide enough information to strongly suggest the current overall operating situation. Build that list of critical variables for each operator area. For each variable or situation on the critical variable list, ensure that the HMI screens have sufficient visible ways to see each become off-normal or abnormal. Usually this means icons, gauges, and dashboards will show them to indicate normal, off-normal, or abnormal. Even where a critical variable has an alarm configured for it, it should be visible to the operator before alarm activation. This is an affirmative requirement for the design of the appropriate HMI screens. No operator screens can be properly designed for information content without fully understanding what weak signal information they must contain.

9.14 Operator Weak Signals Are Tactical (Short-Term)

Until now, you might have thought that weak signals is such a good operator tool that it would be used all the time. Every time a new screen is brought up on an HMI display would be a great chance to look for them. Every time an alarm clears (returns to normal) might be a good time to check for them. Whenever the operator thinks that things are going too well, he looks for them so as not to jinx things. Or, and this is more likely, operators will look for them instead of the stronger signals of regulatory control drift because weak signals seems to be a really good tool. Looking for regulatory drift seems to be much more open-ended and unstructured. Underlying all of this is the nagging possibility that something might be going wrong with the plant and the operator is not seeing what it is. So, it is not much of a stretch to think operators might overuse looking for weak signals. Now is the time to ensure that weak signals are kept in perspective.

Tactical versus Strategic

To be useful, a weak signal must be examined when it will yield proper information. This means that each weak signal *must* be examined within the time frame of change that revealed it. We briefly started this in Section 9.9 with near hits. Now, let us dig deeper. Weak signals can be accumulated or logged without timely processing, only if they remain processable (that is, Section 9.6, step 4 has not occurred). These are referred to as *tactical* weak signals. It follows that *strategic* weak signals are those that persist over a long enough time frame for them to be gathered but not processed immediately. This is the difference between tactical weak signal analysis and strategic weak signal analysis. Tactical analysis must be done almost immediately after the signal is noticed. This is what operators do. The operator will do it as soon as his operation schedule permits. This chapter assumes the weak signals are tactical ones.

Tactical Weak Signals

The operator actions of looking for and processing weak signals must be a separate activity for maintaining situation awareness over and above the usual checking and validation of the effectiveness of maintaining regulatory process control. *Regulatory* here means that controlled variables are at or near set points and measured variables are at or near their desired values. Effective weak signals processing requires separating the responsibility of maintaining regulation from the use of weak signals to identify likely developing abnormal situations. Each occurs in clearly distinct time periods of the operator shift. For the most part, an operator will check and adjust process conditions often and regularly during the shift. Together with his other duties of coordinating maintenance and other activities and preparing reports (including shift handovers), this forms the major activity. Separately and at predetermined times in the shift (see Chapter 10, "Situation Management," for recommendations), the operator will do a weak signals sweep of his unit. This is the active process of looking for weak signals in the places where they are most exposed. By separating the regular duties from weak signals processing, an operator is able to give proper attention to each. This reduces the blurring between "what is going now" and "what is building up to occur later." Both are important. Mixing them can lead to downplaying or missing a short-term problem, overreacting or reacting too early to a potential building problem, or overly relying on the corrective effects of more closely managing a short-term problem as a protection for the longer term ones.

The weak signal question: What is wrong with the current operation and my assessment of everything that I may not be aware of now?

Summarizing briefly, operators are encouraged to ask the *weak signal question*: What is wrong with the current operation and my assessment of everything that I may not be aware of now? In effect, we use the weak signal question to avoid the pitfall of several biases laid out in Chapter 7, "Awareness and Assessment Pitfalls." The reasons why it works well is that it (1) provides an unbiased procedure to look for potential off-normal situations and (2) provides a powerful and structured methodology for evaluating what is found.

Strategic Weak Signals

All weak signals that do not become stale can be accumulated and evaluated significantly less often. Let us take a quick look before we leave this topic. Although strategic weak signals can be quite useful, their use would be to identify infrastructure issues rather than contemporaneous operating problems. Certainly, infrastructure issues can and almost always lead to operating problems. But the weak signal analysis of the operating problem would point to an operational remedy (if one were available), not usually any underlying infrastructure issue. Of course, Monday morning quarterbacking could suggest them. But that would be on Monday—too late for our operator. Some examples of strategic weak signals would be maintenance costs trending up, uneven quality, slightly bumpy production deliveries where smooth ones are best, or the pattern of alarm activations. The underlying causes might be due to training deficiencies, poor alarm system design, reduced quality of maintenance, lax oversight, or any number of other enterprise deficiencies. For the most part, the standard methodology of forward-extrapolation and looking for collaborating evidence after backward-projection should reveal confirmations, because the underlying problems or defects change (unfortunately) very slowly or not at all.

There are interesting illustrations of underlying problems changing rapidly. For example, when a specific support individual who has been falsifying work schedules leaves for another post; the plant shifts from one seasonal operating regimen to another; or budgets change. One cannot always depend on delaying strategic weak signal analysis. Lest this entire thread be taken out of context, our interest and focus for the control room should remain tightly centered on weak signals for the operator's use during shift. Every enterprise should establish and enforce operations protocols and responsible activities and procedures. Failures of these systems and ineffective safeguards are beyond the intention of weak signals for operators.

9.15 The Dependence of Weak Signal Analysis on Model Quality

Operators need to have a thorough understanding of the process, plant, or operation that they are entrusted to manage. This is fundamental. They need to fully know how their plant operates. They need to be able to identify all the different ways it goes astray. We accept this as an essential requirement. The "understanding" we are discussing is called the operator's *mental model*. Let us shorten *mental model* to just *model*. When we take a closer look at how this model works specifically for weak signals, we discover that weak signals reinforce the already existing need for good operator understanding. *Understanding* means the operator has a good mental expectation (the model) of how his plant or operation works. Notice how all of this is fitting together nicely.

Model Fidelity

Model accuracy and appropriateness are a main concern when considering models. Model quality is important. Without it, even an operator who is exquisitely well trained cannot be successful. Moreover, that same model will be used for alarm management and training purposes. Success depends on a quality model of the real-world plant or enterprise. It is this model that engineers use to design a plant. It is the model that safety management personnel use to understand and control for operational risk. It is the model that those designing procedures and training use. The closer the operator's mental model matches this actual model, the better.

For weak signals, it is only the conceptual mental model that the operator is using that is important. This model is the central mechanism by which an operator (1) comes across a weak signal indication of a suspect abnormality, (2) chooses or not chooses to further investigate it, (3) processes it using the weak-signal extrapolation and projection methodology, and (4) identifies it as something of concern or not. If his model is a good one and is sufficiently close to the actual model (and by direct inference the real world), this process can work properly. If it is not so good, then the whole weak signals thing is not going to work. And if it were a tool in his kit, it would be flawed and the operational integrity of the plant he is managing would be at risk.

Identifying Model Inadequacies

Suppose your operator possesses perfect training in using the weak signals technology. All is for naught unless his mental model is a good enough representation of reality. This boils down to an inevitable conclusion that for any plant or enterprise that is being managed by operators who are required to act properly to remediate when sufficient abnormalities are present, the operators must have adequate models for what is going on. Responsible management must ensure that these models are good enough. This presents quite an engineering dilemma. We have found that the work process of alarm rationalization (the activity by which an enterprise arrives at a proper alarm system design) can be a powerful test of the collective understanding of the elements in a plant or operation. The term *collective* means that rationalization is one of the few formal activities that plants use to pool cross-discipline and cross-experience knowledge for safe operation objectives. The other ones are the related HAZOPs, PREOPs, and incident investigations. It is not uncommon for the alarm rationalization to identify issues that were not recognized during the other reviews. Another "test" of model adequacy would be the simulations and training methods that an enterprise must use to train operators to use the weak signals work process. If operators with skill in using weak signals are having difficulty processing candidate weak signals because (1) they cannot seem to see them in the plant operational records (HMI screens, logs, laboratory results, shift communications, etc.); (2) they see them but cannot effectively forward-extrapolate to potential problems or backward-project to potential confirming abnormalities; or (3) they do the projections just fine, but the results are false (which means either the problems projected to were not the right ones, or the evidence backward-projected to was visible but indicated a problem that on investigation did not exist), then we must suspect faulty models.

Weak Signal Work Process

Draw upon a variation of that time-honored *reductio ad absurdum*.[31] In the extreme, consider an operator with a mental model that is a perfect product of a perfect actual model that is a perfect representation of the physical world. If the operator is not capable enough to use the weak signal methodology effectively, weak signals fails as an operational tool. Therefore, enterprises must also expend proper effort to train operators in the effective use of weak signals.

9.16 Getting Weak Signals Working

This brings us full circle back to the weak signals tool. Certainly, the whole technology around weak signals is new. Weak signals were most likely entirely unknown to you before you picked up this book. But when you think about it, isn't this what a savvy operator was trying to do by looking for things that might be going wrong? He was asking, what is wrong with the current operation and my assessment that I may not be aware of now? Aren't weak signals easier to understand than an *ad hoc* invention? Using weak signals makes sense. If you do not do it this way, what other tools can you give the operator instead?

---

31 *"Reductio ad absurdum,"* Merriam Webster, accessed October 2, 2016, http://www.merriam-webster.com/dictionary/reductio ad absurdum.

Let us try a digression. If you take great pride in being aware of how your automobile is working, you will carefully follow most (if not all) of the recommended preventative maintenance service requirements. That is good. It is the baseline of proper care. However, I think that you will readily admit that you are always on the lookout for anything unusual: a slight "pull" in the steering when you back up, a noise that you heard yesterday and are listening for today, or perhaps a new smell. Sure, you would prefer that issues would never be subtle and unusual. But when they are, you think about them. These are weak signals.

Proper Foundation

Weak signal management may not be as big a job as it first seems. Starting from scratch may seem like an uphill climb. But surely, your enterprise is not starting from the beginning. Typically, there are lots of pieces and parts of these tools already in procedures or in general practice by your operators. Now that you have a better understanding of their purpose and how they might be better designed and improved, you will benefit from cleaning things and sharpening them. Consider working this into next year's plan, if you need to spread it out. Otherwise, the longer the delay, the longer the unnecessary stress on the operator and the greater the risk of bad things happening. This is the time to emphasize how every earlier topic in this book fits into the picture.

It should be evident by now that a management culture that prepares and empowers operators to function as highly effective professionals is vital to their trust that what they are doing is important; that they are expected to take leadership in operational effectiveness; and that they can depend on the enterprise to train, guide, and support them. The essential details of proper HMI design and implementation are front-and-center facilitators for operators being able to identify off-normal situations as weak signals. The proper design and implementation of the control room or operation center is key to ensuring that operators can use the HMI as intended and their work processes are not encumbered by unnecessary distraction. Effective training and support procedures and processes are essential for operators to possess the needed depth of understanding to enable them to properly extrapolate weak signals forward and then project backward.

To summarize, if you are using weak signals, but it is not working well as an effective tool for a specific operator or control room, then there are likely serious gaps in one or more of the following: (1) the alarm system, (2) the HMI, (3) fundamentals training, (4) maintenance, (5) operator loading, (6) morale and motivation, (7) policies and procedures, and (8) readiness and fitness for duty.

No Shortcuts for Weak Signal Management

Let us clarify how weak signal management must work in the long haul. Like any useful tool, the more it is used, the more familiar operators are with using it. Consequently, they will know how to look for weak signals. They can appreciate why it is useful to have specific times during a shift to look. They will develop an understanding of the inner workings of their plant or enterprise. Along the way, it might be natural to assume that finding and using some shortcuts could make it all work better. That approach, as tempting as it may be, will damage the entire benefit of weak signals. Weak signals is a protocol-based tool. The benefit is derived from using the tool—not skipping steps. Skipping steps can be tempting. It would work like this: see a potential weak signal, remember what it meant last time, and simply jump to whatever was done last time to handle it. The process must not work that way. Let us examine the reasons why, point by point.

Each Weak Signal Is Always New

Each candidate weak signal must be approached as if it had never been seen before. Even if it looks just like one before, it cannot be the same as before until *after* it has been proven to be the same. The proof is the end result of the work, not assumed at the start. This understanding ensures that the candidate (new weak signal) is properly processed in its own right. Otherwise, it is likely that the new one will be (1) processed too quickly by shortcutting here and there, (2) dismissed because the last time it turned out to be nothing, or (3) handled just like the last time even though this time it could be very different.

Finding Weak Signals

The ways to find weak signals will improve with practice. That is, knowing what to look for in some cases might get sharper. But it must not interfere with the looking. Where to look for them must be forever kept new. No matter how often there were no weak signals in a certain place during the middle of a shift, for example, one must still check there every time looking happens. Let us list the reasons why.

- The process, plant, or enterprise does not stay the same.
- The process, plant, or enterprise may be used differently now than before.
- The raw materials, weather, market conditions, and other factors do not all stay the same.
- The experience of the personnel (including the operator) does not stay the same.

With this number of changes, it is likely that subtle indications of things going amiss might be different as well. Therefore, in order to keep weak signals effective, do not shortcut where to look and for what.

Work to Resolve All Problems Found

Finally, when confirmation is found for any problem or issue, work to resolve each, taking into account the current operating procedures and the current state of the plant, enterprise, and surrounding issues. The same problem that was found months back might not be resolved the same way. Evidence and situation matter. Remember that remediation must follow established procedures, protocols, and all the rest. That methodology is part of the plant or enterprise infrastructure. This book provides additional tools and guidance to do that job better.

Indicators of Weak Signal Management Problems

Just as anything else can fail, weak signals can fail. Just as any tool must be kept sharp and the user continually trained, the same must be done for weak signal management. Some critical indicators for poor weak signal management are as follows:

- Small problems or operational concerns are present without sufficient weak signals available to operators.

- Operators fail to identify or detect weak signals when they are present.

- Operators have difficulty forward-extrapolating or backward-projecting the weak signals.

- Operators' success rate in using weak signals to identify underlying abnormal situations is below 95%.

Everything depends on the infrastructure of the plant or enterprise being in good order. Effective weak signal management can only build on a solid foundation for the entire enterprise. Situation management cannot overcome a lack of proper executive management, missing safety and environmental safeguards, weak financial management, missing training, impaired personnel, maintenance lapses, lost morale, or any other reasonable requirement for responsible enterprise operations.

Weak Signal Overload

Everything unusual can be useful. But not everything useful is useful all the time. On the one hand, weak signals seem to be everywhere. The more "tuned" an operator is to the idea, the more often weak signals are seen. The more success stories your operations team has to tell about finding and using weak signals, the more they will look for them. And in part, this is what keeps a winning team winning. But this can be carried too far. Everything in moderation. Remember concept of scale in Chapter 1, "Getting Started." Weak signal management is only one tool—make sure it is not overused. We should find a comfortable working approach to look for weak signals. There are two easy ways to do that.

Weak Signals at Regular Planned Intervals

The first recommendation is that looking for weak signals is not "job one." Most of their "chair time" should be taken up by their usual duties of coming on shift, managing or coordinating maintenance work, checking on production quality and rates, monitoring personnel access to the plant (including permits), preparing the end-of-shift handover report, and engaging in the task of going off shift. In between, at regular intervals (there are specific suggestions in Chapter 10, "Situation Management"), they would make the rounds of plant operations to ensure that production/operations were going according to plan. During these regular intervals is when the operator would put on his "weak signals hat" and look for and process them. The operator will do as much work as needed to identify them and work on them, but only at clearly identified parts of the shift. This looking and understanding is structured and limited. There should not be any weak signal overload problem here. If there is, this means that too much is going wrong all at the same time. This means significant portions of the ordinary infrastructure are out of line: maintenance is not what it must be, procedures are broken, operators are inadequately trained, or process controls are poorly functioning. Operators are not able to operate the plant effectively. This is a management problem, not so much an operational one.

Weak Signals as They Are Observed

The more likely way weak signals can overload operators is when the operator is overly sensitive to them or over-aggressively looks for them. This means operators are too ready to find them everywhere. The way out is to ensure that operators maintain a perspective. During ordinary monitoring, which they do as an intrinsic part of watching out for things, we should establish an expectation of respecting that things can be a bit out of normal without announcing a problem in the making. There must be a tolerance for variation that is expected and understood as part of normal operation. In weak signals terminology, they would all be defined as less than immature weak signals—thus not meriting any attention.

This could lead us either to a definitions dilemma or a mental calibration situation. To avoid both, we should understand that an essential part of the operator's training must make him competent to understand the differences between truly normal variability and off-normal operation. Any candidate weak signal that appears likely to be beyond normal would be fully processed. Any weak signals that are not, will not be processed. The bottom line is that operators should be able and encouraged to look for anything off-normal even when they are just checking on things and keeping an eye out. They should also know to keep a proper perspective.

Selling Management

How can weak signals not be useful and usable? And remember, this technology is not replacing anything. What would any operator, engineer, technician, or manager require to be able to convince senior management of the essential need for this? There is not any magic here, unfortunately. Managers should readily see the benefits and come to the conclusion that responsible stewardship requires enough tools for the operator to feel comfortable and competent in the job. Lay it out as a simple way to formalize what operators can do as they "look around" for unusual stuff. Now as they find something suspect, they have an efficient tool to work it through. It is a tool they did not have before. It is a tool that can be trained for, practiced, folded into procedures and operations, and used in collaboration, delegation, and escalation.

9.17 Troubleshooting Guide

Things can appear to go amiss while working the weak signals processing procedure. This section will briefly review a few of them.

Finding Too Few Weak Signals

If operators are finding too few weak signals, it suggests several situations: the process is operating extremely well, the operator is not skillful enough to find them, the operator is not looking for them, or the HMI design is too weak to provide them sufficiently. Except for the first situation, you have a real problem. Resolving it has more to do with infrastructure and training than with weak signals. You have work to do.

Finding Too Many Weak Signals

An operator finding too many weak signals means that there really are too many or he is looking too hard. Remember, looking for weak signals is not job one. Certainly, if one is visible, the operator should take a look. But rather than hunting around for them in every nook and cranny, he should keep it under control. On the other hand, if weak signals really are all around, it is not too much of a stretch to suggest that the plant may have significant design issues, operating procedures are not up to the job, operator training is not good enough, there is poor maintenance, or some other significant problem exists. Escalate this issue soon.

Cannot Forward-Extrapolate a Weak Signal

An operator having any difficulty forward-extrapolating a weak signal means that his process understanding is too limited or he has very weak skills. This is basic and needs immediate attention!

Finding Too Many Problems When Forward-Extrapolating

An operator forward-extrapolating a weak signal to many potential problems means that his process understanding is confused or he has very weak skills. One significant reason for this situation is that the operator is attempting the extrapolation to too low a level of significant problems. This is basic and needs immediate attention!

Many Weak Signals Forward-Extrapolate to the Same Problem

An operator finding that forward-extrapolating several weak signals to the same potential problem suggests that that a potential problem exists. Get to work on it!

Cannot Backward-Project a Potential Problem

An operator having any difficulty backward-projecting a potential problem suggests that his process understanding is too limited or he has very weak skills. This is basic and needs immediate attention!

Many Potential Problems Backward-Project to the Same Evidence

When an operator finds that backward-projecting several potential problems points to the same evidence, it suggests a potential problem exists. Get to work on it!

Confusing Evidence Obtained from Backward-Projections

An operator finding that backward-projecting a potential problem yields confusing evidence suggests that the potential problem being projected might not be the correct one. Try a second round. If things are still confusing, this suggests that something is going on but cannot be properly identified. Something is amiss, so it is best to escalate right away!

No Evidence Found from Backward-Projections

An operator finding that backward-projecting a specific potential problem yields no evidence suggests that the potential problem probably does not exist. Either consider it as noise and forget about it for now, or keep an eye out and look again later. Or the operator is not working the process correctly. Or the model is weak.

The Weak Signals Procedure Does Not Seem to Work at All

If the weak signals procedure does not seem to work for you, something basic and important is going on. Stop. Do not force it. Keep your focus and temporarily work out other ways for finding problems.

Bring this situation up to management so they can work this out. The weak signals process can be effective and powerful. Let management figure out how to get it working for you.

The Weak Signals Procedure Does Not Work All the Time

If the weak signals processing procedure does not seem to work well all the time, not to worry. It is just a tool. If the weak signals are so weak that sometimes you find nothing and other times it helps, then things are just fine. However, if sometimes it works fine and other times you get lost, or worse, get a wrong result, something basic and important is going on. Stop. Do not force it. Keep your focus and work out other temporary ways for finding problems.

Bring this situation up to management so they can work this out. The weak signals process can be effective and powerful. Let management figure out how to get it working for you.

The Weak Signals Procedure Is Too Hard to Use

If you, as an operator, are finding the weak signals processing procedure too hard to use, we both have a problem. My problem is that I did not do a good enough job at convincing you that the weak signals procedure should be easy, intuitive, and empowering. I also probably led you astray by all the definitions and extrapolations and projections stuff. So, I ask you to pause, take a deep breath, and let me try to give this a brief fresh start. There are only four moving parts: (1) see clues that something does not look quite right; (2) imagine the worst that the "not quite right" things could be (what we might call the tip of an iceberg); (3) assume that if the iceberg is really there, find other tips hiding in slightly different places; and (4) check each "tip" to find evidence that it is amiss or not. Let's take a closer look.

Seeing Weak Signals

If you are having trouble seeing them, either they are not there or they are not easy to see. Review the earlier section, "Finding Too Few Weak Signals."

(Forward) Extrapolating Weak Signals

If you are having trouble extrapolating them, then your training might need some refreshing. Get together with your supervisor and see what he suggests. Review the earlier section, "Cannot Forward-Extrapolate a Weak Signal."

(Backward) Projecting Potential Problems

If you are having trouble projecting potential problems, then your training might need some refreshing. Get together with your supervisor and see what he suggests. Review the earlier section, "Cannot Backward-Project a Potential Problem."

Finding Evidence of Potential Problems Actually Present

If you take each item you are looking for evidence about (stuck valve, missing sign, etc.) but you cannot seem to locate the right place to look, or you look but nothing wrong is there more often than not; something very basic is amiss. Seek supervisory assistance soon.

9.18 Additional Thoughts about Weak Signals

There are more benefits operators can gain from looking with a different "weak signals" lens. Let us take a brief look to see how they might be useful to you.

Artificial Intelligence for Weak Signal Analysis

Smartphones have applications (apps) that can tell if someone is looking at the screen or not. Search engines have the capability to figure out what is being looked for and suggest other ways of finding it. For example, credit card companies try to figure out whether a given charge is proper or not. Likewise, the work process for weak signal analysis could be a natural candidate for intelligent aids for the operator to use.

Let us explore this for, of all things, a vehicular traffic control example.

1. The operator notices something that appears to be suspicious, for example, the between-vehicle spacing suddenly decreasing near the beginning of an off-ramp.

2. A query to the intelligent aid (IA) would be, "I see a sudden drop in the between vehicle spacing in the sector XA12-42 off-ramp but not in later sectors."

3. The IA responds with likely worst things (forward-extrapolations), such as the lead vehicle in XA12-42 (1) has vehicle problems, (2) the driver might be suddenly physically impaired, (3) the driver spotted debris in the roadway, (4) the driver realized that the wrong exit was taken, or (5) the off-ramp load has suddenly increased.

4. For the options in step 3 that the IA system can backward-project to, it will look for confirmation. One that comes to mind would be to check on whether or not the traffic flow on the main roadway just beyond XA12 has abruptly changed

so that drivers were exiting to avoid being caught in slow traffic. And it can look for other possibilities.

5. The operator follows up the options in step 3 that are beyond the design of the IA system, such as inspecting a camera to see if one car in XA12-42 is trying to leave the exit ramp and get back on the main roadway, if the lead vehicle is not moving, or if vehicles appear to be driving around to avoid a roadway problem.

Other IA systems might work differently, but this illustration should give you an idea of how it might work in this example.

Identifying Gaps in Training and Procedures

Now that you have read this far in this book, you have probably noticed that the approach to all of this is synergy. Sure, there is a lot here. Some is framework content that encourages you to understand and check for needed infrastructure in your enterprise that can form an effective foundation on which to build situation management. The rest are tools and processes to get it all done. Everything should weave together into a strong fabric for doing that job well. We now take a look at how weak signal processing can identify gaps in training and procedures. Finding the gaps is surprisingly direct.

Training Gaps

Important gaps exist in operator training on the basic functionality and underlying technology used in the plants or enterprises they are charged with operating, if operators have difficulty doing (meaning cannot do it well or make errors doing) the following:

- Seeing weak signals that are present

- Effectively forward-extrapolating a candidate weak signal to appropriate worst case potential problems

- Effectively backward-projecting worst case potential problems into likely abnormal situation causes

- Following the established procedures for confirming or disconfirming the abnormal situations identified by backward-projection

These issues expose core problems around the assumptions about what operators need to know and how to teach it in order for them to do their job properly. To a lesser extent, these problems may also indicate that the operator selection process may be a concern.

Procedure Gaps

Important gaps exist in the design or development of proper operating procedures if

- not enough weak signals can be observed (using the HMI, referring to laboratory data, responding to issues raised by field observations or field operators there) for off-normal situations,

- appropriate procedures do not exist for handling the abnormal situations identified and confirmed by backward-projection, or

- following the procedures for handling the abnormal situations confirmed by backward-projections do not remedy the problem.

These issues above expose core problems with the understanding or assumptions about how the enterprise must be viewed and managed in order to keep it functioning properly. Fixing this is the responsibility of operations management and engineering.

Conceptual and Functional Concerns

The core of weak signals is that they must be available (visible) to be detected by operators. The sum total of all the ways operators look into their plants and examine how they are working must be enough to see what they need to see. What the operator cannot see cannot be used for anything. Moreover, the qualifications and shared experiences of operators must build into effective working methods that take advantage of weak signals and assist in their use. Care must be taken to avoid assuming anything, including that "experts" are always right. Finally, looking for weak signals should not be elevated or prioritized over other activities and procedures beyond the enterprise's deliberate plan for specific operator responsibilities.

Weak Signals and Incident Investigations

It is tempting to assign a role for weak signals in the incident investigation process. They do have a place there, but it is probably different from what you are thinking. Missing a *weak signal* by operations is by no means a failure nor can it be assigned as an operator error. Remember, weak signals are subtle, vague, and otherwise indistinct. They might and often do appear to be different to different observers. Therefore, finding abnormal situations by looking for weak signals is not going to be reliable enough to put responsibility on the operators. So, who is responsible for the incident that the operator was not able to see coming? Make no mistake; your enterprise must close this loophole. Here is the message: *if a weak signal (or a few of them) would be the primary way for an operator to learn about an abnormal situation that likely leads to an incident, then it is the responsibility of the engineering and operations teams to design other mechanisms for discovering that abnormal situation.* Those other mechanisms are beyond the scope of this book.

Our topic started with weak signals having a role in incident investigations. And that role is clear. *Any weak signal that is noticed by the operator but improperly processed is a contributing factor to the incident.* While finding a weak signal might not be reliable, once found, any failure to work out its significance is a mistake. A short list of causes might include operator impairment, inadequate training, weak understanding of how the process functions, insufficient detail or errors in procedures, and maintenance problems. Discovering what causes this failure is a proper incident investigation activity. Recommending changes to remedy this failure is a proper incident investigation conclusion.

There is one concluding note about how all of this fits into post-incident activities. It has to do with the extent to which the actual weak signals make it into the preserved record or history of the incident. Most weak signals are subtle, indistinct, unremarkable, and not too remote from normal. As such, no one should expect them to be flagged, annotated, or otherwise noted. They are not going to be visible down the road unless the data history is expansive and of high fidelity. The result is they are not always going to be in the record. Missing them cannot be reliably confirmed or denied. This reinforces the guidance that an operator who might miss them would not generally be at fault for the miss.

Weak Signals Are Not the Only Way

Just in case you may have settled on the idea that weak signal analysis is the only way for an operator to gain early knowledge of trouble, let me quickly suggest that any other effective substitute ways of ensuring that off-normal situations are identified and processed would be just fine. Good operators may have developed their intuition and surveillance protocols into remarkably useful tools, though their proper function must still be evaluated and approved. If this skill varies greatly from operator to operator and can ebb and flow over time or during stressful situations, it will not be adequate. It must be left up to operations supervision and enterprise management to decide if and how such intuition and history are adequate guarantees of future success. If so, how would it be rolled out for all? If not, what will take its place? If nothing is done, how safe are your personnel and surrounding community, and how protected is the enterprise's future? Doing nothing should not be an option.

Skipping over Weak Signal Processing

There are a couple of important situations to clarify before closing out this discussion of weak signals. First, why not simply forget about all of this careful and sometimes tedious weak signal processing and the cases and all the rest, and work out what is wrong directly after seeing a small indication? Why not forget about all the procedures and cases? Just figure out what is the matter. When you do, just fix it. It is quick. It is easy. You have certainly thought about doing it this way. Maybe you even used those thoughts to back away from carefully following the details of this chapter. So push everything onto that back burner and go back to the way you were doing it before. Well, you can. And you will get it wrong more often than not. Every time you get it wrong, the plant is closer to a difficult situation and the operator will have missed seeing it.

The second important situation to clarify is why we do not want to stay local and just look around for more indications. Let us look at the first one first.

Jumping Directly
When you as the operator see something that does not look quite right and you do not carefully check it out, you risk getting it wrong. Sometimes you get it right. When you do, you fix it or decide that it was nothing to worry about. When you are not right, you are going to miss. You are going to miss because most subtle things do not nearly often enough point clearly to what is wrong. When things are not clear, it makes it easy for you to assume. When you assume, you skip over a lot, you jump to a conclusion without evidence or support, and you do it all through the lens of all of those human foibles we talked about so carefully in Chapter 7, "Awareness and Assessment Pitfalls." Remember doubt? Remember biases? Remember the information fork? They are not curiosities. They really can get in the way. And because they can, they often do.

The power of weak signal processing is its ability to work around most of those pitfalls and keep you thinking objectively. It helps you identify problems. You should miss less. And what you do sense as something not quite right, you have a structured way to find out. You no longer go directly from something you see to land on a conclusion. You are not going to make that big jump in a single step. You are going to carefully consider everything reasonable that might be going on from the clue that you got by seeing a weak signal (forward-extrapolation). Then, you go for more evidence (backward-projection). You are looking for evidence (where projection indicates that it should be found) that you might not have even considered looking for before. You did not consider it before because without weak signal processing, you might not think of it as being useful. Now it is. If you do not find evidence, you can be comfortable that nothing much is going on. When you do find evidence, you know that a problem is building up and what it is. Now you can stay on top of things better.

Bypassing Going Step-by-Step
You agree to not try to jump from an observed weak signal directly to a conclusion. So that is settled for now. But something about all the effort and imagining to get weak

signal processing to work may still seem to be overkill. Seeing weak signals is so intuitive. Why complicate this by asking operators to pause and think how to extrapolate and to what? And why ponder how what was extrapolated to and project backward to find enough evidence, when it is so easy just to follow the clues directly? Why not take the weak signal and just look around for more evidence in the neighborhood? If something is going wrong, you might see it as another weak signal. It appears as a clue. If the thing that is going wrong is close by, just looking around should do the trick. Well, it might. And if it does, you are lucky. But what if that same weak signal were just the tip of a bigger problem? That bigger problem might easily be some distance away from the clue you see. If it is, following the nearby pathway might only uncover something limited to the original clue, not the bigger problem. You are likely to fix it and go on to something else. However, a bigger problem just got missed!

Yes, you just solved a local problem. But no, you did not even see the larger one. You just worked symptoms. The underlying root cause still exists. Now you have to wait and hope for another advance warning; otherwise, you have blown a perfectly good chance to find the larger problem by not finding it using the weak signal process. An important and powerful product of using weak signal processing is that it is designed intentionally to uncover important underlying problems, the real problems. The part of weak signal processing that asks you to forward-extrapolate brings to the fore important possibilities that the weak signal might be pointing to. This extrapolation helps to skip over smaller local problems and issues to get to any larger ones. It is an important part of finding problems early. And finding problems early is important for situation management. In the final analysis, situation management is about finding problems in time to be able to mitigate or prevent damage. Weak signals are just a means to that end.

Wrapping It Up

The material in this chapter is nothing short of amazing. It is about finding potential problems, not diagnosing them. Weak signals fills in the huge gap between alarms and nothing. It provides a methodology and work process that builds on operator intuition and observation. It allows the operator to take full advantage of all his process knowledge. It guides the HMI designer to suggest depicting as many indications and variables that he suspects should provide small clues and other knowledge about things that can vary from normal to a bit off-normal. It supplements good operating procedures. The whole idea of exploiting small indications of abnormal now has a purpose and a process. But, as your author, I would be remiss if I were to suggest that weak signals are singularly essential. They are not. The other material in this book is just as relevant, just as important, perhaps even more important, if there was not anything here about weak signals. So please do not get so carried away that you give the other parts less attention. Without weak signals, situation awareness has worked forever. If weak signals were not in this tool kit, the rest of situation awareness still would work. And if you do not like weak signals and do not want to use it, fine. But please develop other ways for your operators to find problems not announced by alarms. Make sure operators have the skills and resources to use those other tools effectively.

> The sum total of all the ways that operators look into their plants and examine them to see how they are working must be enough to find what they need to find.

9.19 Close

At this point you might be either quite impressed with weak signals or ready to brush it off as interesting but not useful. Either way, it is good that you have visited the topic and are still reading. Weak signals is a tool that is only as sharp as you keep it. But it is not magic. I am amazed at the power of weak signals to close the gap between alarms and a frantic search for the abnormal or off-normal to keep things on track. Understanding weak signals and working the process into the operator's tool kit should tip the odds for better operation in favor of success. If you have found this whole topic exciting and potentially useful, let me add a word of caution. The power of situation management clearly can be enhanced by weak signal processing. However, by now you realize that the only way weak signals have even a chance of working is if the rest of the operational infrastructure is strong and effective. Practices and procedures must be competent. Plant design and maintenance must be robust and up to date. Operator training must be deep and useful. And management must be encouraging and enabling. Readers are encouraged to find the most value from this reference book in the entirety of the topics and tools and processes suggested. Weak signals are just an added benefit to help it all work better.

9.20 Further Reading

API RP 1167. *Pipeline SCADA Alarm Management*. Washington, DC: API (American Petroleum Institute), 2009.

EEMUA (Engineering Equipment Materials Users' Association). *Alarm Systems—A Guide to Design, Management and Procurement*. EEMUA Publication No. 191. London: EEMUA, 2007.

Heimer, Olaf, and Norman Dalkey. Inventors of the Delphi Method; Rand Corporation; ca 1950s. Accessed August 2, 2015. https://www.rand.org/content/dam/rand/pubs/research_memoranda/2009/RM727.1.pdf

Hollifield, Bill R., and Eddie Habibi. *Alarm Management: A Comprehensive Guide.* 2nd ed. Research Triangle Park, NC: ISA (International Society of Automation), 2011.

ISA-18.02-2016. *Management of Alarm Systems for the Process Industries.* Research Triangle Park, NC: ISA (International Society of Automation).

NTSB DCA-10-MP-007. "Control Room and Supervisory Control and Data Acquisition (SCADA) Group Chairman Factual Report." Washington, DC: NTSB (National Transportation Safety Board), April 10, 2012.

PHMSA (Pipeline and Hazardous Materials Safety Administration). *Hazardous Liquid Integrity Management Enforcement Guidance Sections 195.450 and 452.* Washington, DC: PHMSA. Accessed September 4, 2018. https://www.phmsa.dot.gov/sites/phmsa.dot.gov/files/docs/Hazardous_Liquid_IM_Enforcement_Guidance_12_7_2015.pdf.

Rothenberg, Douglas H. *Alarm Management for Process Control: A Best Practice Guide for Design, Implementation, and Use of Industrial Alarm Systems.* 2nd ed. New York: Momentum Press, 2018.

Shermer, Michael. *The Believing Brain: From Ghosts and Gods to Politics and Conspiracies—How We Construct Beliefs and Reinforce Them as Truths.* New York: Henry Holt, 2011.

Weick, Karl E., and Kathleen M. Sutcliffe. *Managing the Unexpected: Resilient Performance in an Age of Uncertainty.* San Francisco: Jossey-Bass, 2007.

Part III

Situation Management

10 Situation Management

Never mistake motion for action.

Ernest Hemmingway (Author)

We learn so little from experience because we often blame the wrong cause.

Joseph T. Hallinan (Author)

We must stop all this communication and start having a conversation.

Mark Twain (Author, Samuel Langhorne Clemens)

Now is the time to claim your prize! You did not win the lottery. Sorry. Anyway, relying on chance to keep things going well is not a dependable strategy. There are better ways. Here are some. In this concluding chapter on *situation management*, you will see the many ways you can change "chance" into competence. It is about ensuring that the job operators do will be beneficial. This chapter cements everything together so operators can deliver. It will help operators gain confidence in operations. To get the real prize, management must make changes, followed by operations, engineering, and maintenance. Everyone is part of this team. This team is going to work.

How the team works is everything. How the individual works will lay the framework for how the team works. We rely on a four-step process to keep everything clear and working.

1. **Observe** – See what about the enterprise might need attention and action.

2. **Confirm (or disconfirm) problem** – Understand everything that needs attention. Keep the parts that actually need clarification, attention, or action. Discard those that do not.

3. **Remediate** – Decide what intervention must be done and how to do it to manage everything that needs it. Carry out the needed actions and activities that were decided on.

4. **Confirm (or disconfirm) resolution** – Verify the effectiveness of actions and correct where needed.

The effectiveness and power of taking each step in order comes from knowing *only that step is being thought about and worked on*. Nothing is permitted to spill over into the next step before you are complete with this current step. Keeping everything distinct is effective in ensuring that each step is fully explored and understood in its own right. By not jumping ahead and by being careful, you can minimize those natural tendencies to succumb to biases that always lead down the wrong track; provide clear and structured opportunity to engage dual tools to work, including collaboration and delegation; and ensure enough information is at hand to prepare for the next step.

As you read into this chapter, you will come across many ideas and suggestions for changing the way your enterprise is designed and managed. As you find ideas you would like to use, please keep in mind that the best way to do so would be to design them into the fabric of your existing enterprise. Resist the temptation to add a new procedure, a new committee, or a new list of responsibilities. Doing it that way could be quick and less costly, but it is unlikely that it would be either efficient or sustainable. You want both. Work out how each new idea or concept should fit back into an existing procedure or training program. Design and modify the existing ones to include the new.

10.1 Key Concepts

Situation Management	Situation management is the ability to identify possible and potential threats to good operation of a plant or enterprise; confirm the validity, significance, and extent of the threats; undertake appropriate response to the threats to remediate or, if remediation is not possible, limit the extent of damage; and evaluate the results of those efforts.

Fundamental Concepts for Effective Situation Management	1. Explicitly *identify* what the current situation is. 2. Fully and independently *confirm* that situation is as identified. 3. *Remediate* or resolve only the situation identified and confirmed (no "changing horses in midstream"). 4. **Confirm** or disconfirm that current corrective actions are working.
Effective Situation Management Changes the Rules; New Rules Mean Better Outcomes	When an infrastructure is prepared for the unexpected as if it were "expected," the entire rules of the "game" become operations of competence. Managing shift workflow will significantly improve the operators' ability to detect and manage impending abnormal situations, and at the same time reduce uncertainty and stress.
Knowledge and Experience Should Rule	Operating in a major situation requires the active participation of individuals solely based on their expertise (demonstrated knowledge and experience) regardless of their immediate position in the organization hierarchy.
Design In Rather Than Add On	The best way for sustainable implementation is to design in rather than add on.
Control Room "Condition"	Use of specialized control room conditions can provide the environment where operator ability to focus is enhanced by explicitly managing distraction and simultaneously providing special targeted resources.
Goals versus Risk	When operational risk challenges operational goals, managing risk always comes first.
Escalation Is a Safeguard That Improves with Use	The ability to rapidly deploy competent escalation teams to the control room can make the difference between dangerous operations through inappropriate risk-taking or inability to recognize incipient danger and an orderly transition to a safe operating state. Enterprises will readily deploy escalation teams as often as needed to back up their culture for safe and effective operations.
Situation Management Bottom Line	Successful situation management requires operators to apply specific solutions to problems without specifically experiencing them beforehand. This requires competence training on how situation management tools and concepts are used.

10.2 Introduction

Automation today places the operator well above the minute-by-minute responsibility of watching process variable values and making manual adjustments to maintain proper regulation. Enterprises have sophisticated automation equipment to do that job. Whether it is a single-loop controller that measures a pressure to modulate the position of a control valve or a vast distributed control system that manages an entire plant, operators watch rather than manipulate. Only when proper operation might be at risk is the operator called on to act. In this new world of operations, the most important operator task is to know when things are going well and when they might not be. Every time an early abnormal operating situation is missed, valuable solution time is lost. When the abnormal situation gets worse and is also missed, much more is at risk. Often incidents, some very serious, are the result. And when intervention is called for, the skill of the operators and the performance of the supporting actors can make the difference.

Situation management is about finding and managing problems, not so much about resolving them. Once found, most of the solving is usually well understood. Most trouble comes from not finding the real problem, finding only a part and not finding those other parts, or failing to properly work a solution.

This chapter is about how things come together in the control room. Let us start by remembering the fundamental concepts for effective situation management. Everything in this book supports one of these four core concepts:

1. Explicitly *observe* what the current situation can be or is.

2. Fully and independently *confirm* or *disconfirm* that the actual situation is what you think it is.

3. *Remediate* or resolve only the situation identified and confirmed (no "changing horses in midstream").

    a. Factor in all *threats* present or on the horizon that can shape the solution.

    b. Always stay in *familiar* territory by never letting the process or operation take the operator to a place he has never been before (instead, shut down or safe park).

4. Make sure the resolution effectively *resolves the problem*.

A problem is not considered to be found until its potential risk (if not resolved) is fully understood.

The Situation Management Activity

Situation management is the competent execution of the four fundamental activities. Refer to Figure 10-1 (repeated from the Chapter 1, "Getting Started"). Everything in this book supports these activities.

Successful situation management requires operators to use specific solutions to problems without ever being trained or experiencing them specifically beforehand. That tool kit requires operators to be proficient at detecting early abnormal operations, evaluating those situations to identify all that must be addressed, and successfully addressing each. This is what the entire book is about. It is how you can ensure that what the operator needs gets successfully done. The shared understanding and acceptance of the principles of good operation protocols, competency training, and operational improvement

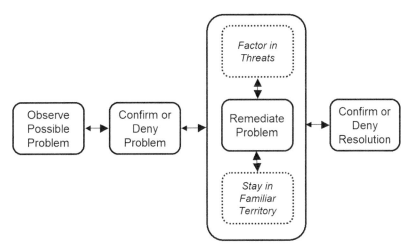

Figure 10-1. Four-part situation management activity process that requires appropriate tools and technology to be able to find all potential problems, specifically confirm or deny their presence, and resolve them if there. Note the importance of factoring in any threats that are near, keeping the remediation relevant, and making sure that everything done is familiar (not invented on the fly).

tools are the front line of success. Because these principles are so important, tools must be carefully designed, often following industry good practice guidelines or conforming to regulatory guidance and requirements. Operators and supervisors must be trained and refreshed in their use. Recommendations from incidents and accidents are carefully examined and folded back appropriately. Technology and equipment providers are ever-improving their offerings. Given proper management resourcing and direction, all of this should work to improve the control of operations.

Step-by-Step Working Process

Let us break this down into a nice, clean working process. Remember, for each bit of information, be sure to use the information fork (see Chapter 8, "Awareness and Assessment Tools") to identify the known parts and flag the unsure and assumptions parts. Only the known parts can be used.

Activity 1: Observe Possible Problem
- **Step 1.** Find all potential and reasonable problems that may be causing the process to be abnormal. *Make no evaluation* about how confident you feel about each or how plausible each might be.

    Here you are just making a list of cautionary clarifications for the step.

Activity 2: Confirm or Deny Problem
- **Step 2.** Find all evidence that tends to *confirm* each problem as being the right problem, and find all evidence that tends to *deny* each problem as being the

right one. Do not attempt to decide which problem might be the one causing the process to be abnormal.

Here you are making lists for each separate candidate problem.

- **Step 3**. Where some evidence confirming and/or denying one candidate problem is shared by another candidate problem found in step 2, then carefully and deliberately look for a larger or different problem that combines the two that share confirming and denying evidence. If you can, add that one to the list of potential problems found in step 2.

Refine the list of problems to the smallest sensible number found.

- **Step 4**. Settle on the Step 3 list of problem(s) (abnormal situation observed and confirmed that needs active resolution).

Make the list and arrange the items in order of operational risk. Make sure that none are moving too fast to manage.

Activity 3: Remediate Problem
- **Step 5**. Identify remediation protocols and procedures applicable to managing each problem.

These must be established and approved procedures as well as trained-for activities.

- **Step 6**. Identify all threats that are outside the immediate problem but can affect either the choice of remediation method or the way the remediation is applied.

Example threats include threat of loss of essential utilities, missing support manpower or other resources, problems in adjacent operating areas that might spill over into this one, and need for escalation or collaboration.

- **Step 7**. Execute the remediation activities making sure that all actions and activities, and all results and changes, are expected and understood. If not, at any time, immediately begin "permission to operate" activities and procedures.

Activity 4: Confirm or Deny Resolution
- **Step 8**. Continue monitoring and remediation activities until the situation is resolved. At any time where remediation activities fail to progress or achieve the needed results, and alternate activities are not a planned part of the remediation, immediately begin "permission to operate" activities and procedures.

At this point, the problem is resolved or the plant activities are in a safe place under established "permission to operate" activities and procedures. No other state is permitted.

10.3 Lessons from Air France Flight 447

In important respects, the tragic crash of Air France Flight 447, an Airbus A330-203, into the South Atlantic Ocean provides a rich experience in the value, *actually the necessity*, of proper operations management in the face of uncertainty—and the intimately related damaging effects of confusion.[1] It is precisely this that situation management is designed to detect and advise. To understand this event, we will use what has been pieced together from the various records of the flight and expert conclusions.

Background

On June 1, 2009, Flight 447 was at cruising altitude after a routine scheduled departure from Rio de Janeiro on its way to Paris. It crashed into the sea some 3.5 hours later with a loss of life of all 228 passengers and crew. The initiating event was inoperable air speed measurements caused by excessive icing of the Pitot tube measurement elements located on the outside skin of the fuselage. The proximate cause was aerodynamic stall caused by a too low airspeed. The too low airspeed was the direct result of pilot action. Predictably, the stall resulted in a complete loss of ability to fly. Although the problem of excessive icing of the Pitot tube airspeed sensors was well documented, Air France failed to upgrade the tubes on all fleet aircraft. This was compounded by the lack of proper training of Air France pilots in Airbus aircraft handling under full manual control. Training was deficient to the point that most current pilots would fail.

We examine the timeline of events leading up to the time of the crash. During an uneventful portion of Flight 447, the captain woke the second pilot for duty and left the cockpit for a routine nap. Later, within a few minutes after the airplane entered turbulent weather, the autopilot and auto-thrust systems disengaged. The airplane had flown into a thunderstorm with significant turbulence. It was one of several storms in the area, so it may have been difficult to avoid without extensive flight path deviation, although most pilots would choose flight path changes to avoid the storms. There was no indication that the pilot of Flight 447 made any effort to avoid this particular storm. The airspeed indicator increased sharply (for reasons unknown at the time, but later learned to be due to the failure of the airspeed sensors). The stall alarm sounded (an antithetical event; had airspeed actually increased, which it did not, the pilot neglected to process the true cause of this alarm). This led the pilot to erroneously assume that the aircraft was flying at too high a speed. The crew made consistent changes in flight surfaces designed to reduce airspeed. The actual airspeed eventually fell below stall, and the aircraft lost all ability to maintain flight. It has been estimated

---

1 "Air France Flight 447," Wikipedia, last modified June 20, 2018, https://en.wikipedia.org/wiki/Air_France_Flight_447.

that its altitude at that time was below 1000 feet above sea level. By this time the sleeping pilot had returned to the cockpit, recognized what was wrong, and attempted to correct it. His efforts were too late. The airplane crashed into the sea.

History of Failures

> Starting in May 2008 Air France experienced incidents involving a loss of airspeed data in flight […] in cruise phase on A340s and A330s. These incidents were analyzed with Airbus as resulting from Pitot probe icing for a few minutes, after which the phenomenon disappeared.
>
> A later report from the BEA [Bureau of Enquiry and Analysis for Civil Aviation Safety; France], released on 29 July 2011, indicated that the pilots had not been trained to fly the aircraft "in manual mode or to promptly recognize and respond to a speed-sensor malfunction at high altitude" nor was this a standard training requirement at the time of the accident.
>
> The report [29 July 2011 BEA report] indicated that the copilots had never been trained to fly the aircraft at high altitude in manual mode with unreliable airspeed indication.[2]

Seventeen days after the crash of Flight 447, all Pitot tubes on all Air France A330s had been replaced with improved ones.

Lessons

No single failure contributed to the fatal results. The sum total of contributing failures is sobering.

- **Failure to assess** – There was no evidence on the voice recorder or through flight data information that the flight deck crew either discussed alternate approaches to the problem or attempted alternate commands. They consistently failed to examine whether the airplane was actually flying at or near a proper speed. Moreover, they consistently placed the flight control surfaces in an improper position for the clearly indicated engine thrust settings. It would have been evident (but was totally unobserved) that no airplane could maintain flight at such a low engine power.

- **Failure to collaborate** – There were three qualified pilots on board. At most times two were in the cockpit. Voice recorder records exhibit no evidence of any discussion of the potential problems and alternative reactions.

- **Failure to evoke clear command and control** – There was no evidence of a clear command-and-control structure by which the pilot was communicating to the copilot observations and conclusions. It was not until the last minute of flight

---

2 "Air France Flight 447," Wikipedia.

that the captain returned to the cabin and realized what was about to happen, but it was too late.[3]

- **Failure to invoke fail-safe operations** – The complete loss of flight speed indication is traditionally managed by placing the aircraft engines and flight surfaces in a safe setting. At that setting, it is almost impossible for the aircraft to fail to maintain altitude. The altimeter was fully functional at all times.

Any further analysis would be mostly speculation. However, it does not take a stretch of the imagination to see the effects of *confirmation bias* and *continuation bias* at work. Reliance on competency training (flying an aircraft with safe settings and using the functioning altimeter and attitude indicator) would have made a critical difference. There was no evidence of even rudimentary exercise of crew resource management. No member of the cockpit staff ever raised a concern about too low a thrust or altitude. Collaboration (early recalling the captain to the cockpit) would have made the difference. It would seem that some of the same tools useful for *situation assessment* need to be utilized during situation management—a successful assessment is determined only by outcome. This section has discussed activities that did not take place in the cockpit (control room) that should have. Later in this chapter we move on to understanding and managing who should be in the control room and the management of distraction to facilitate focus and concentration.

10.4 Operations Safety Nets

There are "safety nets" that should be in place to back up the operators' best efforts. These are introduced by Figure 10-2. Each check mark (√) is a requirement for a part of the net. Your enterprise should take specific steps to make sure they are properly in place. Each x-mark (X) recognizes a problem area that your enterprise should take specific steps to prevent or eliminate.

To have a necessary level of situation management, the operator must be able to manage all abnormal events. In order to manage these events, the operator needs the appropriate tools and the enterprise infrastructure must have competent design. The tools can only be used if the operator is able to observe when they need to be. Only a trained and capable operator can consistently observe, process, and manage. Each of these steps has critical success factors. Each of these steps has problems to be avoided. Success comes from meeting the critical success factors and properly avoiding anything that gets in the way.

---

3 Peter Allen, "Final Words of Air France Passenger Jet Emerge: 'What's Happening?'," *The Telegraph*, October 13, 2011, https://www.telegraph.co.uk/news/worldnews/europe/france/8825264/Final-words-of-Air-France-passenger-jet-emerge-whats-happening.html.

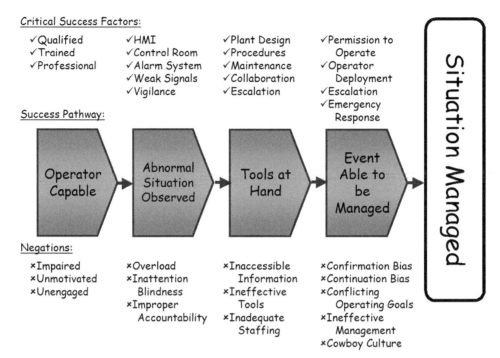

Figure 10-2. Situation management success pathway illustrating both enablers and detractors for everything working together. We need to enhance the success factors and work hard to eliminate the things getting in the way (negations).

Consider that things have been done right. Let us summarize the situation. We have placed a very high importance on making sure that the operator knows about anything that might be going wrong. We have designed and put into place useful and powerful tools and processes. The control room has been carefully designed. The operator interface was built with the needed tools (e.g., graphics design, symbols and relationships, and viewability). At this point the operator is aware. He is examining the process. He will observe the weak signals. He has situation awareness. We have provided the important technology to understand the value of that awareness and make the decision whether or not to further examine any concerns raised. The operator processes all alarms. He handles all strong signals. He weighs the weak signals to decide their type; he observes the process abnormalities. He classifies the weak signals. He seeks confirmation (backward-projection). He makes the decision either to wait or to take action based on that result. He has situation assessment.

If action or additional investigation were the plan, we have provided tools to do that competently without undue risk. The needed decision support tools are in place for the procedures, to collaborate on action, and to gain needed additional aid to execute. He is presumed to be working on a manageable problem. With that plan in operation, the need becomes ensuring that his actions are working. Here is where the important "safety net" is cast. Here is how we design and deploy that net. This is the outcome management part of situation management.

Whenever proper situation management is possible, it will take place so long as the problem at hand is manageable. Effective management will result in the situation eventually returning to some form of normal so the operator can move on to take up another task. When the situation does not respond to being managed, it effectively forces control out of the operations environment into disaster management. In some cases, the control room personnel are involved and have specific responsibilities for that management. In other cases, management passes on those responsibilities to others more skilled in them. The control room provides support only at their direction. The plant is not expected to return to normal operation without remediation work.

We now visit additional tools in the operations kit the operator can use to keep process operations safe and productive without exposing them to undue risk. Many of the tools involve operators and supervisors as a team. The observant reader will note that some material for this chapter was previously introduced in the author's earlier book, *Alarm Management for Process Control*.[4] It remains valuable and will be reviewed briefly in this chapter as an introduction and for completeness. Later, take a look at that book for just about everything you will need to deliver effective alarm technology to your plant.

10.5 Using Experts and Benefiting from Experience

Available experts and reliable experience are to be sought after. We depend on experience to recognize previous problems and recall what to do and how to properly manage. We rely on experts to be able to use their broad knowledge and sharp skills to understand problems that might be different from what we expect or go beyond available experience. Expertise and experience are sometimes found in the same individuals but not always. This is especially true when individuals are on duty in very small groups. The likelihood of operators being both experts in the enterprise technology and experienced in the broad range of operations is important but small. This is not an enterprise deficiency. It is how enterprises and operations are designed and operated.

How experience is used and expertise is relied on must be part of the operational protocol. Like most other things, if left to happenstance, limiting or even dangerous wrinkles develop that get in the way of proper operation.

---

4 Douglas H. Rothenberg, *Alarm Management for Process Control—A Best Practice Guide for Design, Implementation, and Use of Industrial Alarm Systems*, 2nd ed. (New York: Momentum Press, 2018).

Experience

Experience is having enough familiarity with something to be able to call on it with the expectation of it helping with something related going on now. "In my experience" usually means that the person sees something happening, feels that it is familiar, and likely knows how to deal with it. This is all well and good, but this experience is not reliable enough to count on without careful controls. To be sure, this experience goes beyond the casual. We are talking about experience that is going to be used to handle an important abnormal operating condition. Here is a recommended process for being able to rely on experience.

1. Take enough time to identify what "experience" is going to be used (fully frame all parts of it).
2. Take enough time to identify what is currently going on (fully frame all parts of it).
3. Specifically find out everything in the experience that does not fit well with the current situation.
   a. Concentrate on the differences, not the similarities
4. Specifically review each "does not fit" difference identified; any nontrivial one is grounds for dismissing experience and approaching the problem as a new one.

These steps are important safeguards against jumping to conclusions and using a previous approach to work the current closely related but different problem.

Expertise

Having the big picture and possessing the depth of expertise to understand the foundations and operations are essential for a safe and productive enterprise. This holds for the conceptual stage, the design stage, and the training and operational stages. Nowhere is this more important than in managing difficult abnormal situations. The deference to expertise is one of the key success factors of situation management. However, this deference must always be balanced with the good understanding and acceptance of others on the team. One of the ways to do that is to always insist on following crew resource management. Then the knowledge of each team member is factored into the overall situation.

10.6 Safe Conversations

Safe conversations are at the heart of how communication is understood, accepted, and managed by individuals. The starting point is that safety is the underlying requirement for all industrial control room operations. We all want (and surely need) to avoid incidents and accidents that can expose people to harm, the community to environmental insults and damage, and the public to undue disruption of their daily activities. The adage "If it cannot be done safely, then it should not be done at all" is the golden responsibility. But words and intentions do not make policy. And policy does not make safety. People make policy. People's concerted efforts can make the difference for safer operations. Here we explore what it takes to ensure that what is said is what is intended, and what is said is what is heard.

First, a Word about Safety

When safety is everyone's responsibility, it often ends up being no one's. Sure, there are rules. There are guidelines. There are specific safety and safe operation requirements as part of many (but not most, and certainly not all) procedures. They do not naturally fit together tightly. The cracks appear in two places: (1) management does a weak job of designing the requirements, and (2) operators are unable or unwilling to consistently follow them.

1. Management's roles
    a. Set the frame, policy, and rules.
    b. Lead by following the rules (including following them personally).
    c. Audit and amend (frame, policy, and rules).
2. Operator's roles
    a. Fully understand the frame, policy, and rules.
    b. Follow the rules.
    c. Get clarification in case of *any* doubt as to applicability or requirements.
    d. Provide timely upward feedback on all weak, uncertain, or difficult situations or events.
    e. Stop and go safe when you are not sure (or things are not going the way they should and you are not sure why).

If all of this is beginning to sound like you have heard it before, what comes next should help.

Safe Communication

All conversations in the control room have a safety component. Sometimes the issue of safety is right up there—something is going wrong. More often, nothing about safety seems involved. In order to keep this simple, approach all conversations with a bit of structure. How strong might the coffee be or when the weather is expected to shift a bit are casual, so do not get too carried away. The purpose is that both parties must know what the conversation is about, why the conversation is happening, and what each party expects of the other.

Here are the guiding factors:

- Explicitly state what is known about the situation and how it is known (use the knowledge fork).

    This confirms what is known for both parties and anything that must be actively observed and challenged where there is any doubt or discrepancy. Each point must be "read back to the other party" in much the same way that communications are confirmed in voice communication (e.g., between a pilot and an air traffic controller). Each party must ask if there is anything else that must be known that has not been mentioned.

    The text in quotes below each bullet are illustrative narratives as examples.

    Known

    "The outlet flow has been increasing for the past hour. Here are the trends."

    Or:

    "I've received no deliveries of additives this shift and the last delivery was a week ago. Here is the delivery log for the past 30 days."

    Or:

    "All of my logged production values but one are within normal operating ranges. Here is the shift handover report; let's view the current values on the screen."

- Explicitly state what is of concern.

    This sets the reason for the conversation, but not the outcome.

    Concern

    "I don't ever remember the outlet flow dropping off like this for no reason."

    Or:

    "We seem to be running a bit low in the tanks."

    Or:

    "I can find nothing to explain that this is okay or that I might have an unseen problem that must be fixed."

- Explicitly state what is expected of each party.

   This sets the working relationship between the parties. Is confirmation expected or another opinion needed? Is actual operational assistance being requested, including escalation, hands-on assistance, or a handoff?

   Expected

   "Please find the lead pipe fitter and ask him to come here immediately to look for leaks. If you cannot find him right away, can you do it yourself?"

   Or:

   "Could you check on the coordinated delivery board in the field office for the current delivery status when you go on break?"

   Or:

   "What am I missing that could explain what we see here?"

Mindful Conversations

Conversation is one of the most important processes that go on in the control room. There are conversations between board and field operators. There are conversations between one board operator and another board operator, with supervisors, during shift handovers, and more. Mindful conversation is an important culture to create and practice in the control room. According to Rosa Antonia Carrillo:

> While effective safety management systems are part of every successful organization, the most successful ones also pay attention to creating superior communication and collaboration that support the free-flow of information. Absent a trusting relationship among team members and between different teams, there is little chance that information will be shared.

> Mindful conversations are a powerful tool for incident prevention, but they require a productive safety culture to take hold and a strong leader/sponsor. A socially reinforced approach using conversation and relationships cannot succeed within a managerial climate of [inflexible] command and control. People will not freely contribute their ideas and observations if they fear retribution or do not feel valued. Thus, the leader as creator of conversations is an important role in creating a climate for open communication.[5]

It might be helpful to review a checklist for communicating for collaboration and coordination developed by Michael Roberto.[6] This five-step activity is designed to share

---

5 Rosa Antonia Carillo and Neil Samuels, "Safety Conversations—Catching Drift and Weak Signals," in *Professional Safety* (Park Ridge, IL: American Society of Safety Engineers, 2015), 22–32.
6 Michael A. Roberto, *The Art of Critical Decision Making* (Chantilly, VA: The Great Courses, 2009).

a starting point of information with others. The sharing would be for collaboration, delegation, or escalation. Here are the steps:[7]

1. Here is what I think we face.

2. Here is what I think we should do.

3. Here is why.

4. Here is what we should keep our eye on.

5. Now, talk to me.

   Ask the following focus questions below to round out step 5.

   a. What have I missed?

   b. What do you know that I don't know?

   c. Where do we go from here?

An important special-case tool for enabling mindful conversations during emergencies in situations of strong command and control requirements (e.g., an aircraft cockpit) is discussed later in Section 10.14 under the topic "Crew Resource Management." All control room conversations should be mindful. Sure, there is room for the informal conversations about things such as fishing tackle, recommended new tires, or tonight's date, but it goes without saying that they have no bearing on operations. No one should ever be able to confuse informal conversations with mindful ones. Mindful ones have a specific time and a place.

Mirroring, Acknowledging, and Tracking

Let us add an important communication tool before we move on. It involves the three aspects of mirroring, acknowledging, and tracking conversations. These activities are used to ensure that "what is said" is "what is heard." They also build in any essential status follow-up so that the intent of the communicated event is tracked. These activities actively engage the speaker and the listener. You saw a part of this in action in Chapter 3, "Operators," where the arriving operator declared to the leaving operator, "I am now operating this unit," to confirm the fact. What the leaving operator also said (but was not included in Chapter 3) was the acknowledgment of "You are now operating this unit." This form of communication is integral to the command

---

7 Roberto, *Art of Critical Decision Making*.

communications structure of the US Navy; all other branches use the functionality in a less formal way.

The following discussion covers specific behaviors for ensuring the effectiveness of conversations. Each plant or enterprise should clarify their practices, which includes what to do and when to do it. Best practices for effectiveness should be an integral part of the culture.

Mirroring
Mirroring is the specific activity where the recipient (the intended listener) communicates back to the sender (who issued the communication) exactly what he thought he heard. It is a repetition action. But there is an essential ingredient that is part of why it is so effective. Before the recipient can repeat back what he heard, he "translates" it into his own thoughts. This is to say that we hear what we hear because we process it subconsciously through our own personal experiences and biases. Only then can we reply back with what we think we understand. By repeating it back to the sender, the sender gets to hear it and make sure it is what he intended the recipient to hear in the first place. Remember the "in hot water" of the game of *telephone* in Chapter 3? Anything that gets "lost in the communication" can be repaired in the moment when it is most needed.

Acknowledging
An acknowledgment is a specific communication back to the sender that the recipient will perform the requested action (by restating which action is to be performed). Mirroring guarantees that both the sender and the recipient have the same message content. This next step is to ensure that the recipient takes ownership of the intent of the message. If the message was an order or instruction (hereafter termed *instruction*), the recipient informs the sender that the instruction will be followed. This assures the sender that the recipient will perform the requested action and that the action is what he intended. The acknowledgment does not need to be lengthy or couched in any sort of legalistic or formal terms. Just to be sure we understand this part, mirroring alone would not imply acknowledgment of any kind beyond hearing the message.

Tracking
Tracking ensures that the progress or results of actions requested by the sender are communicated back to him in a timely manner. Not all actions require tracking. For those that do, tracking will either be requested at the time or done so according to the requirement of an established protocol.

An Example

Here is an illustrative example of a hypothetical conversation involving the three aspects of mirroring, acknowledging, and tracking. The left-hand column shows notations of the type of activity. The right-hand column gives the communicator and the message content.

Activity	Communicator and Message
	OPERATOR: Pump RT-4157 seems to be getting warmer than I expected it would. Can you check it out? I filled out a work order.
Mirror	TECHNICIAN: You are concerned that Pump RT-4157 seems to be getting too warm?
Mirror	OPERATOR: Yes.
Acknowledge	TECHNICIAN: I am going over to the rundown tank to free up a level float. As soon as I am done, I will have a look at your pump.
Track	TECHNICIAN: It looks like the pump is running OK. But I noticed that the temperature RTD (a type of temperature-sensing element) had been removed to keep it from breaking when the pipe fitters went in to check the flange nuts of the bypass line. It was sitting on the steam line, so that is why the reading was not right. I repositioned it and the local indicator seems to read properly.
Mirror	OPERATOR: I see now that the temperature indicator is fully functional. My readings look reasonable.
Acknowledge	TECHNICIAN: I am done. I will close out the work order for this job.
Acknowledge	OPERATOR: Yes, the work is complete.

10.7 Operational Drift

Almost everything changes. We understand that small changes are a part of everyday life. A Reuben sandwich somehow seems to taste a bit different each time we order one. But it is a favorite, and you allow very little to get in the way of eating it every time you lunch at the deli around the corner. All in all, small changes are understood as the stuff of real life. We usually do not know they are there, or if we do, they quickly fall under the radar. Yet, small changes in the control room can be a big problem if they lead to important things we do not want to miss. A few of the sticky problem areas are the following:

- Short-term progress seems more useful than long-term goals
- Conflict between making production (and profit) versus operating completely safely
- Number and quality of the messages about operational activities and safety

- Forgiving small irregularities in favor of waiting until they became strong enough to call for attention

An illustration about a frog and a stove bears telling. As you read this piece, please remember that it is an illustration, not a prescription for injury to small animals.

> Take a frog and place it in a pan of cool water. Then gradually heat the water. If it is done very slowly, by the time the frog actually senses the danger of the situation, the water is already too warm for it to be able to jump out. On the other hand, take that same frog and place it into that very same pan, now with very hot water, and the frog will immediately jump out and thereby save itself.

So, drop the image; keep the lesson. The lesson is that there are ways to reduce small drifts from leading to larger issues. We want to recognize the difference between small variations around something basically constant and small variations that progress. The former is just noise. The latter is operational drift. Operational drift works like this: Something breaks a little bit. It does not get fixed right away. A bit more stops working. It does not get repaired. We stop depending on it to work properly and start to "baby" it and work around its growing list of problems. All the while, production might slip a bit. Lab reports on quality suggest it is moving closer to a limit. But we keep going. Pretty soon the sales team understands that they probably should not remind the buyer about quality as much as before. We have drift.

Drift also comes from procedures being field corrected because they were hard to follow or seemed to miss parts some of the time. Down the road a bit, operators find other small ways that seem to work better—a change here, a skipped step there, all seemingly to help get the job done. Until something happens. The initial procedure was designed with that problem in mind. The drifted procedure now is not strong enough to manage that problem.

Operational drift always results from unfixed small problems. Those problems can come from anywhere. No matter where they come from, if not fixed, we will drift. Effective organizations manage operational drift by doing the following:

- Ensure effective communications (say what you mean, and mean what you say).
- Recognize all small problems and fix them. For example:
  - Small process problems get explicit engineering or operational attention.
  - Weak procedures get revised and management of changes (MOCs) done.
  - All broken equipment gets fixed right away.

- Build procedures and protocols that are clear and form an understandable and rational structure, not just a basket full of stuff.

- Conduct periodic reviews specifically designed to spot drift and fix it.

Tenets of Operation

Chevron Oil (that is a legacy name) experienced an operational mishap that should not have happened. We skip the details here, but what Chevron did to make sure it did not happen again is important to know. What the enterprise quickly found out during a root cause investigation was that although it had all the rules in place to prevent the situation it experienced, there were too many of them. It was too hard to know and follow all the rules. Little by little, Chevron slipped a bit here and a bit there. Eventually, it became so used to slipping that it stopped watching out for the problem. Then something awful and expensive happened. The way out, and this is where we benefit by buying this lesson wholesale (instead of paying full price, retail), was for Chevron to set a few very high-level rules that allowed no drift. They called the rules "Tenets of Operation."[8] Others refer to similar ones as "rules to live by." No slippage is allowed. Adhering to them provides a reliable safety net for the big things. Of course, all the rest of the rules are still in place. Slippage is not encouraged. Where slippage does manage to enter, it will not be so costly.

Here are the basic Tenets of Operation that Chevron developed. The first three are key principles and are the foundation for everything.

- Do it safely or not at all.

- There is always time to do it right.

- If you don't know what's going on, stop.

The next ones are still basic and high level but add more substance. The tenets address a wide range of behaviors. The key word in the tenets is *always*.[9]

- Always operate within design and environmental limits.

- Always operate in a safe and controlled condition.

- Always ensure safety devices are in place and functioning.

---

8 "Tenets of Operation," Chevron Corporation, accessed March 23, 2013, http://www.chevron.com/about/operationalexcellence/tenetsofoperation/.
9 Chevron, "Tenets of Operation."

- Always follow safe work practices and procedures.
- Always meet or exceed customers' requirements.
- Always maintain integrity of dedicated systems.
- Always comply with all applicable rules and regulations.
- Always address abnormal conditions.
- Always follow written procedures for high-risk or unusual situations.
- Always involve the right people in decisions that affect procedures and equipment.

Consider these for your enterprise. Use them so that you might avoid falling prey to unintended operational protocol drift. The most important contribution of Chevron's work led to the concept of *permission to operate*. It is discussed in detail later in this chapter (Section 10.15).

10.8 The Restricted Control Room

At times during operating situations it can be extremely important to manage the behavior of people in the control room. Unnecessary distractions are eliminated. Each individual is placed and ready to handle the abnormal situation. Needed resources are at hand. The restricted control room and the even more managed sterile control room (discussed in the next section) are special constructs for these situations.

Control Rooms in Normal Operations

The control room or operation center is not a social center in any way. A control room is a purpose-designed, purpose-built, and purpose-used facility. Yet, your control room can be a very busy place. During normal operations there is a lot going on. At other times operating conditions are moving so slowly that nothing seems to be happening for hours at a time. It all depends. Operators are doing their thing: watching the board, talking on the phone, consulting other operators, and more. If more than one operator is sited there, each operator will be managing his assigned area; or they may be jointly operating the same area. Each has his own schedule to follow and own way of doing things. An operator will take brief breaks, sometimes asking another operator to "keep an eye out on his board." He will take meals, so food may be delivered or in process of preparation.

Supervisors also attend to a myriad of duties and tasks. Visitors may be present. Other enterprise personnel may be observing, consulting, or just present for a brief check-in. There may be background music playing. Relief operators may be arriving or departing. Personnel may be engaged in replacing materials that were used during a safety or operational drill. Authorized personnel will visit as part of their duties and responsibilities: field operators checking in, maintenance personnel coordinating, materials being refreshed or removed, and all manner of other direct support activities going on. All of this happens with the understanding that any distractions to the interacting operator and other uninvolved operators are minimal. But there are situations where even these normal distractions can interfere. At times they can interfere in a big way. This will need to be carefully managed.

It is important to note that no activity should be permitted to take place in the control room for which the control room has not been explicitly designed and tasked. It is also important that no individual be in the control room that the control room operations protocol has not explicitly permitted to be there. Of course, this is often situationally dependent. This is a given. The enforcement of these practices should be the responsibility of all those present, including supervision (who might not necessarily be present at the time).

The Control Room in Abnormal Operating Conditions

In a well-managed control room, most activity would be acceptable during normal operations. However, should any part of the control room be engaged in significant nonroutine operations of any kind, it is important to alter the entire "background" to ensure that those charged with management of the operation are able to devote full attention to that task. There are two specific protocols designed to manage unnecessary distraction in the control room. The first is the *restricted control room* discussed here. The second is the *sterile control room* discussed in the following section. The restricted control room is designed to shift the internal functioning of the control room into a situationally supportive environment to respond to a significant operational situation. This mode is designed to support operational activities that are preparatory or reactive to some significant condition or situation. Examples include a major weather event coming or arrived, unusual stresses in operation that might be caused by significant curtailment or increase in product demands, and a state of emergency due to security issues. During these situations, the enterprise functions according to preestablished plans and protocols. The control room environment is positioned to support that activity.

Please keep carefully in mind that the configuration of a restricted control room is only that. It does not specify duties or procedures for managing the plant or enterprise. Those are to be found in the specific instructions for them. Those specifics are designed by each plant or enterprise based on their requirements and operating practices.

Restricted Control Room Conditions

A restricted control room condition might require the following:

- All visitors are out (following accepted security protocols, of course).
- All scheduled personnel are in attendance; if assigned supervision personnel are not available, predetermined backup individuals must step in.
- All essential personnel are at their designed station(s).
- All operations are conducted with heightened attention.
- All nonessential activity in the field, including maintenance, construction, and survey, is expedited for early completion or delayed (always in a safe manner).
- Where access to special or detailed documentation and other information that is not normally readily available is needed, a specific individual will open access and prepare it for use.
- Where appropriate, issue a declaration of a control room situation code (see section 10.21).

Initiators of a Restricted Control Room Condition

This is a partial list of initiator situations. When any one exists, a transition to a restricted control room status is recommended. The specific initiator would be set by the enterprise.

- Severe weather arrival within the current shift or already arrived
- Significant change in operational modality requiring more than "due care"
- Medical emergency in the control room
- Medical emergency in the controlled part of the enterprise (outside the control room) for which control room operators must accommodate operation and/or protect personnel and/or facilitate rescue or recovery
- Loss of the minimum reasonable number of operational personnel needed in the control room
- Declaration by any authorized individual of the need to be a restricted control room

Termination of a Restricted Control Room Condition

Ending the restricted control room condition is by determination of the authorized individual declaring the condition, or by consensus of the attending operators. All suspended activities are permitted, but the necessary protocols for them will be followed.

10.9 The Sterile Control Room

There are situations where even a restricted control room is insufficient to position personnel and restrict unnecessary activity. This situation can benefit by being declared a *sterile control room*. The configuration of a sterile control room is only that. It does not specify duties or procedures for managing the plant or enterprise. Those are always governed by the protocols, procedures, and knowledge and skill of the operations team. Being able to do their work in a special control room configuration amplifies their ability to concentrate and bring to bear the needed resources. The state of being in *sterile* conditions is more restrictive than *restricted* conditions. It is designed to move the control room into an operational state that simultaneously eliminates all nonessential distractions and ensures essential participants are present.

Sterile Conditions

The requirements of a sterile control room are in addition to restricted control room requirements. They are as follows:

- All conditions for a restricted control room are activated.
- All background activities are stopped: food is placed on temporary hold; music is turned off; nonessential telephone calls and radio calls are ended and all future ones screened.
- All activity in the field that is not vital, including maintenance, construction, and survey, is terminated (rendering them safe). Unless maintenance personnel are required to ensure the safety of the holding of field activities, all are removed from the operating area. They are to remain at the ready in case of need.
- Any operators in training or newly assigned must have their backup take over their activities.
- Operators whose duties are split between the control room and the field must have both the control room and field manned by separate qualified individuals.
- Operator relief due to shift change is delayed pending a staffing determination. This is not a time for "make do."
- All nonessential movement within the control room is ended.
- All nonessential conversation is ended.
- All essential conversation is conducted in a muted fashion designed for clear understanding and minimal distraction to others. The recipient confirms each exchange via voice-back repeat.

Initiators of a Sterile Control Room Condition

This is a partial list of initiator situations in the control room. When any one of them exists, a transition to a sterile state is recommended. The actual initiator situations would be set by the enterprise.

- Any operation (planned or unplanned) that has a potential to risk life or disability, serious economic loss, significant environmental exposure, or other significant risk to the enterprise or surrounding community that enhances the risk above that expected in normal operation
- Activation of any single highest priority alarm
- Activation of multiple second-highest-priority alarms in different process areas
- Partial loss of operational control (power outage, substantial communications loss, severe weather, etc.)
- Security (cyber or physical) attack or breach
- Declaration by any authorized individual of the need to be a sterile control room

Termination of a Sterile Control Room Condition

It is important that the ending of any specific situation or condition that caused the control room to initiate sterile operation in the first place is not sufficient to allow the control room to end sterile operating conditions without further care and attention. Thus, care must be taken to ensure that all relevant aspects of operation within the control room and the enterprise are sufficiently understood and determined to be within reasonable or normal ranges. Unless there is confirmed reason to believe that all initiators of a restricted control room are absent, a return to a restricted control room would be necessary. Supervisory personnel would normally determine this.

10.10 Lessons of Defensive Operating

Recall the five defensive operating principles from Chapter 1, "Getting Started":

1. **Aim high** – Maintain an overall perspective for what is going on.
2. **Get the big picture** – Keep alert to the other aspects surrounding what is going on.
3. **Keep your eye moving** – Make sure that all of the small details are accounted for.

4. **Leave yourself an out** – Always have a backup plan.

5. **Make sure they see you** – Ensure the rest of the operation, including upstream, downstream, customers, and suppliers, knows what you are doing and any issues/problems you are encountering.

Added to the list above, the specific supporting principles below are followed:

- Use extra caution when making operational changes.
- Leave some space.
- Pay attention to current surroundings/situations.
- Trust but verify.
- Always yield.

Following these is a good way to reduce operating risks. Whenever the ability to maintain operation increases the risk of operating properly, managing risk must always come out on top.

10.11 Managing Everyday Situations

We now explore how to progress the shift or duty cycle to balance the ever-present job of "looking for trouble" with everything else, including brief downtime. We will lay out what and how operator resources are used (and balanced) to manage production and other activities to remain abreast of what is going on and be able to fix what is amiss.

Fictional Illustration

Imagine that you have been tasked to determine whether or not a certain television network program might be using material that may not be sufficiently original. The nature of the originality is an important aspect in an ongoing ethics investigation. You were a writer for a similar type of program that had aired for several years, but is now off the air. You were hired for the present work because of your background and experience. You have worked with the writers in question on an earlier project, but it has been a while. The program series in question aired for 2 hours each Saturday morning. Ten weeks of programs are to be examined. Interestingly enough, one of the stipulations for your work is that you are allowed to view each program only once, without interruption or pause of any kind. You will be observed to make sure. Your responsibility is to identify specific potential unoriginal material but to offer no proof. Your list will be given to a team of archive experts who have wide access to programming to conduct the actual investigations. They are extraordinarily expert at finding specific material but have no ability to identify what to look for. You are providing the "what to look for" part.

You accepted the job and are now working out a plan of attack. On the surface, you are not really looking forward to 20 hours of staring at television recordings. You need the money but wonder how in the world you are going to stay focused enough to do this? One way would be to sit down, watch the recordings, and jot down notes. You tried doing this using recorded tapes from another show series. The first hour, you found nearly a dozen things that bothered you. After the third hour, you found yourself not paying enough attention. In the end, you decided that straightforward viewing was going to provide way too many "suspect" items or you were going to miss the few that might have been there. Clearly, you need to find a better way.

You have decided to look for two situations:

Situation 1: Material that is familiar enough for you to be able to categorize as likely not being original.

Situation 2: Some material catches your attention as possibly not original, but your feeling is vague. It just does not seem to jibe with your sense of what those writers would produce.

When you tested this approach, you discovered that it was working quite well. The Situation 1 items "jumped out." And the Situation 2 items were few but clearly nagging. Months later, you were advised that one Situation 1 item was verified and five Situation 2 items were verified. Interestingly enough, it was not the Situation 1 item, but the totality of the Situation 2 items that was sufficiently persuasive to a jury to turn the case to decide the work was stolen.

You know that the story above is fiction. And you know just finding potential problems is only a part of the job. But looking for subtle stuff in the midst of a lot of other stuff for long periods of time is difficult. We ask operators to do it on every thankfully, uneventful shift. We ask them to watch over the plant (or operation) to find and identify anything going wrong that might spell trouble. What if they adopted a similar plan to our video expert's? If they did, Situation 1 items would be alarms. An alarm is clear, it is not going to be missed, and what to do has already been adequately identified and planned. That leaves all the rest. So what if the operator looks for Situation 2 items? They would be weak signals.

Let us work out ideas for doing this during the course of normal shifts.

Operator Intervention Caution

First, let us visit the foundation for having an operator in the control room. The operator is there to keep an eye on everything and, if needed, step in and adjust or correct what might be going amiss. Notwithstanding all the training, proficiency, motivation, and need for taking action, the situation requiring operator action is abnormal! "Operator intervention always opens plant disturbances to potentially dangerous events."[10] This is why there is the need for clear operating instructions and procedures. This concern reinforces the need for a sufficient level of detail for alarm response sheets. It is the basis for competency training as opposed to skills training.

Operators Must Not Be Innovators

Innovation ordinarily is a high-value skill. We readily recognize the importance of new and better solutions to existing problems and situations. Every day there are new products out there, better services available, and more interesting ways to enjoy leisure activities. Sure, the "tried and true" always has value. It forms a basis for what makes us comfortable. Yet it is the new and novel that often captures our interest and our hearts. So what part should operator innovation play in the control room? The answer is none.

Innovation is a change in the existing way of doing things. It challenges our mind's eye to explore doing or thinking about things in new, often untried, ways. By not permitting operator innovation, we prevent an operator from going "outside the box" to look for reasons or explanations or work-arounds. You do not want operators attempting to resolve problems or concerns by looking for something novel or attempting to work them out on their own. Abnormal situations and operational problems can surely be difficult to see and challenging to manage. However, the search for a proper resolution cannot be "trial and error." There are three basic and governing reasons for this principle: (1) the time to see, understand, decide, and act on a problem is very limited; (2) safe and reliable remediation activities need to understand the totality of operation and take into proper balance all aspects in a preplanned approach; and (3) operators and other coordinating players need to be functioning together and have practiced the activity enough to properly master it.

There is little room for optional approaches. The stakes are too high to afford mistakes. This truth forms the backbone of responsible industrial process management. This requires the availability to the operator of sufficiently effective tools and procedures. While the discovery of cause, hopefully root cause, for any given abnormal situation can involve great skill, a cool head, and a steady hand, once discovered, remediation must follow established plans and safeguards to ensure effectiveness. Operators are not innovators. Any time an abnormal operating situation goes unmanaged and the operator has

---

10 Gregory Hale and Luis Duran, "Human Factors and the Impact on Plant Safety," *Control Engineering* (January 2014): 32–39.

exhausted the available tools, we must not permit creative impulses to take over in the hope that something useful might be found. In its place, proper escalation and collaboration (Section 10.16) or ceasing operations (Section 10.15) are the only options.

Operator Duties

Operators are tasked with ensuring the operation under their care performs as required. Requirements can be varied, of course. They generally consist of the following items:

- Orient to the current status and objectives of the operation
- Prepare reports including log entries
- Arrange for nonroutine testing and sampling of production (including product quality, environmental quality, and resource utilization)
- Evaluate results of routine testing and sampling and adjust operation appropriately
- Arrange for, monitor, and confirm maintenance activities
- Originate, authorize, and accept receipt of materials used in the operation
- Manage delivery or shipment of production operation results (produced chemicals, electrical power, railcar movements, etc.)
- Effectively monitor the totality of the operation and make appropriate adjustments
- Engage in ancillary aspects of the operation as needed (alternate pumps to even out wear, visitor visits, etc.)
- Prepare operation to conclude or transfer to relief operator

Following the Rules

Having lots of rules can be useful. Having too many can be problematic. To be useful, rules must be complete, accurate, and available when needed. Operators will use them when they are available and they have been adequately trained to use them. However, and this is a big *however*, no rule can be followed blindly. Every rule requires "thoughtful compliance."[11]

> Personnel are expected to follow rules and procedures. However, **personnel are also expected to think** about what will happen if the established rules and procedures are applied to the current situation. If they believe the risks of implementing [following]

---

11 CCPS (Center for Chemical Process Safety), *Conduct of Operations and Operational Discipline for Improving Process Safety in Industry* (New York: Wiley, 2011).

the rules and procedures are unacceptable they are expected to stop and seek advice from other knowledgeable people.[12]

The concept of thoughtful compliance is extended to include most of the situation management tools and processes in this book. All are meant to either augment the operator's ability to do his work or safeguard against losses. However, there is an important additional tool that belongs in the "kit." Each enterprise must write the rules for following established rules. This is specific guidance for how to look for, understand, and manage all conflicts between following basic rules and anything that does not appear to be doable or safe. The best way to do this is by setting out a general protocol for following the rules. These would involve deciding the following:

- How to know that a rule is available

- When the rule is to be followed

- How to manage parts of the rule that do not seem to apply or do not seem to be doable

- What to do when actions appear to be needed but are not in the rules

- How to know that following the rule should be terminated and how to do it

The actual work for meeting the needs of this activity is left to the enterprise.

Flow of Operator Activities

A traditional aspect of how a shift lays out is illustrated by Figure 10-3.

The parts of a shift include coming on shift (orientation), logging and reporting, coordination with testing and maintenance, monitoring and adjusting, and going off shift (handover). Typically, these duties account for around 30% of operator time during the shift. The rest of the time (see the red area in Figure 10-3) is loosely organized and generally not planned or structured. The operator just uses it to make sure things are okay. Because this time is so loosely organized, operators will sometimes devote too much effort to one task for a while and then leave it for longer than desirable, or the other way round. Either way, it is a source of stress and uncertainty.

General Flow of Operator Activities for Situation Management

Rather than leave this largely unstructured, and by doing so leave the operator's responsibility to discover where things might be going wrong to the ad hoc, use the

---

12 CCPS, *Conduct of Operations and Operational Discipline*.

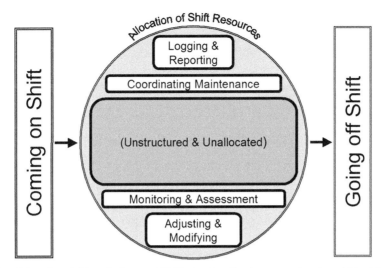

Figure 10-3. Traditional activities during a shift illustrating the large unstructured block(s) of time where the operator has to invent working processes within them. This can result in operators having their own methods or operators adopting unproven "common denominator" methods without proper enterprise involvement.

new tools. This is not about operator loading. It is about the operator's need to have tools to keep an eye on things during the bulk of the shift—to help operators "keep a finger on the pulse." Let us see how the tools of situation awareness and situation assessment are included in operator duties in a planned and effective manner. The new situation management model of how a shift lays out is depicted in Figure 10-4. The figure refers to either a 12-hour shift (to the left) or an 8-hour shift (to the right). Only one is followed, of course. For a 12-hour shift, the operator would conduct a weak signal review of the entire operating area four additional times during the shift, for a total of six. The first one would be part of the initial shift handover. Another would be an hour later (hour 1). This would be repeated for hours 4, 8, and 11 into the shift. The final review (the sixth) would be in preparation for the end-of-shift handover.

This arrangement provides a new operational practice opportunity. Using it enables a regular workflow that balances structured activities with informal aspects in a way that ensures good monitoring with time for the other support activities. The approach should enable operators to feel confident that their efforts for "keeping an eye on things" are responsible, appropriate, and adequate. Here is a good time to recall the guidance for actively looking for weak signals (see Chapter 9, "Weak Signals"). Of course, this does not mean that between these shift reviews, the operator is freed of all monitoring and management. This process is just to ensure that fewer things slips through cracks. At the same time, it replaces the stress of needing to constantly look for things that rarely occur. This benefit should not be underestimated.

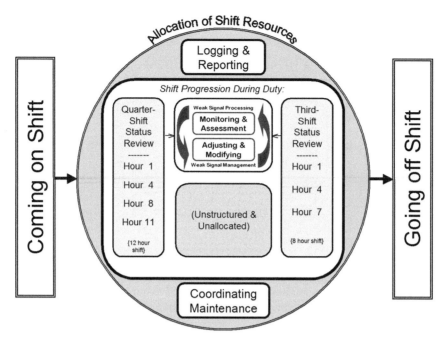

Figure 10-4. Recommended new situation management structure for the shift illustrating the clearly structured block(s) of time to manage operator responsibilities. This provides operators with approved enterprise methods for maintaining the proper balance between the various roles needed for good operation and for ensuring they keep on top of both the short-term regulatory performance and early developing problems that may impact things later.

Other shift activities include the conventional and necessary duties and responsibilities of the operator. But now, there is a regular and predictable review of the operational status using the technology of weak systems. At regular intervals, we have tasked the operator to perform what amounts to in-shift operational assessments. See "Operator Ownership Transfer at Shift Change" in Chapter 8, "Awareness and Assessment Tools," for how this might work. These tasks include the following:

- Review all alarms that have not returned to normal.

- Review all ongoing maintenance activities in progress.

- Review all key performance indicators (KPIs).

- Search for and process all weak signals on a schedule. (Both 8-hour and 12-hour shifts are shown.)

- Prepare appropriate documentation regarding this assessment.

Each item in the preceding list is explained in the following sections.

Review Alarms

Review all alarms that have not returned to normal. Work out any additional steps necessary to fully resolve that abnormal situation. Proper alarm management design ensures that the abnormal situation causing the alarm to activate will not lead to an out-of-hand situation. In this step, we take the action to understand and work out any residual issues. If the situation required maintenance follow-up, that would be attended to, and so on for other related items. Also included here would be *standing alarm* and *stale alarm* remediation. These would be done in accordance with the requirements set out in the site alarm philosophy.

Review Maintenance

This includes ensuring that appropriate work orders have been requested (or written, as the case may be), ensuring that any active work is monitored (proper permits, safe working activity, etc.), checking on suspect equipment, and returning to service activities.

Review Key Performance Indicators

There will be KPIs related to production values or flow rates, quality, costs and resource usages, and more. Those suggesting production modifications or other activities should be noted and scheduled as soon as the assessment is completed.

Weak Signal Assessment

This is a full examination for and analysis of all weak signals not already identified during the intervening portion of the shift. Refer to Sections 6.4 and 6.5 in Chapter 6, "Situation Awareness and Assessment," for the complete work process and procedures. See also Section 9.16 for a strategy of looking that fits into the plan.

Document and Annotate

Prepare appropriate log entries and add to the shift handover report being prepared for the end of the shift. It is important to include notes on procedures used that may need improvement and any other resources that did not properly and adequately assist the operations activities.

10.12 Doubts and Concerns

You would be missing an important tool in the kit if, after reading Section 7.7 on doubt in Chapter 7, "Awareness Assessment and Pitfalls," you concluded that an operator would try his best to avoid all uncertainty and doubt. Without trying to get too deep into definitions, if your operator finds himself wondering about something (in the plant or operation, of course) he should drop a "flag" on the situation (Chapter 9, "Weak Signals"). Our analogy to sports means that his concern should be investigated. It is a form of weak signal. If the operator was unable to do so due to urgent time constraints or he is not able to figure it out enough to resolve the concern, escalation would be in order.

10.13 Delegation

Control room delegation is quite a different aspect from the delegation of responsibility covered in Chapter 2, "The Enterprise." Here, the operator delegates the specific responsibility for following or remediating operational situations to another qualified individual. A delegation also might be from a supervisor to the operator or another operator or individual. The activity of delegation is an immediate and tactical one. That is, there is a problem to follow or resolve and the main responsible individual must pass it on to another. This form of delegation is used only because the responsible individual does not have enough resources (usually time, but could be specific experience or something else) to do the job himself. Delegation is a formal process. The following list is how it must work:

1. Identify the delegate.
2. State the reasons for the delegation.
   a. Identify what the delegate is being asked to do.
   b. Explain why the specific delegate has been selected.
3. State the requirements for the delegation to continue.
   a. What, if any, are the specific parameters of what the delegate is to ensure happens?
   b. Maintain control of this situation until (state specifically what would require the passing back to the delegator).
4. Require the consent of the delegate for the delegation.
5. Require all operational changes to be cleared by the delegator before implementation.
6. Formally end the delegation by delegator or delegate initiative.
   a. Transfer all information needed by the delegator from the delegate to effect proper transition.
   b. Require formal acknowledgement by the other party that delegation has ended.

Keep in mind that delegation transfers the task(s) (eyes, ears, hands, etc.) to another, but the responsibility for the conduct and outcome remain with the delegator (operator or supervisor). This should make sense because the delegated task is only part of a larger situation. Only the delegator will have the broader picture sufficient to judge what he and others might see and be doing.

10.14 Collaboration

Collaboration is a vital and powerful decision support tool. Use it effectively to encourage interactive, interpersonal conduct to bring out the best efforts of working teams while preserving individual responsibility and authority. Collaboration should be put in place by every enterprise where operational failure can expose personnel to significant injury, risk significant financial loss, or risk unacceptable environmental exposure and/or damage. Doing it right involves the entire operational team. Each player brings his part to the table. Of course, this "bringing more to the table" must be useful instead of confusing, overly time-consuming, or of little value.

All control room collaboration must be under the control of the operator. Regardless of how much knowledge, recommendations, and feedback may come from the others in the collaboration activity, only the operator has the full picture and responsibility. The bottom line is that this is about setting the frame for a common basis for understanding leading to a shared basis for action. What follows are suggestions for how this might work.

Use the Knowledge Fork

We need to bring information, opinions, and observations from several individuals into proper focus. Make sure that you use the knowledge fork to do that. Recall the figure from Chapter 8, "Awareness and Assessment Tools," repeated below as Figure 10-5.

Figure 10-5. Knowledge fork with the three "tines" describing how to qualify every bit of information available and intended to be used for problem identification and remediation (if needed). These classifications ensure that decisions are made from only known information. Anything unclear or assumed must be verified to the level of "known" to be used.

Every observation, every bit of data, and every other item that is put on the table during collaboration must be clear. Only *known* information can be relied on. The rest must be verified or ignored.

Caution

The reader is strongly advised to refrain from concluding that any material presented here is intended in any way or under any circumstance to undermine, or to suggest a lack of respect for or challenge to, your existing protocols or authority regarding traditional and accepted command and control. Such established roles and authorities are well settled and for long-standing purpose. The material here is best understood to suggest important considerations to keep in mind as each member of a team carries out his respective duties. Its purpose is to expand the conversation and understanding of what proper team coordination is and how it might be used. We begin with an illustration.

Air Florida Flight 90 Crash

This is the story of the fatal crash of Air Florida Flight 90 (see Figure 10-6)[13]. Flight 90 was a Boeing 737-200 on a regularly scheduled flight from Washington National Airport to Tampa (Florida) International Airport. Of the 74 passengers and 5 crew members, only 5 survived. Four motorists on the bridge were also killed. At exactly 1 minute and 1 second after 4 p.m. on Wednesday, January 13, 1982, during takeoff, Air Florida Flight 90 struck the 14th Street Bridge across the Potomac River in Washington, DC. The plane's takeoff weight was above the maximum limit (due to unmanaged icing). The copilot was aware that something was wrong, but the pilot would not listen. The dramatic failure of the crew to share vital information was the final link in a disastrously long chain of failures leading to the crash. The clarity of these deficits is derived from situational reconstruction of crew conversations from the cockpit voice recorder.

Cockpit Errors Immediately before Takeoff

What follows may be considered extensive detail, especially because few of us are pilots. However, the long list of failures should drive home the critical message: no single safeguard, even a lengthy list of safeguards, is of any value unless each is used to its fullest. To make the message even more forceful: the longer the list of safeguards, the less likely any user will properly work through all of them diligently enough to ensure that each has fulfilled its intended purpose. It is as if a long list of safeguards is a safeguard in and of itself. History says all too often that it is not.

---

13 "In 1982, tragedy struck the Potomac," Washington's Top News, January 13, 2017, https://wtop.com/dc/2017/01/in-1982-tragedy-struck-the-potomac/slide/9/.

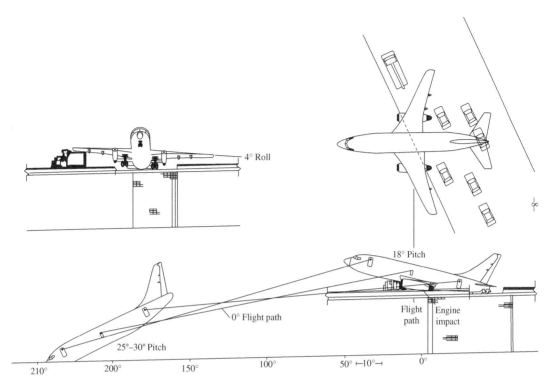

Figure 10-6. Air Florida Flight 90's 1982 crash into the Potomac River Bridge. The cause of the crash was too much ice on the plane. The accumulation was due to a series of pilot errors on the ground. The lack of knowledge of the problem was due to the pilot not listening to the copilot, who observed a problem during takeoff and attempted to notify the pilot.

The weather was well below freezing with a moderate, steady snowfall. During deicing, the ground crew failed to place the proper covers over the airspeed measuring ports and engine air intakes. The deicing compounds were incorrectly prepared and applied by two different operators to opposite sides of the aircraft. The application equipment was not calibrated to the nonstandard mixtures and had no mix monitoring, resulting in a faulty application. Ice buildup on the taxiways was extensive to the point that it prevented normal towing from the gate. The pilot (against all regulations) attempted to assist the tow by using the aircraft engines in a reverse-thrust mode. Due to various delays, Flight 90 finally reached its proper takeoff position more than three-quarters of an hour after deicing. This excessive delay rendered the earlier deicing completely ineffective. The pilot failed to order additional deicing despite the visual appearance of substantial ice buildup on the wings, including leading surfaces (the part affecting proper lift). Nor were the engines' deicing systems engaged.

Cockpit Errors during Takeoff

Analysis of the cockpit voice recorder (CVR) determined that, during the departure checklist, the copilot announced and the pilot confirmed that the plane's own engine anti-icing system was turned off. This system uses heat from the engines to prevent sensors from freezing, ensuring accurate readings [including air speed]. Adding to the plane's troubles was the pilot's decision to maneuver closely behind a DC-9 that was taxiing just ahead of their aircraft prior to takeoff, due to their mistaken belief that the warmth from the DC-9's engines would melt the snow and ice that had accumulated on Flight 90's wings. This action—which went specifically against flight manual recommendations for icing situations—actually contributed to additional icing on the 737. By sitting behind the preceding aircraft, the exhaust gases melted the snow on the wings. During takeoff, instead of falling off the plane, this slush mixture then froze on the wing's leading edges and the engine inlet nose cone.[14]

The formal cause of the crash was ruled pilot error.

> The pilots failed to switch on the engines' internal ice protection systems, used reverse thrust in a snowstorm prior to takeoff, and failed to abort the takeoff even after detecting a power problem while taxiing and visually identifying ice and snow buildup on the wings.[15]

Both the pilot and first officer were inexperienced, having a combined chronological age of 65 years. After the crash, airlines began enacting policies to ensure that at least one older, more seasoned cockpit crewmember was on board at all times.

> As the takeoff roll began, the First Officer noted several times to the Captain that the instrument panel readings he was seeing did not seem to reflect reality (he was referring to the fact that the plane did not appear to have developed as much power as it needed for takeoff, despite the instruments indicating otherwise). The captain dismissed these concerns and let the takeoff proceed. Investigators later determined that there was plenty of time and space on the runway for the Captain to have aborted the takeoff, and criticized his refusal to listen to his first officer, who was correct that the instrument panel readings were wrong. The pilot was told [by air traffic control] not to delay because another aircraft was 2.5 miles out (4 km) on final approach to the same runway.[16]

Crew Resource Management

Crew resource management (CRM, also called cockpit resource management), the process of collaborative decision support, is a tool that balances effective command and control with appropriate consideration of subordinates' concerns. It does so in a powerful and effective way. CRM recognizes that different members of a closely knit team can have different expertise, different responsibilities, see different information, and reach different interpretations of shared situations. Its power derives from the ability to interject critical information from subordinates into the command consideration process during the moments of greatest need. The US National Transportation Safety

---

14 "Air Florida Flight 90," Wikipedia, last modified June 11, 2018, http://en.wikipedia.org/wiki/Air_Florida_Flight_90.
15 "Air Florida Flight 90," Wikipedia.
16 "Air Florida Flight 90," Wikipedia.

Board (NTSB) and the Federal Aviation Administration (FAA) derived the genesis of CRM from the 1978 fatal crash of United Airlines Flight 173. While in the air, this flight "… ran out of fuel while the flight crew were troubleshooting a landing gear problem."[17]

"In addition, they [the FAA] also began reappraising the traditional unwritten rule that the captain had ultimate authority on a flight and could not be questioned. From that point onward, first officers were encouraged to speak up if they believed a captain was making a mistake."[18]

In the Control Room

CRM in the control room is the protocol, procedures, and training to enable all participants to actively engage in situation awareness, situation assessment, and decision management activities. Effective CRM does not undermine vested authority, nor should it be interpreted to suggest any individual's lack of ability or inattention to duty or discernment, but just the opposite. CRM recognizes that the whole can be greater than the sum of its parts. We want to take advantage of this. Here are the key aspects.

- Each *member* of an operations team has a duty not to be distracted by a singular problem that might interfere with his responsibility to remain focused on the full scope of his specific responsibilities using the full extent of his training, responsibility, and experience.

    o *Exception: There are critical times when delegation to an individual of a specific, narrowly focused task or responsibility is crucial to detecting or following something of vital importance. At this time, that individual has no other responsibility but to be hypervigilant to the assigned task.*

- Each *member* of an operations team has a duty to search for and identify any operational parameter or condition that may contribute to a problem or suggest that the present team's understanding might be inaccurate, incomplete, or incorrect.

- Any *subordinate member* of an operations team, having identified any abnormal operational aspect (see the preceding bullet), has the affirmative responsibility to respectfully and accurately bring the identified operational aspect to the attention of the primary member of the operations team.

- The *primary member* of an operations team has the specific affirmative responsibility to hear, in the moment, all subordinate members' concerns and factor them into a proper operational frame.

---

17 "Crew resource management," Wikipedia, last modified March 3, 2018, http://en.wikipedia.org/wiki/Crew_Resource_Management.
18 "Air Florida Flight 90," Wikipedia.

Worst Case First

Group discussion between informed, competent participants is an often-used process for gaining collective knowledge and concerns. One of the tools employed to shape the breadth of thought and to overcome the participants' biases is to start with the problem and identify the worst cases first. The group is tasked to assume that the problem, issue, or situation will play out in a terrible way. Sure, it may seem to be leading with pessimism, but the actuality is far from it. Of course, we all hope for the best. And often that is exactly what does happen. But hoping for the best is not a plan. Doing so and relying on it will have the opposite effect. By not planning, you actually plan for disaster.

The discussion group formally asks each participant to put forward what they can visualize as the most awful result or outcome. It is understood that the "most awful" is not fanciful. It means that no one is expected to evoke the childhood story "For want of a nail the kingdom was lost" as a model.[19] Participants will need to remain serious but stay "likely." Staying "likely" includes remote yet entirely possible outcomes. In fact, these are the situations that the method wants to find.

There is one additional part of this tool that is considered to be important. The order of the participants starts with the most junior and least experienced member. This way, the first to voice his worst-case scenario must do so without knowledge of what his peers and superiors are thinking. And so on up the chain. Working the discussion this way has two important benefits. First, experience, by its nature, favors what is known or what is expected. This is not a problem when it comes to action. There we need all of the experience and competence available. But when it comes to understanding or possibilities, experience can be an opaque veil. The second benefit is that it avoids a whole host of biases—not the stuff of prejudice, but the human nature ones discussed in Chapter 7, "Awareness and Assessment Pitfalls." By not hearing from anyone with more experience or more insight, or actual "stories" about their history, the junior participant must rely on his own ability to think and attempts to visualize what might be possibly going on. Each participant voices his considerations. No consideration is examined until all are on the table, as this would interfere with others having the benefit of an unbiased forum.

10th Man Doctrine

An interesting illustration helps to underpin the intent of this topic. Not surprisingly, it reinforces the entire concept of CRM. Here is the situation: A group of individuals is charged with coming up with a collective decision for which there may be vitally

---

[19] "For Want of a Nail," Wikipedia, last modified April 12, 2018, http://en.wikipedia.org/wiki/For_Want_of_a_Nail.

important consequences for not getting it right. Each participant is an expert. After discussion and deliberation, the matter was polled and the entire group was of one mind. Each of the participants was ready to conclude that the current decision was the one they would support without reservation. This is a nice place to be if true. Such unanimity is very comforting. It reinforces acceptance of the decision. It emphasizes the conclusion that the decision is right.

Not so fast. Actually, it is the very uniformity that should be discomforting! No issue of importance is ever so tidy. No carefully considered decision is without doubt or so certain. Not that uniformity cannot be achieved for difficult decisions, but it should not come readily. (This is a foundation of the American criminal jury system. Refer to Section 7.7 on doubt.) There is also a real-life occurrence that drove this home. After the 1973 Yom Kippur War in which Israel found itself completely blindsided by the attack, it developed a methodology to reduce the untoward effects of groupthink.[20] It established a special branch of intelligence to take the "devil's advocate" position on everything. The movie *World War Z* somewhat fictionalizes this as the 10th man doctrine.[21] Basically, and oversimplifying a bit, if a group of, say, 10 decision makers all reach the same conclusion, one experienced member of the group (against his personal convictions) is given the explicit task of going "outside the box" to find all of the reasons the chosen decision is not the right one, no matter how improbable the disagreement might seem. Therefore, the "experienced team member" will find all of the reasons why the seemingly unchallenged decision might, in some small but important way, not be the correct option. The results are brought back to the rest of the team. The result is more discussion, perhaps more investigation, and ultimately a better decision.

Triangulation

Just so we get this started on the right foot, collaboration assiduously avoids anything to do with confirmation. Sure, the results of a collaboration might (or might not, depending on the truth of it all) be a confirmation of an initial suspicion. But it must never start out to be. Ask "What do you make of this observation," not "Is this really happening?" Seek assistance to understand. Then seek options to manage a confirmed situation. Seek extra eyes, ears, hands, knowledge, and experience.

One of the ways collaboration can work is by an activity called *triangulation*. Triangulation provides a second way to view or see something from a different

---

20 "World War Z Movie: Do the Israelis Really Have a Tenth Man Doctrine?" Quora, accessed October 10, 2014, http://www.quora.com/World-War-Z-2013-movie/Do-the-Israelis-really-have-a-10th-man-doctrine.
21 Armi Legge, "What World War Z Can Teach You About Critical Thinking," *Complete Human Performance*, accessed September 11, 2015, https://completehumanperformance.com/2015/09/11/world-war-z/.

perspective.[22] The word *triangulation* derives from noticing that an object viewed from two vantage points (view one, view two, and the object) forms a triangle. In astronomy or geography, we use geometry to determine the location, including distance away. In situation management, we use the dual perspectives to bring out differences that help provide a better understanding of what is seen, what is happening, or what to do about it. How do we use triangulation in the control room? There are two general approaches: by *job or task* or by *experience*. First we lay out the eight triangulation questions.

Triangulation Questions

The operator may have or suspect something amiss in an area of his plant. There are others available he can ask for assistance. Here are the questions to ask and answer. There are two methods of applying this tool: (1) job or task triangulation and (2) experience triangulation. Each method uses all questions. The operator asks:

1. Do you (the collaborator) see *anything unusual* or out of the ordinary where you are?

2. Are you having *any problems* doing what you are doing?

3. What are the ways that what you are doing may affect my area of the plant?

    a. Are you *seeing everything* that you need to see to confirm that what you are working on is properly functioning?

    b. Are you seeing *all of the indications* that what you are doing is working properly?

    c. Do you see *anything amiss or unusual* that might suggest that something might not be going as expected?

    d. Are you *unable to see* anything that could clarify or expose any concerns?

    e. Do you see anything that you *cannot fully and simply explain*?

4. Assume that you are doing my (the operator's) job:

    a. How would what I am doing affect what you are doing?

    b. Do you see *anything unusual* or out of the ordinary in my area?

    c. Do you see everything that you would *expect to see* to ensure that my area is properly functioning?

    d. Do you see anything that you would *not expect to see* going on in my area?

---

22 R. Yin, *Case Study Research: Design and Methods*, 2nd ed. (Thousand Oaks, CA: Sage Publications, 1994).

5. Are you aware of *anything going on in the larger plant* that might affect what you and I are doing?

6. Are you aware of anything planned to happen in the larger plant that *is not happening* or is not going to plan?

7. What do you (*collaborator is asked*) know that I do not know?

8. What do I (*operator asking himself*) know that you do not know?

Job or Task Triangulation
This approach takes advantage of situations where two (or more, but let us settle on just two for now) individuals (one is an operator) are working on different jobs in the same plant (or enterprise or operational area; we will just use *plant* from now on). The particular jobs are different. However, the general plant knowledge to do both jobs is similar and the specific jobs being worked on sort of overlap each other, perhaps a bit or even substantially. When the operator finds himself with a problem or issue, he asks the other individual (the collaborator) for his input. Remember, because the operator is responsible for the plant, he would clearly know that the other individual (e.g., field operator or maintenance technician) is in his area and what he is doing. In this situation, the operator asks the eight triangulation questions.

Experience Triangulation
The *job or task* triangulation discussed in the preceding section involved the collaborator already being involved in related activities close by. *Experience triangulation* involves the operator actively seeking out someone with experience in his area but who may not be actively working in or near his plant. It could be another operator in the control room (or available via electronic linking). If it is around shift change, it could be the relieving or the arriving operator. It might be the process engineer, safety engineer, trainer, or others. As before, the operator asks the eight triangulation questions.

Processing the Answers to the Triangulation Questions
First, make sure there is enough time to work all of this out. If not, rely on permission to operate.

Triangulation, and for that part collaborating, is not to be taken lightly. This is not about confirmation. Something is amiss, and the operator must know what. The answers (or lack of answers) for each of the eight questions need review. Use what is known. Remember the information fork. When all is in place and there is enough time to work carefully through it all, assemble a picture of what is going on and what the options are for dealing with it. If it still is not clear, escalate. You are not going to magically figure this out.

Red Teams

Red teams are a form of collaboration. It is really a team form of the 10th man doctrine. Ambiguity in situation awareness and assessment is a very difficult thing to work with. When there is ambiguity, everyone works hard to ignore it so it will go away. Yet it is almost always a useful indicator of imperfect, closed, or biased thinking. This will not do, of course. A useful way to ensure that important thinking is well founded is to try and find the gaps or holes in it. Red teams are a way.

> The British Armed Forces and other organizations deploy so-called red teams to accomplish this [way to find holes]. The red team is a parallel task force, made up of senior leaders and support staff, whose only mission is to collect and synthesize information to prove that the current plan is wrong and needs to be changed. This team plays the role of the loyal opposition, in the spirit of Alexander the Great, who would periodically ask himself how much evidence it would take for him to abandon the current plan.[23]

10.15 Permission to Operate

An operator's assigned role and purpose is to operate. The enterprise only makes money when it produces a salable product. To produce, it must operate. It is only natural that there is a strong inclination toward operating. For the most part, this works. When it does not, the results are often extremely costly. Careful studies of abnormal situations have determined that a large percentage of incidents are directly attributable to operators not realizing in time that their operating problem was not going to be contained short of a catastrophic event.[24] In the press of the upset, the innate desire to "manage the problem" crowds out the more important decision of whether or not continuing attempts to manage have a proper chance to succeed. By the time the operator did fully realize the shocking reality, it was almost always too late. It was too late either because he did not have enough time to do what it would take to return to a safe operating mode or because he could not determine what was wrong or what to do. In all cases, the result is the same: equipment damage, production losses, personal injury, environmental degradation, or worse. This topic is one of the most important in all of situation management. It identifies how far an operator should be expected to go in his goal of managing any abnormal situation. And it is as powerful as it is effective. Simply stated, as long as the operator retains an effective responsibility to manage the situation, he may continue to do so. However, once that responsibility is withdrawn, he must cease all attempts at good operation restoration. This functionality is called *permission to operate*.

---

23 Paul J. H. Schoemaker and George S. Day, "How to Make Sense of Weak Signals," *MIT Sloan Management Review* 50, no. 3 (Spring 2009), http://sloanreview.mit.edu/article/how-to-make-sense-of-weak-signals/.
24 Ian Nimmo, "The Operator as IPL (Independent Protection Layer)," *Hydrocarbon Engineering* (September 2005).

First a few key points:

"Know when to hold 'em" and "know when to fold 'em."	The short-term operational goal of a sitting operator must be attainable. By trying to "fix things" when fixing is not possible, he is forfeiting the responsible opportunity to successfully manage.
Planning for implementation of permission to operate	All operations and operational procedures must have a pre-designed and pre-engineered exit point for moving to a safe state. This must be an integral part of the design process.
Retaining permission to operate	In order to maintain permission to operate, the operator must have a firm and accurate understanding of the problem, the operator must be making continual positive remediation or control progress, and no additional threats can appear.
Loss of permission to operate	When permission to operate is lost and neither the shutdown nor safe park has been attained, operations will presumptively transition to disaster management mode.
Alarm response execution	When the operator has completed all documented steps and procedures specified in the alarm response sheet and the abnormal situation has not resolved, permission to operate must be lost.

If the decision to operate in the face of uncertainty is made properly and sufficient technology is in place to implement that decision, a significant reduction in operating losses can be realized. Please keep in the forefront that each individual plant or enterprise will be shaping specific guidelines and policy for operators. This discussion provides grist for the mill of decision-making and policy. The various examples illustrate the range of alternatives and amplify the power of the approach, nothing more. This section reviews the process by which a decision to operate in the face of uncertainty is raised to the level of a strategic tool. When the important supporting procedures and technology are in place to facilitate such a decision, a significant reduction in operating losses can be realized.

Before we get too far into things, it can be helpful to see how unexpressed default operating conditions affect how we think about this. Most industrial plants (chemical, oil refining, foods, etc.) are expected to be running unless they are not. This is not a play on words. What we mean is that when they are operating, they are expected to be kept running. The default state is to keep running. At the other extreme are plants that expose significant risk if they get into problems when they run. Think about nuclear power plants. For these, the default state is to not operate. Only when and as long as safe operation is assured may they be operated. For the former, permission to operate will ensure that any "default" thinking or predisposition will not unduly get in the way of safe operation.

BP Texas City operators[25] were vainly attempting to manage their abnormal situation for more than 10 hours before the tragic explosion.

---

25 US Chemical Safety and Hazard Investigation Board, "Urgent Recommendation (BP Texas City Explosion and Fire, March 2004)," news release, August 17, 2005.

Milford Haven was in serious upset without the operators understanding what was wrong and what to do about it for more than 5 hours before disaster struck. After the investigation, the Health and Safety Executive specifically identified the failure to understand when to continue operating or not as an important proximate cause of the incident:

> "#5—[Provide] clear guidance [to the operator] on when to initiate controlled or emergency shutdowns."[26]

Management's Role

Here we lay out the options for management to consider to establish an effective policy for plant operators having permission to operate within their area of responsibility. The purpose of this discussion is twofold: first, to make the reader aware of an important part of the infrastructure for operating a production enterprise that heretofore was never thought of explicitly; and second, to put forth several options and recommendations to assist the development of an explicit policy. This is one of the primary prerogatives of senior management—determine the rules of the game.

Until now, we did not pay much attention to a decision of this type. No part of the business plan had a line item in it for this. No part of the engineering design package investigated what the operator should do when the process became abnormal and success at coping was at issue. Yes, regulation and safe operation considerations provide for emergency shutdown criteria and the equipment to support that requirement. But operating procedures ignore the subject of activities prior to a challenge to this last line of defense. Once management settles the issue of permission to operate, all operating procedures and support technology need to be in place. All affected personnel need to be fully informed and appropriately trained.

Operating Situations

Not all operating situations and modalities are the same. In this section, we will add some structure to them. This structure will be used later to define situation boundaries to assist in the implementation of the operator's permission to operate.

Operating in Uncertainty

Generally, operators are quite proficient in the day-to-day management of their production units. The automated controls and other equipment maintain production. When variations appear, the operator responds appropriately and normal operation

---

[26] "The Explosion and Fires at the Texaco Refinery, Milford Haven, 24 July 1994," Health and Safety Executive, http://www.hse.gov.uk/comah/sragtech/casetexaco94.htm.

is quickly restored. In this manner, effective process management is maintained by experience, procedures, problem-solving skills, and collaboration. There are a couple of basic ways this comfortable situation becomes problematic.

- Problem incidents begin with a "whimper" or perhaps a small "bang" somewhere, then for a while progress slowly, then take a turn and get complicated, often without the operator knowing, and suddenly get out of control.

- Problem incidents begin with a big "bang" and quickly become (known by the operator or not) out of control.

Unique Events

Most of the seriously damaging incidents never happened before in your enterprise. Nor were they envisioned. For such unique events, experience and specific procedures are of little use. Situation-based training will not be of much use either. Industrial best practices now rely on three key aspects to respond effectively to unprecedented events:

1. Competency-based operator training, where abnormal situation recognition and management are one of the core competencies

2. Abnormal operation detection and identification tools

3. Abnormal situation management procedures and technology

Explosive Events

Some damaging events seem to happen almost without notice and move extraordinarily quickly into a catastrophic state. In this situation, we must rely on either an operator quickly identifying the situation and initiating an emergency shutdown or the automated activation of emergency shutdown protection equipment. In both cases, it is imperative to ensure that the shutdown procedures and equipment are adequate and effective.

Surprisingly enough, however, such an event almost never arises without some sort of prior symptom. If the right sort of notice was taken, it could likely be effectively managed. Therefore, industrial best practices here fall into three key areas:

1. Comprehensive and effective equipment health monitoring programs

2. Recognition of the role that operating stresses place on both personnel and equipment

3. Technology to detect both subtle as well as incipient faults in the process and related equipment

Operational Modes

Figure 10-7 depicts the various operational modes, plant states, critical support systems, operational goals, and required activities for success.

The figure illustrates the three general operational modes all operators face: normal, abnormal, and emergency. For example, when the operational mode is *normal* and the actual plant state is *normal*, the critical systems (those that directly support the operational goal) are the decision support system, process equipment (pumps, switchers, etc.), control equipment (e.g., controllers), and plant management systems. For the same plant state and operational mode, the operator's goal (the rules of operation supported by procedures and all the rest of the support infrastructure) is to keep the operation normal. The operator is expected to devote all reasonable resources to that end.

Let us move to the abnormal mode and state. In the *abnormal* mode, the plant can either be *abnormal* or it can be *out of control*. If the plant state is abnormal and out of control, the operator's required goal is to bring the plant to a safe state. No longer is he permitted to even think about trying to make things normal. That activity is off the table. Permission to operate is now extremely simple to state. Looking inside the red box of the figure, ask the question: At any moment in time is the plant operating above or below the dashed red line? If below the dashed red line, we expect the operator to use all best efforts to restore operation to normal. If above the dashed red line,

Operational Modes:	Plant States:	Critical Systems:	Operational Goals:	Plant Activities:
Emergency	Disaster	Area Emergency Response System	Minimize Impact	Fire Fighting
	Accident	Site Emergency Response System		First Aid
				Rescue
Abnormal	Out of Control	Physical and Mechanical Containment System Safety Shutdown, Protective Systems.	Bring to Safe State	Evacuation
	Abnormal	Hardwired Emergency Alarms DCS Alarm System	Return to Normal	Manual Control & Troubleshooting
Normal	Normal	Decision Support System Process Equipment. DCS, Automatic Controls Plant Management Systems	Keep Normal	Preventative Monitoring & Testing

Figure 10-7. Plant state versus operational modes and goals illustrating the clear change of operator goals. As the situation worsens, the operator changes the objective from trying to return to normal to bringing the plant to a safe state (perhaps shut down if that is the only appropriate state).

the operator has an affirmative duty to cease attempting to restore good operation and immediately move operation to a safe place. This is not optional. This is a duty that is expected by requirement from senior operations management. It is part of the operations rulebook, is trained for, and will be supported when it is used.

Management support must go all the way to accepting that there may be situations in which the operator had good reason to invoke the permission to operate protocol, but later on when adequate time permitted an evaluation, it was revealed that the problem was not unmanageable. Any "Monday morning quarterbacking" will be done, not to put more pressure on operators, but to improve what can be done to better equip the operators to know the situation and to improve the procedures.

By now, you can tell that the real problem occurs when the plant is in the abnormal mode of operation. The basic issue is simple. The issue centers on how to detect and appreciate the significance of whether the plant state is abnormal or out of control. Specifically, is the current operation above or below the dashed red line in Figure 10–7? If all operators had a sign that read "Out of Control" every time their plant was abnormal and out of control, it is unlikely that they would ever try to keep it running in the vain hope of a miracle. Such a sign is really a stretch. *As a surrogate, we use the concept of permission to operate to provide this functionality.*

Definitions
The following definitions are used throughout the remainder of this chapter:

- **Normal region of operation** – The collective operating situations where (1) there are no critical alarms active and (2) the plant critical operating conditions (safety, environmental, and financial) are not violated.

- **Conservative region of operation** – A region of operation that sets limits so that it is unlikely that a manageable upset will go outside the normal region of operation.

- **Upset** – Any process deviation likely to affect production quality or rate, or place equipment, personnel, or environmental limits at risk.

- **Manageable upset** – Any upset that the operator understands and has every reason to believe he can control and thereby return the process to normal operation in a timely manner and without excessive risk.

- **Unmanageable upset** – Any upset that is not a manageable upset.

How Permission to Operate Came to Be
There is an interesting story about how the concept of key operating principles was developed. We started it earlier in Tenets of Operation. During the normal course of

manufacturing, a mistake was made in producing one of its retail products. The product was designed for a niche market use. When it was used, this defective product severely and irreparably damaged users' equipment. The mistake in production was not detected, so the product shipped. Before the general word got out, many millions of dollars of customers' equipment were damaged and needed replacement, which was, of course, paid for by the petrochemical company.

The company's management decided to uncover the root cause of this costly mistake in the full belief that (1) if it were resolved at only the symptom level, a similar problem would repeat again and (2) understanding of the fundamentals of the failure might lead to important lessons that could have an enterprise-wide impact. Both were true.

What the company did was a case study in really good forensic examination. It collected every available record of incidents, accidents, and such, covering about 5 years. If the reports were complete, it recorded the causes and recommendations, taking care to ensure that the lessons learned were detailed and sufficiently deep to uncover root causes. If the reports were incomplete, the company made attempts to revisit them and complete the investigations. All of this work involved a significant amount of effort. The next, and most crucial, step was to categorize the incidents to look for common causal issues. The company found them. It found a list of more than a dozen operating actions that almost always were the root cause of serious operating mistakes. If those erroneous actions had not been taken, most of the incidents would not have happened! Permission to operate was the most important way of preventing these erroneous actions and so forms the foundation of this topic. The others were discussed earlier in this chapter.

How Permission to Operate Works

Operators are empowered to operate the unit(s) under their responsibility by the authority from management that we shall call *permission to operate*. Until permission is withdrawn, the operations staff will employ their best efforts, utilizing plant procedures, training, and their other considerable abilities, to ensure safe, productive, and responsible operation. When permission is withdrawn, current efforts to keep the plant running will cease, and all future efforts will be directed to placing the plant in a safe state. Of course, you know there will not be a watchman looking over the operator's shoulder and calling the shots: okay to operate, watch out, and not okay to operate. Rather, permission to operate will be a structured judgment process that operators, supervisors, and plant staff can understand and apply.

Most plants today have no such understanding. Most plant managers have never thought about this concept. Most operators have no idea how far they should go and that a policy can be thought out in advance that actually provides understandable rules and procedures for this. But now that you have heard about it, does it not sound about the only reasonable way to approach this issue?

To implement this tool, management will establish criteria by which operators can unambiguously ascertain whether they have (constructive) permission to operate their unit. Once conditions change so they no longer have permission to operate, the operations team must take all necessary steps to cease operations by moving their units to predetermined safe operating states. Again, simply stated, if operations are below the dashed red line in Figure 10-7, the operator may use his best efforts to return the process to normal. If operations are above the red dashed line, the operator must direct all his action to moving things to a safe state. The methodology presented here is about how to make the red-dashed-line decision.

Permission to Operate

Countless incident post-audits have concluded that most catastrophic incidents happened *after* the operating personnel were unable to ascertain what was going on.[27] Put another way, once an operator was no longer able to identify what was going on, rarely were his actions successful in restoring normalcy or averting an event. We can now very simply state what permission to operate is:

> The operator has permission to operate the plant under his control so long as he fully understands what the process is doing. Once lost, the operator has an affirmative duty to immediately and with direct purpose safely cease operation.

Using this guideline, management will be able to change the operator's operating charter from de facto "just keep things running" to one of providing explicit permission by defining the circumstances under which they believe the operator can keep things running successfully. This has a profound effect on the operator's responsibility and authority. So long as the rules give the operator permission to operate, he is presumed to have the authority to operate and will use all his skill and knowledge in the pursuance of his task. However, as soon as that permission is lost, the operator must, and without fault (lest he not do so the next time), cease operating and move the plant to shutdown or an alternate safe state. Management therefore assumes total responsibility for the rules of this decision and the results that come from the operator

---

27 US Chemical Safety and Hazard Investigation Board, "Urgent Recommendation."

using them in good faith. Even if it may be later proved that the operator "should have known what might be happening but did not actually know" or the plant was actually not in upset, the operator is to be commended for his action, because he reasonably thought it to be so and acted under authority of his charter.

Withdrawn Permission to Operate

At the moment permission is lost (operator has lost understanding of what is going on), the operator must execute the preplanned process of moving the operation sufficiently away from the unsafe or unmanageable condition. This also means that the ability to change the operating modality from operating to safe state must be an integral part of the plant design. In addition to proper protocols and procedures, controls and equipment must be in place for the operator to command the change, monitor the progress, and remediate any defects along the way. This capability is part and parcel of good engineering. The choice of moving to a shutdown state or a safe park state is an operational one depending on which is available for use.

Alternate Methods for Having Permission

There are a few basic alternatives that can be used to operationalize the "permission" decision. Management will choose or otherwise specify which, if any, should be used.

De Facto Decisions

- **Carte blanche** – Permission to operate is granted if and only if the operator is reasonably sure that he understands what is going on and is quite confident of his ability to manage. He has no bravado, just honest self-awareness. The decision is simple and direct. It is up to the operator to manage.

- **In usual upset** – Permission to operate is granted if and only if the operator is reasonably certain that the plant is in a usual upset state, regardless of whether or not he understands specifically what the current upset is, and he is confident he will be able to manage it successfully. If anything about this upset is unusual, permission to operate is withdrawn. Again, this is simple and direct. It is a bit more subjective than carte blanche, but still clear and unambiguous.

- **Manageable unusual upset only** – Permission to operate is granted if and only if the operator is reasonably certain that though the plant might be upset, he has demonstrable reasons to know or expect the upset to be manageable. If he determines or begins to suspect that the upset might not be manageable, then permission to operate is withdrawn. The only manageable upsets are those that are routinely seen and for which procedures that are expected to work exist. This case is quite a bit more subjective and requires substantial operator competence to employ.

Operating Modality Decisions

The plan for predesigning the operator's role based on plant situation was illustrated in Figure 10-7. Recall that the figure shows three operational modes: normal, abnormal, and emergency. For each mode, the operator's goal is identified. For example, if the plant state is normal, then the operator's goal is to keep it normal. In the abnormal mode, the plant can either be abnormal or it can be out of control. If the plant state is abnormal or out of control, the operator's goal is to bring the plant to a safe state. Notice that now the critical plant infrastructure systems involve, in order of escalating danger, the process control system (PCS) alarm system, hardwired emergency alarms, protective systems, safety shutdown equipment, and eventually any physical or mechanical protection and containment.

The following is a list of alternative definitions for having permission to operate. See Figure 10-8. Generally, plant management will select one of these or develop something else that is compliant with their operating philosophy. When the defining rule for having permission to operate is violated, the operator must, without delay or need for further decision, supervision, or other authority, move the plant to a safe state or shut down.

- **Normal operation definition 1** – The plant is operating within the conservative region of operation and there are no unmanageable upsets.

- **Normal operation definition 2** – The plant is operating outside the conservative region of operation but within the normal region of operation and there are no upsets.

For example, if the plant adopts definition 2 for normal operation, whenever the plant moves outside the normal sphere of operation and there is an upset, the operator

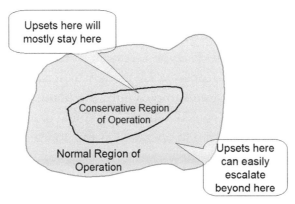

Figure 10-8. Regions of operation boundaries showing that the closer the plant is operating to the desired state, the more likely it is to remain there. As the plant begins to operate further away, the more likely the problem can get worse.

is required to discontinue normal operation. This means that his permission to operate has been challenged. He must make a specific determination to ensure that conditions allow him to maintain operation (that is, at this moment in time, he retains *permission to operate*). Otherwise, *permission* is lost and he must render things safe.

Safe Operating Limits

It is easy to say that we require the operator to ensure that he keeps the plant within safe operating limits. That is his goal. It is the reason we have him. It is the way alarms are designed. It is why weak signals can be so useful. It is why you developed operating procedures and policy. It is everything. And everything in this book is there to support it. Achieving safe operation is a result. There is no magic. Keeping the plant within safe operating limits requires a relentless pursuit of operations improvement. Please make this an essential part of your enterprise culture.

Managing the Operator's Permission

So long as the operator maintains permission to operate, he is free to incorporate all proper means at his disposal to keep things running. We now explore how to actually manage this idea of permission to operate. For this to be useful, you will need ways for the operator to know and understand when to continue and when to stop. These are suggestions, not necessarily requirements. Each plant will develop the specific details as part of an alarm philosophy or other operational guidelines. In this section, we see some specific ways for you to identify situations where a clear decision can be arrived at to know whether the operational charter for the operator will be maintained or not. Previously, we discussed criteria for the operator to use that identified the plant-operating regimes and to make a determination about continuing to operate or not. However, in practice, this is not enough. We need a second way, a check if you will, to ensure that the operator's charter to operate is still in force.

Four separate categories are discussed. Any item from any category that withdraws the operator's charter to operate is enough to terminate his permission to operate. Remember that these are examples only. Their purpose is to provide clear, understandable alternatives for consideration. Any single one being true is enough to withdraw the operator's permission to operate. It is fine to use as many of the types as needed, or your plant may develop new ones that suit it.

Qualifying Abnormal

Let us assume that the plant is now abnormal. The following situations will identify what is meant by continuing to be abnormal. If any one or more of these exist, then we must conclude that the plant remains abnormal, regardless of what opinion the operator might have.

- **Situation 1** – Additional alarms activate in the same or a related plant area.
- **Situation 2** – Few, if any, existing alarms in the same or a related plant area have returned to normal within a reasonable time.
- **Situation 3** – One or more additional pieces of equipment (valve, pump, sensor, or anything else not trivial) fail or are newly discovered as having failed earlier.

No Help at Hand
Difficult abnormal situations where there is not a second experienced plant person at the operator's "left hand" in the control room (or virtually with dependable technology) should automatically suggest that the operator is either already in trouble or soon to be. Consider this to be sufficient reason to withdraw permission to operate.

Observer Evaluation
We continue this discussion by looking through the eyes of a mythical observer in the control room who might be side-by-side with the operator and is making observations about what is seen. Of course, we know that such an observer may not actually be there, so this is a "thought" discussion. The format is question and answer. This methodology will ask the questions. Our observer is responsible for the truth of the answers. The answers will determine whether the operator retains permission or not.

- **Question 1.** Has the operator tried more than one approach to turn the abnormal situation around and it is still abnormal?
    - Yes—Discontinue operation.
    - No—Continue to operate.
- **Question 2.** Does the operator appear to be able to explain what is likely to happen to the plant in the next 10 to 20 minutes?
    - Yes—Continue to operate.
    - No—Discontinue operation.
- **Question 3.** Is the operator using a predetermined, documented approach to work the abnormal situation?
    - Yes—Continue to operate.
    - No—Discontinue operation.

- **Question 4.** Has the situation remained quite abnormal (even if it does not appear to be gravely abnormal) for longer than a reasonable time?
  - Yes—Discontinue operation.
  - No—Continue to operate.

Operator Self-Evaluation

Now we move to the operator himself. This part is self-examination. Each question is expressed in the third person but would be understood as being read and answered by the operator as applying to him personally.

- **Question 1.** Does the operator feel too harried to take a careful, methodical approach to understand, evaluate, and manage the situation?
  - Yes—Discontinue operation.
  - No—Continue to operate.

- **Question 2.** Does the operator feel that there is enough time to slow down and think things out?
  - Yes—Continue to operate.
  - No—Discontinue operation.

- **Question 3.** Has the operator tried the same or similar things to regain control more than once in this event without clear success?
  - Yes—Discontinue operation.
  - No—Continue to operate.

Putting It All Together

You get the idea. Where there is uncertainty, either explicitly by the operator not knowing or understanding, or indirectly from what is happening to the operator, the operator should close things down. Figure 10-9 illustrates the decision process as a flowchart.

10.16 Escalation

The ability to rapidly deploy competent escalation to the control room can make the difference between dangerous operations through inappropriate risk-taking or the inability to recognize incipient danger, and an orderly transition to a safe operating state. Enterprises readily use it as often as needed to back up their culture of safe and effective operations. The more often it is used whenever needed, the

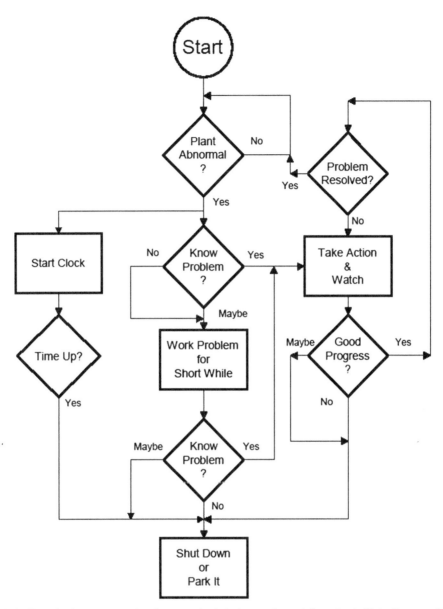

Figure 10-9. Permission to operate diagram depicted as a formal flowchart. Note the parallel "clock" that ensures that working on a problem resolution does not take too long without a change in tactics.

better the operation becomes across the board. Available escalation protocols form an effective safeguard for operator limitations or response to powerful operations faults. It also provides an excellent opportunity for training, mentoring, and coaching. Operational responsibility for the activity that has been escalated transfers to the escalated individual. *However, during the shift, even if operating escalation has taken place, the shift operator is responsible for ensuring that someone is always in operational charge. Always.*

The one vital situation where escalation is not appropriate is when a plant or operation is in grave danger and the time required to escalate exceeds even the time to shut down or render safe. Escalation takes time. That time is not available. For these situations, the operator is the frontline responsible individual. The enterprise will rely on its specific protocols for permission to operate for proper response. Also remember to follow the communication protocols in section 10.6.

Communication and Collaboration

Communication, collaboration, and escalation are three of the more important interpersonal interactions in the control room. The operator is at the center. Please take the time and effort to fully understand the differences and distinctions between each of these modes of interaction. A significant number of terrible incidents and disasters could have been prevented or reduced in severity had the participants understood the proper form, purpose, and steps for use.

Communication

> Communication (from Latin *commūnicāre*, meaning "to share") is the activity of conveying information through the exchange of ideas, feelings, intentions, attitudes, expectations, perceptions or commands, as by speech, non-verbal gestures, writings, behavior and possibly by other means such as electromagnetic, chemical or physical phenomena and smell. It is the meaningful exchange of information between two or more participants (machines, organisms or their parts).
>
> Communication requires a sender, a message, a medium and a recipient, although the receiver does not have to be present or aware of the sender's intent to communicate at the time of communication; thus communication can occur across vast distances in time and space. Communication requires that the communicating parties share an area of communicative commonality. The communication process is complete once the receiver understands the sender's message.[28]
>
> A *communication* has the following attributes:
>
> - Content with enough significance (or potential importance) to merit a communication.
>
> - A *sender* (for our purposes, usually an individual) and a *recipient* (usually another individual). This is not something that is sought by a recipient, such as a search result on the Internet or in a database.
>
> - A referenced and understandable *message* (what or where in the enterprise does this concern relate to; its importance).

---

28 "Communication," Wikipedia, last modified June 20, 2018, http://en.wikipedia.org/wiki/Communication.

- o Specific purpose of the message
- o Specific thing (part of plant, situation, etc.) that the message relates to
- o Specific message content clearly expressed
- o Delivered via a reliable vehicle with reasonable traceability
- o Understood by the recipient
- If communication is required (e.g., by procedure or by standing operations requirements), then it must be formally archived as to content, sender, recipient(s), and timing.

The following are a few examples of communications:

- Verbal messages from one operator to another, or from one operator to any other member of the operations team including field, maintenance, and supervision
- Entries in reports, logs, or other operational annotations (recipients implied)
  - o Shift handover reports
  - o Operator logs
- Written (hard copy or soft) documents with a specific requirement for an individual to note receipt, acceptance (or refusal), or consent (or refusal to grant consent)
  - o Hot work permits, entry permits
  - o Work orders generated and/or received

Both the sender and the recipient have a duty to properly understand the communication. Therefore, it is incumbent on the plant or enterprise to clearly identify any communications that must be classed as a communication. Identifying communications can be as formal or informal as desired by the enterprise. All others are simply remarks.

Collaboration

Collaboration is working with others to do a task and to achieve shared goals. It is a recursive process where two or more or organizations [including individuals] work together to realize shared goals, (this is more than the intersection of common goals seen in co-operative ventures, but a deep, collective determination to reach an identical objective)—for example, an endeavor that is creative in nature—by sharing knowledge, learning and building consensus. Most collaboration requires leadership, although the form of leadership can be social within a decentralized and egalitarian group. In particular, teams that work collaboratively can obtain greater resources, recognition and reward when facing competition for finite resources.[29]

Collaboration is therefore a process of cooperatively working together toward an agreed upon purpose. It is unnecessary for the purpose to be shared in that all collaborating parties stand to gain. While the collaborating parties each need to provide resources and share information, and work cooperatively on the collaborated activity, there is no expectation that the collaboration effort be uniform or even fairly distributed. Nor does the activity of collaboration need to fully occupy each collaborating party. For example, an air traffic controller managing a flight with significant mechanical problems need not have other controllers or supervisors fully dedicated to assist. Rather, the collaboration may take the form of ensuring other flights remain away from anticipated flight paths and emergency landing options, or assisting with contacting emergency aircraft maintenance and engineering resources to assist when located.

In order to collaborate, all parties need to share a common vocabulary, understand the overall framework of the situation to the point that their collaboration efforts are productive, share appropriate information, and respect the professionalism of all participants. Collaboration may be offered by other participants or be required by existing protocol. In the absence of existing protocol to the contrary, collaboration activities that affect other parties to the point that another party must take actions that are not desired by that party, the collaboration should change or cease at the request of the affected party.

Escalation

Escalation is a pre-trained and proceduralized activity for the operator to recognize that the current (potentially abnormal) situation may be beyond his own ability to understand, lay plans for remediation, and/or execute proper management for remediation. The operator will formally seek aid using structured, preplanned operations policies and procedures. Escalation is an essential safeguard for responsible operation because fully qualified, well-trained, and engaged operators at times face situations that exceed their strengths to manage. This is not a subtle suggestion that the program to develop operators is somehow inadequate. Rather, this is a recognition that the sheer complexity of plants and operations as well as the diverse ways they can get into trouble may be beyond any single human's ability to encompass. It is

---

29 "Collaboration," Wikipedia, last modified June 14, 2018, http://en.wikipedia.org/wiki/Collaboration.

therefore essential that control room protocols address how the enterprise expects this to be handled.

First and foremost, escalation is not "passing the buck" to someone else when what is going on gets annoying to the operator. Escalation is a formal process. Specific things happen when escalation is used as part of a planned program of getting help. Escalation only works when those who use it, use it when needed. Once called into play, it must be used effectively. These plans must be carefully laid out by the enterprise in policies and procedures. Those specific procedures and the requisite infrastructure to support them to ensure workability are beyond the scope of this discussion.

There is a clear difference between seeking reactions (communicating), seeking advice and possibly assistance from others (collaboration), and escalation. The difference is that before escalation, the operator maintains full and exclusive control over this problem and the rest of his responsibility. While the same individuals may be consulted or asked to "keep an eye" on something, those activities are not considered escalation. Such activities are part and parcel of a cooperative workplace with mature and capable individuals working there. You would want these activities to occur in a comfortable and effective way.

The bottom line of escalation is that it must follow a formally determined process. For example, when the operator decides to escalate his situation to another, he would formally announce, "I am escalating the situation [and describe the situation explicitly] to you."

Depending on the enterprise protocols, escalation may also be declared by someone else besides the operator, though this would be rare. An operationally qualified supervisor may intervene (invoke the escalation) and take over the processing of a specific situation. In even more rare situations, any operationally qualified individual other than a supervisor may intervene when it is clear and completely self-evident that the operator is unable to function.

Escalation versus Transfer of Responsibility

Only specific problems or highly ambiguous situations are escalated. An escalation will be as clear as operator knowledge and awareness facilitates. Escalated things include: help with the process or authority around being able to discover what might be going on, how best to confirm and/or test the facts or issues, and what should be considered and eventually acted on to remedy the problem. The operator retains full operational authority and responsibility. Any intervention activities recommended by the escalated person should either be executed by the operator or executed with the specific knowledge and permission of the operator. For example, it would not be good protocol for the escalating operator to give blanket permission to do whatever it takes to resolve the situation. This would effectively remove both the knowledge of any problem and the remediation activity away from the responsible operator. This situation would override any benefit.

The bottom line is that operational control is separate from escalation. There are some situations that benefit from a transfer of control during some escalations. These are discussed next.

If the Escalated-to Person Is Oriented to the Current Situation

The person who is escalated to would formally acknowledge, "I accept the escalation. I am now in control." Or for another situation, "I accept the escalation. You remain in control, I will assist."

This formality would continue until escalation is ended.

If the Escalated-to Person Is Not Oriented to the Current Situation

The person who is escalated to would formally acknowledge, "I accept the escalation. You remain in control; I will assist." Later, as the situation and/or operating protocols permit, control may be assumed by declaring, "I am transferring control from you to me."

This formality would continue until escalation is ended.

Escalation Resources

We will clarify what escalation brings to the table. First of all, it means that an additional person or people will be directly involved in the current operational situation. They bring the following to the table:

- Additional manpower for examining documents, following protocols, tracking external effects
- Broader or more in-depth experience
- Added authority to manage or track operational risk
- Extra set of hands (only for operationally qualified personnel)

This means that when an enterprise wishes to use escalation, it must be clear to all what is being added to the operational situation and who will do what.

Specifically, there must be clear command-and-control authority at every step of the way.

Direction of Escalation

Escalation is normally thought of as upward in the enterprise hierarchy. After all, to move upward is what the word *escalate* means. However, for purposes of situation management, we use the more abstract aspect of the term to include looking for added resources upward in the organization, sideways (peer to peer), or downward in the organization. Therefore, when the operator finds himself in a situation in which he feels escalation must be utilized, he should identify the resource and initiate the process.

Escalation may also be initiated (so long as it is covered by enterprise policy) by supervision, or in rare situations by peers or subordinates. This discussion will not digress into any sort of subterfuges or legal ramifications that might be evoked here. A simple example or two should clarify the intent. Suppose another operator observed the *operator* having what may appear to be a serious medical issue. I am not talking about the operator lying on the floor or in the midst of a seizure. Rather, the situation is somewhat less dramatic but nonetheless apparent. The peer or subordinate (if qualified) may request escalation if the operator is cognizant and can understand and agree, or may declare the situation escalated if the operator lacks sufficient awareness. If the peer or subordinate was not sufficiently qualified, then he would follow procedures to escalate on behalf of the operator to someone who was available and qualified.

It should be noted that none of this discussion is meant to supplant or contravene operating protocols for fire, explosion, insurrection, or other crisis or disaster situations.

First Duty of Escalation

Immediately when the operator is either considering escalation or activates escalation, it must be determined that there is sufficient time left to manage the current situation once escalated. This must be done by the escalating operator before activating escalation, or done by the escalated-to individual before any activities at managing the situation are attempted. There must be no realistic opportunity for things to go horribly wrong because a decision to continue operating by escalating was made instead of outright revoking the operator's permission to operate (see Section 10.15).

704   Situation Management for Process Control

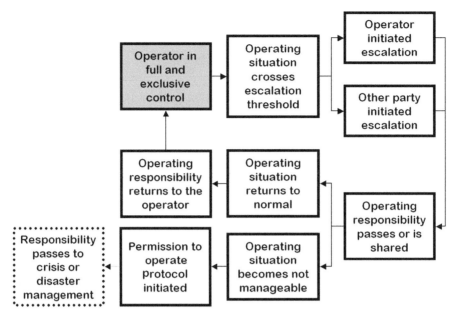

Figure 10-10. Escalation process framework illustrating the initiation events and approved transfer of responsibility that can take place to ensure that the enterprise brings to bear the right skills for the right operating situation.

The Process of Escalation

An illustration of the flow process of escalation is shown in Figure 10-10. Your escalation process would identify and explain when, why, and how each step would be carried out.

When to Escalate

Enterprises must establish clear, unambiguous rules for when escalation is required. These situations should recognize that there are times when escalation should be mandatory and there are times when the operator, without fault or blame, should be able to elect to escalate before a specific situation might require it.

Escalation should be available for supervisors to initiate. This might be obvious because it is well accepted that this is an inherent part of their core duties. However, this action must not be reflexive or authoritative. There is no room for "I know better so I'm taking charge now."

Escalation should be available for others on the operational team to initiate. This will not be used to "show up the operator," but rather as a carefully crafted tool that recognizes that there may be times when the individual closest to the front line may be so engaged that a larger picture is missed. This is at the core of crew resource management.

Let us visit a few situations where escalation can be an important tool.

Crossed Operational Area Boundaries
Escalation is required when the effects of abnormal operation cross an operational area boundary. As we saw in Section 3.4, an operational area boundary cross may be between adjacent operator areas or from an operator area to the enterprise or to outside the boundaries of the enterprise. In all these situations, the effects of operation are beyond the ability of the operator to manage due to lack of control "handles" and much more. Explicit protocols must be used.

Costa Concordia Example
On January 13, 2012, while negligently navigating too close to the Italian coast, the cruise ship *Costa Concordia* struck an underwater rock, resulting in a severe hull breach that soon led to sinking and capsizing with a loss of life of 32 passengers.[30] The onboard ship emergency response operations were severely inadequate, but we leave any additional discussion up to history to fully reveal and report. For our purposes, before the ship sank, the command-and-control responsibility rested with the ship's crew, especially the captain. Once the ship had sunk, the situation crossed the boundary, putting civil authorities in command. During the incident and rescue process, the ship's captain prematurely abandoned the ship. Civil authorities, under their mandate, had the power and legal authority to order the captain's return to the ship, which they did emphatically.

The plan was for the captain to manage onboard activities because he was trained and would use existing command-and-control procedures as well as equipment. The civil authorities would manage things for everyone who had left the ship or required more extensive rescue. We add for clarity that the captain refused to return to the ship; the onboard evacuation and rescue operations were crucially inadequate and unnecessary lives were lost.

Current Situation is Overdemanding
Operating at times of stress or unusually abnormal situations can be daunting. The enterprise should have developed a specific and clear policy for ensuring that escalation is considered for these situations. Section 10.19 covers a general protocol for enhanced operational focus and resources for situations having a reasonable expectation to be manageable. Section 10.15 specifically covered this within permission to

---

[30] "Costa Concordia Disaster," Wikipedia, last modified June 19, 2018, http://en.wikipedia.org/wiki/Costa_Concordia_Disaster.

operate for situations that are unlikely to be successfully managed by any attempts to continue operation.

Escalation Design

Enterprises provide specific protocols and training for conducting escalation. Covered topics include the following:

- When escalation might be a good idea
- When escalation would be required
- Who can request escalation and how
- When escalation is a collaboration and when is it a transfer of operational authority
- When escalation is ended and how is it done
- What the individual responsibilities are, including who will document the escalation and return to normal, including reports, MOCs, further repair or remediation, and lessons learned

Escalation in Perspective

Escalation is a tool, not an end result. Using it can permit a plant or operation to bring additional resources for managing. All existing enterprise policies, procedures, safeguards, requirements, and public responsibilities remain the same before, during, and after escalation.

10.17 Escalation Teams

> An emergency is a situation that requires immediate attention, but is usually small [or limited] in scale. . . . An emergency can turn into a disaster if left unchecked. However, not all disasters are preceded by an emergency.[31]

An abnormal situation is one in which it is no longer possible to continue operations (production, manufacturing, etc.) using the normal procedures. If left unmanaged, it can become an emergency that can endanger safety, financial integrity, and environmental containment.

---

31 "The Difference Between a Crisis, Emergency, and Disaster," Lighthouse Readiness Group, accessed March 26, 2016, http://lighthousereadiness.com/lrg/difference-crisis-emergency-disaster/.

Operational Modes:	Plant States:	Critical Systems:	Operational Goals:	Plant Activities:
Emergency	Disaster	Area Emergency Response System	Minimize Impact	Fire Fighting First Aid
	Accident	Site Emergency Response System		Rescue
Abnormal	Out of Control	Physical and Mechanical Containment System Safety Shutdown, Protective Systems,	Bring to Safe State	Evacuation
	Abnormal	Hardwired Emergency Alarms DCS Alarm System	Return to Normal	Manual Control & Troubleshooting
Normal	Normal	Decision Support System Process Equipment, DCS, Automatic Controls Plant Management Systems	Keep Normal	Preventative Monitoring & Testing

Figure 10-11. Plant state versus operational modes and goals identifying the situation management escalation region (orange dotted box) in which the operator changes the objective from trying to return to normal to bringing the plant to a safe state (perhaps shut down if that is the only one appropriate). When sufficient time is available, a purpose-defined escalation team can best manage this mode of transition.

Let us place this situation within our existing operations framework diagram. Whenever the operational situation is within the orange dotted box (Figure 10-11), timely escalation should be considered to provide the next level of situation management. A plant operating within the orange dotted box is considered to be "abnormal not in control." It is highly unlikely that any operational changes could restore the plant to normal operation. Left unmanaged, the plant will likely endanger personnel, produce heavy equipment damage, cause large financial losses, and threaten environmental containment. Unless the operator acts immediately under the situation of loss of permission to operate, the abnormal situation management escalation team (called *escalation team*) takes over in this situation. If time permits, the team will assist the operator to render the situation safe. The timing of when they begin operation is covered under the topic "The Escalation Activity" later in this section.

The roles of the escalation teams directly relate to a core deliverable of this book: the ability of an enterprise to provide the effective infrastructure to successfully manage operations. These teams are called into action when the abnormal situation goes beyond the expected capability of the operators. These teams are manned from a predetermined list. As soon as an escalation is declared, the team gets together. They can use special rooms outfitted to support this situation. The rooms may have other uses for emergency, crisis, or disaster management. The team functions as long as needed and performs activities in support of predetermined plans governed by the particulars of the current situation. Once the abnormal situation has passed, the team disbands. Although there may be reports and follow-up investigations done by certain members of the team, for the most part, the team responsibility ends with the end of the event. They also have important roles in addition to managing abnormal situations. This section covers those roles.

It is unlikely you have this team at the ready, fully trained, with a known room or facility within which to operate. You have emergency, crisis, and disaster management teams because you understand that some situations become serious enough that managing them requires a level of expertise and authority way beyond operator control. You have explicit protocols under which they function and their composition. This section introduces the additional concept of escalation management teams. Their function is to be able to augment or take over the operator's role with enough experience and resources to prevent the abnormal situation from becoming a full-blown emergency. Where an emergency cannot be avoided, their efforts will reduce the seriousness of the impacts. The appropriate manning of these teams and their roles during operation will be left up to the specific enterprises.

Escalation Team Composition

Enterprises will want to take advantage of the special capabilities of the individuals comprising the team. And those special capabilities are extensive. There are members who have the most experience with the design, operation, and risks of improper operation of the plant or enterprise they support. There are experts in the detailed equipment design and operation. There are experts in the handling of extreme changes in the operating environment that cause or contribute to the emergency. And there are experts in understanding human limitations during severe conditions. They have the ability to engage in the performance efforts needed for success.

Most of the members of the emergency management team are senior in their respective fields. Therefore, they are composed of an appropriate number of the following:

- Senior plant management
- Operating superintendents and senior operations leaders
- Equipment experts in design and operation
- Maintenance supervisors and senior members of the maintenance team
- Senior logistics personnel
- Senior safety, health, and environmental personnel

The Escalation Activity

It is important to understand when operational control should transfer from the operator to the escalation team. The operator is always there. The escalation team is not. It will need to be assembled, and doing so will take some time. We do not want to call out the team unless there is a reasonable likelihood that we should. On the other hand, we do not want to be caught flat-footed by not having them at the ready to do their job when it turns out that we need them. Each enterprise should lay down explicit direction for this to happen properly. What follows are some suggested considerations. Attention is required to provide enough lead time after notification so that they can actually be assembled (either physically or virtually) into a functioning team. Here are some of the triggering events:

- Any single highest priority process alarm is activated.

- Two or more independent second highest priority processes alarms are activated.

- The operator decides to initiate a shutdown of a significant portion of his operational responsibility to manage an out-of-control abnormal situation.

- The operator requests the team (likely where the operator has significant concern about the operational conditions being improper but shutting down has its own important risks).

- The supervisor requests the team (where the operational situation is likely to undergo significant stresses, such as severe weather or significant operational demands, that threaten the integrity of the plant or impact essential product delivery).

Once the escalation team has been called, it is the joint responsibility of the team to decide if and when to take some or all operational control from the operator. Usually, the operator joins the team with specific functions under the direction of the team lead. In addition, the act of assembling the team should be sufficient notice to the enterprise to alert the emergency, crisis, or disaster management teams to the situation and prepare to interact with them as needed and prescribed by enterprise protocols.

Frontline Coaching and Mentoring

At this point the escalation team is in the control room and prepared to do their thing. Here is the perfect opportunity to coach and mentor the operator. Before that can be done, the team members need to reassure the operator that the callout of the team is positive and responsible. It should in no way be considered a defect in operator actions or skill. The fact that the operator or someone else was concerned enough to call the team is due to the overriding desire for safe operation and protection of the community. It is for that reason that the escalation team must reassure the operator that this safeguard is the right thing to do.

As they go about the work of readiness or taking charge, the escalation team will be actively engaged in "thinking out loud." This means that each member of the team shares his thoughts and plans with the operator. Where the team acts collectively, someone interprets their actions for the group and the operator. Each question, each concern, each search for status and other information is accompanied by an introduction before it is done and afterwards an explanation as to what was found and how it fits or not. Everything unexpected is voiced as it happens. All anomalies are discussed with what they might mean and what will be done next. What-if questions work well. Following this process ensures that the team functions as a team. It also exposes the operator to the entire process to build experience and confidence. At the end of the day, the enterprise has the benefit of the best operating support it can have, and the operator has the opportunity to learn firsthand about how it is all done.

Readiness Evaluation Role

There are other important ways to take advantage of the abnormal situation management escalation teams for synergetic purpose-improving situation management capabilities. The first topic is readiness evaluation. Escalation teams have the ideal makeup to plan for and conduct readiness evaluations, not only for themselves as a team, but also for the enterprise as a whole. Readiness evaluation is not about evaluating the emergency management planning or plans. That is done outside of readiness. Readiness evaluation is knowing how prepared the support and operating personnel are before the emergency appears. The better they are at their jobs, the less likely an emergency is to appear. And if and when one does, they are better prepared to recognize one and to support the crisis management teams when they take over. Readiness evaluation might include the following:

- Extent and effectiveness of the plant or enterprise design for operation
- Adequateness of the control room design to support undistracted and effective operations
- Effectiveness of operator (and other important support personnel) fatigue assessments and management programs
- Effectiveness of the design of the human-machine interface (HMI) to support situation awareness, maintaining good operation, and emergency operation

- Extent and effectiveness of the operating procedures to guide the operator to properly manage

- Extent and effectiveness of the shift handover process to capture on-shift knowledge, pass it on to the next shift, and assist the next shift to assume control and successfully manage

- Extent and effectiveness of the plant or enterprise maintenance and support to keep everything fully operational and ready

Try to resist the temptation to just delegate this to others. That is where most plants and industrial operations are now. Just to be clear, it is not possible to delegate experience and judgment. It is not possible to delegate the benefits of seeing senior enterprise members walking the walk and talking the talk.

Training

Escalation management teams probably have the best understanding of the basic operation of all equipment and process elements in the plant or operation. Who better is best suited to "invent" operational irregularities and abnormal situations? Who better is best suited to test and coach weak signal processing? Good weak signal processing requires in-depth understanding in order to

- recognize weak signals from operating data and displays,

- connect each weak signal to its worst likely situation (forward-extrapolation),

- list the likely early indicators of the worst likely situation, assuming it to exist (backward-projection), and

- find all early indicators of the worst likely situation that are actually present (evidence).

Who better than the members of the emergency management team or a special escalation team would be best able to take near hits (near misses) and use them for effective training scenarios? And at the same time, these team members would be coaching and training operators, shift supervisors, and other midlevel management and support leaders.

Abnormal Situation Management Process Model

This is perhaps something new. Effective abnormal situation management requires a process model that is very different from the one used for normal operator control and management. The control model shows what the operator must do to keep the plant running well in order to make good product and do it efficiently. That model would have lots of inputs (which operators would set or watch) and lots of outputs (which operators would try to keep in good places for proper production). An emergency management process model should not center on how to manage an emergency or direct the team. Those topics are well-documented in the literature. The type of model here should expose the "fingers" and "toes" of what the emergency can wreck if it works some of its untoward ways. As such, an abnormal situation management process model should depict the exposures to fire, explosion, excessive pressures and temperatures, and such, as well as exposures to environmental releases. It should have lots of inputs (which escalation teams would use to prevent or mitigate serious problems) and lots of outputs (e.g., explosivity, toxicity, and danger to safety). It should, of course, be related to the actual plants or processes, but not care about efficiency, quality, or anything else about normal production. This model should use the dangers to be managed as outputs. The model should be used to know what to watch for, what handles to use to mitigate or manage, and what turning the handles would likely do.

Just to be sure we understand the situation, these models do not now exist. It is not a practice to develop them either as part of the plant design process or afterwards. Yet without them, how can the escalation team adequately predict where the not-in-control abnormal situation is likely to end up? Without reasonable models, the team members are forced to rely on their individual understanding of what parts of the plant are likely to break and how. Members will have their own versions. Without a shared model, the team will have to develop one on the fly or accept one of the individual models as their collective working one. Either way, time is lost and chances for error elevated.

It is recommended that abnormal situation management process models are developed as part of the normal hazard evaluation process after design. Specifically, a good time to do it would be as an integral part of the HAZOP analysis, consequent quantitative risk analysis, and layers of protection analysis. And just to remind the reader, the results of those processes are also needed for proper assignment of alarm priority and other alarm response protocols. The actual design and specifics for these models are beyond the scope of this chapter and book.

Weak Signals for Abnormal Situation Management

Abnormal situation management process models available for the emergency team to use should be quite suitable for weak signal analysis by the team. Like any other model, they should be able to depict small challenges to process integrity before they become full on emergencies. If the unfolding time frames for these emergencies were slow enough (and history is full of terrible incidents that unfolded rather slowly in time), weak signals would likely be present and could be used for situation awareness similarly to what the operator does during normal operation.

10.18 Command and Control

The previous section on escalation may have had the effect of seeming to blur the boundaries between command and control in the control room. Actually, those boundaries are reinforced by that discussion. Take a closer look. At every juncture, unless the operator explicitly agreed and relinquished operational control to another, that responsibility remained with the operator. Yes, there are a few very special cases where the decision was removed from the operator without requiring agreement. One was where the supervisor (who must be in the direct chain of command with operators) ordered the change. The other was where the operator was functionally unable to understand and conduct duties due to incapacity or absence. Both are very clear exceptions.

An important situation still remains for discussion. That situation is when the operator in charge is given a difficult operational order by a responsible supervisor in the direct chain of command. It is generally recognized that the sitting operator is the single individual responsible for all hands-on operational management. *Hands-on* means taking actions. It also includes responsibility for the consequences of not taking actions when actions should have been taken—in formal words: *errors of commission* and *errors of omission*. What might happen during a shift to affect this operator responsibility? If the operator feels that the specific operational order is contrary to the legal, safe, and responsible operation, what is the proper course of action? To the author's knowledge, there are no statutory requirements that address the situation (violation of law is the exception, of course). What follows is an explanation of the circumstances and possible suggestions. Every enterprise must identify the necessary situations and put policies and procedures in place for appropriately managing them.

There are two situations. The first is where the supervisor is not formally qualified to be an independent operator. The second is where the supervisor is formally qualified.

Supervisor Not Fully Qualified to Be an Operator

To be clear here, the supervisor is expected to be fully trained and qualified in a supervisory role. This is prudent but not the concern here. Rather, the concern is that the supervisor is not fully trained and qualified to actually sit in the chair as an operator for the specific plant or operation in question. We take a lead from the US Department of Transportation Pipeline and Hazardous Materials Safety Administration (often referred to as PHMSA) regulations.[32] The only individual allowed to make operational changes from a control room or to order, from the control room, others to make operational changes under the operator's authority must be a fully qualified operator for the specific control room. Therefore, for all operations managed from a control room, no unqualified individual may make, or order the operator to make, any operational change (or to prohibit the operator from making an operational change). They may suggest changes but may neither expect nor require the changes to be followed. This prohibition extends from the lowest-level supervisor to the highest-level enterprise authority.

This would seem to be a prudent policy to adopt by all enterprises. The only exception, and this is an important exception, is the supervision's authority to safely and immediately order a cease of operations. This is retained by anyone in proper authority.

Supervisor Is Fully Qualified to Be an Operator

To be clear here, the supervisor is fully trained and qualified to operate the plant or enterprise for which an operational order is given. What must the operator do if he feels that the operational order is unsafe or will result in unlawful operation, or will likely lead to unsafe or unlawful operation? I cannot prescribe a remedy here, yet this situation is all too real. It must be adequately, formally, and fully covered in the enterprise policies and procedures.

A few suggestions might be useful for consideration. Once the suspect order is given or the operator thinks the order is given, the operator does the following:

1. The operator restates the order to the supervisor and waits for confirmation from the supervisor that the operator got it right.

2. The operator (or anyone else on the operations team) invokes the principles and tools for crew resource management (CRM) by respectfully questioning the order and suggesting why the reasons following the order will or may result in unsafe, unlawful, or other significant damaging consequences; and suggesting alternates that would be acceptable to both the operator and the supervisor.

---

32 49 CFR 192.631, *Control Room Management, Transportation of Natural and Other Gas by Pipeline; Minimal Federal Safety Standards* (Washington, DC: PHMSA [Pipeline and Hazardous Materials Safety Administration]); 49 CFR 195.446, *Control Room Management, Transportation of Hazardous Liquids by Pipeline; Minimal Federal Safety Standards* (Washington, DC: PHMSA [Pipeline and Hazardous Materials Safety Administration]).

3. The operator formally states that he objects to following the order because it is unsafe, is unlawful, or will cause significant damage. Said objection should be witnessed and documented.

4. If the operation is unlawful, the operator has a legal responsibility to not follow it. No supervisor or enterprise manager has the authority to ask, or much less order, anyone to commit a criminal act.

5. If the operation is unsafe or damaging, the operator may, at his option,

    a. request a relief from operational responsibility (meaning that he would not be operating and that someone else will follow the order), or

    b. follow the order with the documented full understanding that it is done under protest and that the ordering individual accepts full and complete responsibility for all outcomes including any errors made in good faith by the operator.

All enterprises are advised to visit this situation and adequately plan for it in a responsible way.

10.19 Alternate Safe Operating Modes

When permission to operate is lost, the operator must move or place the operation to a safe condition or state. Usually, this means shutting down all or part. Here are some of the practical ways to employ the concept.

Operator-Initiated Shutdown

The standard emergency shutdown is familiar to all current industrial manufacturing plants. In one form, the US Occupational Safety and Health Administration mandates its presence. It is currently designed as a last-resort control measure. We call it an emergency shutdown (ESD). Operator-initiated shutdown is one small but significant step back from that. For this, much more care is taken to do things in an orderly, controlled, but direct manner. It is a predesigned, pre-implemented capability that the operator can initiate from the PCS that will start a largely automated sequence of shutting down the unit. This is important. We need to make sure that this "solution" does not place unnecessary risk on the enterprise. Implementing this capability usually requires the replacement of most manually operated field components with ones that can be operated remotely. A few manual ones can be left, but the "people action" must be carefully coordinated. When implemented correctly, this will ensure that the plant will be in a much less damaged shutdown state than if the ESD system had done it. This should eliminate or significantly reduce equipment damage. As a result, it should facilitate a faster restoration of production when the plant is ready to restart. Of course, those situations that could not be properly managed by this controlled shutdown will be caught by the ESD, thus ensuring process safety integrity foremost.

Automated Shutdown

There is an actual enterprise strategy that nicely introduces this topic. Unfortunately, the document reference to this has been lost so there is no convenient record. There was a small chemical plant that faced the problem of managing upsets in a very straightforward, logical, yet unusual way. The management recognized that there might be a very dramatic payout for exploiting the operator load differences between normal operation and upset or abnormal operation. Studies have traditionally shown that during normal operation, operators are utilized about 25% of their time. But during upsets, they are loaded to 300% of their ability. Because operator load is not elastic, this means that up to two additional operators might be needed during upsets. This flexibility is not economical for a small plant. The solution was to invest in enough instrumentation and controls equipment to fully automate a reliable shutdown and start-up. As soon as the operator determined that the plant was unusually abnormal, rather than attempting to work out what might be amiss and try and fix things, the operator instead initiated the automatic shutdown process. The plant was then quickly and properly shut down.

The result was minimal production losses with only a rare equipment problem. Any equipment problems that did happen were almost always due to the abnormal situation that caused the situation rather than to a flawed shutdown. Once the system was shut down, the operator would attempt to diagnose the problem and effect a repair or call in others to assist. Most problems could be promptly remedied. Once this was done, the operator would initiate the automated start-up process, and the unit was back online in a timely manner. The operator did the job well and, at the end of the day, saved the plant quite a bit of money. In the final analysis, the combined savings in manpower and avoided equipment damage far outweighed the expense of the automation and its associated upkeep. Not surprisingly, during normal operation, the plant actually ran better due to the added instrumentation and operational flexibility.

Alternatives to Shutdowns

Shutting down can be harsh. Done carefully, it is time-consuming. It is expensive, and it is fraught with opportunity for more trouble with equipment damage and poor product. But it does the job. In the next sections are options for doing the job, but in a less impactful way. They are only used if they can be effectively done.

Safe Alternatives

The recurring principle of operating is *not to do something that does not have a procedure, has not been effectively learned, and has not been practiced*. Operating is not inventing. But there is a broad line between invention on the one hand and effective use of knowledge on the other. We examine this next. We expect operators to use their understanding of process operation to safely plan and execute operational actions that are true to their understanding and abilities but not necessarily formally found in procedures. We rely on their ability to understand a situation, evaluate alternatives, and implement a chosen one. It is one of the important reasons why we have operators. But to be allowed to exercise this level of responsibility, we will need a few basic guidelines. Operators may exercise their experienced operational initiative only if all the following are true:

1. The current abnormal situation is clearly and fully understood.
2. All external factors are either not limiting or accounted for.
3. The proposed solution (in this order)
   a. has less or equivalent operational risk than any other safe choice,
   b. will not violate existing specific or implied safe operating requirements, and
   c. has a better chance of being effective than any other safe choice.
4. After implementing the proposed solution, the overall operating situation will not expose the plant to increased operating risk or operational stress later on.

Example Illustrations

It is not possible to fully cover this topic to the degree that plants and enterprises can take the results and directly make policy. Rather, you will need to understand what is suggested here and plan what your enterprise should potentially implement. Here are some examples to think about as you begin.

USS Carl Vinson. [W]hen Dick Martin, the first captain of the [aircraft] carrier Carl Vinson, found himself in an intense storm off the coast of Virginia in 1983 [with planes in the air needing to land soon], the winds were so strong that he improvised by driving the carrier at 10 knots in reverse in order to reduce the speed of the winds across the deck and allow the aircraft to land more safely.[33]

---

33 Karl E. Weick and Kathleen M. Sutcliffe, *Managing the Unexpected: Resilient Performance in an Age of Uncertainty*, 2nd ed. (San Francisco: Jossey-Bass, 2007).

Mann Gulch Fire. There was an out of control forest fire in Mann Gulch in Montana. Dodge was firefighting team leader. To save himself he headed into the fire, found a small grassy patch, and set it on fire. It saved his life. This maneuver had never been done before; he used principles from other situations and applied it on the spot.[34]

Safe Park

Full automation for start-up and shutdown is normally not considered for a complex processing unit. Moreover, many are highly integrated with neighboring units. Consequently, having an option that is capable of providing additional time for the operator to work problems and at the same time greatly reduce exposure to the danger and damage of continuing to operate would be very valuable. That option is *safe park*. Safe park is a methodology that designs appropriate ways for production units to move to a much safer and more reliable operational mode to buy time and at the same time to avoid the problems of an extensive shutdown. It provides a valuable way station between full operation and shutdown can buy time for remediation, reduce the risk of unwanted consequences, and permit a smoother and more efficient return to normal operation when it is okay to do so. For a safe park, a set of process operating states are identified that can be expected to render the process in a place that is safe and somewhat operable, though far from productive. The objective is to stabilize the process and avoid the necessity for complete shutdowns. The constructive result is an ability to forestall the impending consequences likely to come as a result of the abnormal situation in progress—that is, to slow down, or freeze time. Time is the most valuable commodity that an operator can have during an abnormal situation. If enough time is available, almost any situation can be managed. Moving to a safe state can be extremely effective when employed for downstream units during an upset in the upstream ones. This process is also quite effective for stabilizing an upset unit itself under certain conditions. These safe states are carefully identified and controls put into place to assist the operator. Operators following a carefully scripted procedure do the transition manually.

For each needed safe park process unit, a carefully worked-out scenario is developed that permits it to be parked instead of shut down. Simple examples are placing distillation towers in total recycle, putting pumps on low power with sufficient kickback to avoid operational problems, and placing a furnace on low fire but in a way that controls thermal stresses and inventoried product problems. Other examples from transportation systems would be to side-track a train off the main track when another

---

34 "Mann Gulch Fire," Wikipedia, last modified April 21, 2018, https://en.wikipedia.org/wiki/Mann_Gulch_fire.

train may be attempting to share it but that train is in an unknown position. Air traffic controllers will use this (to airplane travelers' frustration) to place all airplanes in a holding pattern around a potential airfield until a weather problem or another emergency is sorted out. Note that some "parking" cannot be done forever. It is a tool, not a solution.

The particular unit will govern the specific parking details, of course. Any additional needed instrumentation and controls are added. Operating procedures should be expressly designed for this activity, and the corresponding protocols need to be in place.

Keeping Perspective

By establishing the criterion for understanding the plant's operating status, have we replaced one dilemma with another in the guise of improving operations? Have we simply transferred the problem? The answer is a definite *no*. The reason is basic and important. If the operator ever fails to understand what his process is doing—no matter what the reason, no matter what additional technology could have been provided to help him to see what is wrong and find a prompt and effective solution—at this very moment he does not know what is happening. Using the permission to operate model, he is obligated to cease operations and move to safe or shutdown state.

The more effective the operator is, the better job he will do at foreseeing the abnormal and managing it before things get out of hand. There are two key areas where technology can assist in the application and effectiveness of the operator. The first is in the area of detection and warning of abnormal situations. That is the stuff of situation awareness in Chapter 6, "Situation Awareness and Assessment." The second is in effectively placing the process in a safe state or shutdown. That is what was covered here.

10.20 Managing Major Situations

There are operational situations that are so difficult or require such concentrated activities that the normal personnel roles for them can greatly benefit success if they are modified. The next tool we add to the operations kit is situationally based, dynamic operator deployment. This is a way of shifting operator roles during major abnormal situations. These role changes will have a profound effect on both the operator's understanding of the situation and his ability to manage it effectively. Here our operator has not lost permission to operate yet; however, the situation is likely to exceed his personal ability to keep on top of it. Before it does, and knowing that it might, we build additional tools for it. Let us turn our attention to what determines a major situation.

Major Situations

Normal situations are the stuff of everyday operations. Things stop unexpectedly so the operator must find out what happened and why, and then figure out and execute a resolution. Things break. Raw materials do not always stay the same. Products might need to be changed or modified. What characterizes these normal situations are their somewhat limited nature, their familiarity, and their ready management. This is not to say that they are trivial. Their impacts vary from the barely noticed to a nasty shutdown that might have been avoided had maintenance been better. So while we are not too happy when they come, we accommodate them without undue add-on risk.

Major situations are different, much different. They risk extensive plant damage, personnel injury, and environmental degradation. They come very infrequently, and when they do, they are never the same. Yet, properly managed, their impact and cost can be managed as best as reasonable. Here is an effective way to do that management. You should consider adding these or similar procedures and technology to your situation management kit.

The first item of business usually is to declare a *restricted* or *sterile control room*. Next, the control room personnel are reorganized as discussed next.

Operator Redeployment

We first show a single control room operator, a single "chair," to explain the components in the figures to follow. Call him the console operator. Figure 10-12 illustrates this. Our operator is located at his operating console that happens to have five video display units. He also has the usual accompaniments of telephone, plant radio, and such; these are assumed present but not depicted in the figure.

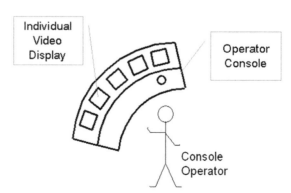

Figure 10-12. A single control room operating "chair" showing operator console and typical video displays.

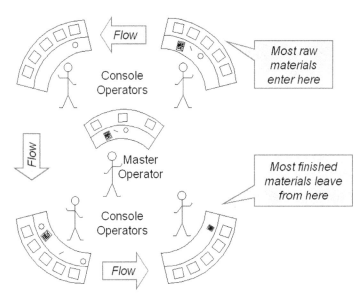

Figure 10-13. Control room with four console operators and a supervisor illustrating an arrangement where each operator, in turn, manages a part of the plant before sending product downstream to the next operator.

Let us put our single operator into a central control room with three others and a supervisor. Figure 10-13 shows a typical plant arrangement. The basic and primary materials flow is shown progressing from the upper-right operator counterclockwise through all four stations. The primary exporting is done from the plant managed by the operator of the lower-right station. Each operator has his own area of responsibility with separate activities. The operators' interaction with each other is normally limited to product coordination, though they might also relieve one another for short breaks. In some plants, these operators rotate responsibilities on a scheduled basis to ensure cross-training, if that is needed.

Let us suppose that the second operator (the upper-left chair in Figure 10-13) experiences a problem in a part of his responsibility. That problem is depicted in Figure 10-14 as a red screen on the display at the left end of his console. We know that screens and displays are not dedicated to specific plant equipment, but let us just assume this case for illustration.

Let us take a closer look. We presume that the problem is significant and involves about 20% of his plant. It is significant, because it is in red. Remember the graphical color conventions. Its extent is about 20% because only one display is involved. Remember that this example is figurative. At this time several important operational events happen in quick, preplanned, and proceduralized steps. The first is shown in Figure 10-15. The last is shown in Figure 10-16.

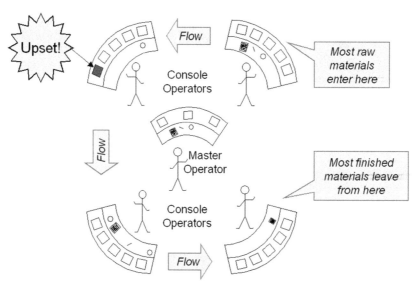

Figure 10-14. Control room with four console operators and a supervisor depicting a serious upset at the second (in the flow of production) operator position.

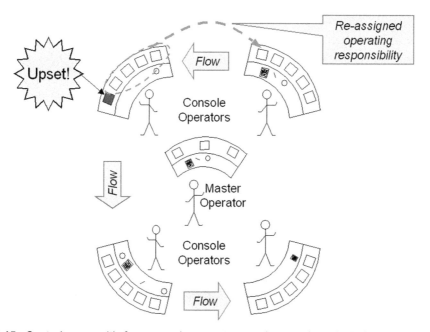

Figure 10-15. Control room with four console operators and supervisor depicting a serious upset at the second (in the flow of production) operator position and showing the (temporary) assignment of his operating responsibility for his non-upset units to the first upstream operator.

The procedure begins:

Step 1. The remaining 80% of the console operator's operational responsibilities are passed to the control and management of the nearest upstream operator. In our case, it is

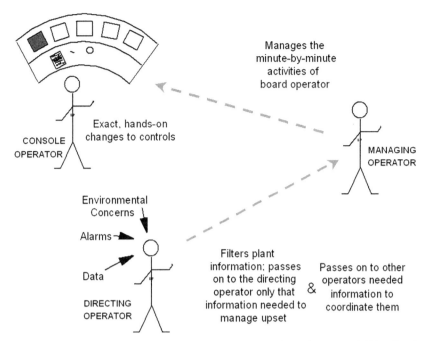

Figure 10-16. The upset operating situation showing the assigning of responsibilities for the escalation team. The console operator is now only responsible for all hands-on changes and eyes-on monitoring of the upset unit. The managing operator is responsible for planning the proper operator responses and then asking the console operator to execute each step one at a time. The directing operator is responsible for filtering out all plant situations except for what is needed to understand the current operating risks.

passed to the closest operator upstream in the material flow path—upper right. It is passed there for two reasons. First, this operator will be the least influenced if the current problem is not managed effectively. Second, this operator may be able to make changes in his operation to better enable the affected operator to better manage. If there were not an upstream operator available, the responsibilities are passed to the operator farthest downstream.

Step 2. Our console operator remains on station. However, his focus is narrowed to include only the limited portion of his plant that is upset. For the moment, his only task is to locate the appropriate displays and control "handles" that might be needed for management. No independent actions are taken. Only at the direction of the managing operator (explained in step 3) will he perform clear and distinct "hands-on" control and other state changes to plant equipment that are possible from the control room. His counterpart in the field will do the same for field equipment.

Step 3. Another position, say for example that of the master operator (Figure 10-16), steps in. This post may be filled by the operating supervisor for the shift, by a senior floating operator, or by any other prequalified operations individual. This managing operator will, for the entire course of the upset, be responsible for the minute-by-minute operational tactics to be used to manage this situation. The managing operator will direct the console operator to perform single tasks, one at a time. Only the current task will be told to the console operator. When that task is complete and confirmed, he will be told the next task. This procedure is continued as long as needed.

Remember, only one task at a time is on the agenda for the console operator. It is up to the managing operator to decide what to do and in what order.

For example, the managing operator might instruct the console operator to perform the following tasks in sequence, one at a time, each announced by itself by the managing operator:

1. Put the reflux flow controller in manual and set the valve output to 20% flow.

2. Stop the pump on the rundown tank.

3. Reduce the steam flow to the reboiler to keep the temperature in the upper tray area around 165°F.

The tasks continue until the current specific tactical plan of the managing operator has been carried out. If escalation or additional delegation is needed, the *managing operator* makes appropriate requests to the *directing operator*. It is the responsibility of the directing operator to determine the nature of the need, locate appropriate resources, assign responsibilities, and manage them. As a result, the directing operator may eventually be supervising two or more managing operators at the same time.

Continuing with our procedure:

Step 4. A directing operator (Figure 10-16) steps in to assist the managing operator by providing all external situational information needed to fully understand the rest of the plant and its state. His job is to be aware of the entire rest of the plant and only pass on to the managing operator what is needed to manage the immediate and specific threat posed by the upset event.

Many plants will quickly suggest that they do in fact use this concept during upsets—not always, but enough to suggest to themselves that this might not be anything new. That should be comforting to those of you reading about this for the first time. Good, but not good enough. When those same plants, claiming that this is a practice, are asked if it is part of the formal plant procedures, they will say no. When asked if it is practiced to build skills and competencies, they will again say no. When asked if there are any other ways that this is incorporated into the operational infrastructure, they will again respond no. I leave it to you to draw the lesson.

You might be interested to know that operator redeployment is an actual working procedure for US nuclear power plants to manage upsets. It is one form of an *emergency management escalation team* discussed earlier. It has been used with considerable success. It is proceduralized. It is practiced. It is part of the fabric of operations.

10.21 Control Room Situation Codes

This section does not document established protocol or design. It is new to the control room. The concept is borrowed from medical operations used in hospitals.

It is essential for everyone in the control room to know rapidly and without ambiguity that "things as usual" have changed and changed in a big way. Declaring a "code" is the broadly recognized method used by hospitals and other emergency management facilities. This instant realization cuts through everyone's current activities and mental models of what is going on. Everyone knows that at this moment and forever until managed, they are facing a clear and present danger. Knowing this, and having clearly defined roles, tools, and protocols, enables the operations team to make the transition to a heightened state of awareness and initiate purpose-built processes for response. Control room situation codes may be combined as an integral part of a restricted control room or sterile control room discussed earlier.

One of the significant impediments for a team to realize that their world has undergone a change extensive and risky enough to require a dramatic change in its management is the lack of a clear "tipping point" that all can immediately relate to. A code declaration is a proven protocol used by hospitals to do just that.[35] An example is a *code blue* at a specified location. A code blue means that there is an individual (patient, visitor, or otherwise) in the hospital at the specified location who has stopped breathing (and is therefore turning blue). The preplanned response involves the following actions: (1) all preassigned individuals immediately move to the specified location; (2) one or more of these individuals finds and delivers the required predefined and prepared "cart" of equipment (crash cart) and other equipment to the specified location; and (3) the preassigned individuals now function as a team, each performing required activities under the preassigned team leader, until the event is over (announced by a *code clear*). Once code blue is declared, everyone in the immediate area of the problem knows that whatever they have been working on is now not the problem to work on.

---

35 Malcolm Gladwell, *The Tipping Point* (London: Little, Brown and Company, 2000); "Hospital Emergency Codes," Wikipedia, last modified June 16, 2018, https://en.wikipedia.org/wiki/Hospital_emergency_codes.

Whatever activity was underway or planned for the near future is shelved. No one is thinking about how to approach the old situation. This code is the current reality.

The following is an example of a code blue description:

"Code Blue" is generally used to indicate a patient requiring resuscitation or in need of immediate medical attention, most often as the result of a respiratory arrest or cardiac arrest. When called overhead, the page takes the form of "Code Blue, (floor), (room)" to alert the resuscitation team where to respond. Every hospital, as a part of its disaster plans, sets a policy to determine which units provide personnel for code coverage. In theory any medical professional may respond to a code, but in practice the team makeup is limited to those with advanced cardiac life support or other equivalent resuscitation training. Frequently these teams are staffed by physicians (from anesthesia and internal medicine in larger medical centers or the Emergency physician in smaller ones), respiratory therapists, pharmacists, and nurses. A code team leader will be a physician in attendance on any code team; this individual is responsible for directing the resuscitation effort and is said to "run the code."[36]

A classic illustration of individuals not recognizing that their current situation had changed from routine to *anything but routine* is the story of the 1949 Mann Gulch fire blowup in Montana.[37] The general story goes like this:

The fire started when lightning struck the south side of Mann Gulch at the Gates of the Mountains, a canyon over five miles long that cuts through a series of 1,200 foot cliffs. The place was noted and named by Lewis and Clark on their journey west in 1805.

The Mann Gulch fire was a wildfire reported on August 5, 1949, in a gulch located along the upper Missouri River in the Gates of the Mountains Wilderness, Helena National Forest, in the state of Montana in the United States. A team of 15 smokejumpers parachuted into the area on the afternoon of August 5, 1949, to fight the fire, rendezvousing with a former smokejumper who was employed as a fire guard at the nearby campground. As the team approached the fire to begin fighting it, unexpected high winds caused the fire to suddenly expand, cutting off the men's route and forcing them back uphill. During the next few minutes, a "blow-up" of the fire covered 3,000 acres (1,200 ha) in ten minutes, claiming the lives of 13 firefighters, including 12 of the smokejumpers. Only three of the smokejumpers survived. The fire would continue for five more days before being controlled.

Most of the team was inexperienced. They had never fought a serious fire. Under the lead of experienced firefighter Wagner Dodge, the team had classified this as a "10 o'clock fire." Meaning that it was routine and they would have it out by 10 o'clock

---

36 "Hospital Emergency Codes," Wikipedia.
37 Norman McClean, *Young Men and Fire* (Chicago: University of Chicago Press, 1993).

the next morning. Dodge stopped to eat lunch as the first activity after parachuting down. To the men, this act further reinforced the situation as a minor one. Conventional firefighting activities began a few minutes later. Within a very short time after that, it became clear to Dodge that the fire was anything but routine. He ordered his men to drop their gear and run for their lives. However, the crew was unable to recognize that this routine fire of a few minutes earlier was now out of control and threatened their lives. The mere act of "dropping their gear" meant that they were abandoning the only role that they knew: to fight fires. Some did. Most could not, even though it meant that they were significantly slowed down. They could not change their ingrained initial assessment nor could they shed their identifying "badges" signified by the tools.

Dodge headed into the fire, found a small grassy patch, and set it on fire. He ordered his men to lie down in the already burned area. They did not. He did and the fire passed around and over him. He emerged completely unscathed. Two others managed to outwit the fire by finding a rock fissure. All the remaining crew perished.[38]

Every member of the Mann Gulch firefighting team could have been saved. They were not saved due to their own inability to accept that their initial evaluation was wrong until too late. Once they did, their raw fear kicked in. Instead of listening to their experienced leader's orders, they ran. The problem was their inexperience with rapidly changing operating paradigms. Control rooms have similar issues. One is the need to change operating modes in a controlled and effective manner. If they were to use codes, it would be clear to all that operations must change immediately and to what. Everyone would be aligned. Already in this chapter, you have seen many tools and protocols that can form the "pieces parts" of what an operating mode might be composed of. Codes are another tool. Within each category (its color), you will need to add the appropriate process and procedures to get it working at your site. Now to the codes and the individual parts of each.

Code Gray

Code gray is a situation where there is a significant threat to the integrity of a facility that comes from outside. Specifically, before the situation exists, the enterprise or plant is operating properly and faces little or no internal operational threats. The threat announced by a code gray is likely to cause a significant change in operations including curtailment, disruption, or danger. It must be recognized and operations modified to accommodate it. Examples include severe, disrupting weather or natural disaster on the near horizon; civil unrest likely to directly affect operations; impending loss of essential utilities like electrical power and natural gas, or sewer malfunction;

---

38 "Mann Gulch Fire," Wikipedia.

impending significant loss of ability to store or ship product or dramatic increase for producing product; rampant disease or risk of disease exposure that reduces availability of sufficient operations personnel; and more. Note that a code is quite different from one where these situations might develop gradually. These should be part of the enterprise strategic planning and implementation activity to design. A code must be preplanned, of course. But to qualify as a code, its appearance would be largely unexpected and the response tactical.

Code Orange

The difference between a *code orange* and a code gray is that for a code orange, the direct appearance of trouble is within the plant or facility itself. The situation would always be unexpected. There is imminent significant danger when it happens. Responses primarily require deep understanding of the particular inner workings and function of the plant and the ready availability of proper equipment and staff. Response activity requires careful coordination of personnel within the enterprise and effective communication with personnel outside. Example initiator events would be outbreak of fire; explosion; loss of containment of flammable, toxic, or excessive materials; failure of equipment that exposes significant operational stress; presence of unauthorized personnel likely to initiate dangerous operation or threaten safety of personnel; and more.

Ending Codes

Closing down a code properly is nearly as important as initiating one. Each code must have a clear ending procedure and process by which the plant or enterprise operation is declared to be inoperative, the plant or operation will go through a carefully choreographed plan to be returned to good operation, or good operation has already been restored as a result of the code response.

10.22 Operability Integrity Level for Online Risk Management

Chapter 2, "The Enterprise," introduced operability integrity level, which could be an efficient measure of the overall risk posed by an operating entity (plant, power grid, etc.). *Operational integrity* is the inherent capability of an enterprise to be responsive to proper operation. *Responsive to proper operation* meaning whether or not an enterprise is capable of being appropriately managed by automatic controls and other instrumented mechanisms supplemented by human individuals who monitor and intervene where necessary.

Suppose one could measure and then track the enterprise operational risk as it might change from hour to hour. When it was within an acceptable level, the operator would be devoting his efforts and time to identify and manage specific situations and

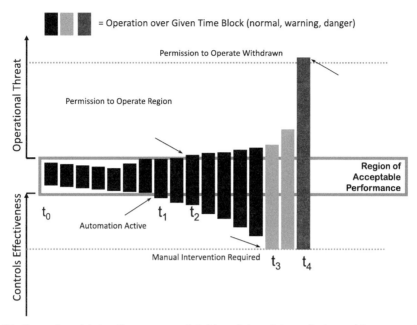

Figure 10-17. Operational integrity measure "dial-type" trend icon that provides operators with an early notification of building abnormal operation (eventually indicating that the plant is not going to be adequately managed if kept operating).

events he was able to see and control. On the other hand, if it were to increase above an acceptable risk level, then no matter what the operator may or may not see or be working on, prudence would dictate that operations cease or be sufficiently curtailed so that continuing to operate would pose no substantial risk going forward.

As an illustration of visual iconography or dashboards, Figure 10-17 depicts how the hour-by-hour measure of operational integrity might change. Operational integrity is a composite demonstration. Therefore, it is illustrated as simultaneous departures from both normal control and normal threats that are shown on a backdrop of how close the current operation is to being in trouble to the point of requiring specific operator intervention. Intervention includes adjustment to automatic controls, plant equipment configuration, and such. In case of a loss of proper operational ability, there would be the operational requirement for the operator to cease normal operations entirely and either "park" (Section 10.19) or shut down the operation.

At time t_0, the process is operating within acceptable performance. That is, the automatic controls are performing adequately and there is no operational threat (safety challenge, product quality or production rate concerns, environmental exposure, etc.). No explicit operator action is needed nor expected at this time. This good operation continues until t_1. At time t_1, the plant has become less than acceptably managed by the controls, but there is no operational threat. Something is amiss. An example might be that the process is emitting some pollutant into an effluent stream. The situation worsens at the next time step. Such a problem does not in and of itself suggest that the plant is in any danger of becoming out of reasonable management by the operator. The plant may continue to degrade without necessarily challenging the inherent ability of the operator to manage things.

At around t_2, the plant enters an operational region that can eventually lead to the inability of the operator to manage. But at this point, there is no real operability issue. What is ahead of t_2 is in the future. From t_2 until t_3, the situation continues to deteriorate with the plant becoming less and less responsive to the automatic controls. At time t_3, the appropriate plant controls should be placed in manual (if they are not already there by operator discretion). Operability continues to degrade. At time t_4, the plant has crossed the permission to operate line, so the situation is bad enough to be presumed to be unable to respond to any reasonable manual action. It must be "parked" or shut down in order to provide adequate situation management.

Looking forward, this type of information is being explored by technology providers who are building sophisticated process performance and risk exposure models for control, optimization, and production integrity.

10.23 Managing Biases, Overcoming Pitfalls

Now is the time for Chapter 7, "Awareness and Assessment Pitfalls," with all the pitfalls of situation awareness and situation assessment to catch up with us. That was the part about how important mental models are. That was the discussion of doubt and the stuff about uncertainty. We plowed through biases, the myth of multitasking, how our cultural heritage shapes our thinking, and the nature of our different personalities. Your dedication to understanding all of this and your kind patience must have overcome a natural desire to "throw in the operational towel" in the face of what must seem to be unbendable obstacles in the way of your success in the control room and keep reading.

You did not stop. Or maybe you jumped over the bad stuff and cut to the chase. I am thinking you have opened most of these chapters. And when you did and now take a brief step back, you are able to see the big pieces falling into place. You have seen how a carefully done control room with a competent HMI can provide important overview and framework information to maintain good overall situation awareness. You have seen how the alarm system and weak signals used together can enable operators to attend to existing problems as well as see emerging ones before they mature. And in this chapter, you have seen how high-level tools like managing situations, permission to operate, safe park, and escalation provide both a working methodology and a safety net to bridge the ever-present gaps.

This book was designed to help you build an understanding of what is available when you approach significant operational improvement and offer ways of doing that job effectively. By now you should be aware that all the tools and technology were designed only for the purpose of aiding you to build a strong control room. Overcoming the foibles of human nature and substituting good tools was the objective of it all.

10.24 Safety and Protective Systems

Safety and protective systems are installed to provide proper protection in case the actions of man cannot or do not adequately prevent serious problems. Their design, choice, installation, use, and maintenance are critical but lie outside the scope of this book. What is of importance to you is how they fit into the situation management picture. There are two brief points to make.

1. The safety system must not be used as a default state to justify inadequate or insufficient situation management. The fact that some safety component, such as a high-pressure relief valve, is in place should never be used to avoid appropriate controls or monitoring of the associated operation. *All challenges to the safety system must be as a result of failure of other planned remediation processes and activities.*

2. In individual cases where human intervention is likely to make matters worse due to a lack of resources, a failure to understand, or other reason sufficient to strongly suggest that initiating or continuing with manual operation (including attempts to shut down) is ill-advised, it is prudent to refrain from manual intervention and allow the safety system to protect.

10.25 Key Performance Indicators

How well the operator's activities are suited for effectively managing daily operations can easily be measured. What follows are some potential ways to measure them.

- **Alarm activations** – Each process alarm signals an aspect(s) of the operation that has gone astray to the point where operator intervention is required to remedy it.

- **Safety equipment activations** – Every activation (which is good in that it works to protect persons and other equipment) indicates an aspect(s) of the operation that has gone seriously astray and is currently beyond operator intervention to remedy.

○ *The list includes relief valve activation, automatic power shutdown, overspeed trip for rail transit cars, automatic load shedding, and much more.*

- **Poor operational performance** – Enterprises are operated for a purpose, and a failure to deliver implicates one or more critical aspects gone wrong.

- **Lack of proceduralized operations** – Appropriate operations require preestablishment of proper, effective procedures to operate over normal, abnormal, and crisis situations.

- **Failure to follow procedures** – Operations must be conducted with utilization of approved methodology.

- **Conducting operations for which no adequate training has been provided** – Good operation must be managed by experience.

- **Unavailable supervision and/or escalation personnel** – Experience must be brought to the operation by appropriate personnel; it is generally impossible to provide the needed breadth of experience and knowledge to a single duty operator.

We can see the degree of effective process management by tallying up the occurrences of events for the indicators above. Effective process management is a necessity. Monthly evaluations are recommended for all indications. Add or modify any that your enterprise finds useful. Not measuring them misses an important continuing improvement opportunity. The detrimental aspects of not having it include the following:

- Diminished time available for other duties
- Unnecessary operator stress
- Reduction of confidence
- Increased propensity for operational errors including omissions and failure to act
- Loss or decrease of job satisfaction

Self-confidence and pride are some of the most important benefits life has to offer. Let us not diminish them in the workplace.

10.26 Miracle on the Hudson

On January 15, 2009, at about 3:25 p.m. EST, an Airbus A320 designated as US Airways Flight 1549 departed New York LaGuardia Airport on a routine flight to Charlotte, North Carolina. The command pilot was Chesley B. "Sully" Sullenberger. The first officer was Jeffrey B. Skiles. Both were very experienced pilots. Within 2 minutes after takeoff and at an altitude of about 2000 feet, the aircraft encountered a bird flock and

Figure 10-18. US Airways Flight 1549 after a successful safe landing in the Hudson River with no loss of life.
Source: Greg Lam Pak Ng.[39]

immediately sustained a strike in both engines, rendering them inoperable. The crew's successful efforts during the remaining 4 minutes of flight relate an amazing tale of situation awareness, situation assessment, and situation management (see Figure 10-18).

The takeoff was under the physical control of the first officer. Immediately after the bird strike, the pilot took operational and physical control of the aircraft. The first officer immediately recognized the event implications, located the quick reference handbook (QRH; section on loss of thrust on both engines), and called out and followed its procedures. These procedures are credited in part to the successful outcome. One activity was to note that the aircraft was below the optimal engine relight speed and the critical minimum altitude, so restarting the engines was going to be a long shot. Sullenberger immediately rejected that option. The plane successfully ditch-landed on the river without extensive structural damage, which allowed flotation for a time. The cabin crew's professional efforts were also instrumental in all passengers and crew escaping without substantial injury or any loss of life.

> The NTSB concluded its investigation on May 4, 2010. It determined the probable cause of the accident to be "the ingestion of large birds into each engine, which resulted in an almost total loss of thrust in both engines." The accompanying statement credited the accident outcome to the fact that the aircraft was carrying safety equipment in excess of that mandated for the flight, and excellent cockpit resource management among the flight crew. Contributing [positive] factors were the good visibility and fast response from the various ferry operators.

---

39 Greg Lam Pak Ng, "Plane crash into Hudson River," flickr, January 15, 2009, https://www.flickr.com/photos/22608787@N00/3200086900.

Captain Sullenberger's decision to ditch in the Hudson River was validated by the NTSB.

Both pilots identified issues potentially impacting the safety of the passengers and crew. They fully evaluated and quickly addressed the corrective measures needed. Each pilot adhered to his role and responsibilities while the captain planned the landing—wings exactly level, nose slightly up, survivable descent rate; touch down just above minimum flying speed but not below it. All of these things needed to happen simultaneously.

The National Transportation Safety Board (NTSB) report states that the cockpit voice recorder data indicated that the communication between the captain and first officer were excellent. In addition, the flight crew made only pertinent callouts to the air traffic controller in order to manage the workload and focus on safety.

In the CRM training, everyone learns to articulate a discrepancy between what is happening and what should be happening. After the bird strike, the captain quickly realized there was an emergency and took control of the aircraft. Although the Air Traffic Controllers were clearing the way for Flight 1549 to land at LaGuardia Airport, the captain analyzed the existing conditions, questioned the guidance to land at LaGuardia and instead made the decision to land on the Hudson, which was the safer option.

The NTSB report cites four major factors contributing to the survival of all 150 passengers and 5 crew members:

- The decisions and "crew resource management" of the flight crew
- The airplane itself, which was equipped with forward slide/rafts although these were not required on this flight
- The performance of the cabin crew in expediting the evacuation of the airplane
- The proximity and rapid arrival of emergency responders.[40]

Before this amazing story is closed, it must be told that this was not a perfect response. It is important to hear this lest we assume (or worse, require) perfection as a necessary expectation for success. Errors made include the following:

- The pilot failed to use the QRH section on water-ditching the aircraft.
- The pilot failed to announce a brace for emergency water landing, instead announcing brace for emergency landing (this was potentially a contributor to passengers not immediately seeking flotation devices)

---

40 "US Airways Flight 1549," Wikipedia, last modified June 11, 2018, http://en.wikipedia.org/wiki/Miracle_on_the_Hudson; United States Nuclear Regularity Commission (US NRC), "US Airways Flight 1549: Forced Landing on Hudson River," *Safety Culture Communicator* (August 2011), http://pbadupws.nrc.gov/docs/ML1122/ML11228A218.pdf.

- The pilot failed to rig the aircraft for emergency water landing (thus not closing off all hull openings that could potentially flood the fuselage).

- The flight attendants failed to prevent a passenger from opening a rear emergency exit (thus accelerating the flooding of the fuselage).

- More than half of the passengers evacuated the aircraft (most to the wings) without their emergency flotation devices. Because this aircraft was actually configured for over-water flight (unusual for this flight itinerary), in addition to seat cushions there were life vests below the seats. Few were able to retrieve them; most had trouble inflating them once leaving the aircraft.

10.27 Achieving Situation Management

Situation management is the successful culmination of situation awareness, assessment, decision, action, and resolution. The interactive, stepwise process to manage situations is clear and powerful. There are no shortcuts, not much room for error, and much to be gained from success. This chapter knits it all together into an understandable whole. The benefits are sufficient to empower practitioners to deliver effective management to enterprises under their care. From a health-care provider sitting at a patient's side to an operator on an offshore petroleum platform, effective understanding of situation awareness, proper design of the workspace tools and environment, and successful decision management will make the difference. Be it life or death, or financial success or ecological disaster, responsible individuals have a clear effect on successful outcomes. The proper tools, training, protocols, and individual fitness for duty spell the difference.

Consider the following illustration. Suppose the plant has a sudden and very serious abnormal operation situation. Clearly, the control room and operator(s) will need to do things quite different from normal. Here is what an illustration might look like. The following operational states and condition tools might be utilized:

1. Operator detects a significant abnormal situation (e.g., the activation of a critical alarm or the detection of a significant spill for a yet unknown cause).

2. Control room is set to "restricted condition" (Section 10.8).

3. Situation is escalated after the full impact of the situation is observed (Section 10.16).

4. Escalation team takes responsibility (Section 10.17).

5. Escalation team decides the problem is serious and initiates a code orange (Section 10.21), which automatically invokes the following:

   a. Control room is set to "sterile condition" (Section 10.9).

   b. Operational structure is redeployed (Section 10.20).

   c. Operational safeguards are up and running.

      i. Active collaboration is initiated (Section 10.14).
      ii. Permission to operate evaluations are continually done (Sections 10.15 and 10.19).

6. Operation continues until the abnormal situation has passed (Section 10.21).

The preceding list is how one situation might play out. It started with recognition by the operator or a watchful supervisor that something serious appears to be going on. Prompt evaluation suggests that the problem is so significant that it requires additional assistance. That added assistance determines that a proper mode of response is to dramatically change the operational protocols and responsibilities to manage it. This change in protocol directly invokes other supporting protocols. All of this continues until the situation is managed or effects are stabilized or, in the extreme case, the situation is out of the scope of operational management.

Other situations will use different tools and processes drawing from the tools and technology for situation management both here and elsewhere. There is no secret to achieving situation management. There are no shortcuts to getting there. All is available. To get it, you will need to understand the individual tools and processes outlined in this book and elsewhere. One by one, you select them, understand them, and bring them into your enterprise. Then you knit them together into your way of operating. The whole becomes the sum of the parts. Successful situation management is about delivering sustainability: sustainability of the tools and technology laid out in this chapter, and sustainability of the entire enterprise infrastructure. Situation management is the structure you will build to provide it. Successful situation management is the effective utilization of that entire structure. In the end, this is less about hardware and software and more about thinkware and peopleware. It is less about policies and more about procedures, processes, protocols, and culture.

A large part of this book centers on what operators can do and how the operations team supports success. But we must not close this without a review of the other engineering support roles.

Control Room Management Choices

How you choose to design your control room, operator interface, data and information content of operator displays, procedures for operation and decision making, and specific tools and processes for situation management does not matter at all, *so long as they are effective and sustainable.*

Why is there all this attention to situation management? Because there are many ways to do each part of the job, how the choices are made matters. Individually, most ways to do the job are reasonably good. Advice and standards are everywhere. All have some benefit. Enterprises choose what to do and how to do it based on their understanding of benefit, their confidence in being able to implement, and the fit into individual and corporate style and culture. This reference book is filled with frameworks, technology, processes, and recommendations because long-term effectiveness is the goal. Why make so many choices so carefully? The answers are simple and powerful. There are three points of explanation for them:

1. The pathway to effective situation management requires that specific choices be made for each individual component part (alarm system, operator interface design, escalation, etc.)

2. The proper design and implementation of each individual component part requires a deep knowledge of its technology and best practices.

3. Not all designs for all parts will fit back together into an effective whole (most leave important gaps that are hard to detect until it is too late).

To overcome these limitations, you will need to design specifically for effective sustainable situation management. It cannot be a hoped-for result. It must be an overriding requirement. To help you do it well, this book provides a robust framework including useful specific options that cover the job of effective control room management for operators. Let us review this as if it were a thoughtful exercise in reverse engineering.[41] Start with the requirement that the operator in charge must be able to manage his plant or operation. To manage requires sufficient information, the ability to effectively use that information to make appropriate operational changes, and proper safeguards when a given situation threatens to move beyond the operator's

---

41 "Reverse Engineering," Wikipedia, last modified June 2018, https://en.wikipedia.org/wiki/Reverse_engineering.

ability to manage. It ends with who we would like to be sitting in the chair—we call him the professional operator. The anchor points for all of this are the subject and content of this book. Taken together, you will know why each component makes the design choices it does. You can use each component and technology point to get comfortable that when everything is put together, it fits and works.

The Process Engineer

Traditional process design works out how to get equipment, materials, and people into the right setting to make whatever the enterprise wants to produce. It relies heavily on proven design principles and experiences. The usual result is a customized plant to do the job—customized because each enterprise has its own way of doing things, its own markets, and its own business constraints. Customizing is usually cost-efficient because the overall cost of a plant typically far outreaches any reasonable tailoring necessary. During the work of design, the primary emphasis is on getting a design that works efficiently. Situation management asks the process designer to visit this work with extra attention paid to operability. He must better understand what is normal. He must fully anticipate how his design can move to the edges of normal and slip into the beginnings of abnormal. He must identify how operators can identify this movement in efficient and understandable ways. And, of course, he must design for and provide sufficient safeguards to ensure that even if operators fail to notice and correct, the plant will not cause harm. He also must ensure that this information is fully integrated into the control room management team.

The Controls Engineer

The controls engineer, over and above his work to assist the process engineer with the design and controls, is responsible for communicating to the operator where the plant is operating now, how exposed the operation is to abnormal situations, and the extent of the problem. His tools are the choice of how to control, where to move the disturbances to, and what to provide on the HMI. The primary information for this task will come from the planned-for and identified ways the plant can get into trouble. He will do this in close consultation with the designer and the operations team.

For example, in Chapter 5, "Human-Machine Interface," you saw a lot of detail on how to lay out display screens. It included specific navigation requirements. It included much coverage on ensuring operating data gets to the operator in ways that he can understand and assimilate. The specifics on icons, dials, gauges, and dashboards showed the powerful ways that they could aid operators in seeing and understanding data. Chapter 7, "Awareness and Assessment Pitfalls," reminded us that human nature is an ever-present force. When it is understood and accounted for, operators are better able to keep proper focus on the job at hand. The alarm system, discussed in Chapter 8, "Awareness and Assessment Tools," is used to alert the operator to every abnormal situation that requires operator action to manage. Weak signals technology, discussed in Chapter 9, "Weak Signals," assists the operator in finding the rest of the operating situations that do not have alarms but could lead to abnormal operation.

The Operations Engineer

Given an effective plant design, an effective controls system, and an effective operator interface (HMI), the operations team must fully lay out how the plant is to utilize operators to ensure all goes well. They will do this by setting operational expectations regarding the role of procedures, the extent of training, the management of fatigue and workload, and the way operational anomalies must lead to coordination of the operations and management teams. The purpose of this book is to provide the insight and tools to do that job well. The material has been designed to provide choices that are powerful and effective in their own right. When they are put together into a whole, they provide industry best practices, or leading-edge options where best practice gaps exist. They are not the only ways. Other ways exist. Those other ways must pass the test of providing what is necessary while at the same time fitting together as a unified whole. There is no room for hidden defects or unknown conflicts. Be creative but not inventive. Be brave but not foolhardy. Be effective, not lucky.

10.28 Training, Practicing, Evaluating, and Mastering

If you have stuck with the materials in this book, even if you exercised the fundamental reader's prerogative of picking and choosing what to explore, a lot of useful ground has been covered. Some, hopefully a lot, of the ideas and concepts are going to be useful to you professionally and to your enterprise functionally. This section is about all that you plan to incorporate into your enterprise. Your operators need to know and master everything that makes it into their kit. This activity requires as much planning and dedication as everything else he uses. It needs training, practicing, evaluating, and eventual mastering. The training here is over and above what is considered as operator training introduced in the "Operator Training" section of Chapter 3, "Operators." Here we visit specialized considerations to train for now that we have more tools. Practice these as part of an active program of skill building and performance reinforcement. Make searching for errors a basic part of near-miss and incident evaluation. Let us see how this might work for the important tools in this book.

Weak Signals

Mastering weak signals can be a powerful opportunity to train operators to better understand the inner workings and detailed behavior of the plants and processes under their management. This level of knowledge is essential for effective operation. The discussion was introduced in "Additional Thoughts about Weak Signals" in Chapter 9, "Weak Signals," so here is a brief reminder. Operators and supervisors who regularly engage in (hypothetical or tabletop) processing of weak signals should be able to easily and consistently forward-extrapolate them to all reasonable important possibilities of things going wrong, and just as capably be able to backward-project them to places to look for both confirming and denying evidence. Any hesitation or missteps here would be understood to represent gaps in their basic understanding of how their processes and plants operate.

Collaboration

Structured, purposeful collaboration is an important control room tool. It is an important aspect of crew resource management as well as the tools of worst case first, 10th man doctrine, and triangulation. Practice these as role-play and skill-sharpening activities.

Escalation

Section 10.16 covered this topic functionally. Sure, the operator will escalate when it is needed. But escalation must work. Now it is time to plan for rolling escalation into the daily plan. Do a role-play or tabletop simulation. Ensure that the operator knows how to do it and the individual escalated to is ready. Make sure that the electronic links to screens and supporting documentation work and everyone knows how to use them. Make sure that both sides know exactly how to communicate, ask questions, and provide support and advice.

Permission to Operate

Permission to operate (Section 10.15) provides a vital operational safeguard. Alone it stands out as a critical operational infrastructure activity. Almost every near-hit (near-miss, if you prefer to be traditional) and minor incident provide ample situations for practicing how operators would understand whether or not they had "permission" or not. Practice and coach extensively. It is that important.

Other Tools and Support

As you review how you have added situation management tools and practices into your control room, work out training plans to build proficiency and maintain effectiveness.

10.29 Closing Thoughts on Situation Management

This book has focused on control rooms and similar operation centers. They are the backbone of managing the production part of industrial enterprises. The product can be a chemical commodity or the movement of people by a commuter rail system. The problems and operational challenges are similar. The principles and tools are similar. A key one is structured "looking around" by operators. We all have a tendency to look without seeing (Chapter 7, "Awareness and Assessment Pitfalls"). In this chapter we saw useful structured ways of looking. Shift handovers (both going on shift and going off) provide a procedure and a process for going through the operating area and understanding the current status. Everything important is examined. Weak signal processing provides a powerful tool for uncovering the off-normal to see if there is more to it. It is integrated into the operator's tool kit as a periodic assessment activity. Weak signals processing is not a lifestyle. If something was noticed by looking around, confirm it first and then work it. But looking for problems mostly at specific points during a shift adds the essential element of structure to be able to successfully "look to see."

I end this book by urging you to put as much of this into practice as your professional responsibility permits.

10.30 Further Reading

Endsley, Mica R., Betty Bolté, and Debra G. Jones. *Designing for Situation Awareness: An Approach to User-Centered Design*. Boca Raton, FL: Taylor & Francis, 2003.

Shermer, Michael. *Skepticism 101: How to Think like a Scientist*. The Great Courses. Chantilly, VA: The Teaching Company, 2013.

Appendix: Definitions of Terms, Abbreviations, and Acronyms

"What's in a name?"

William Shakespeare

abnormal. The state of not being normal. See also *normal*.

abnormal situation. Any unexpected or unintended situation that differs from what is expected.

absolute alarm. Standard construction for most alarms. The present condition of a potential alarmed variable is compared against the preset alarm activation point. When crossed, the alarm becomes active.

Dynamic modification of the alarm activation point and/or priority is permitted.

See also *enhanced alarming*.

accident. An unforeseeable and unexpected turn of events that causes loss of value, injury, and increased liabilities.

ACK: acknowledge.

ACK'd: an alarm that has been acknowledged. See *acknowledge*.

acknowledge. The first operator action that indicates the recognition of an alarm to the alarm-managing portion of the control system.

action. The activity of doing something tangible (usually physical).

activate. An alarm that becomes "in alarm" by the monitored entity passing over the alarm activation point from the normal into the abnormal condition. See also *alarm activation point*.

active state. The status of a configured alarm being in alarm (the process or condition value on the alarm side of normal or within the alarm deadband).

adequate resources. Management is required to provide operators with accurate information and the training, tools, procedures, management support, and operating environment necessary for the operator to take action to help prevent accidents and minimize commodity losses.

advanced alarm. See *enhanced alarm*.

alarm. An audible or visible means of indicating to the operator an equipment or process malfunction or abnormal condition requiring a response.

alarm acknowledge. See *acknowledge*.

alarm activation. The state in which an alarm becomes active (enters the state of being in alarm).

alarm activation point. The threshold value or discrete state of a process variable that triggers the alarm into the active state.

alarm class. This is a misleading term. Dividing alarms into classes is not useful for operators or for designing alarm systems in general. Alarm priority should be used to assign a level of importance to alarms. See *alarm priority*.

alarm deadband. The range through which the alarmed variable must be varied away from the alarm activation point in order to clear the alarm.

alarm flood. The situation when the number of alarm activations exceeds the operator's ability to process them.

alarm group. The set of alarms associated within the alarm system. See also *plant area model*.

alarm historian. The database that contains the long-term record of all alarm (and other) activities associated with alarms. See also *alarm summary*.

alarm horn. Any type of audible sound (including voice response commands and warnings) that is initiated when an alarm activates and alerts the operator to the presence of an active alarm. The alarm horn will silence when the operator silences or acknowledges the alarm.

Some plants use different sounds and/or tones to distinguish between various operator positions and/or alarm priorities. This practice must be carefully designed to avoid confusion and unnecessary distraction during times of stress.

alarm indication. All means of indicating to the operator that an alarm has been activated.

alarm limit. Archaic term. See *alarm activation point*.

alarm log. The historical record of alarm indications and actions.

alarm management. The processes and practices for determining the need for, documenting, designing, monitoring, and maintaining alarm systems.

alarm message. The text message that is normally used to convey identification information about an alarm when it activates. This is an important part of the alarm summary.

alarm philosophy. The guiding document for design or redesign of the entire alarm system.

alarm points. The process conditions that are configured to be alarms. These are usually specific physical entities (pressure, temperature, purity, etc.).

alarm priority. The attribute of an alarm that defines a specific level of importance to the alarm to be used by the operator in deciding which alarms to work and in which order.

This is the primary way enterprises manage operational risk when alarms activate.

alarm rationalization. See *rationalization*.

alarm readiness state. The physical state of an alarm with regard to whether it is expected to activate when the alarm activation point is passed.

alarm response manual. The set of all alarm response sheets. See *alarm response sheet*.

alarm response sheet. A document (sheet) that contains all information about a given alarm and fully documents every operational aspect. It is extremely useful to both the operator and the alarm system designer.

The format may be a printed or online form, which may include dynamically assembled basic process control system (BPCS) screen information relevant to the alarm and plant states.

alarm set point. This is a misleading term. See *alarm activation point*.

alarm summary. See *alarm summary display*.

alarm summary display. A graphic human-machine interface (HMI) display that lists alarm indications over a period of time. The list includes the status of all active alarms and all recently cleared alarms.

In general, the BPCS manufacturer preconfigures the display and only facilitates minor customer customization.

alarm summary manual. See *alarm response manual*.

alarm summary sheet. See *alarm response sheet*.

alarm system. The entire construction (the collection of hardware, software, procedures, and protocols) that detects the alarm state, conveys that information and other supporting information to the operator, and archives all alarm information.

alarm trip point. Archaic term. See *alarm activation point*.

alert. A message or other visible and/or audible means of indicating to the operator a plant situation or equipment condition that does not require a response.

Alerts are useful in replacing old alarms that were used only for the purpose of notifying the operator of equipment status or condition. See also *message* and *notification*.

audit. The activity of examining status or performance by comparing the current conditions against the requirements. The audit results usually include recommendations for improvement.

auto shelve. An advanced alarm or enhanced alarm function that detects the need for shelving an alarm and does so without operator intervention, according to a pre-planned procedure.

automation. The technology and equipment designed to operate mostly without undue operator attention.

backward-projection. The activity of identifying places to look for evidence of the most impactful implications that resulted from forward-extrapolation. See also *forward-extrapolation* and *weak signals*.

bad actors. Any alarm (usually thought of as a significant number of alarms) that activates too often to permit the proper operator actions. Taken as a class, these alarms result in a significant distraction to the operator who must acknowledge them, and because they can block or hide more useful alarms, even very important ones.

base load. Operator loading for all basic tasks and responsibilities when the plant or operation is functioning normally. See also *operator load*.

BATNEEC: best available technology not entailing excessive cost.

best practice. A recommended practice that is considered the most appropriate by those in the field of activity. See also *recommended practice*.

BPCS: basic process control system. See *primary controls platform*.

bypass. To manually modify a function to prevent its activation. The term is used to imply that some mechanical method (e.g., using jumper wires) has been employed to ensure that the affected alarm will not activate. See also *cutout, disable, inhibit,* and *suppress*.

(This term is not normally used to describe alarm readiness states.)

chattering alarm. An alarm that repeatedly and rapidly transitions between the alarm state and the return to normal state, generally indicating that nothing is wrong with the process. This type of alarm is typically associated with discrete or digital alarms. See also *bad actors, cycling alarm,* and *nuisance alarm.*

checklist. A checklist is a specialized tool that is usually part of a larger procedure or protocol. It consists of a specific list of required items to check, the order to check them in, and the condition to be observed or confirmed against.

circadian rhythm. A daily cycle of inherent biological activity that affects the organism (operator) in a way that affects performance.

classification. The process of separating rationalized alarms into categories based on the type of consequences.

clear. See *cleared alarm.*

cleared alarm. An alarm that has returned to normal.

code gray. A condition in which the personnel in a control room are aware of a significant threat to the integrity of the facility that comes from *outside* of the plant or enterprise boundary.

code orange. A condition in which the personnel in a control room are aware of a significant threat to the integrity of the facility that comes from *within* the plant or enterprise and is a result of it being severely abnormal or heading toward that status.

collaboration. The formal process of seeking qualified assistance and guidance from others.

competency training. Foundational training that melds core skills and a deep understanding of underlying concepts and processes.

communication. The sending or receiving of information between two or more entities; the term does not imply the success or failure of the activity.

configure. To modify the control system parameters by setting appropriate values for a pre-structured database provided by the equipment manufacturer (BPCS, process control system—PCS, or programmable logic controller—PLC).

confirm/confirmation. The ability of operators (or others) to identify sufficient evidence to appropriately reach a decision that the fact or situation is true. See also *backward-projection, disconfirm,* and *weak signals.*

consequential alarm. An alarm that activates often enough after one or more specific other alarms activate that it may be considered unnecessary when the predecessor alarm(s) activates. See also *nuisance alarm.*

console. The interface for a single operator to monitor the process, which may include multiple displays and other annunciators and communications equipment. See also *display, operator area,* and *screen*.

control room. A room for operators and related personnel where a plant or operation is managed (monitored and controlled).

control room code. Specific notification to all control room personnel that a significant threat to operations must be figured into the current operating situation. See also *code gray* and *code orange*.

control room management. The sum total of policies, procedures, and protocols used to guide and inform the roles and performance of all personnel in a control room as they engage in assigned responsibilities.

This guidance is over and above the specific procedures and information (e.g., alarm response manual, procedure for swinging compressors, and such) used by those personnel for proper and effective operation.

control system. See *BPCS, PCS,* and *PLC*.

controller. The primary person responsible for ensuring that process parameters are maintained within limits in an operation center such as air–traffic control or pipeline control. See *operator*.

creed. A deeply held expression of dedication to a belief.

cutout. To automatically prevent the transmission of the alarm indication to the operator through a designed function; this is done on a temporary basis. See also *bypass, disable, inhibit,* and *suppress*.

cycling alarm. An alarm that repeatedly transitions between the alarm state and the return to normal state, generally indicating that nothing is wrong with the process. This type of alarm is generally associated with analog-type alarms. See *bad actors, chattering alarm,* and *nuisance alarm*.

DCS: distributed control system.

decomposition. The task of separating an entity into smaller components for examination and operation while retaining the ability to put everything back into a properly functioning entity.

delegation. The act of redirecting the authority and responsibility for managing to another (qualified) individual. Delegation may be downward (to a subordinate) or upward (to a superior).

deviation alarm. An alarm that uses a difference between two plant aspects (analog values, computed values, etc.) and compares the difference to an allowable value.

A deviation alarm can be an excessive difference (high deviation, which is either positively high or negatively low) or an insufficient difference (low deviation). See *mismatch alarm*.

digital alarm. An alarm that is activated based on a logic variable being either on or off, zero or one. Most often, this alarm arises from a process switch.

disable. To manually or automatically remove the alarm function from a process indication. See also *bypass, cutout, inhibit,* and *suppress*.

disaster. A sudden calamitous event bringing great damage, loss, or destruction.

disaster chain. The safeguards and protections designed to prevent or significantly reduce the impact of each specific abnormal operational situation.

A broken disaster chain means that at least one of the safeguards or protections functioned as designed and the worst-case situation was avoided. See *disaster*.

disconfirm. The ability of operators (or others) to identify sufficient evidence to appropriately reach a decision that the fact or situation is not true (false). See also *confirm*.

discrepancy alarm. See *mismatch alarm*.

discrete alarm. See *digital alarm*.

display. A single piece of hardware that is capable of depicting a visual graphical rendition normally associated with operator consoles. Any number of screens may be viewed (serially or by combining views) on a given display, up to the limit of the hardware and software controlling them. See also *screen*.

duty of care. See *standard of care*.

eclipse. An alarm that becomes redundant or otherwise unnecessary once another alarm has activated. See also *nuisance alarm*.

enforcement. The process of resetting the alarm system to the approved configuration state that is specified as the base or norm. It may be done in any manner consistent with the alarm philosophy.

enhanced alarm. See *enhanced alarming*.

enhanced alarming. Any method of modifying alarm parameters or information to ensure that the alarm maintains its relevance to the current operating conditions. This also includes the provision of advice and information to the operator regarding the interpretation and management of alarms.

enterprise. The general term for a manufacturing plant or a distribution or control operation (pipeline, rail system, etc.)

escalate. See *escalation*.

escalation. The formal process of seeking assistance and guidance from predefined individuals, usually those having more experience and authority.

ESD: Emergency shutdown. The initiation of formal action using physical devices to take the equipment in a plant or system to a safe state.

evidence. The facts and other proven observations that are used to decide the truthfulness or falseness of a belief.

extrapolation. See *forward-extrapolation*.

failure. The inability to properly operate or function normally.

fatigue. The inability of an individual to perform his normal function usually due to excessive sleepiness or as a result of excessive stress.

fault tolerance time. See *process safety time*.

flag. Two or more weak signals in the same plant area at the same time.

To specifically identify or note.

forward-extrapolation. The activity of identifying the most impactful implications that may result from an observed small indication or weak signal being significant or important.

fundamental. An essential component, without which the whole concept is not adequately realized or properly understood.

GEP: good engineering practice. Proven and accepted engineering practices. See *RAGAGEP*.

grouping. The process of identifying a number of different alarms with the same response.

guideline. A standard or principle used to make a judgment or determine a policy or course of action.

handback. A handover that passes back to the originator of the handover. See also *shift handover*.

handover. See *shift handover*.

hazard. A condition, event, or circumstance that could lead to or contribute to an unplanned or undesirable event. See also *HAZOP* and *pre-op safety review*.

HMI: human–machine interface. Equipment used by operators to observe, understand, and manage their responsibilities. HMIs are usually composed of a number of display screens and other audible or viewable information and command entry equipment. See also *display*.

HAZOP: hazard and operability study. A systematic review of all hazards and abnormal operating modes to determine the potential dangers of operation.

horn. See *alarm horn*.

identify. The successful act of an operator locating and understanding the current operational situation.

impairment. The inability of an operator to properly engage in all activities for which he is charged. This can be caused by fatigue, stress, illness, injury, unusual physical limitation, addiction, or excessive distraction.

incident. An occurrence, other than an accident, associated with the operation that affects or could affect the safety of operations.

independent layer of protection. Mechanical, electrical, chemical, or other equipment installed for the express purpose of preventing or significantly reducing the effects of abnormal operation. Each "layer" is functionally independent of the others so as to eliminate one from contributing to the failure of any other.

indication. See *small indication*. See also *weak signal*.

inhibit. To manually or automatically prevent the transmission of the alarm indication to the operator, usually on a temporary basis. See also *bypass, cutout, disable, and suppress*.

initiating event. A malfunction, failure, or other condition that can cause an alarm activation.

intelligent fault detection. See also *enhanced alarming*.

IPL: independent protection layer. See *independent layer of protection*.

key concept. A short phrase explaining an important bit of information or an interpretation of information contained in the chapter.

knowledge fork. A mental tool to sort all information (and data) into that which is known, unclear, or assumed. Only known information is useful.

latching alarm. An alarm that remains in an alarm state after the process has returned to normal and requires an operator action beyond acknowledgment before it will clear.

layer of protection. See *independent layer of protection*.

log. An official report (usually made during a shift) containing a record of events and observations for archiving, remediating, and preparing reports, among other related activities.

logic-driven alarm. Any form of calculated or recipe-determined alarm designed to alert the operator that certain situations are present that require action to remediate. Refers to all forms of calculations including statistical and inferred.

long arm of the operator. A term used to signify when an operator requests that others at a remote location observe, adjust, or engage in another hands-on activity. All such requests are done according to the strict authority of the operator who takes full operational responsibility for them.

major event. An abnormal situation in which the plant or production unit is disturbed to the point of seriously jeopardizing its operational integrity. See also *alarm flood*.

MOC: Management of Change. An approved and systematic way to deal with all aspects of making changes to the operations or equipment of a plant or enterprise.

message. A timely communication (excluding alarms) containing information. Alternate terms for messages include alerts and notifications.

mismatch alarm. Used for discrete values (e.g., switches), valves that do not open or close properly, or motors that do not start or stop when they should. See also *deviation alarm*.

mobile operator. An operator who has the freedom and responsibility to move beyond the confines of a control room to operate.

model. An effective yet simplified functional description of a plant or enterprise—or a part of either—used to understand the processes, predict vulnerabilities, and determine how to interact with individual elements or the plant or enterprise as a whole.

monitor. To actively observe values or situations at appropriate intervals for evaluation to be able to implement remediation as necessary to maintain proper function.

near hit/near miss. An abnormal situation that came close (even very close) to but did not result in injury, incident, or accident, even though it could have.

normal. An alarm state that indicates the particular point is not in alarm; however, the alarm must not have been inhibited or otherwise prevented from activation.

notification. Alternate term for message. See also *alert*.

nuisance alarm. An alarm that transitions to the alarm state but does not require action from the operator. See *bad actors*.

observe. To watch or perceive for the purposes of conducting qualification evaluations using on-the-job performance; observations must include the interaction of the evaluator and qualification candidate to ensure that the candidate's knowledge of the specific task requirements and procedures (and the reasons for key task steps) is adequate to ensure the continued safe performance of the task.

operate. To start, stop, and/or monitor a device or system.

operating mode. The particular situation under which a plant is operating (e.g., startup, reduced production rates, and unusual substitute utilities). Generally, it is presumed that the various operating modes are sufficiently different to affect how it is managed and perhaps how the alarm system must work.

operational integrity. The inherent capability of an enterprise to be responsive to proper operation.

operator. The primary person responsible for ensuring the process parameters are maintained within limits. See *controller* (alternate use).

operator area. The specific portion of the plant that a single operator has been assigned to operate.

operator console. The HMI equipment used by an operator to manage the aspects of production where operational parameters are observed and managed.

A console is typically composed of many displays, communications devices, and other related and supporting equipment.

operator interface. See *HMI*.

operator load. The amount of unavailable time that an operator has, usually expressed as a percentage. An operator load of 85% means that only 15% of his time is open for any unexpected task. See also *base load*.

operator readiness. The state or condition of an operator that refers to his unimpaired ability to conduct his duties.

operator response time. The time between the annunciation of the alarm and when the operator takes the (presumed) correct action in response to the alarm. See *SUDA*.

operator, board. An operator in a control room. See also *operator, inside*.

operator, field. An operator whose primary (or even sole) responsibility is to work in the field or actual plant area(s) outside the control room. Also called *outside operator*.

operator, inside. An operator whose primary working location is an operator console, usually in a sheltered area that may be within but separate from the physical plant. See *operator, board*.

operator, outside. See *operator, field*.

operator's creed. A creed specifically for operators. See *creed*.

out of service. A term used to designate a plant or part of a plant that is not actively being used for production.

Usually, an out-of-service plant has been specially prepared to ensure that dangerous situations or conditions are not present, but this cannot and should not be presumed. Therefore, out-of-service plants may have active alarms and alarms that are not suppressed or otherwise inhibited or bypassed.

overload. Any operator whose load exceeds 100%. See also *operator load*.

ownership. The act of being responsible for the proper function or action.

PI: proportional-integral controller.

P&ID: piping (or process) and instrumentation diagram.

PCS: process control system. See *PLC*.

permission to operate. A formal protocol by which an operator will make a determination that justifies continuing to operate. A loss of permission to operate requires that the equipment be immediately rendered safe (usually, shut down).

personal tools. An invention by an operator of a way of doing things outside of existing procedures, protocols, or responsible practices.

PFD: process flow diagram or probability of failure on demand. See also *P&ID*.

PHA: process hazard analysis. A systematic, organized process used to identify and assess the impact of potential hazards of an activity that handles or uses highly hazardous materials.

philosophy. See *alarm philosophy*.

PHMSA: Pipeline and Hazardous Materials Safety Administration, a part of the US Department of Transportation (DOT).

pick list. A list of the only possible entries that must be selected to answer a specific query (usually from a machine-generated interaction).

plant. A coordinated or physically contiguous collection of areas including an operation and an enterprise.

plant area. The basic part of a plant that is managed by one operator position. It may be operated by more than one physical operator so long as the equipment and controls interface only contains all the pieces for which the operator is responsible. See also *plant area model*.

plant area model. A construction (explicit, or generally shared understood representation) of a plant area wherein the plant is divided into constituent parts in a structured way that enables the identification and control of alarms as a group with respect to related alarms, consequential alarms, and others.

Also used for operational background and associations in weak signals.

plant enterprise. The largest entity in a corporate structure, generally consisting of multiple plants located at one or more sites.

plant equipment. Individual items in a plant module.

plant module. Collections of plant equipment; the smallest general entity for which alarm readiness states are managed.

plant response time. The time it will take a plant to respond to an operator action that is designated solely to avoid an abnormal state of operation.

plant site. A single, physical geographical location for manufacturing.

plant unit. See *unit*.

PLC: programmable logic controller; one instantiation of a BPCS. See *BPCS*.

pre-op (or pre-operation) safety review. A formal process by which an enterprise reviews the design, construction, and procedures of a manufacturing plant (or a portion thereof) to ensure that it would be safe to operate. "Safe to operate" normally refers to sufficient financial and environmental integrity.

primary controls platform. The combined hardware and software that manage the enterprise. This platform usually contains the configuration of alarms and has the ability and authority to generate alarms. See also *BPCS, PCS,* and *PLC*.

prioritization. The process of assigning a specific level of importance (based on operational risk) to an alarm. See also *alarm priority*.

priority. See *alarm priority*.

procedure. An established and approved way (normally written) of performing a task or meeting a responsibility.

process response time (PRT). The time that a process takes to become adjusted enough to avoid unwanted consequences due to an upset after an operator (or automated) change has been made.

process safety time (PST). The amount of time between an abnormal process event and the inception of serious consequences. Effective management of the abnormal event, if done in time, can usually prevent the inception of serious consequences.

If the process response time is less than the process safety time, an incident can usually be avoided through proper operator actions.

projection. See *backward-projection*.

protocol. The enterprise plans, procedures, rules, or recommended practices for a given activity.

proximate cause. The apparent or first cause. See also *root cause*.

PSAT: pre-start-up acceptance test. See also *pre-op safety review*.

PSSR: pre-start-up safety review. See *pre-op safety review*.

qualified. To be qualified, an individual must be able to properly perform assigned task(s) and recognize and react appropriately to any abnormal operating conditions (AOCs) that may be encountered. The individual should be capable of doing this whether the condition arises as a direct result of his work performance (e.g., if it is specific to the task being performed) or not (e.g., if it is generic in nature but still observable because the individual is present on site).

qualified operator or controller. A qualified individual who has been evaluated and can (1) perform assigned tasks and (2) recognize and react to AOCs.

RAGAGEP: recognized and generally accepted good engineering practice.

rate-of-change alarm. An alarm indicating that the monitored value is changing at either too high a positive rate or too low a negative rate. *Positive* and *negative* refer to direction of change.

Rate-of-change is not applied to values that are *not* changing fast enough.

rationalization. The process of accepting a potential alarm using the alarm philosophy, determining and documenting the rationale, and specifying all design requirements for the alarm (including the response sheet, activation point, and priority).

readiness. See *operator readiness*.

recommended practice. Procedures that come directly from the available pool of resources and is considered the appropriate way to do things.

redundancy logic. Using multiple methods to determine that an alarm condition exists and in doing so, activating only a single appropriate alarm.

redundant alarm. Two or more alarms that, if activated, will indicate the same process condition. Therefore, all alarms might be removed except the best indicator of the abnormal condition.

Where removal of the redundant alarms is not possible, redundancy logic can be used to limit the activations to the first alarm.

regulation. A rule, order, or requirement issued by a governing authority that one must comply with according to the law.

related alarm. An alarm that activates often enough either before or after one or more specific other alarms activate. A related alarm differs from a consequential alarm in that any related alarm may activate either before or after the other alarm rather than solely after it. See also *nuisance alarm*.

release. To manually prevent the transmission of the alarm indication to the operator through a managed list or shelf, usually on a temporary basis. However, an alarm on release is not reactivated until it has been cleared. See also *shelve*.

remediate. An operator taking the proper action to remedy or reduce the impact of an abnormal situation.

remote alarm. A console alarm from a remotely operated facility that often mirrors a local alarm at that facility.

report. A document or orally transmitted account of information.

response time. See *plant response time*.

responsible engineering authority. An experienced and respected individual who is the main enterprise contact for collective operating experience and technical knowledge.

restricted control room. The operating condition in a control room that imposes restrictions for personnel and operations in order to reduce nonessential operations and distractions for operators. See also *sterile control room*.

retriggering. The process of automatically causing an alarm (regardless of its acknowledge status) to re–alarm. This process activates the horn, adds an entry into the alarm summary, and otherwise indicates the presence of an active, unacknowledged alarm.

This is *not good practice* and should not be used.

return/return to normal. The alarm system indication that an alarm condition has cleared.

risk assessment. Any form of assessment that provides useful information about the probability and impact of any event or activity.

root cause. The most basic cause(s) that can reasonably be identified as being the cause(s) of the incident or other abnormal operation, and if remedied would likely prevent the problem in the future.

RP: See *recommend practice*.

RTN: return to normal. See *normal*.

safe operating limits. A set of values (or targets) for operating variables (set points, etc.) that generally ensure safe operation. Safe operating limits can be important safeguards against operational distraction.

safe park. A predesigned process condition for significantly curtailing operations, which provides an effective measure of safety. Operators can use this process condition to manage an abnormal situation that must not remain. Under appropriate situations, safe park may be considered as an acceptable alternative to a shutdown. See also *permission to operate*.

safety. Protection from danger to personnel and equipment, financial loss, and environmental degradation.

safety related. Any operational factor that is necessary to maintain plant integrity or that could lead to the recognition of a condition that could impact the integrity of the plant, or a developing abnormal or emergency situation.

SCADA: supervisory control and data acquisition (control system).

screen. The entire content visible on a display. See also *display*.

shelve. To manually prevent the transmission of the alarm indication to the operator through a managed list or shelf, usually done on a temporary basis.

shift. An assigned period of time that an operator (or supervisor) remains on duty.

shift handover. The act of transferring knowledge and responsibility from one individual (operator, controller, supervisor, etc.) to another, usually a relief or replacement.

SIF: safety instrumented function. Equipment designed and intended to reduce or eliminate a specific hazard by automatically bringing the situation to a safe state. See also *safe park* and *permission to operate*.

SIL: safety integrity level. These equipment categorization levels provide universally understood safety performance ratings based on equipment design, legal codes, and accepted practices.

simulation. The use of a model of an entity (either formal or ad hoc) to create useful information about the original entity.

SIS: safety instrumented system. The collective set of safety instrumented functions forming the equipment-based operational management that brings hazardous operations to a safe state. See also *permission to operate*, *SIF*, and *safe park*.

situation management. The competency, ability, and willingness of the human operator to properly and successfully manage the enterprise or activity under his charge.

skill. A demonstrable ability to perform a given task well, arising from talent, training, or practice. See also *competency*.

skills training. Training to provide a skill. See *skill*. See also *competency*.

small indication. A barely noticeable awareness of something not quite right or not quite wrong that cannot be confirmed or readily explained without further examination.

One way of examining the meaning of the small indication is to treat it as a weak signal. See also *weak signal*.

stale alarm. See *standing alarm*.

standard. Established criteria or a set of rules provided by an approved entity governing acceptable practices.

standard of care. A requirement that a person or enterprise acts toward others with the watchfulness, attentiveness, caution, and prudence expected of a reasonable person in the circumstances. If a person's actions do not meet this standard of care, then the person is considered negligent.

standing alarm. An alarm that remains in the alarm state for an extended period of time (well beyond any usefulness to the operator). See also *stale alarm*.

station. A single HMI within the operator console. An operator console may have a number of stations. See *display*.

sterile control room. The operating condition in a control room that imposes additional restrictions on personnel and operations beyond the restricted control room in order to eliminate all nonessential operations and distractions for operators. See also *restricted control room*.

strong signal. A clear indication of operations gone amiss. It does not include why the situation exists or what to do to remedy it.

SUDA: see, understand, decide, and act. The steps an operator takes when responding to an alarm activation. The activity begins with the operator seeing and recognizing an alarm activation and continues through all activities needed for the operator to make what is anticipated as the proper corrective action to remedy the abnormal condition. See also *SUDA time*.

SUDA time. The time it takes for the entire SUDA activity to occur.

supervision. The act of providing direction, guidance, and feedback. See also *supervisor*.

supervisor. An individual designated to provide supervision.

suppress. To automatically (based on designed logic) prevent the indication of the alarm to the operator when the base alarm condition is present. See also *bypass, disable,* and *inhibit*.

tag. A unique identifier that is used for BPCS data, such as temperatures and flows. Displays, alarms, logs, and other aspects are usually organized by their tag.

template. A specific model that is used as a reference or starting point for understanding items similar to the model.

tenents of operation. See *permission to operate*.

time to manage fault. The amount of time required for the operator to become aware of an abnormal event, understand it, and take action to manage it and the process to respond in a way that prevents serious consequences. So long as the time to manage fault is less than the process safety time, serious consequences can be thought of as being amenable to operator action. See *process safety time* and *process response time*. See also *SUDA*.

transformational analysis. The activity of graphically mapping the flow of operations and the complexity of operation at each flow step.

trouble point. When this value is crossed, the process variable becomes abnormal. Alarms are configured to initiate operator intervention to prevent the crossing.

UNACK: Unacknowledged.

unacknowledged. An alarm in the alarm state that has not been acknowledged by the operator.

unit. A general term usually used to identify a major portion of an operator area that performs one or more important functions in a somewhat self-contained way.

upset. The state of not being normal to the point of exposing risk to safe and responsible operation.

user alert. See *alert*.

user notification. See *alert*.

weak signal. A seemingly random or disconnected piece of information that at first appears to be background noise but can be recognized as part of a significant pattern when it is viewed through a different frame or connected with other pieces of information.

Credits

The author wishes to credit these individuals and organizations for providing source material that was used in this book.

ABB

Africa Studio/Adobe Stock

Alex White/Adobe Stock

Alvaro German Vilela/Alamy Stock Photo

arquiplay77/Adobe Stock

Associated Press

BP

CAISO (California Independent System Operator corporation) for the background photograph on the front cover

cigdem/Shutterstock.com

Columbia Pictures

Daniel Simons

DILBERT © 2008 Scott Adams. Used By permission of ANDREWS MCMEEL SYNDICATION

Dive and Discover

Emerson

EnerSys

Ergon Refining

EUMETSAT

Eyematrix/Adobe Stock

Freepik

Greg Lam Pak Ng

Harris Shiffman/Adobe Stock

Honeywell ASM

HSE

Kiro 7

Kyodo News

metamorworks/Adobe Stock

Momentum Press

National Commission on BP Deepwater Horizon Oil Spill and Offshore Drilling

National Transportation Safety Board

neilb2200/Adobe Stock

neirfy/Adobe Stock

PAS Global LLC

Phil/Adobe Stock

racksuz/Adobe Stock

Reuters/Glab Garanich

SAGE Publications, Inc.

StockCharts.com

Twitter user @Missxoxo168

User Centered Design

US Chemical Safety and Hazard Investigation Board

viappy/Adobe Stock

Winsted Control Room Console Company

YuliaB/Adobe Stock

ZPAS

My thanks and apologies to any others whose identity was unknown or who I was unable to locate and thus escaped attribution.

Index

Note: Page numbers followed by f or t indicate figures or tables, n refers to chapter footnotes.

A

ABB, operator effectiveness program, 53
Abnormal, 526
Abnormal operation region, 35–37
Abnormal situation, 472
    definition, 134
    plant actually abnormal, 135
    plant actually normal, 135
    weak signals, 518–519
Abnormal Situation Management (ASM) Consortium, 228
Abstract mimicry, screen, 278
Accident(s), 5, 136. *See also* Incident(s)
Accident-prone behavior, 455
Accommodation, 247
Acknowledging, communication, 657, 658
Acoustic, operator interaction, 278
Action notification, 507
Active lifetime, 509
Adams, Richard, 538
Administration, role in operations, 50
Advanced Operation Assistance Solutions, Yokogawa, 55
Advanced regulatory control, 108
Advanced technology control center(s), 239–241
Affecting operator, 160
Affordance, 245
Air and Space Operations Center, 236
Airbus 330-203, 552
Airbus A330, 421

Aircraft
    pilot training, 275–277, 276f
    See also Plane crashes
Air Florida Flight 90
    cockpit errors before takeoff, 676–677
    cockpit errors during takeoff, 677–678
    crash, 676–678
    crash into Potomac River Bridge, 677f
Air France Flight 421, 551–552, 647–649
AkzoNobel, 127
Alarm(s)
    abnormal situation identification, 477f
    definition, 506
    event data, 198
    excessively complicated dashboard for nuisance, 324f, 324–325
    line of defense, 40
    management, 66
    multitasking and, 453–454
    nuisance example gauge, 311–313
    operating region showing, 40f
    operation regions, 35–37
    overview display screen showing, 338f
    precedence of operator activity, 618
    process, 14f, 15f, 16f
    role of system, 472
    secondary display page showing, 341f
    something is wrong for sure, 40
    system, 402
    weak signals and, 617, 618
    See also Notification(s)

Alarm activations, weak signals from, 618–619
Alarm Management for Process Control, 498
Alarm rationalization, weak signals, 611
Alarm response sheet, 485–489, 486f
   abnormal situation, 486
   advanced alarm considerations, 489
   automatic actions, 488
   causes, 487
   configuration data, 485
   confirmatory actions, 487
   consequences of not acting, 487–488
   header information, 485
   manual corrective actions, 488
   online example, 489, 490f
   safety-related testing requirements, 489
Alarm system, 474
   alarm activation point, 491–493, 493f
   alarm fundamentals, 477–478, 478f
   alarm priority, 494–497
   alarm rationalization step-by-step, 497–499
   alarm response sheet, 485–489, 486f, 490f
   anatomy of, 478–479, 479f
   configuration metrics, 499–500
   management, 480–481
   metrics, 499–501
   operator alarm loading, 501–503
   performance metrics, 500–501, 501t
   philosophy, 479–480
   process abnormal, 475
   process fault, 475
   process normal, 474
   process trouble point, 489–490, 491f
   rationalization, 482–485
   time management, 491, 492f
Alignment failure, 465
Amagasaki rail crash, 425–427, 464
   aerial view of crash site, 426f
   setting up the fear, 426–427
   takeaway, 427
American National Standards Institute (ANSI), 43
American Petroleum Institute (API), 47
Analog, device technology, 277
Anchoring bias, 445–446
Annotation, 376
Announcements, HMI, 379–380
Ansoff, H. Igor, 520
Antonia, Rosa, 60
Aristotle, 461
Artificial intelligence, weak signal analysis, 631–632
Asiana Flight 214, 423–425, 447
   July 2013 crash, 423, 424f
   setting up the roles, 424
   takeaways from incident, 424–425

Asimov, Isaac, 527
Assessment situation, awareness and, 473–476
Audience of book, 10–11, 56, 58
Auditing, 514
Automated shutdown, 716
Automation, 643
   bias, 421–422
   ceasing operations during significant upsets, 114
   complacency, 420–421
   dangers from, 418–422
   generation effect, 422
   operating periodically without, 113–114
   plant, 112–115
   procedural, 110–112
   proper level of, 114–115
   selective, 113
   substitution myth, 419–420
   See also Selective automation
Awareness, 416
   alerts, messages and notifications, 506–511
   biases, 441–448
   concepts, 415
   definition, 392, 511
   doubt, 432–436
   geography of thought, 456–464
   inattention blindness, 448–450
   institutional culture versus individual responsibility, 464–469
   looking without seeing, 418
   making decisions, 436–441
   making mistakes, 416–418
   myth of multitasking, 452–454
   partial information, 450–451
   personalities, 454–456
   See also Notification(s); Situation awareness

B

Backward-projection
   confusing evidence from, 629
   illustration, 578f, 579–580
   no evidence from, 629
   potential problems, 631
   Texas City disaster, 615
   work process, 573–575, 574f
Bandwagon effect, bias, 446
Bank of England, 333, 334f
Baseline operational measures, 557–558
BATNEEC (best available technology not entailing excessive cost), 46–47
Beliefs, 416
Belonging, 215f, 216
Bender Treater, 351–352

Best available technology not entailing excessive cost (BATNEEC), 46–47
Best practice, 44
Beuthel, Carsten, 345
Biases, 441–448
 anchoring bias, 445–446
 automation, 421–422
 bandwagon effect, 446
 confirmation bias, 442–444
 continuation bias, 445
 diffusion of responsibility, 446–447
 halo effect, 446
 post hoc ergo propter hoc, 447–448
 "what then" question, 448
Boardroom, process safety beginning in, 87–88
Boeing 777-200ER airplane, 423
Boston Transit System, 369, 370f
BP, 93
 Toledo Refinery, 79
 Transocean *Deepwater Horizon* and, 467–469
BP Texas City disaster, 99–100, 403
 critical failure, 143f, 143–144, 147, 468
 operator's permission to operate, 685
 retrospective weak signal analysis of, 616t
 splitter process flow diagram, 614f
 weak signal case study, 612–613, 614f, 615
Brafman, Ori, 439
Brafman, Rom, 439
Brazerman auction, 439–440
Build-on-demand trends, 330
Bullemer, Peter, 273
Bystander effect, 447

C

Canadian Standards Association (CSA), 43
Carr, Nicholas, 419
Carrillo, Rosa Antonia, 567, 655
Carte blanche, permission to operate, 691
Cautions, conflict with protocols or statutory requirements, 9
 design and safety notice, 8–9
Center for Operator Performance, Wright State University, 54
Challenger Space Shuttle, 465–467
 prelaunch command and control management failures, 467
 prior-to-launch technology management failures, 466–467
Charles Schwab (SCHW), 80
Checklists, 182–183, 602
Chemical impairment, 169
Chemical Safety Board (CSB), 133
Chernobyl, 167, 167f
Cherry, Kendra, 162

Chevron Corporation, 217
Chevron Oil, tenets of operation, 660–661
China syndrome, 537–538
China Syndrome, The (movie), 537–538, 538f
Churchill, Winston, 223, 517
Circadian clock, 165, 166f
Circadian rhythm, 165, 166f
Clemens, Samuel Langhorne, 641
Coaching, escalation teams, 709–710
Code blue, 725–726
Code gray, 727–728
Code orange, 728
Coffman, Bryan S., 520, 567
Coherent view, 285
Collaboration, 402, 675–684
 10th man doctrine, 680–681
 Air Florida Flight 90 crash, 676–678
 control room, 740
 in control room, 679
 crew resource management (CRM), 678–679
 escalation in control room, 699–700
 red teams, 684
 shift handover, 186
 triangulation, 681–683
 using knowledge fork, 675f, 675–676
 weak signal management, 595–596
 work space, 243–244
 worst case, 680
Collaboration center(s), 237–239
 bring-your-own workstations, 238, 239f
 configurations, 238
 functional requirements, 237–238
 importance of, 238–239
 physical requirements, 237
Collective, term, 623
Color, 261
 blindness, 297–298
 display screen, 293, 294f, 295f
 excessive use of, 382f
 improved screen, 383f
Command and control
 control room, 713–715
 supervisor fully qualified as operator, 714–715
 supervisor not fully qualified as operator, 713–714
Commercial enterprise, term, 79
Communication
 escalation in control room, 698–699
 handover, 199
 hypothetical, 658
 mindful, 655–656
 mirroring, acknowledging and tracking, 656–657, 658

Communication (*Continued*)
  mobile operator, 212–213
  safe conversations, 652–658
  shift handover, 186
  word about safety, 653
Competency, training, 176–177
Complacency, automation, 420–421
Concept design, 261
Concept of scale, 74
Confirmation, situation management activity, 24, 25, 25f
Confirmation bias, 430, 442–444, 543, 649
  illusion of skill, 444
  loss of scale, 443–444, 444f
  myside bias, 442
  See also Biases
Consensus, weak signal management, 595–596
Content change management, 375
Continuation bias, 445, 649
Continuous improvement, 69
Contradictions, weak signals, 592
Control (magazine), 7
Control loop
  flow control example, 106–107
  magic of, 105–107
  moving disturbances, 107
  process control, 105–107
  temperature control example, 105–106
Control measures, difficulty of, 554–555
Control room(s), 223–224, 233–235
  in abnormal operations, 662
  access management, 231
  advanced technology control centers, 239–241
  architect perspective, 256–257
  architectural aspects, 235, 236f
  automation complacency, 420–421
  automation in, 643
  building style, 251
  code blue, 725–726
  code gray, 727–728
  code orange, 728
  collaboration, 675–684, 740
  command and control, 713–715
  console design, 252–253
  controls engineer, 738–739
  crew resource management, 679
  dangers from automation, 418–422
  delegation, 674–675
  design, 225, 226–227, 249–253, 257
  design circa 1950, 226f
  design circa 2014, 227f
  design considerations, 252
  design evolution, 234–235
  directing operator, 724
  environmental controls, 225, 229–230
  escalation, 740
  future of, 255–256
  information support tools and technology, 230
  key concepts, 224–225
  layout, 251–252
  life cycle, 253, 256–257
  location, 250–251
  management choices, 737–738
  managing operator, 724
  MBTA Boston Transit System, 369, 370f
  meteorological, 228f
  mobile, 253–254
  mobile operator, 210–211
  in normal operations, 661–662
  observer evaluation, 695–696
  operational conditions showing current primary structure in, 474f
  operational conditions showing extended awareness structure in, 476f
  operational support, 231
  operations engineer, 739
  operator redeployment, 720–725
  operator self-evaluation, 696
  permits, 232–233
  personnel, 233
  physical protection and security, 229
  principles and ergonomics, 252
  process and operational controls, 230
  process engineer, 738
  remoteness of, 234
  requirements, 229–233
  restricted, 661–663
  role of, 254
  scope, 228, 233
  security, 251
  situation codes, 725–728
  special operating situations, 232
  sterile, 664–665
  upset operating situation, 723f
  USS Seawolf submarine, 240f
  video walls for, 366, 368–370
  visitors, 233
Control room management, 19, 37
  equipment, 21
  essential components of, 20–22
  framework of, 19f
  maintenance, 21
  management pillar, 20
  operation interface, 21–22
  operator training, 21
  procedures, 21
Control room operators, 161. *See also* Operator(s)
Controls engineer, 738–739

Controls platform, formal notifications by, 510
Conversations
  mindful, 655–656
  safe communication, 653–655
Cooperation, situation awareness, 404, 405
Costa Concordia (cruise ship), 705
Costs
  best available technology not entailing, 46–47
  understanding, 59–60
Coveralls, 70
Crew resource management (CRM), 162, 170, 455
  collaboration, 678–679
  operators, 714
  training, 734
Crisis management, 12
Critical failures, 137–146
  BP Texas City, 143f, 143–144, 147
  chains, 138–140
  Deepwater Horizon, 144–145, 145f
  Milford Haven at Texaco refinery (UK), 140–142, 141f, 147
  Olympic pipeline in WA, 146, 146f
  Piper Alpha, 142–143
  See also Incident(s); Plane crashes
Critical variables, weak signals and, 619
Crossover(s), 159
  basic procedure framework, 159
  management, 159–161
  Piper Alpha lesson, 160
Cues, HMI, 378–379
Culture, 79–80, 416
  alignment failure, 465
  geography of thought, 456–464
  handling and reporting problems, 463–464
  individuality, 463
  institutional, versus individual responsibility, 464–469
  logic and reason, 460–462
  norms and conventions, 457–460
  visual perception, 459–460
  world map delineating East, West and blended frameworks, 457f

D

Darley, John, 163
Dashboards, 318–325
  better, 325
  clear and effective, 325f
  definition, 318
  design fundamentals, 326–327
  design fundamentals for, 325–329
  deviation diagram, 320–322
  excessively complicated, 324f, 324–325
  overloaded, 323f, 323–324
  pipeline system, 319–320
  salience requirements, 328–329
  social media users, 319
  unity of presentation, 322–325
Data versus information, 261
Da Vinci, Leonardo, 413
Day, George, 519
Dead reckoning, navigation, 302
Deception of two reasons, 430–431
Decisions
  Brazerman auction, 439–440
  loss aversion, 438–440
  making, 436–441
  over many shifts, 437–438
  short-term versus long-term, 436–438
  sixth sense, 440–441
  within a shift, 436–437
Decomposition, 127
  basics, 118
  concept, 78, 117
  illustration of typical boundaries, 118f
  input and output classification questions, 119f
  key repeated elements, 120–121, 121f
  key repeated subsystems of, 121–122, 122f
  looking for abnormal situations in repeated elements, 123–125
  looking for abnormal situations in repeated subsystems, 125–127
  structure of, 120–123
  subsystem boundary attributes, 118–119
  subsystem internal attributes, 119–120
  system, 123, 123f
  underlying situation management, 120–127
  using transformational analysis for, 130–131
Deepwater Horizon
  BP and Transocean, 467–469
  critical failure, 144–145, 145f
  during the incident, 468
  operation prior to incident, 468
  postscript, 468–469
De facto decisions, permission, 691
Defensive driving, Smith System, 32
Defensive operating, 32–35
  five principles of, 33–34
  lessons of, 665–666
  supplementary guidance, 34–35
  term, 32
Delegation
  of authority, 92
  control room, 674–675
  of responsibility, 92–93
Deming, W. Edwards, 77, 153

Design
  concept of, 225
  considerations, 252
  console, 252–253
  control room, 226–227, 249–253, 257
  evolution, 234–235
  high-reliability organizations, 88
  safety notice and, 8–9
  user-centered, 244–249
Development supervision, role in operations, 50
Deviation diagram
  dashboard example, 320–322
  operating area, 565f
Dials and gauges
  design types, 327–328
  nuisance alarms, 311f, 311–313
  pipeline nominations tracking, 313–318
  resolution-based display screens, 328
  risk-based display screens, 327–328
Diffusion of responsibility, bias, 446–447
Digital
  device technology, 277
  smart devices, 559–560
Direct enter search, navigation, 302
Directing operator, 724
Direct measurements or observations, weak signals from, 542–545
Direct strong signals, 398–399
Disaster, 137. *See also* Critical failures; Incident(s)
Disaster management region, 531
Display screen(s)
  attention-based principles, 268
  building effective screens, 385–386
  dashboards, 318–325
  dials and gauges, 311–318
  Engineering Equipment and Materials User Association (EEMUA) principles, 269–270
  glyphs, 305–308, 308f
  icons, 308–310
  ISO 9241 principles, 271–272
  large and small displays, 353–365
  memory principles, 268–269
  mental model principles, 268
  mimicry concepts, 278
  nomenclature for, 263–265
  paper versus electronic screens, 370, 372–377
  perception principles, 267–268
  principles of design, 267–272
  Wickens' 13 principles of, 267–269
  See also Off-workstation (OWS)
Display screen design, 292–295
  color, 293, 294f
  color blindness, 297–298
  construction, 281
  display complexity and minimum view time, 297
  dynamic page assembly, 295–296
  examples, 286–290
  flash, 291–292
  hierarchy, 282–284, 283f
  overview level, 283f, 283–284, 335–339
  overview of, 281–290
  principles of, 267–272
  procedure tracking, 344f
  screen arrangement and layout, 294–295, 296f
  screen structure, 281, 282–291
  secondary level, 283f, 284, 339–343
  segmentation, 284–285
  tertiary level, 283f, 284, 343–345
  visibility and viewability, 290–291
Display screen examples, 334–345
  overview identifying components, 337f
  overview page, 335–339
  overview showing alarms, 338f
  overview showing operator's responsibility, 336f
  secondary page, 339–343
  secondary page for Riser/Regenerator, 340f
  situationally based secondary pages, 340, 342–343
  subordinate secondary pages, 339–340
  tertiary page, 343–345
Display screen organization
  geographically based, 290, 290f
  hierarchy, 282–284, 283f
  responsibility-based, 286f, 286–287
  risk-based, 287f, 287–288
  similarity-based, 289, 289f
  situation-based, 304–305
  task-based, 288, 288f
Documentation preparation, 198–199
Dodge, Wagner, 726–727
Doubt, 432–436
  dealing with uncertainty, 434–435
  lingering, 435
  managing "truths," 435–436
  possible, 433
  probable, 433
  reasonable, 433–434
  shadow of a, 434
Douglas, Michael, 537
Doyle, Sir Arthur Conan, 517, 541
Ducommun, Jesse C., 153
Due diligence, 44
Dutch Safety Board, 133
Duty of care, 47
Dynamic page assembly, 295–296

E

EEMUA (Engineering Equipment and Materials User Association), 228, 269–270
Effective screens, 385–386
Electronic, device technology, 277
80:40 rule, 431–432, 432f
Electronic formats
   content change management, 375
   documents, 374–375
   following the thread, 374–376
   personalization and annotation, 376
   pros and cons, 372–373
   readability, 373–374
   related content, 375–376
Elements, 261, 263, 264f
Ellis, Graeme, 38n7, 61, 61n16
Email, informal notifications, 510–511
Emergency, alarm priority, 494, 496t
Emergency shutdown (ESD), 715–716
Emerson, 54
Emerson Automation Solutions, 382, 382f, 383f
Emotional impairment, 169
End of Eternity, The (Asimov), 527
Endsley, Mica, 378
Engaged handover interaction, 199–200
Engineer(s), 10–11, 59
Engineering Equipment and Materials User Association (EEMUA), 228, 269–270
Engineers' Creed, 155
Enough displays, 304
Enterprise(s), 77–78
   abnormal situation, 134–135
   accidents, 136–137
   automated plant, 112–115
   capabilities, 86–89
   commercial, 79
   communication between silos, 85–86
   component organization, 82–85
   control loop, 105–107
   critical failures, 137–146
   crossover management, 159–161
   culture, 79–80
   decomposition, 117–120
   decomposition underlying situation management, 120–127
   delegation of responsibility, 92–93
   hazard, 134
   key concepts, 78
   knowing what is right, 73–75
   learning from experiences, 133–134
   near miss, 135–136
   operational boundaries and responsibilities, 157–164
   operational integrity, 93–99
   plant area model, 115–117
   preparation, 199
   process hazard management, 146–148
   responsible engineering authority (REA), 103–105
   safety, 99–103
   selective automation, 107–112
   short-term versus long-term, 89–92
   silo concept, 82–86
   suits and coveralls, 70
   transformational analysis, 127–132
   understanding, 81
Enterprise design, concept, 78
Enterprise scale points, definitions, 98
Environmental controls, 225, 229–230
Epicycle, 74
Equipment, plant area model, 115–117
Equipment reliability indicators, 560–561
Escalation, 696–706
   collaboration, 699–700
   communication, 698–699
   control rooms, 740
   Costa Concordia cruise ship example, 705
   design, 706
   direction of, 703
   first duty of, 703
   operator recognizing, 700–702
   process of, 704, 704f
   resources, 702–703
   transfer of responsibility, 701–702
   when to escalate, 704–706
Escalation teams, 706–713
   abnormal situation management process model, 711–712
   activity, 709
   composition, 708
   frontline coaching and mentoring, 709–710
   readiness evaluation role, 710–711
   training, 711
Essential decomposition, concept, 78
Esteem, 215f, 216–217
European Agency for Safety and Health at Work (EU-OSHA), 133
European Organisation for the Exploitation of Meteorological Satellites, 228f
Evaluating effectiveness, HMI, 380–383
Evaluation, 261
Event notification, 507
Everyday situations
   fictional illustration, 666–667
   flow of operator activities, 670, 671f
   operator duties, 669
   operator intervention caution, 667–668

Everyday situations (*Continued*)
   operators following the rules, 669–670
   operators not being innovators, 668–669
   situation management, 666–673
Evidence for confirmation
   backward-projections, 578f, 579–580
   confirming evidence, 575–576, 580
   degrees of evidence, 577–578
   disconfirming evidence, 575–576, 580
   weight of evidence, 581
Expanse, work space, 243
Experience
   benefiting from, 651–652
   learning from, 75–76
   operating, 67
Expertise, situation management, 651–652
Explicit, 38

F
Face-to-face handover, 200–201
Failures. *See* Critical failures
Fan charts, 332–333, 334f
Fate, logic versus, 462
Fatigue
   causes of, 167
   circadian rhythms, 165, 166f
   impairment, 169
   incidents and, 165–167
   managing, 168
   understanding, 164–167
   See also Operator readiness
Fault-tolerance clock, 31
Federal Aviation Administration (FAA), 679
Few, Stephen, 318
Field operator, 162
   handover, 202–203
   handover timelines, 202f, 203f, 207f
Field shelters, 235
Flag, 198
Flags, weak signals as, 598–600
Flash
   cycle, 291, 292f
   screen display, 291–292
   visibility, 291
Flow, control loop example, 106–107
Focus, navigation, 303
Fonda, Jane, 537
Formats, 261, 263, 264f, 332
Forward-extrapolation
   problems of, 629
   proper, 587
   weak signals, 630
   work process, 570–573, 571f, 572f

Four-corners tool, 409–411, 410f
Foxboro Company, 259, 320
Franklin, Benjamin, 463
Frontline supervisors, role in operations, 49–50
Functional silo, 84–85
Fundamental guide, 261
Fundamental tools, situation management, 24–30

G
Galilei, Galileo, 517
Gauges. *See* Dials and gauges
General observations, weak signals from, 534–537
Generation effect, 422
Geography of thought
   eye flow on HMI page in West and East, 458f
   handling and reporting problems, 463–464
   individuality, 463
   language construction, 460
   logic and reason, 460–462
   logic versus fate, 462
   nailing down the issue, 462
   normal visual flow directions, 457–459, 459f
   norms and conventions, 457–460
   simplicity as truth, 461–462
   three postal codes, 464
   visual perception, 459–460
   world map delineating East, West and blended frameworks, 457f
Geometry, work space, 242
Georgia System Operations, video wall, 370, 371f
Gibran, Kahlil, 221
Gladwell, Malcolm, 598
Glass Cage, The (Carr), 419
Glass cockpit, 370
GlobalSantaFe Corporation, 467
Glyphs, 305–308
   as navigation buttons, 307, 307f
   social media, 308, 308f
   tertiary page, 345f
   tool tips combined with, 307f
   traffic sign, 306f
Goals, short-term versus long-term, 89–92
Godell, Jack, 538
Good engineering practice, 65–69
   alarm management, 66
   maintenance, 67–69
   operating experience, 67
   operators on the job, 66
   providing adequate information, 66
   roles and responsibilities, 66
   training, 67
Google, 373

Gould, Stephen Jay, 77
Graphics symbol library, 280–281
Gropius, Walter, 223
Ground rules, cautions and, 8–10
Guerlain, Stephanie, 273
Guideline(s), 44–45
    notifications, 509–511
    permission to operate, 691–692

H

Hallinan, Joseph T., 391, 641
Halo effect, bias, 446
Handover. *See* Shift handover
Hazard(s), 134
    definition, 134
    process hazard management, 146–148
Hazardous chemicals, OSHA 29 CFR 1910.119 for process safety management of, 45–46
Head-up displays, 363–365
    automobile example, 364f
    eyeglasses version of, 365f
Health and Safety Executive (HSE), United Kingdom, 44, 133, 686
Heat exchanger
    heat efficiencies of, 346–347
    mass data display, 346f
Helicobacter pylori, 428
Hemmingway, Ernest, 641
Herd mentality, 446
Hewlett, Bill, 80
Hewlett-Packard (HPQ), 80
Hierarchy
    display screen, 282–284, 283f
    geographically based organization, 290, 290f
    responsibility-based screen organization, 286f, 286–287
    risk-based screen organization, 287f, 287–288
    similarity-based organization, 289, 289f
    task-based organization, 288, 288f
High-reliability organizations (HROs), 81, 88–89
HMI. *See* Human-machine interface (HMI)
Holmes, Sherlock, 399
Honeywell, 54, 368, 369f
Hopkins, Andrew, 391
Hotton, Donald, 538
Human-machine interface (HMI), 5–6, 17, 47, 57, 64, 211, 259–260, 262–263, 274–281
    communication, 212
    components of, 277–278
    contemporary control room, 227f
    design philosophy, 278–279
    Emerson, 54
    evaluating effectiveness, 380–383
    fire, gas, safety instrumented systems and security systems, 377
    graphics library, 280–281
    high-performance, 53
    key concepts, 261
    mass data displays, 345–349
    multivariate process analysis, 349–353
    navigation, 298–305
    nomenclature for display screens and components, 263–265
    operation interface, 21–22, 27
    principles of display screen design, 267–272
    principles of workspace design, 272–274
    requirements for operator screens, 265–266
    Rockwell Automation, 54
    Schneider Electric, 55
    simple control design, 235, 236f
    sound, audio and video, 377–380
    style guide, 279–280
    trend plots, 329–334
    video walls, 365–370
    wartime story setting a stage, 275–277
    weak signals, 517–518
    See also Display screen

I

Icons, 308–310
    design fundamentals, 326–327
    design fundamentals for, 325–329
    progress, 310, 310f
    salience requirements, 328–329
    temperature, 309f, 309–310, 310f
Icons, dials, gauges and dashboards (IDGDVs), 564–565, 567
IDCON, root cause analysis, 148–149
Identification
    work process, 569–570
    situation management activity, 24, 25f
Illusion of skill, 444
Immature situation, 525, 526, 530, 531f
Immature weak signals, 532, 585
Impairment
    categories of, 169
    managing, 169
    understanding, 168–169
    See also Operator readiness
Imperial Chemical Industries (ICI), 127
Implicit, 38
Impulse control, 406
Impulsive behavior, 455
Inattention bias, 448–450
Inattention blindness, 448–450

Incident(s), 5, 137
  Amagasaki rail crash, 425–427, 464
  BP and Transocean *Deepwater Horizon*, 467–469
  Challenger Space Shuttle, 465–467
  costs of, 60
  fatigue and, 165–167
  literature, 133
  Mann Gulch fire, 718, 726–727
  near miss, 135–136
  Piper Alpha, 142–143
  "Swiss cheese" model of, 140f
  weak signals and investigating, 633–634
  See also Critical failures; Plane crashes
Independent protection layer (IPL), term, 150
Indirect measurements or observations
  baseline operational measures, 557–558
  difficulty of control measures, 554–555
  equipment reliability indicators, 560–561
  instrument condition monitoring, 558–560
  key performance indicators, 561–562
  mass and energy balances, 546–551
  operational plausibility values, 551–554
  statistical process control information, 555–557
  weak signals from, 545–562
Indirect strong signals, 397
Individuality, 463
Inferential control, 110
Inflation report, Bank of England, 333, 334f
Inflections, weak signals, 592–593
Information
  in-shift handover emulation, 506
  intuition and "raw," 565–566
  mobile operator, 212–213
  partial, 450–451
  receiving, 505
  support tools and technology, 230
  transferring for shift handover, 504f, 504–505
  verifying and taking ownership, 505–506
Information fork, 434
Information silo, 84
Infrastructure, plant (enterprise) area model, 115–117
Inherent safety/complexity, definitions, 98
Insecurity, 406
Inspector(s), note to, 62
Instrument condition monitoring, 558–560
  basic sensor validation, 558–559
  smart field devices, 559–560
Instrument landing system (ILS), 423–424
Instrument panel, aircraft, 276f
InTech (magazine), 111–112

Integrity, operational, 63
International Association of Oil and Gas Producers (OGP), process safety, 86–87
International Electrochemical Commission (IEC), 43
International Society of Automation (ISA), 47
International Standards Organization (ISO), 43, 228
  ISO 11064 (Ergonomic Design of Control Centres), 249, 252, 256
  ISO 11064-5 (graphical displays), 263, 264f
  ISO 9241 design principles, 271–272
Intervention actions, effectiveness of, 17–18
Introduction to Human Factors Engineering, An (Wickens), 267–269
Intuition, 261, 406, 565–566
iOps command center concept, 54
Irritability, 455

J
Jamieson, Greg, 273
Japan, Amagasaki rail crash, 425–427, 464
Joint operation, 199

K
Key performance indicators (KPIs), 561–562, 731–732
Key repeated elements, 482
  alarm management, 611
  decomposition, 120–121, 121f
  looking for abnormal situations in, 123–125
Key repeated subsystems, 482
  alarm management, 611
  decomposition, 121–122, 122f
  looking for abnormal situations in, 125–127
Kletz, Trevor, 3
Knowledge fork, 472, 473, 473f, 675–676
Kourti, Theodora, 349
Ko Wen-je, 29
Kroc, Ray, 81

L
Latané, Bibb, 163
Layers of protection, 150–151
Leadership
  culture, 79–80
  process safety beginning with, 87–88
  situation awareness, 404–405
  talking the talk, 81
  walking the walk, 80–81
Leadership principles, 61

Lehrer, John, 440
Lemmon, Jack, 537
Liao Chien-tsung, 29
Life cycle, of weak signal, 608–609
Liu Tse-chung, 29
Logic
   reason and, 460–462
   versus fate, 462
Logic-tight compartments, 427–429
Logs, 196, 198, 511
Long-arm operations, 154, 213–214
Loss aversion, 438–440
Loss of scale, confirmation bias, 443–444, 444f
Loss of view, 383–385
Lowell, Elliott, 537

M

McDonald's (MCD), 81
MacGregor, John, 349
McMaster Advanced Control Consortium, 349
Maintenance, 21, 154
   good engineering practice, 67–69
   handback, 208
   shift handover for, 206, 208
Major situations
   managing, 719–725
   operator redeployment, 720–725
   See also Everyday situations
Malfunction, clear notice of, 558
Management
   alarm, 480–481
   control room, 20
   control room choices, 737–738
   fatigue, 168
   impairment, 169
   leadership principles, 61
   note to senior, 59–62
   operator workload, 171–172
   process hazard, 146–148
   role in permission to operate, 686
   role in situation management, 26
   role of, 47–48
   silo, 85
   See also Situation management; Weak signal management
Management of Change (MOC), 104, 375
Manager(s)
   bottom line, 5
   industrial setting, 38
   role of, 4
Managing operator, 724
Mann Gulch fire, 718, 726–727
Marshall, Barry, 428
Maslow's needs hierarchy diagram, 215f

Mass and energy balances, 546–551
   energy balance gauge, 550f, 551f
   mass balance gauge, 548f, 549f, 550f
   plant with area for calculation of balance, 547f
   showing imbalances to operator, 547–551
Mass data displays, 345–349
   departure from normal/expected value mode, 347
   departure from steady-state value mode, 347
   format in advanced control monitoring display, 348f
   heat exchanger example, 346f
Mastering, situation management, 739–740
MBTA Boston Transit System, 369, 370f
Measurement evaluation, operational integrity levels (OiL) as, 97–99
Mechanical engineering, 83f, 84–85
Mental model(s), 422, 622
   deception of two reasons, 430–431
   80:40 rule, 431–432, 432f
   expected roles, 423–425
   failure avoidance, 425–427
   good is not really good enough, 431–432
   logic-tight compartments, 427–429
   remembering, 431
   surrogate models, 429, 430
Mentoring, escalation teams, 709–710
Messages, role of, 472
Meteopole, Toulouse, France, 366, 367f
Metrics
   alarm configuration, 499–500
   alarm performance, 500–501
Meyer, Danny, 81
Milford Haven
   critical failure, 140–142, 147
   HSE report, 546
   operator's permission to operate, 686
Mindful organization, 81–82
Miracle on the Hudson, 732–735
Mirroring, communication, 657, 658
Mistakes
   looking without seeing, 418
   making, 416–418
Mobile operator(s)
   control room mobility, 210–211
   information and communication, 212–213
   large geographical mobility, 211–212
   plant area mobility, 211
   protocols and processes, 213
   requirements for mobility support, 212–213
   safeguards, 212
   tools, 212–213

Model(s)
  accuracy, 622
  fidelity, 622
  inadequacies, 622–623
  mental, 422–432
  model-based reasoning, 109, 109f
  plant (enterprise) area, 115–117
  surrogate, 429
  weak signal analysis, 566
  See also Mental models
Model reference adaptive control (MRAC), 109–110, 110f
Moore-Ede, Martin, 164, 167
Morton Thiokol, 466, 467
Motivation
  basics of, 215–217
  belonging, 216
  esteem, 216–217
  physical needs hierarchy diagram, 215f
  self-actualization, 217
Multitasking, 452–453
  alarms and, 453–454
  myth of, 415, 452–454
Multivariable process control (MPC), 110
Multivariate process analysis, 349–353
Myside bias, 442
Myth of multitasking, 415, 452–454

N

National Transportation Safety Board. *See* US National Transportation Safety Board (NTSB)
Navigation, 281, 298–305
  definition, 298
  navigating cycle, 299–302
  product of, 303–305
  purpose of, 298–299
  tablet icons, 362f
  tools, 302–303
Navigation tools
  dead reckoning, 302
  direct enter search, 302
  focus, 303
  table lookup, 302
  targets, 303
  yoking, 303
Near hits, 609
Near miss, 135–136, 609
News & Observer, The (newspaper), 29
New York Times (newspaper), 28
Nielsen, Jakob, 271
Nikkin Kyoiku, retraining program, 427
Noise, of weak signals, 605

Nomination, 564
  gauge, 564f
  process of, 313
Nominations gauge
  design structure of, 314–315, 315f
  explanation of symbols, 314f
  flow changes, 315–318
  natural gas and petroleum pipelines, 313–318
Noncontiguous shifts, 196
Nonpharmacological addiction, 169
Normal, 526
Normal operation region, 35–37
Notification(s)
  action, 507
  in combination, 508–509
  event, 507
  formal, 510
  general design and implementation guidelines, 509–511
  informal, 510–511
  logs and, 511
  properties of, 508–509
  situation, 507–508
  sorted, 509
  as weak signals, 506–507
NTSB. *See* US National Transportation Safety Board (NTSB)
Nuisance alarms, example gauge, 311f, 311–313

O

Observation, possible meaning, 534, 535–536
Occam's razor, 74
Occupational Safety and Health Administration (OSHA), 45–46, 133, 715
Off-normal, 526, 530, 531f
Off-normal operation region, 35–37
Off-workstation (OWS)
  ability to focus, 359–360
  auditing, 360
  benefits of, 362–363
  display screen visibility, 359
  formats, 361–362
  illustrations of OWS large displays, 360–361
  large displays, 356–358
  OWS small displays, 361–363
  physical analog mimic panel, 359f
  physical design of large OWS displays, 358–359
  problems of, 363
  requirements for large OWS displays, 358–360
  size, 353t

Operating displays
  weak signals from comparison, 544–545
  weak signals on, 543, 544f
Operating observations, weak signals from, 537
Operating situations, 686–687
  explosive events, 687
  operating in uncertainty, 686–687
  strong signals, 396–399
  unique events, 687
Operating values, deviation diagram, 565f
Operation(s)
  causes of poor, 71–73
  defensive principles, 665–666
  frontline supervisor's role, 49–50
  human operator in control room, 48–49
  management's role in, 47–48
  plants and, 156–157
  safety nets, 649–651
  tenets of, 217–218, 660–661, 689
  vendors, 53–55
Operational area
  boundaries and responsibilities, 157–158
  control room coordination with, 161–162
  crossing from one, to another, 159–160
  crossing to enterprise, 160
  crossing to universe, 160–161
Operational area division of responsibility, transformational analysis, 131
Operational drift, 658–661
Operational integrity, 63, 93–99
  equipment readiness, 96
  equipment suitability, 96
  levels (OiL), 95, 97t
  management competency, 95
  management effectiveness, 95–96
  measuring dial-type trend, 729f
  OiL as measurement evaluation, 97–99
  online risk management, 728–730
  plant complexity and inherent stability/safety, 96
  plant operability components, 95–96
  proper operation of enterprise, 94
  record, 96
  safety (SIL), 95
  staffing competency, 96
  staffing quality, 96
  staffing readiness, 96
Operational modes, 688–689
  definitions, 689
  plant state versus, 688f
Operational plausibility values, 545, 551–554
  model, 553, 554f
  plant showing identification of, 552f, 553f
Operational safety, 102

Operational success, key ingredients for, 70–71
Operational supervision, role in operations, 50
Operation center(s), 236–237
  concept of, 225
  environmental controls, 229–230
  information support tools and technology, 230
  permits, 232–233
  personnel, 233
  physical protection and security, 229
  process and operational controls, 230
  requirements, 229–233
  special operating situations, 232
  support, 231
  visitors, 233
Operation center controllers. See Operator(s)
Operation display, key operating parameters, 545f
Operation regions
  alarms, 40
  condenser and its, 124f
  illustration identifying, 394f
  for key repeated subsystem, 126f
  nested nature of, 39
  normal, off-normal and abnormal, 35–37
  overall setting, 39–42
  pump and its, 124f
  strong signals, 41
  weak signals, 41–42
Operations engineer, 739
Operator(s), 10–11
  active versus passive monitoring, 401
  alarm loading, 501–503, 502f
  alarms over weak signals, 618
  alarm system as line of defense, 40
  alertness, 172–173
  automation bias, 421–422
  basics of motivation, 215–217
  boundaries, 157–164
  creed, 155
  dealing with uncertainty, 434–435
  definition of, 155
  do list, 220–221
  don't list, 220
  duties, 669
  duty period plan and schedule, 52
  effectiveness pillars, 53
  flow of activities, 670, 671f
  following the rules, 669–670
  general flow of operator activities, 670–673
  goals, roles and culture, 214–222
  good engineering practice, 66
  human, in control room, 48–49
  improving performance, 173

Operator(s) (*Continued*)
  interface, 21–22
  intervention of, 4–6, 667–668
  key concepts, 154
  long arm of, 213–214
  managing permission to operate, 694–696
  mobile, 210–213
  modality decisions, 693–694
  not being innovators, 668–669
  note to, 58
  objectives of, 218–222
  operational success, 7–8
  operations and, 155–157
  overview showing area of responsibility, 336f
  ownership transfer at shift change, 503–506
  permission to operate, 684–696
  principles of effective, 33–34
  qualification, 178–181
  redeployment, 720–725
  responsibilities of, 13–16, 51–52, 62, 162–164
  role of, 4, 393f
  self-evaluation, 696
  situation awareness, 6
  situation management by, 11–18, 153, 222
  success of, 12
  term, 3
  training, 21, 173–178
  See also Shift handover
Operator interface, 402. *See also* Human-machine interface (HMI)
Operator of the Future initiative, Honeywell, 54
Operator readiness
  circadian rhythms, 165, 166f
  improving operator performance, 173
  managing fatigue, 168
  managing impairment, 169
  managing overload, 171–172
  operator alertness, 172–173
  understanding fatigue, 164–167
  understanding impairment, 168–169
  understanding overload, 170–171
Operator screens
  requirements for, 265–266
  See also Display screen(s)
Operator tools, 181–185
  checklists, 182–183
  procedures, 183–184
  protocols, 183
  reports, 185
  simulators, 184–185
Organization(s), mindful, 81–82
Outcome based, 43
Outside operator, term, 161

Overload
  management, 171–172
  supervision, 171
  understanding, 170–171
  work complexity, 170–171
  workload, 170
OWS. *See* Off-workstation (OWS)

P

Pacific Gas and Electric, diversion of funds, 403
Page, 261, 263, 264f
Parametric situation, 525, 526, 530, 531f
Parametric weak signals, 532–533, 585
Pareto Optimum, 431
Pathologic mental impairment, 169
Pay It Forward (Hyde), 220
Pay it forward illustration, 221
Peopleware, 65, 156
Perfect process understanding, 78
Performance, auditing, 514
Perishable skills, concept, 78
Permission to operate, 440, 684–696, 740
  alternate methods for having, 691–694
  BP Texas City operators, 685
  de facto decisions, 691–692
  diagram of formal flowchart, 697f
  functionality, 684
  guideline, 691–692
  how it came to be, 689–690
  how it works, 690–691
  management's role, 686
  managing operator's, 694–696
  observer evaluation, 695–696
  operating modality decisions, 693–694
  operating situations, 686–687
  operational modes, 688–689
  operator self-evaluation, 696
  operator's role, 684–686
  qualifying abnormal, 694–695
  safe operating limits, 694
  withdrawn, 692
Permits, control room, 232–233
Personalities
  accident-prone behavior, 455
  quiet ones, 455
  situation management points, 456
Personalization, 376
Personal safety, 101
Personal tools, 177–178
Personnel protective equipment (PPE), 101
PG&E San Bruno Pipeline, institutional failure, 87
Philosophy, HMI design, 278–279
Physical limitations, illness or, 169

Physical mimicry, screen, 278
Pilot training, 275–277, 276f
Pipeline, explosion at PG&E San Bruno, 87
Pipeline and Hazardous Materials Safety Administration (PHMSA), 47, 179, 180
Pipeline nominations tracking, example gauge, 313–318
Pipeline system, dashboard example, 319–320
Piper Alpha, 100
   critical failure, 142f, 142–143
   doubt, 432
   lesson learned, 160
   maintenance for, 206
Plane crashes
   Air Florida Flight 90, 676–678
   Air France Flight 447, 447, 421, 551–552, 647–649
   Asiana Flight 214, 423–425
   Miracle on the Hudson, 732–735
   TransAsia Flight 235, 28–30
   United Airlines Flight 173, 679
   US Airways Flight 1549, 732–734
Plant area model, 115–117
Plant operability components
   equipment readiness, 96
   equipment suitability, 96
   management competency, 95
   management effectiveness, 95–96
   plant complexity and inherent stability/safety, 96
   record, 96
   staffing competency, 96
   staffing quality, 96
   staffing readiness, 96
Polarity, 464–465
Pop-up, 345
Pop-up trends, 330
Portable document format (PDF), 374
Positive materials identification (PMI), 21, 104
Possible doubt, 433
Post hoc ergo propter hoc, bias, 447–448
Preoperation safety review (PREOP), 518
Prescriptive, 43
Pressure symbol, 280f
Pre-startup safety review (PSSR), 519
Pride and exceptionalism, 405
Principal component analysis (PCA), 350
Probable doubt, 433
Procedural automation, 110–112
Procedure(s), 183–184
Procedure gaps, weak signals, 633
Process control system (PCS), 105–107, 115
Process engineer, 738
Process hazard management, 146–148

Process safety
   beginning in boardroom, 87–88
   International Association of Oil and Gas Producers (OGP), 86–87
Process safety management, OSHA 29 CFR 1910.119, 45–46
Process trouble point, alarm system, 489–490, 491f
Production management, See-Understand-Decide-Act (SUDA), 30–32
Progressive abnormal situations, 617
Projection on least squares (PLS), 350
Protection layers of, 150–151
Protocols, 183
   conflicts with, 9
   mobile operator, 213
Proximate cause, 149–150
Psychological impairment, 169

Q
Qualified operator(s), 178–181
   glossary of, 179–181
   message of, 181
   resources, 179–180
   roles and responsibilities, 188
   See also Operator(s)
Quality of operation, concept, 78
Quick reference handbook (QRH), 733, 734

R
RAGAGEP. *See* Recognized and generally accepted good engineering practice (RAGAGEP)
Ratio control, 108
Rationalization
   alarm, 482–485
   step-by-step, 497–499
Raw information, 565–566
Readiness evaluation, situation management, 710–711
Reading suggestions, 56–57, 76, 151, 222, 386–387, 411, 470, 515, 637–638, 741
Reason, logic and, 460–462
Reasonable doubt, 433–435
Recognized and generally accepted good engineering practice (RAGAGEP), 45, 47, 103
Recommended practice, 44
Red teams, collaboration, 684
Reductio ad absurdum, 623
Regulator(s), note to, 62
Relative prime responsibility, 403
Remediation, situation management activity, 24, 25f, 25–26
Remembering, 431

Reports, 185, 197–198
Resistance, 455
Resolution, work process, 581–582
Resources, escalation, 702–703
Responsibility
  delegation of, 92–93
  maintaining, 162–164
  term, 158
  transfer of operational, 199
Responsible engineering authority (REA), 103–105
Responsive to proper operation, 728
Restricted control room, 661–663
  conditions, 663
  initiators of, 663
  termination of, 663
  See also Control room(s)
Retention, work space, 244
Retrospective weak signal analysis, Texas City disaster, 612–615, 616t
Riser/Regenerator, secondary display page, 339, 340f, 341f, 342
Risk assessment, using transformational analysis for, 129–130
Ritchie, Hugh, 29
Roberto, Michael, 655
Rockefeller Standard Oil, 93
Rockwell Automation, 54
Root cause/root cause analysis
  definition of, 148
  explanatory case for, 148–149
  weak signals, 610–611
Ruled-in problem, 399
Ruled-out problem, 399
Ruskin, John, 223
Russian nuclear power plant, 248–249

S

Safeguards, mobile operator, 212
Safe operating limits (SOLs), 104, 694
Safe operating modes
  alternatives to shutdowns, 716–718
  automated shutdown, 716
  Mann Gulch fire, 718, 726–727
  operator-initiated shutdown, 715–716
  safe park, 718–719
  USS Carl Vinson (aircraft carrier), 717
Safe park, 718–719
Safety, 9
  characteristics of, 100–101
  components of, 101–102
  conduct of personnel, 101
  definition, 99
  delivering, 103

equipment and operational design, 102
high-reliability organizations, 88
importance of, 78
inherent complexity and, 98
International Association of Oil and Gas Producers (OGP) on process, 86–87
management and, 38, 59–61
performance of activities, 102
personal, 101
process safety time, 31f
responsibility and, 23
safe communication, 653–655
word about, 653
Safety in numbers, 446
Safety instrumented systems (SISs), 377
Safety integrity level (SIL), 95
Safety nets, operations, 649–651
Safety notice, 8–9
Salas, Hector, 538
Salience, requirements, 328–329
Salk, Jonas, 471, 527
Salk vaccine, 527
Samuels, Neil, 60, 567
San Francisco International Airport, 423
Saving face, 463
Schneider Electric (Foxboro, Invensys), 55
Schoemaker, Paul, 519
Schwab, Chuck, 80
Security systems, 377
See-Understand-Decide-Act (SUDA), 30–32, 35, 492, 492f
Segmentation, display screen, 284–285
Selective attention, 448–450
Selective automation
  advanced basic control, 108
  advanced control, 108–110
  basic process control system (BPCS), 107
  concept of operator as coordinator, 111f
  models, 108–109
  MRAC (model reference adaptive control), 109–110
  procedural automation, 110–112
Self-actualization, 215f, 217
Senior management, note to, 59–62
Sensor validation, 558–559
Shadow of a doubt, 434
Shift assessments, 472
Shift change, operator ownership transfer at, 503–506
Shift handover, 154, 185–186, 198–201, 402, 741
  arriving operator, 185, 194, 503
  beginning the operator's shift role, 189–190
  ending of operator's shift role, 190–191
  engaged interaction, 199–200

face-to-face, 200–201
field operators' handover, 202f, 202–203, 203f, 207f
functional components of, 191–196
information and decision flow for, 504f, 504–505
information content of, 208–210
in-shift handover emulation, 506
leaving operator, 185, 194, 503
logs, 196, 198
for maintenance, 206, 208
noncontiguous shifts, 196
overlapped operation, 195–196
physical presence, 193–194
preparing full status report, 192–193
reasons for shift changes, 186–189
receiving information, 505
reports, 197–198
for supervisors, 204f, 204–206, 205f, 207f
taking operational control and ownership, 194
timelines, 188f, 202f, 203f, 204f, 205f, 207f
timing activities, 186f
tool tracking handover process, 201f
verifying information and taking ownership, 505–506
Shift progress reports, 402
Shirley, Richard S., 259
Short-term versus long-term
concept, 78
story of two enterprises, 89–92
Shutdown(s)
alternatives to, 716–718
automated, 716
operator-initiated, 715–716
Silo(s), 82
agricultural, 82f
communications between, 85–86
conceptual term, 82–84
functional silo, 84–85
information silo, 84
management of, 85
mentality, 82
pictorial design of mechanical engineering silo, 83f
Simplicity, 73
Simulations, 113
Simulators, 184–185
Situation assessment, 391, 649
question, 392, 406–407
region, 531
Situation awareness, 6, 64, 249, 391, 471, 472, 472–476, 531
accepting reality, 404

active versus passive monitoring, 401
alarms and weak signals for, 617
alarm system, 474, 477–503
assessment situation, 473–476
auditing, 514
cooperation, 404, 405
definition, 399–400
design and implementation, 513
interactive flow to situation management, 400f
intuition and raw information, 406
job capability, 513
key performance indicators, 561–562
knowledge fork, 473
leadership, 404–405
operational situations, 396
ownership, 403
problematic situations, 395–396
process of, 400
psychology of, 402–406
question, 392
relative prime responsibility, 403
requirements, 512
selling management, 513
tools, 401–402
triple package, 405–406
usability, 513
weak signals, 519
See also Alarm system
Situation-based display screens, 304–305, 340, 342f, 342–343
Situation codes of control room, 725–728
code blue, 725–726
code gray, 727–728
code orange, 728
Situation management, 4, 6, 11–18, 64–65, 260, 644, 741
abnormal, process model, 711–712
achieving, 735–739
activity, 644–645, 645f
alarm system, 22
basic four-activity process, 24–26, 25f
benefiting from experience, 651–652
changing the game, 35–39
collaboration, 675–684
command and control, 713–715
components of control room management, 20–22
contribution of alarms to, 503
cornerstones of, 4
decomposition underlying, 120–127
delegation, 674–675
diagram, 529f, 531f
doubts and concerns, 673–674

Situation management (*Continued*)
  escalation, 696–706
  escalation teams, 706–713
  essential components of, 22–23
  fitting process, 16–18
  framework for, 19f
  fundamental tools for, 24–30
  general flow of operator activities, 670–673
  importance, 23
  interactive flow from solution awareness to, 400f
  key concepts, 642–643
  key performance indicators, 672, 673, 731–732
  last opportunity, 5
  layers of protection and, 150–151
  lessons from Air France flight 447, 647–649
  limitations of, 75
  loss-of-view lesson, 383–385
  managing biases, 730–731
  managing everyday situations, 666–673
  managing major situations, 719–725
  operability integrity level for online risk management, 728–730
  operational drift, 658–661
  operational setting, 394–395
  operations safety nets, 649–651
  operator roles and responsibilities, 393f
  overcoming pitfalls, 730–731
  permission to operate, 22–23
  premise, 78
  preparation, 51
  reality in function, 50–52
  recommended structure of, 672f
  region, 531
  responsibilities of, 12–13
  safe conversations, 652–658
  safety and protective systems, 731
  "situation," 393–395
  step-by-step working process, 645–646
  straightforward, 13–16
  structure of, 18–24
  success pathway, 650f
  suits and coveralls, 70
  training, practicing, evaluating and mastering, 739–740
  TransAsia crash, 28–30
  using experts, 651–652
  weak signal analysis, 22
  weak signals for, 520, 529–531
  weak signals for abnormal, 712–713
  See also Communication; Decomposition
Situation notification, 507–508
Sixth sense, 392, 416, 440–441, 511
Skiles, Jeffrey B., 732

Skills training, 176
Sleep deprivation, 167
Sleep disorders, 167
Smart field devices, 559–560
Smartphone screens, 361, 362f
Smith, Karen, 256–257
Smith System, defensive driving, 32
SMS (short message service), informal notifications, 510–511
Social media
  dashboard example, 319, 319f
  glyphs, 308, 308f
Sohio, 93
Solenoid valve icon, 281f
Sound, HMI, 377–378
Space
  concept of, 241–242
  See also Work spaces
Sparkline graphs, 562–563, 563f
Sparkline trend, 330, 333–334
Spatial introduction, work space, 242
Standard, 43
Standard Oil Company, 93
Statistical process control (SPC)
  gauge, 555, 556f, 557f
  information, 555–557
Statutory requirements, conflicts with, 9
Sterile control room
  conditions, 664
  initiators of, 665
  termination of, 665
  See also Control room(s)
Stimulant drug use, 167
StockCharts.com, 331, 331f
Stomach ulcers, 428
Strategic part, weak signal extrapolation, 610
Strategic weak signals, 620, 621
Stress, impairment, 169
Strong signals, 396–399, 402
  direct, 398–399
  indirect, 397
  operating region showing, 41f
  something is wrong but not sure what, 41
  weak signals among, 605–606
Structural operational issues, transformational analysis for identifying, 131–132
Structured message, 509
Stumbling stone, 221
Style guide, screen design, 279–280
Substitution myth, automation, 419–420
SUDA (See-Understand-Decide-Act), 30–32, 35, 492, 492f
Suggestions for reading, 56–57
Suits, 70

Sullenberger, Chesley B. "Sully," 732–734
Supervision
   permission to operate, 171
   role in operations, 49–50
Supervisor(s)
   handover timelines, 204f, 205f, 207f
   note to, 59
   shift handover for, 204–206
Surrogate models, 407–409
   model test, 430
   from our immediate surroundings, 408
   from personal history or folklore, 409
   sources for surrogates, 408–409
Sustainability, 56, 92
Sustainable enterprise, 63
Sway (Brafman), 439
Symbols
   graphics library, 280–281
   icons, 308–310
   nominations gauge explanation, 314f
   pressure, 280f
   solenoid valve, 281f

T

Table lookup, navigation, 302
Tabletop simulations, 113
Tactical part, weak signal extrapolation, 609–610
Tactical weak signals, 620–621
Tactile, operator interaction, 277
Tags, plant area model, 116–117
Targets, navigation, 303
Task return point, 431
Teamwork, four-step process, 642
Technician(s), 10–11
Technologist(s), 59
Temperature, control loop example, 105–106
Templates, weak signals, 611–612
Tenets of operation, concept, 217–218
Tenth (10th) man doctrine, crew resource management, 680–681
Texaco Milford Haven
   alarms on previous unit, 540
   incident, 140–142, 141f, 539–540
   knock-out (KO) pot, 539
   stuck valve, 539
   weak signal example, 538–540
Texas City disaster
   retrospective weak signal analysis of, 616t
   weak signal case study, 612–613, 614f, 615
Theater Air Control System (TACS), 236
Things that go bump in the night, 606–607
Thinkware, 65
Three Mile Island, 167, 167f, 537–538
Time, 4
   power of, 30–32
   process safety, 30–32
Time-to-manage clock, 31
Tipping point
   after the, 598
   before the, 597
Tools, mobile operator, 212–213
Tracking, communication, 657, 658
Training, 154
   aircraft instrumentation, 275–277, 276f
   competency, 176–177
   escalation team, 711
   gaps and weak signals, 632
   good engineering practice, 67
   on-the-job (OJT), 177–178
   operator, 173–178
   personal tools, 177–178
   process understanding, 178
   situation management, 739
   skills and, 174–175
TransAsia crash, 28f, 28–30
Transfer of responsibility, escalation versus, 701–702
Transformational analysis, 127
   for decomposition, 130–131
   diagram, 128f, 130f
   identification of structural operational issues, 131–132
   for operational area division of responsibility, 131
   process, 127–129
   for risk assessment, 129–130
Transformative situation, 525, 526, 530, 531f
Transformative weak signals, 533, 585
Traveler's immunity, 416–417
Trend/trend plot, 329–334
   build-on-demand trends, 330
   complex trends, 331, 331f
   components, 332, 332f
   continuous trends, 330
   fan charts, 332–333, 334f
   pop-up trends, 330
   single, 329f
   sparkline charts, 333–334, 334f
   superimposed time-related trends, 333f
   weak signals from, 562–563, 563f
Trial and error, 43, 49
Trial by fire, 275
Triangulation, 681–683
   experience, 683
   job or task, 683
   questions, 682–683
Triple package, situation awareness, 405–406
Troubleshooting guide, weak signals, 628–631

Truth, 261
Tufte, Edward, 333
Twain, Mark, 471, 641
Two-cycle weak signal analysis, 600–601, 604
Two-reasons trap, 430–431

U

Union Square Café, 81
United Airlines Flight 173, 679
United Kingdom, Health and Safety Executive (HSE), 44, 133, 686
United Nations, 464
Unmanageable problem, 530
US Airways Flight 1549, 732–734
US Americans with Disability Act, 252
US Department of Transportation (DOT), 47, 187
US Department of Transportation Pipeline and Hazardous Materials Safety Administration (PHMSA), 714
User-centered design, 244–249
 affordance, 245
 compensation, 247
 environment, 245–246
 human factors, 245, 246f
 implementability, 248
 mixed technology, 248–249
 scaling, 246–247
 understandability, 247–248
 unified feel, 248
US Federal Aviation Administration (FAA), 134, 425, 463
US National Aeronautics and Space Administration (NASA), 236, 466–467
US National Transportation Safety Board (NTSB), 133, 678–679, 733, 734
US Occupational Safety and Health Administration, 45–46, 133, 715
USS Seawolf submarine, 240f

V

Vaccine, polio, 527
Vendors
 ABB, 53
 Emerson, 54
 Honeywell, 54
 Rockwell Automation, 54
 Schneider Electric, 55
 Yokogawa, 55
Video, HMI, 380
Video and animation, 261
Video walls, 365–370
 as augmentation to workstation HMI, 368–370
 as background, 365, 366f
 for control rooms, 366, 368–370
 conventional, 365–366
 futuristic display, 368, 368f
 main HMI, 366, 368
 as view into process, 367f
Viewability, screen display, 290–291
VigilantPlant initiative, Yokogawa, 55
Visibility, screen display, 290–291
Visitors, control room, 233
Visual, operator interaction, 277
Visual perception, 459–460

W

Walk the walk
 talking the talk, 81
 walking the walk, 80–81
Wal-Mart (WMT), 80
Walton, Sam, 80
Wang, Thomas, 29
Warren, Robin, 428
Weak signal(s), 402, 517–521, 524–541
 abnormal situation management, 712–713
 actively looking for, 591–593, 606–607
 from alarm activation, 618–619
 among strong signals, 605–606
 announce, 526–528
 artificial intelligence for analysis, 631–632
 building and displaying, 541–566
 categories of, 532–533, 597
 categories of trouble indicators, 588, 589f, 590f
 characteristics of, 541–542
 checklists and, 602
 China syndrome as example, 537–538
 classifying, 585
 clues something might be wrong, find out, 41–42
 collaboration and consensus, 595–596
 concepts, 533–534
 contradictions, 592
 critical variables and, 619
 design for, 587
 from direct measurements and observations, 542–545
 expectations interfering with, 540–541
 extrapolations, 609–611
 extremes, 593
 finding, 528, 625–626, 628–629
 as flags, 598–600
 gaps in training and procedures, 632–633
 from general observations, 534–537
 hunches and intuition, 540–541
 incident investigations and, 633–634
 inflections, 592–593
 intuition and "raw" information, 565–566

key concepts, 521–522
life cycle, 608–609
mastering, 740
model quality for analysis, 622–623
models for analysis, 566, 604
notifications as, 506–507
observation by experts, 595
off-normal operations, 526
from operating observations, 537
operating region showing, 42f
persistence of, 603, 607–609
prove true or prove false, 594–595
relationship between alarms and, 615, 617–619
retrospective case study, 612–615
role of icons, dials, gauges and dashboards (IDGDBs), 564–565
seeming to lead nowhere, 603–605
for situation management, 529–531
skipping over processing, 634–636
special case mapped to specific problem, 593–594
templates, 611–612
Texaco Milford Haven as example, 538–540
from trend plots, 562–563
troubleshooting guide, 628–631
two-cycle analysis, 600–601
without escalation, 596–598
working, 623–628
Weak signal extrapolations
alarm rationalization, 611
near hits, 609
near miss, 609
root cause analysis, 610–611
strategic part, 610
tactical part, 609–610
what-if and HAZOP, 610
Weak signal management, 566–587
accentuating negative and eliminating positive, 602
backward-projection, 573–575, 574f
before and after, 586–587
collaboration and consensus, 595–596
evidence for confirmation, 575–581
extrapolation and projection, 586
flowchart, 569f
forward-extrapolation, 570–573, 571f, 572f
identification, 569–570
indicators of problems, 626
model quality and, 622–623
no shortcuts for, 625–626
overload, 626–628
proper foundation, 624

recapping steps, 582–585
resolution, 581–582
selling, 628
work process, 567–569, 569f, 623
Webb, Amy, 592
Wells, Kimberly, 538
What-if, weak signal extrapolation, 610
What then question, bias, 448
Wickens, Christopher, 267–269
Wilkins, Maurice, 111–112, 112n26
Window, normal visual flow directions, 457–458, 458f
Windows, 261, 263, 264f, 345f
Wittgenstein, Ludwig, 3
Workload, 261
Work process, 567–569
backward-projection, 573–575, 579–580
evidence for confirmation, 575–579
forward-extrapolation, 570–573
identification, 569–570
recapping the steps, 582–585
resolution, 581–582
steps of, 567–568
tying up loose ends, 589–591
Work spaces
collaboration, 243–244
concept of space, 241–242
design of effective, 241–244
expanse, 243
geometry, 242
principles of design, 272–274
retention, 244
spatial introduction, 242
See also User-centered design
Workstation displays, 354–356
closely spaced, 356f
configuration options, 355f
entire plant floor status, 358f
head-up displays, 363–365
with large off-workstation (OWS), 357f
size, 353t
See also Off-workstation (OWS)
World War Z (movie), 681
Wright State University, 54

Y

Yoking, 376
construction details, 343f
navigation, 303
situationally based secondary page, 342f
Yokogawa, 55
Yom Kippur War (1973), 681